T0323671

PROBLEMS IN STRUCTURAL
INORGANIC CHEMISTRY

PROBLEMS IN STRUCTURAL INORGANIC CHEMISTRY

SECOND EDITION

Wai-Kee Li

Hung Kay Lee

Dennis Kee Pui Ng

Yu-San Cheung

Kendrew Kin Wah Mak

Thomas Chung Wai Mak

DEPARTMENT OF CHEMISTRY
The Chinese University of Hong Kong

OXFORD
UNIVERSITY PRESS

UNIVERSITY PRESS

Great Clarendon Street, Oxford, OX2 6DP,
United Kingdom

Oxford University Press is a department of the University of Oxford.
It furthers the University's objective of excellence in research, scholarship,
and education by publishing worldwide. Oxford is a registered trade mark of
Oxford University Press in the UK and in certain other countries

Published in the United States of America by Oxford University Press
198 Madison Avenue, New York, NY 10016, United States of America

British Library Cataloguing in Publication Data
Data available

Library of Congress Control Number: 2018944464

ISBN 978–0–19–882390–2 (hbk.)
ISBN 978–0–19–882391–9 (pbk.)

DOI: 10.1093/oso/9780198823902.001.0001

Printed and bound by
CPI Group (UK) Ltd, Croydon, CR0 4YY

PREFACE TO THE FIRST EDITION

The present publication serves as a problem text for Part I (Fundamentals of Chemical Bonding) and Part II (Symmetry in Chemistry) of our book *Advanced Structural Inorganic Chemistry* co-authored with Gong-Du Zhou, which was published by Oxford University Press in 2008. It may also be used as a supplement for a variety of inorganic chemistry courses at the senior undergraduate level. A brief description of the origin of the problems collected in this book is given below.

In the early 1970s, four young faculty members in our department, Thomas C. W. Mak, K. Y. Hui, O. W. Lau, and W.-K. Li, initiated a major revision of the inorganic chemistry curriculum. The effort led to their co-authorship of a book entitled *Problems in Inorganic and Structural Chemistry*, which was published by The Chinese University Press, Hong Kong, in 1982. It consists of problems that they designed for teaching inorganic chemistry courses, either as take-home assignments or as examination questions. Four years later, a Chinese edition was published by Science Press in Beijing, China. The translation was done by chemistry faculty members of Peking University, the target readership being the growing number of advanced undergraduate and beginning postgraduate students in mainland China.

Three decades have elapsed since the publication of that book. Of the original authors, K. Y. Hui retired in 1996 and now resides in Vancouver, whereas O. W. Lau retired in 2003 and sadly passed away the following year. The undersigned now hold the title of Emeritus Professor of Chemistry and manage to remain active in teaching and research. Thus this book represents a compilation of problems that the two of us have used in our courses, including "Chemical Bonding", "Advanced Inorganic Chemistry", "X-Ray Crystallography", etc., during the past 40 years.

Since examination papers and assignments usually include elementary as well as more advanced types of questions, the present collection does betray this unevenness in regard to the degree of difficulty. Should the students wish, they may first attempt the more elementary problems before proceeding to the more challenging ones.

The questions are organized into nine chapters. However, the classification is done in a somewhat arbitrary way since, in some cases, one "pigeon-hole" appears to be as appropriate as another. Nevertheless, an attempt has been made to group together questions of a similar nature for the sake of unity and consistency.

In the course of putting together the manuscript of this book, we received indispensable assistance from two junior faculty members in our department: Dr. Yu-San Cheung and Dr. Kendrew K. W. Mak. In addition to manuscript preparation and proof-reading, they have essentially solved all the problems one more time to ensure the validity of the solutions. Hence, their names deservedly appear as co-authors of the book.

After the publication of *Advanced Structural Inorganic Chemistry*, five book reviews have appeared in the literature. One of the reviewers lamented that "there are no end-of-chapter problems. I routinely assign these types of problems, and lack of them makes it difficult to require this text for my courses." As mentioned above, we have kept a record of problem assignments and examination questions in structural inorganic chemistry for decades, but did not include them in the textbook mainly because its length already exceeds 800 pages. So we sincerely hope the appearance of this book fills a need in the chemistry community, and university teachers and students everywhere will find the problems collected here helpful, instructive, and sometimes challenging.

Wai-Kee Li
Thomas Chung Wai Mak
DEPARTMENT OF CHEMISTRY
The Chinese University of Hong Kong
June, 2012

CONTENTS

Contents

ABOUT THE AUTHORS

Wai-Kee Li obtained his B.S. degree from the University of Illinois in 1964 and his Ph.D. degree from the University of Michigan in 1968. He joined The Chinese University of Hong Kong (CUHK) as a Lecturer in 1968 and retired as Professor of Chemistry in 2006. He then held the title of Emeritus Professor of Chemistry and remained active in teaching and research. His keen interest in theoretical and computational chemistry led to over 200 research papers in international journals. He had taught a variety of courses in general, physical, and inorganic chemistry and won the Vice-Chancellor's Teaching Award twice. He was a co-author of two textbooks in both English and Chinese editions. Unfortunately, when preparation of the present book manuscript was at its final stage, he passed away due to illness in January 2016.

Hung Kay Lee obtained his B.Sc. and Ph.D. degrees from CUHK in 1990 and 1995, respectively. After working as a postdoctoral scholar at California Institute of Technology, he joined the Chemistry faculty at CUHK as an Assistant Professor in August 1997, and he is now an Associate Professor in the Department. He is engaged in teaching of undergraduate chemistry courses of various levels in the Department such as Advanced Inorganic Chemistry, Bioinorganic Chemistry and Environmental Chemistry. His current research is focused in two areas: (1) the chemistry of low-coordinate d- and f-block metal complexes supported by sterically bulky non-cyclopentadienyl ligand systems, and (2) synthetic and structural studies of metal complexes of biological relevance. He has so far published more than 70 publications in international journals.

Dennis Kee Pui Ng obtained his B.Sc. and M.Phil. degrees from CUHK in 1988 and 1990, respectively, and was conferred a D.Phil. degree by the University of Oxford in 1993. After receiving a postdoctoral training in the California Institute of Technology for one year, he returned to his alma mater to take up a Lecturership in 1994. He is currently an Associate Vice-President, the University Dean of Students, and a professor at the Department of Chemistry. His research interests lie in the chemistry of various functional dyes, particularly phthalocyanines and boron dipyrromethenes, focusing on their biomedical applications and supramolecular chemistry. He has so far published more than 180 publications in international journals.

Yu-San Cheung obtained his B.Sc. (1992) and M.Phil. (1994) degrees from CUHK, and Ph.D. (1999) from Iowa State University. His research interests in his postgraduate studies included computational chemistry and laser spectroscopy, and he has published about 30 papers in international journals. In July 1999 he joined CUHK and he is now a Senior Lecturer in the Department of Chemistry. In this capacity he is in charge of all the undergraduate physical chemistry laboratory courses in the department.

Kendrew Kin Wah Mak obtained his B.Sc. and Ph.D. degrees in Chemistry from CUHK in 1994 and 1998, respectively. He joined his *Alma Mater* in 1999 and he is now a Senior Lecturer in the Department of Chemistry. He is actively engaged in teaching practical organic chemistry and developing new strategies for promoting science education.

Thomas Chung Wai Mak obtained his B.Sc. (1960) and Ph.D. (1963) degrees from the University of British Columbia. After working as NASA Postdoctoral Research Associate at the University of Pittsburgh and Assistant Professor of Chemistry at the University of Western Ontario, he returned to Hong Kong in June 1969 to join CUHK, where he is now Emeritus Professor of Chemistry and Wei Lun Research Professor. His research interests lie in inorganic synthesis, chemical crystallography, supramolecular assembly and crystal engineering, with over 1100 publications recorded in webofknowledge.com and a h-index of 64. He was elected as a member of the Chinese Academy of Sciences in 2001.

Electronic States and Configurations of Atoms and Molecules

1

PROBLEMS

1.1 (i) How many electrons can be placed in

  (a) a shell with principal quantum number n?
  (b) a subshell with quantum numbers n and ℓ?
  (c) an orbital?
  (d) a spin-orbital?

  (ii) How many microstates are included in a term defined by quantum numbers L and S? Explain clearly your deductions.

1.2 Discuss the possible electronic configurations of elements 118 and 154. Predict their chemical behavior by extrapolation of known properties of their lower homologs in the periodic table. For simplicity, an abbreviated notation such as $[Rn]6d^17s^2$ may be used to describe the electronic configuration of element 89 (actinium). Using the $(n + \ell)$ rule as a guide, work out the electronic configuration of element 103 (lawrencium) and proceed from there.

REFERENCE: G. T. Seaborg, Prospects for further considerable extension of the periodic table. *J. Chem. Educ.* **46**, 626–34 (1969).

1.3 Atomic volume, defined as the volume in cm^3, occupied by one mole of an element, was shown by Lothar Meyer in 1869 to be a periodic function of the atomic weight (more correctly the atomic number). With reference to the atomic volume curve shown below, rationalize the following features on the basis of your knowledge of the electronic structure of atoms.

  (i) The alkali metals occupy the maxima.
  (ii) The alkaline earths correspond to a marked decrease in volume.
  (iii) The gaseous and volatile elements lie on the rising portions of the U-shaped curves.

Problems in Structural Inorganic Chemistry. Second edition. Wai-Kee Li, Hung Kay Lee, Dennis Kee Pui Ng, Yu-San Cheung, Kendrew Kin Wah Mak, and Thomas Chung Wai Mak. © Oxford University Press 2019. Published in 2019 by Oxford University Press. DOI: 10.1093/oso/9780198823902.001.0001

(iv) Each transition series corresponds to a gradual decrease toward the trough of a U-shaped curve.

(v) The atomic volumes of corresponding elements of the second (Y to Sb) and third (Lu to Bi) long periods of the periodic table are strikingly similar.

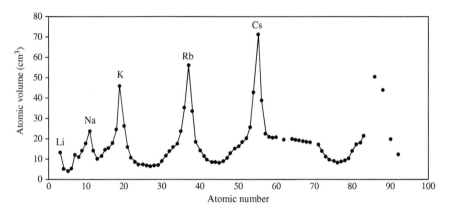

REFERENCE: R. N. Keller, The lanthanide contraction as a teaching aid. *J. Chem. Educ.* **28**, 312–7 (1951).

1.4 Without writing out all the microstates, derive all the Russell-Saunders terms arising from configurations p^2, d^2, and f^2.

1.5 (i) Consider the electronic configuration $2p^1 3p^1$. Derive all the microstates arising from this configuration. Based on the distribution of the microstates, construct all the spectroscopic terms (L and S values only) made up of these microstates.

  (ii) Taking advantage of the terms arising from the configuration $2p^1 3p^1$, derive all the terms (again L and S values only) arising from the configuration $2p^1 3p^1 4p^1$.

1.6 Derive all the Russell-Saunders terms arising from configuration p^3.

1.7 Derive all the Russell-Saunders terms arising from configuration d^3.

1.8 The ground configuration of Cr is $3d^5 4s^1$.

  (i) What is the ground term according to Hund's rule?

  (ii) How many spectroscopic terms can be derived from this configuration? Do not write out all the microstates!

1.9 Deduce the ground electronic state for all possible configurations (d^n, $n = 1, \ldots, 9$) in an octahedral complex, without applying formal group theory techniques.

1.10 One of the excited configurations for Zn^{2+} is $3d^9 4p^1$. Derive all the Russell-Saunders terms arising from this configuration. Can you say anything about which term has the lowest energy?

1.11 Part of Hund's rule states: "For a configuration less than half filled, the lowest state is the one with the smallest J value; if the configuration is more than half filled, the lowest state is the one with the largest J value." Why is the half filled configuration not specifically dealt with?

1.12 The following paragraph is found in a certain supplementary text. Read it carefully and point out any error it may contain.

"The next most stable configuration is $ns^1 np^1$. [The author here is discussing the spectra of the alkaline earths, which have the ground configuration ns^2.] In this case $\ell_1 = 0$ and $\ell_2 = 1$, so $L = (0 + 1), (0 + 1) - 1$; that is, there are two terms arising from this configuration – one of P ($L = 1$) and another of S ($L = 0$) type. The total spin can be either 1 or 0, depending on whether the two electrons have the same or opposite spins. The value $S = 1$ gives rise to triplet terms, and $S = 0$ gives rise to singlet terms."

1.13 The ground electronic configuration of Li is $1s^2 2s^1$. When lithium is heated in a flame, bright red light is emitted. The red color is due to light emission at a wavelength of $6710 \, \text{Å}$. Also, no light emission is observed at longer wavelengths.

   (i)   What is the frequency (in Hz) of the light? The speed of light is $3.00 \times 10^8 \, \text{m s}^{-1}$.

   (ii)   What is the wavenumber (cm^{-1}) of the light?

   (iii)   Suggest an electronic transition responsible for the emission at $6710 \, \text{Å}$. In your answer, you need to include the electronic states involved (each state is defined by its L and S values). Also, you need to specify the electronic configurations from which the states arise. [Hint: The 1s electrons are not involved.]

1.14 Assume that Hund's rule holds for all the systems considered in this problem.

   (i)   Consider the first 36 elements in the periodic table, for which the L–S coupling scheme is applicable. Among these elements, which one has $^7S_3$ as its ground term? Note that this ground term has to arise from the ground electronic configuration of the element.

   (ii)   Again consider the first 36 elements of the periodic table. What two other elements or their (positively charged) ions also have $^7S_3$ as their ground term? Note that this ground term does not have to arise from the ground electronic configuration of the element or ion you just selected.

1.15 (i) Some chemists have suggested using the weighted-average energy of all the terms arising from a given electronic configuration to represent the energy of the configuration. For example, the ground configuration of Ti is . . . $3d^2 4s^2$. If we ignore the J states arising from spin-orbit interaction, the terms derived from this configuration are ^3F, ^1D, ^3P, ^1G, and ^1S. The energies of these states are $E(^3F)$, $E(^1D)$, $E(^3P)$, $E(^1G)$, and $E(^1S)$, respectively. The aforementioned chemists propose that the energy of the configuration $3d^2 4s^2$, $E(3d^2 4s^2)$, is:

$$E(3d^2 4s^2) = [21E(^3F) + 5E(^1D) + 9E(^3P) + 9E(^1G) + E(^1S)]/45.$$

Explain why the coefficients for $E(^3F)$, $E(^1D)$, $E(^3P)$, $E(^1G)$, and $E(^1S)$ are 21, 5, 9, 9, and 1, respectively. Also, why is the number in the denominator 45?

(ii) For the carbon atom in its ground state with configuration $1s^2 2s^2 2p^2$, it has multiplets ^3P, ^1D, and ^1S, with a total of five electronic states: $^3P_0$, $^3P_1$, $^3P_2$, $^1D_2$, and $^1S_0$. These five states have relative energies 0.0, 16.4, 43.5, 10197.7, and 21648.4 cm^{-1}, respectively.

   (a) Calculate the weighted-average energy for each of the three multiplets.
   (b) With the results obtained in (i), calculate the weighted-average energy for the $1s^2 2s^2 2p^2$ configuration.

1.16 For an electronic configuration $s^1 d^1$, specify all the energy states due to (i) $L–S$ coupling and (ii) $j–j$ coupling. Also, draw a correlation diagram for the electronic states arising from the configuration $s^1 d^1$ employing $L–S$ and $j–j$ coupling schemes.

1.17 Consider the excitation of a 4s electron of the Ca atom to (i) the 4p and (ii) the 3d subshells. Deduce the terms and levels for these two excited configurations, and interpret the observed emission spectrum of Ca shown below in terms of transitions between these levels. Selection rules: (a) $\Delta S = 0$, (b) $\Delta L = 0, \pm 1$, (c) $\Delta J = 0, \pm 1$ (except for $J = 0 \rightarrow J = 0$).

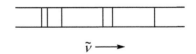

$\tilde{\nu} \longrightarrow$

REFERENCE: G. Herzberg, *Atomic Spectra and Atomic Structure*, Dover, New York, 1944, pp. 77–8.

1.18 Specify quantum numbers L, S, and J for each electronic state you derive in this problem.

 (i) The ground electronic configuration of magnesium is $1s^2 2s^2 2p^6 3s^2$. Write down the electron state(s) arising from this configuration.

 (ii) The first excited configuration for Mg is $\ldots 2p^6 3s^1 3p^1$. What is/are the electronic state(s) arising from this configuration? Write down the allowed electronic transitions between the ground state in (i) and the state(s) you have derived for the $3s^1 3p^1$ configuration. Refer to the selection rules stated in Problem 1.17.

 (iii) Another excited configuration for Mg is $\ldots 2p^6 3p^1 3d^1$. What are the states arising from this configuration?

1.19 In high-school chemistry classes, we saw the bright yellow flames in the flame test for sodium compounds. In spectroscopic terms, these flames correspond to the spectral lines measured at 5890 and 5896 Å in the emission spectrum of the sodium atom.

 (i) Write down the ground term (including the J value) of Na.

 (ii) If the observed spectral lines correspond to the electronic transitions between the ground state you have just derived and the states arising from the first excited electronic configuration of Na, what are the electronic states (including the J value) arising from this excited configuration?

Note that the sodium atom is responsible for the yellow flame, even though sodium exists as Na^+ in most of its compounds. In a flame, we are often burning hydrocarbons (such as butane in our Bunsen burners) and, under very high temperature, these compounds break up easily into various radicals. The radicals are highly unstable and they tend to lose their unpaired electrons. These electrons are then captured by Na^+ cations to form sodium atoms, which emit yellow light upon burning. In short, there is a very "violent" situation in a flame, where all kinds of (unstable) species are found.

1.20 The energy levels for states arising from the ground configuration $1s^2 2s^2 2p^2$ of carbon are shown below (not to scale). Write down the allowed transitions among the five states given in the diagram. For each transition, specify the two states involved as well as the energy difference between these two states.

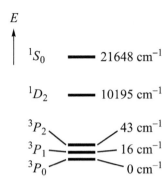

1.21 The energy level diagram shown below illustrates how light is produced by the He–Ne laser.

(i) Write down the electronic configurations which give rise to the three states (^1S, ^3S, and ^1S) of helium and the ^1S state of neon, all shown in the energy level diagram. These four states are labeled (A), (B), (C), and (D), respectively, in the diagram.

(ii) Derive the electronic state(s) arising from the configurations $2p^53s^1$, $2p^54s^1$, and $2p^55s^1$ of neon. The J values of the states are not required.

(iii) Derive the electronic state(s) arising from the configurations $2p^53p^1$ and $2p^54p^1$ of neon. The J values of the states are not required.

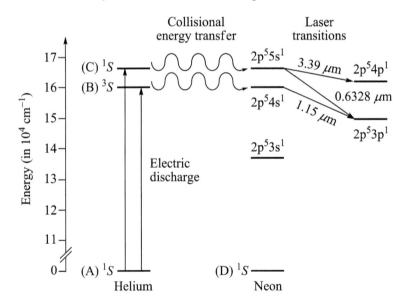

1.22   Derive the states for the following configurations:

   (i)   A D_{5d} molecule with configuration $(e_{1g})^1(e_{2g})^1$;

   (ii) A $C_{\infty v}$ molecule with configuration $(1\delta)^3(2\delta)^1$.

   Given: $\cos 2\phi = 2\cos^2\phi - 1$.

1.23   Derive the states for the following molecules:

   (i)   A C_{3v} molecule with configuration $(e)^2$;

   (ii)   A $C_{\infty v}$ molecule with configuration $(\pi)^2$;

   (iii) A T_d molecule with configuration $(t_1)^3$;

   (iv)   An O_h molecule with configurations $(t_{2u})^2$;

   (v)   An I_h molecule (such as C_{60}) with configuration $(h_g)^5$.

   Given:

   For (i), (ii), and (iv), the characters χ of singlet and triplet states for operation $\boldsymbol{R}$ are:

$$\chi(\text{singlet}) = \frac{1}{2}[\chi^2(\boldsymbol{R}) + \chi(\boldsymbol{R}^2)],$$

$$\chi(\text{triplet}) = \frac{1}{2}[\chi^2(\boldsymbol{R}) - \chi(\boldsymbol{R}^2)].$$

   For (iii), the characters χ of doublet and quartet states for operation $\boldsymbol{R}$ are:

$$\chi(\text{doublet}) = \frac{1}{3}[\chi^3(\boldsymbol{R}) - \chi(\boldsymbol{R}^3)],$$

$$\chi(\text{quartet}) = \frac{1}{6}[\chi^3(\boldsymbol{R}) - 3\chi(\boldsymbol{R})\chi(\boldsymbol{R}^2) + 2\chi(\boldsymbol{R}^3)].$$

   For (v), the characters χ of doublet, quartet, and hextet states for operation $\boldsymbol{R}$ are:

$$\chi(\boldsymbol{R}: \text{doublet}) = \frac{1}{24}[\chi^5(\boldsymbol{R}) - 2\chi^3(\boldsymbol{R})\chi(\boldsymbol{R}^2) - 4\chi^2(\boldsymbol{R})\chi(\boldsymbol{R}^3)$$
$$+ 6\chi(\boldsymbol{R})\chi(\boldsymbol{R}^4) + 3\chi(\boldsymbol{R})\chi^2(\boldsymbol{R}^2) - 4\chi(\boldsymbol{R}^2)\chi(\boldsymbol{R}^3)];$$

$$\chi(\boldsymbol{R}: \text{quartet}) = \frac{1}{30}[\chi^5(\boldsymbol{R}) - 5\chi^3(\boldsymbol{R})\chi(\boldsymbol{R}^2) + 5\chi^2(\boldsymbol{R})\chi(\boldsymbol{R}^3)$$
$$+ 5\chi(\boldsymbol{R}^2)\chi(\boldsymbol{R}^3) - 6\chi(\boldsymbol{R}^5)];$$

$$\chi(\boldsymbol{R}: \text{hextet}) = \frac{1}{120}[\chi^5(\boldsymbol{R}) - 10\chi^3(\boldsymbol{R})\chi(\boldsymbol{R}^2) + 20\chi^2(\boldsymbol{R})\chi(\boldsymbol{R}^3) - 30\chi(\boldsymbol{R})\chi(\boldsymbol{R}^4)$$
$$+ 15\chi(\boldsymbol{R})\chi^2(\boldsymbol{R}^2) - 20\chi(\boldsymbol{R}^2)\chi(\boldsymbol{R}^3) + 24\chi(\boldsymbol{R}^5)].$$

1.24   Benzene, with D_{6h} symmetry, has six π molecular orbitals and half of them are filled. Its ground electronic configuration is $(1a_{2u})^2(1e_{1g})^4(1e_{2u})^0(1b_{2g})^0$. Thus the ground state is simply $^1A_{1g}$. This question deals with the doubly excited configuration of $(1a_{2u})^2(1e_{1g})^2(1e_{2u})^2$, which is considerably more complex than the ground configuration.

(i)   Derive the states arising from the configuration $(1e_{1g})^2$.

(ii)   Derive the states arising from the configuration $(1e_{2u})^2$.

(iii)   Derive the states arising from the configuration $(1e_{1g})^2(1e_{2u})^2$ by combining the results of (i) and (ii).

1.25   (i)   List the symmetry operations of the group D_3 and combine these operations with the operation $\boldsymbol{R}$ (rotation by 2π) to obtain the symmetry operations of the double group D_3'. Make sure that the list prepared has the properties of a group.

(ii)   How many classes are there in D_3'? What are the operations in each class? [Hint: Make a comparison of the groups O and O'.]

1.26   Apply the formula $\chi(\alpha) = \dfrac{\sin(J + \frac{1}{2})\alpha}{\sin \frac{\alpha}{2}}$ to calculate the characters of the representation

for a state defined by $J = 2\frac{1}{2}$. Then reduce this representation into the symmetry species of the D_4' group, which is the double group for D_4.

The character table of double group D_4'

D_4'		E	R	C_4	C_4^3	C_2	$2C_2'$	$2C_2''$
$(h = 16)$		$(\alpha = 4\pi)$	$(\alpha = 2\pi)$	$C_4^3 R$	$C_4 R$	$C_2 R$	$2C_2' R$	$2C_2'' R$
Γ_1	A_1'	1	1	1	1	1	1	1
Γ_2	A_2'	1	1	1	1	1	-1	-1
Γ_3	B_1'	1	1	-1	-1	1	1	-1
Γ_4	B_2'	1	1	-1	-1	1	-1	1
Γ_5	E_1'	2	2	0	0	-2	0	0
Γ_6	E_2'	2	-2	$\sqrt{2}$	$-\sqrt{2}$	0	0	0
Γ_7	E_3'	2	-2	$-\sqrt{2}$	$\sqrt{2}$	0	0	0

1.27   (i)   With the aid of the character table of the double group O' and by means of the formula, $\chi(\alpha) = \dfrac{\sin(J + \frac{1}{2})\alpha}{\sin \frac{\alpha}{2}}$, determine the characters of the representations $J = \frac{1}{2}, 1\frac{1}{2}, 2\frac{1}{2}, ...,$ or $6\frac{1}{2}$. Reduce the representations obtained into the irreducible representations of the O' group.

The character table of double group O'

O'	E	R	$4C_3$	$4C_3^2$	$3C_2$	$3C_4$	$3C_4^3$	$6C_2'$
$(h = 48)$	$(\alpha = 4\pi)$	$(\alpha = 2\pi)$	$4C_3^2R$	$4C_3R$	$3C_2R$	$3C_4^3R$	$3C_4R$	$6C_2'R$
$\Gamma_1 \quad A_1'$	1	1	1	1	1	1	1	1
$\Gamma_2 \quad A_2'$	1	1	1	1	1	-1	-1	-1
$\Gamma_3 \quad E_1'$	2	2	-1	-1	2	0	0	0
$\Gamma_4 \quad T_1'$	3	3	0	0	-1	1	1	-1
$\Gamma_5 \quad T_2'$	3	3	0	0	-1	-1	-1	1
$\Gamma_6 \quad E_2'$	2	-2	1	-1	0	$\sqrt{2}$	$-\sqrt{2}$	0
$\Gamma_7 \quad E_3'$	2	-2	1	-1	0	$-\sqrt{2}$	$\sqrt{2}$	0
$\Gamma_8 \quad G'$	4	-4	-1	1	0	0	0	0

(ii) Consider the ground term ^4F of configuration d^7 in a crystal field with O_h symmetry. Determine the states arising from this term when both crystal field and spin-orbit interactions are included. In order to check the results obtained, do this problem by considering both of the following two cases: (a) strong field and small spin-orbit coupling and (b) large spin-orbit coupling and weak O_h field.

1.28 Consider the term ^4G arising from the d^5 configuration of a transition metal under the influence of an octahedral crystal field. Deduce the states arising for the cases:

(i) strong crystal field interaction and small spin-orbit coupling, and
(ii) large spin-orbit coupling and weak crystal field interaction.

The table given below should be useful.

L or J value	States in O or O' group	L or J value	States in O or O' group
0	Γ_1	$3\frac{1}{2}$	$\Gamma_6, \Gamma_7, \Gamma_8$
$\frac{1}{2}$	Γ_6	4	$\Gamma_1, \Gamma_3, \Gamma_4, \Gamma_5$
1	Γ_4	$4\frac{1}{2}$	$\Gamma_6, 2\Gamma_8$
$1\frac{1}{2}$	Γ_8	5	$\Gamma_3, 2\Gamma_4, \Gamma_5$
2	Γ_3, Γ_5	$5\frac{1}{2}$	$\Gamma_6, \Gamma_7, 2\Gamma_8$
$2\frac{1}{2}$	Γ_7, Γ_8	6	$\Gamma_1, \Gamma_2, \Gamma_3, \Gamma_4, 2\Gamma_5$
3	$\Gamma_2, \Gamma_4, \Gamma_5$	$6\frac{1}{2}$	$\Gamma_6, 2\Gamma_7, 2\Gamma_8$

1.29 The d^9 complex $[CuCl_5]^{3-}$ has a square pyramidal structure (C_{4v} symmetry). Deduce the states arising for the cases

(i) strong crystal field interaction and small spin-orbit coupling, and

(ii) large spin-orbit coupling and weak crystal field interaction.

Use the Bethe notation for the states derived. Useful formula: $\chi(\alpha) = \dfrac{\sin(J + \frac{1}{2})\alpha}{\sin\frac{\alpha}{2}}$.

The character table of double group D'_4

D'_{2d}	E	R	S_4	S_4^3	C_2	$2C'_2$	$2\sigma_d$
			S_4^3R	S_4R	C_2R	$2C'_2R$	$2\sigma_dR$
C'_{4v}	E	R	C_4	C_4^3	C_2	$2\sigma_v$	$2\sigma_d$
			C_4^3R	C_4R	C_2R	$2\sigma_vR$	$2\sigma_dR$
D'_4	E	R	C_4	C_4^3	C_2	$2C'_2$	$2C''_2$
$(h = 16)$	$(\alpha = 4\pi)$	$(\alpha = 2\pi)$	C_4^3R	C_4R	C_2R	$2C'_2R$	$2C''_2R$
Γ_1 A'_1	1	1	1	1	1	1	1
Γ_2 A'_2	1	1	1	1	1	−1	−1
Γ_3 B'_1	1	1	−1	−1	1	1	−1
Γ_4 B'_2	1	1	−1	−1	1	−1	1
Γ_5 E'_1	2	2	0	0	−2	0	0
Γ_6 E'_2	2	−2	$\sqrt{2}$	$-\sqrt{2}$	0	0	0
Γ_7 E'_3	2	−2	$-\sqrt{2}$	$\sqrt{2}$	0	0	0

SOLUTIONS

A1.1 (i) (a) $2n^2$; (b) $2(2\ell + 1)$; (c) 2; (d) 1.

(ii) $(2L + 1)(2S + 1)$. This value is obtained through summing up of the arithmetic series $[2(L + S) + 1], [2(L + S - 1) + 1], \ldots, [2(L - S) + 1]$.

A1.2 Element 103, Lr : $[Rn]7s^2 5f^{14} 6d^1$ (usually expressed as $[Rn]5f^{14}6d^1 7s^2$).

Element 118　: $[Rn]7s^2 5f^{14}6d^{10}7p^6$, an inert gas.

Element 154　: $[Rn]7s^2 5f^{14}6d^{10}7p^6 8s^2 5g^{18} 6f^{14} 7d^2$; this is a metal in Group 14, its principal oxidation state being $+4$.

A1.3 (i) The alkali metals correspond to the start of a new external shell and hence an expanded volume.

(ii) In going from an alkali metal to the next alkaline earth metal, the addition of an electron in the same ns subshell does not compensate entirely for the effect of the increased nuclear charge.

(iii) These elements correspond to the filling of the np subshells. Their tendency to form covalent and van der Waals solids accounts for the increased atomic volumes.

(iv) In a transition series, successive d electrons are added to an incomplete inner shell. Thus, with an ever-increasing nuclear charge but no compensating increase in the distance of the entering electrons from the nucleus, the atom shrinks.

(v) Filling of the 4f subshells from ^{58}Ce to ^{71}Lu results in a continuous diminution in atomic size known as the "lanthanide contraction". Consequently the atomic volumes (and more importantly the chemical properties) of corresponding elements in the second and third long periods are remarkably similar, starting with the pair ^{40}Zr and ^{72}Hf and continuing as far as ^{47}Ag and ^{79}Au.

A1.4 For p^2, the L values for the terms are 2, 1, and 0; the spins are either singlet or triplet. Of the six possibilities, the ones excluded by the Pauli principle may be conveniently crossed out in the following manner:

$$\cancel{^3D} \quad {}^3P \quad \cancel{^3S}$$
$$^1D \quad \cancel{^1P} \quad {}^1S$$

leading to terms ^3P (ground), ^1D, and ^1S for this configuration. It is noted that the ^3D term clearly violates the Pauli principle.

Similarly, for d^2,

$$\cancel{^3G} \quad {}^3F \quad \cancel{^3D} \quad {}^3P \quad \cancel{^3S}$$
$$^1G \quad \cancel{^1F} \quad {}^1D \quad \cancel{^1P} \quad {}^1S$$

with ^3F as the ground term. For f^2,

$$\sout{^3I} \quad ^3H \quad \sout{^3G} \quad ^3F \quad \sout{^3D} \quad ^3P \quad \sout{^3S}$$
$$^1I \quad \sout{^1H} \quad ^1G \quad \sout{^1F} \quad ^1D \quad \sout{^1P} \quad ^1S$$

with ^3H as the ground term.

A1.5 (i) Short-cut: For this system consisting of inequivalent electrons, the exclusion principle is "automatically" obeyed. Hence the terms arising from this configuration can be written readily, i.e., without specifying all the microstates.

$$\ell_1 = \ell_2 = 1, \ L = \ell_1 + \ell_2, (\ell_1 + \ell_2) - 1, (\ell_1 + \ell_2) - 2, \dots, |\ell_1 - \ell_2| = 2, 1, 0.$$
$$s_1 = s_2 - \frac{1}{2}, \ S = s_1 + s_2, (s_1 + s_2) - 1, (s_1 + s_2) - 2, \dots, |s_1 - s_2| = 1, 0.$$

So the terms arising from configuration $2p^1 3p^1$ are: $^3D, \ ^3P, \ ^3S; \ ^1D, \ ^1P, \ ^1S.$

(ii) $^3D \oplus p^1 \Rightarrow {}^4F, \ ^4D, \ ^4P; \ ^2F, \ ^2D, \ ^2P; \qquad ^1D \oplus p^1 \Rightarrow {}^2F, \ ^2D, \ ^2P;$

$^3P \oplus p^1 \Rightarrow {}^4D, \ ^4P, \ ^4S; \ ^2D, \ ^2P, \ ^2S; \qquad ^1P \oplus p^1 \Rightarrow {}^2D, \ ^2P, \ ^2S;$

$^3S \oplus p^1 \Rightarrow {}^4P, \ ^2P; \qquad\qquad\qquad\qquad ^1S \oplus p^1 \Rightarrow {}^2P.$

So there are 21 terms all told.

A1.6 The microstates for this configuration may be tabulated in the following manner:

M_L \\ M_S	$1\frac{1}{2}$	$\frac{1}{2}$	$-\frac{1}{2}$	$-1\frac{1}{2}$
2		$(1^+ \ 1^- \ 0^+)$	$(1^+ \ 1^- \ 0^-)$	
1		$(1^+ \ 1^- \ -1^+)(1^+ \ 0^+ \ 0^-)$	$(1^+ \ 1^- \ -1^-)(1^- \ 0^+ \ 0^-)$	
0	$(1^+ \ 0^+ \ -1^+)$	$(1^+ \ 0^+ \ -1^-)(1^+ \ 0^- \ -1^+)$ $(1^- \ 0^+ \ -1^+)$	$(1^- \ 0^- \ -1^+)(1^- \ 0^+ \ -1^-)$ $(1^+ \ 0^- \ -1^-)$	$(1^- \ 0^- \ -1^-)$
-1		$(0^+ \ 0^- \ -1^+)(1^+ \ -1^+ \ -1^-)$	$(0^+ \ 0^- \ -1^-)(1^- \ -1^+ \ -1^-)$	
-2		$(0^+ \ -1^+ \ -1^-)$	$(0^- \ -1^+ \ -1^-)$	

The terms formed by these 20 microstates are ^4S (ground), ^2P, and ^2D.

A1.7   The microstates for this configuration may be tabulated in the following manner:

M_L \ M_S	$1\frac{1}{2}$	$\frac{1}{2}$	$-\frac{1}{2}$	$-1\frac{1}{2}$
5		$(2^+2^-1^+)$	$(2^+2^-1^-)$	
4		$(2^+2^-0^+)(2^+1^+1^-)$	$(2^+2^-0^-)(2^-1^+1^-)$	
3	$(2^+1^+0^+)$	$(2^+1^+0^-)(2^+1^-0^+)$ $(2^-1^+0^+)(2^+2^--1^+)$	$(2^+1^-0^-)(2^-1^+0^-)$ $(2^-1^-0^+)(2^+2^--1^-)$	$(2^-1^-0^-)$
2	$(2^+1^+-1^+)$	$(2^+1^+-1^-)(2^+1^--1^-)$ $(2^-1^+-1^+)(2^+2^--2^+)$ $(1^+1^-0^+)(2^+0^+0^-)$	$(2^+1^--1^-)(2^-1^+-1^-)$ $(2^-1^--1^+)(2^+2^--2^-)$ $(1^+1^-0^-)(2^-0^+0^-)$	$(2^-1^--1^-)$
1	$(2^+0^+-1^+)$ $(2^+1^+-2^+)$	$(2^+0^+-1^-)(2^+0^--1^-)$ $(2^-0^+-1^+)(2^+1^+-2^-)$ $(2^-1^+-2^+)(2^+1^--2^-)$ $(1^+1^--1^+)(1^+0^+0^-)$	$(2^+0^--1^-)(2^-0^+-1^-)$ $(2^-0^--1^+)(2^+1^--2^-)$ $(2^-1^+-2^-)(2^-1^--2^+)$ $(1^+1^--1^-)(1^-0^+0^-)$	$(2^-0^--1^-)$ $(2^-1^--2^-)$
0	$(2^+0^+-2^+)$ $(1^+0^+-1^+)$	$(2^+0^+-2^-)(2^+0^--2^+)$ $(2^-0^+-2^+)(1^+0^+-1^-)$ $(1^+0^--1^+)(1^-0^+-1^+)$ $(1^+1^--2^+)(2^+-1^+-1^-)$	$(2^+0^--2^-)(2^-0^+-2^-)$ $(2^-0^--2^+)(1^+0^--1^-)$ $(1^-0^+-1^-)(1^-0^--1^+)$ $(1^+1^--2^-)(2^--1^+-1^-)$	$(2^-0^--2^-)$ $(1^-0^--1^-)$
-1	$(-2^+0^+1^+)$ $(-2^+-1^+2^+)$	$(-2^+0^+1^-)(-2^+0^-1^+)$ $(-2^-0^+1^+)(-2^+-1^+2^-)$ $(-2^+-1^-2^+)(-2^--1^+2^+)$ $(-1^+-1^-1^+)(-1^+0^+0^-)$	$(-2^+0^-1^-)(-2^-0^+1^-)$ $(-2^-0^-1^+)(-2^+-1^-2^-)$ $(-2^--1^+2^-)(-2^--1^-2^+)$ $(-1^+-1^-1^-)(-1^-0^+0^-)$	$(-2^-0^-1^-)$ $(-2^--1^-2^-)$
-2	$(-2^+-1^+1^+)$	$(-2^+1^+-1^-)(-2^+1^--1^+)$ $(-2^-1^+-1^+)(-2^+-2^-2^+)$ $(-1^+-1^-0^+)(-2^+0^+0^-)$	$(-2^+1^--1^-)(-2^-1^+-1^-)$ $(-2^-1^--1^+)(-2^+-2^-2^-)$ $(-1^+-1^-0^-)(-2^-0^+0^-)$	$(-2^--1^-1^-)$
-3	$(-2^+-1^+0^+)$	$(-2^+-1^+0^-)(-2^+-1^-0^+)$ $(-2^--1^+0^+)(-2^+-2^-1^+)$	$(-2^+-1^-0^-)(-2^--1^+0^-)$ $(-2^--1^-0^+)(-2^+-2^-1^-)$	$(-2^--1^-0^-)$
-4		$(-2^+-2^-0^+)(-2^+-1^+-1^-)$	$(-2^+-2^-0^-)(-2^--1^+-1^-)$	
-5		$(-2^+-2^--1^+)$	$(-2^+-2^--1^-)$	

The terms formed by these microstates are ^4F (ground), ^4P, ^2H, ^2G, ^2F, $2\,^2D$, and ^2P.

A1.8   (i)   According to Hund's rule, the ground term should have the S value as large as possible and hence both d and s shells are half filled, with all six electrons being unpaired. So $S = 3$ and $L = 0$, resulting in ^7S for the ground term.

(ii)   There are 16 terms arising from d^5. These terms have non-zero S values. So adding one s electron to the system, there should be 32 terms.

A1.9

Free ion	High-spin complexes	Low-spin complexes
$d^1, {}^2D$	$t_{2g}^1, {}^2T_{2g}$	—
$d^2, {}^3F$	$t_{2g}^2, {}^3T_{1g}$	—
$d^3, {}^4F$	$t_{2g}^3, {}^4A_{2g}$	—
$d^4, {}^5D$	$t_{2g}^3 e_g^1, {}^5E_g$	$t_{2g}^4, {}^3T_{1g}$
$d^5, {}^6S$	$t_{2g}^3 e_g^2, {}^6A_{1g}$	$t_{2g}^5, {}^2T_{2g}$

Free ion	High-spin complexes	Low-spin complexes
$d^6, {}^5D$	$t_{2g}^4 e_g^2, {}^5T_{2g}$	$t_{2g}^6, {}^1A_{1g}$
$d^7, {}^4F$	$t_{2g}^5 e_g^2, {}^4T_{1g}$	$t_{2g}^6 e_g^1, {}^2E_g$
$d^8, {}^3F$	$t_{2g}^6 e_g^2, {}^3A_{2g}$	—
$d^9, {}^2D$	$t_{2g}^6 e_g^3, {}^2E_g$	—

Comments:

(a) "Obvious" cases: d^1, d^9, high- and low-spin d^5, low-spin d^6, and low-spin d^7. Many others can be derived by applying the following rule: high-spin complexes with configurations d^{n+5} and d^n have the same orbital ground state.

(b) The ground term for d^2 complexes (configuration t_{2g}^2) is ${}^3T_{1g}$ [which can be derived by the steps shown in A1.23(iv)]. Also, for d^3 (atomic term ${}^4F$) complexes, since a F term splits into A_{2g}, T_{1g}, and T_{2g} in an octahedral field, and with a half filled configuration t_{2g}^3, the ground term should be the orbitally non-degenerate ${}^4A_{2g}$. All others may now be readily obtained.

(c) Ground terms for tetrahedral complexes may also be readily written down: the ground state of a d^n tetrahedral complex is the same as that of a d^{10-n} octahedral complex. The subscript "g" should however be dropped, since tetrahedral complexes lack an inversion center.

A1.10 The terms arising from configuration $3d^9 4p^1$ are identical to those from $d^1 p^1$: ${}^3F$, ${}^3D$, ${}^3P$, ${}^1F$, ${}^1D$, and ${}^1P$. Since Hund's rule is only applicable to systems with equivalent electrons, nothing definite can be said as to which is the ground term. If Hund's rule held, ${}^3F$ would have the lowest energy. As it turns out for Zn^{2+}, ${}^3P$ does.

A1.11 Hund's rule is only applicable for the determination of ground term. For half filled configurations, the L value of the ground state is always equal to zero. For such a term, there is only one J value: $J = L + S, L + S - 1, \ldots, |L - S| = S$, as $L = 0$.

A1.12 The last term in the formula,

$$L = \ell_1 + \ell_2, (\ell_1 + \ell_2) - 1, (\ell_1 + \ell_2) - 2, \ldots$$

is $|\ell_1 - \ell_2|$, which happens to be equal to $\ell_1 + \ell_2$ in this case with $\ell_1 = 0$ and $\ell_2 = 1$. Hence, there is only one value for L: $L = 0 + 1 = |0 - 1| = 1$. So we only have P terms for $ns^1 np^1$. When spin is taken into account, the terms for this configuration are ^1P and ^3P.

A1.13 (i) $\nu = c/\lambda = (3.00 \times 10^8 \text{m s}^{-1})/(6710 \times 10^{-10} \text{ m}) = 4.47 \times 10^{14}$ Hz.

(ii) $\tilde{\nu} = 1/\lambda = 1/(6710 \times 10^{-10} \text{ m}) = 1/(6710 \times 10^{-8} \text{ cm}) = 1.49 \times 10^4 \text{ cm}^{-1}$.

(iii) $(1s^2 2s^1)^2 S \rightarrow (1s^2 2p^1)^2 P$.

A1.14 (i) Chromium, with ground configuration $3d^5 4s^1$, will have $^7S_3$ as its ground term.

(ii) Ions Mn^+ (ground configuration $3d^5 4s^1$) and Fe^{2+} (excited configuration $3d^5 4s^1$) can also have $^7S_3$; Co^{3+} is also acceptable.

A1.15 (i) The energy of the configuration is weighted with the numbers of microstates (which are also the degenerancies) of the terms. For example, for $^3F, L = 3, S = 1$, the numbers of microstates is $(2L + 1)(2S + 1) = 21$. The denominator is the sum of the numbers of microstates, which is 45 in this case.

(ii) (a) The weighted-average energy for ^3P:

$$[1(0.0) + 3(16.4) + 5(43.5)]/9 = 29.6 \text{ cm}^{-1}.$$

Similarly, $E(^1D) = 10197.7$ and $E(^1S) = 21648.4 \text{ cm}^{-1}$.

(b) The weighted-average energy for the $1s^2 2s^2 2p^2$ configuration is

$$[9(29.6) + 5(10197.7) + 1(21648.4)]/15 = 4860.2 \text{ cm}^{-1}.$$

A1.16 (i) $s^1(\ell = 0, s = \frac{1}{2}) \oplus d^1(\ell = 2, s = \frac{1}{2}) \Rightarrow {}^3D, {}^1D$.

$^3D(J = 3, 2, 1) \Rightarrow {}^3D_3, {}^3D_2, {}^3D_1$.

$^1D(J = 2) \Rightarrow {}^1D_2$.

(ii) $\ell_1 = 0, \ell_2 = 2, s_1 = s_2 = \frac{1}{2}, j_1 = \frac{1}{2}, j_2 = 2\frac{1}{2}, 1\frac{1}{2}$.

$j_1 = \frac{1}{2}, j_2 = 2\frac{1}{2}, \Rightarrow J = 3, 2$.

$j_1 = \frac{1}{2}, j_2 = 1\frac{1}{2}, \Rightarrow J = 2, 1$.

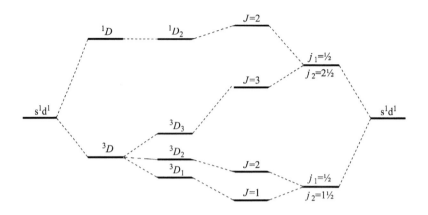

A1.17 For Ca, the ground configuration is $4s^2$.

Excited state configuration	Term	Levels
$4s^1 4p^1$	^3P	$^3P_2, {}^3P_1, {}^3P_0$
$4s^1 4p^0 3d^1$	^3D	$^3D_3, {}^3D_2, {}^3D_1$

The transitions corresponding to the lines in the emission spectrum are illustrated below.

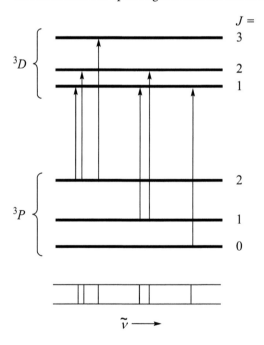

A1.18 (i) The ground term for Mg is $^1S_0$.

(ii) States arising from configuration $3s^1 3p^1$ are $^3P_2$, $^3P_1$, $^3P_0$, and $^1P_1$. Allowed transition from the ground state to these excited states is $^1S_0 \rightarrow {}^1P_1$.

(iii) For configuration $\ldots 2p^6 3p^1 3d^1$, $\ell_1 = 1$ and $\ell_2 = 2 \Rightarrow L = 3, 2, 1$. It follows that $s_1 = s_2 = \frac{1}{2} \Rightarrow S = 1, 0$. So, there are 12 states arising from the configuration: $^3F_4$, $^3F_3$, $^3F_2$, $^3D_3$, $^3D_2$, $^3D_1$, $^3P_2$, $^3P_1$, $^3P_0$, $^1F_3$, $^1D_2$, $^1P_1$. Again, the only allowed transition from the ground state of Mg to the states arising from configuration $3p^1 3d^1$ is $^1S_0 \rightarrow {}^1P_1$.

A1.19 (i) Ground configuration of Na atom: $\ldots 3s^1 \Rightarrow (L = 0, S = \frac{1}{2}, J = \frac{1}{2}) \Rightarrow {}^2S_{1/2}$.

(ii) First excited electronic configuration of Na atom: $\ldots 3p^1 \Rightarrow (L = 1, S = \frac{1}{2}, J = 1\frac{1}{2}, \frac{1}{2}) \Rightarrow {}^2P_{1 1/2}$, $^2P_{1/2}$, with the latter being the ground state of the configuration, according to Hund's rule. In other words, the emission lines at 5890 and 5896 Å may be assigned to transitions $^2S_{1/2} \rightarrow {}^2P_{1 1/2}$ and $^2S_{1/2} \rightarrow {}^2P_{1/2}$, respectively.

A1.20 $^3P_2 \rightarrow {}^1D_2$, $\Delta E = 10152\,\text{cm}^{-1}$;
$^3P_1 \rightarrow {}^1D_2$, $\Delta E = 10179\,\text{cm}^{-1}$;
$^3P_1 \rightarrow {}^1S_0$, $\Delta E = 21632\,\text{cm}^{-1}$.

Note that all these transitions are spin-forbidden (refer to the selection rule stated in Problem 1.17).

A1.21 (i) (A) He: $1s^2$, $^1S_0$. (B) He: $1s^1 2s^1$, $^3S_1$. (C) He: $1s^1 2s^1$, $^1S_0$. (D) Ne: $1s^2 2s^2 2p^6$, $^1S_0$.

(ii) Terms arising from configuration $2p^5 3s^1$, $2p^5 4s^1$, and $2p^5 5s^1$ are ^3P and ^1P.

(iii) Terms arising from configuration $2p^5 3p^1$ and $2p^5 4p^1$ are ^3D, ^3P, ^3S, ^1D, ^1P, and ^1S.

A1.22 (i) The direct product of E_{1g} and E_{2g} is $E_{1g} \times E_{2g} = E_{1g} + E_{2g}$. Hence the terms arising from the configuration $(e_{1g})^1 (e_{2g})^1$ are $^1E_{1g}$, $^1E_{2g}$, $^3E_{1g}$, and $^3E_{2g}$.

D_{5d}	E	$2C_5$	$2C_5^2$	$5C_2$	i	$2S_{10}^3$	$2S_{10}$	$5\sigma_d$
E_{1g}	2	$-\eta^-$	$-\eta^+$	0	2	$-\eta^-$	$-\eta^+$	0
E_{2g}	2	$-\eta^+$	$-\eta^-$	0	2	$-\eta^+$	$-\eta^-$	0
$E_{1g} \times E_{2g}$	4	§	§	0	4	§	§	0

$\eta^\pm = (1 \pm \sqrt{5})/2$.
$^\S \eta^+ \eta^- = -1 = -(\eta^+ + \eta^-)$.

(ii) The terms arising from the configuration $(1\delta)^3(2\delta)^1$ are the same as those from $(1\delta)^1(2\delta)^1$. The direct product $\Delta \times \Delta$ is $\Sigma^+ + \Sigma^- + \Gamma$. Hence the terms arising from the configuration $(1\delta)^3(2\delta)^1$ are: $^3\Sigma^+, ^1\Sigma^+, ^3\Sigma^-, ^1\Sigma^-, ^3\Gamma, ^1\Gamma$.

$C_{\infty v}$	E	$2C_\infty^\phi$	$2C_\infty^{2\phi}$	...	$\infty\sigma_v$
Σ^+	1	1	1		1
Σ^-	1	1	1		-1
Δ	2	$2\cos 2\phi$	$2\cos 4\phi$		0
Γ	2	$2\cos 4\phi$	$2\cos 8\phi$		0
$\Delta \times \Delta$	4	$4\cos^2 2\phi = 2\cos 4\phi + 2$	$4\cos^2 4\phi = 2\cos 8\phi + 2$		0

A1.23 (i)

C_{3v}	E	$2C_3$	$3\sigma_v$	
A_1	1	1	1	
A_2	1	1	-1	
E	2	-1	0	

R	E	C_3	σ_v	
R^2	E	C_3	E	
$\chi(R)$ in E	2	-1	0	
$\chi^2(R)$ in E	4	1	0	
$\chi(R^2)$ in E	2	-1	2	
χ (singlet)	3	0	1	$= A_1 + E$
χ (triplet)	1	1	-1	$= A_2$

Therefore, $(e)^2 \Rightarrow {}^3A_2, {}^1A_1, {}^1E$.

(ii)

$C_{\infty v}$	E	$2C_\infty^\phi$	$2C_\infty^{2\phi}$	$\cdots$	$\infty\sigma_v$	
Σ^+	1	1	1		1	
Σ^-	1	1	1		-1	
Π	2	$2\cos\phi$	$2\cos 2\phi$		0	
Δ	2	$2\cos 2\phi$	$2\cos 4\phi$		0	
R	E	C_∞^ϕ	$C_\infty^{2\phi}$	$\cdots$	$\infty\sigma_v$	
R^2	E	$C_\infty^{2\phi}$	$C_\infty^{4\phi}$	$\cdots$	E	
$\chi(R)$ in Π	2	$2\cos\phi$	$2\cos 2\phi$		0	
$\chi^2(R)$ in Π	4	$4\cos^2\phi = 2 + 2\cos 2\phi$	$4\cos^2 2\phi = 2 + 2\cos 4\phi$		0	
$\chi(R^2)$ in Π	2	$2\cos 2\phi$	$2\cos 4\phi$		2	
χ (singlet)	3	$1 + 2\cos 2\phi$	$1 + 2\cos 4\phi$		1	$= \Sigma^+ + \Delta$
χ (triplet)	1	1	1		-1	$= \Sigma^-$

Therefore, $(\pi)^2 \Rightarrow {}^3\Sigma^-, {}^1\Sigma^+, {}^1\Delta$.

(iii)

T_d	E	$8C_3$	$3C_2$	$6S_4$	$6\sigma_d$	
A_1	1	1	1	1	1	
E	2	-1	2	0	0	
T_1	3	0	-1	1	-1	
T_2	3	0	-1	-1	1	
R	E	C_3	C_2	S_4	σ_d	
R^2	E	C_3	E	C_2	E	
R^3	E	E	C_2	S_4	σ_d	
$\chi(R)$ in T_1	3	0	-1	1	-1	
$\chi^3(R)$ in T_1	27	0	-1	1	-1	
$\chi(R^2)$ in T_1	3	0	3	-1	3	
$\chi(R^3)$ in T_1	3	3	-1	1	-1	
χ (doublet)	8	-1	0	0	0	$= E + T_1 + T_2$
χ (quartet)	1	1	1	1	1	$= A_1$

Therefore, $(t_1)^3 \Rightarrow {}^4A_1, {}^2E, {}^2T_1, {}^2T_2$.

(iv) Short-cut: Firstly, the center of inversion is ignored and we consider the case for the configuration $(t_2)^2$ for an O molecule:

O	E	$6C_4$	$3C_2(= C_4^2)$	$8C_3$	$6C_2$	
A_1	1	1	1	1	1	
E	2	0	2	-1	0	
T_1	3	1	-1	0	-1	
T_2	3	-1	-1	0	1	
R	E	C_4	C_2	C_3	C_2	
R^2	E	C_2	E	C_3	E	
$\chi(R)$ in T_2	3	-1	-1	0	1	
$\chi^2(R)$ in T_2	9	1	1	0	1	
$\chi(R^2)$ in T_2	3	-1	3	0	3	
χ (singlet)	6	0	2	0	2	$= A_1 + E + T_2$
χ (triplet)	3	1	-1	0	-1	$= T_1$

Therefore, $(t_2)^2$ in $O \Rightarrow {}^3T_1, {}^1A_1, {}^1E, {}^1T_2$.

For $(t_{2u})^2$ in O_h: parity of center operation $i = u \times u$ (for two electrons)

$$= g \Rightarrow {}^3T_{1g}, {}^1A_{1g}, {}^1E_g, {}^1T_{2g}.$$

(v) As in (iv), the center of inversion is ignored and we consider the case for the configuration $(h)^5$ for an I molecule:

I	E	$12C_5$	$12C_5^2$	$20C_3$	$15C_2$	$\eta^{\pm} = (1 \pm \sqrt{5})/2$
A	1	1	1	1	1	
T_1	3	η^+	η^-	0	-1	
T_2	3	η^-	η^+	0	-1	
G	4	-1	-1	1	0	
H	5	0	0	-1	1	
R	E	C_5	C_5^2	C_3	C_2	
R^2	E	C_5^2	C_5	C_3	E	
R^3	E	C_5^2	C_5	E	C_2	
R^4	E	C_5	C_5^2	C_3	E	
R^5	E	E	E	C_3	C_2	

$\chi(R)$ in H	5	0	0	-1	1
$\chi(R^2)$ in H	5	0	0	-1	5
$\chi(R^3)$ in H	5	0	0	5	1
$\chi(R^4)$ in H	5	0	0	-1	5
$\chi(R^5)$ in H	5	5	5	-1	1
χ (doublet)	75	0	0	0	3	$= 2A + 3T_1 + 3T_2 + 5G + 7H$
χ (quartet)	24	-1	-1	0	0	$= T_1 + T_2 + 2G + 2H$
χ (hextet)	1	1	1	1	1	$= A$

Therefore, $(h)^5$ in $I \Rightarrow {}^6A, {}^4T_1, {}^4T_2, 2{}^4G, 2{}^4H, 2{}^2A, 3{}^2T_1, 3{}^2T_2, 5{}^2G, 7{}^2H$.

For $(h_g)^5$ in I_h: parity of center operation $i = g \times g \times g \times g \times g$ (for five electrons) $= g$.

Hence, $(h_g)^5 \Rightarrow {}^6A_g, {}^4T_{1g}, {}^4T_{2g}, 2{}^4G_g, 2{}^4H_g, 2{}^2A_g, 3{}^2T_{1g}, 3{}^2T_{2g}, 5{}^2G_g, 7{}^2H_g$.

A1.24 (i)

D_{6h}	E	$2C_6$	$2C_3$	C_2	$3C_2'$	$3C_2''$	i	$2S_3$	$2S_6$	σ_h	$3\sigma_d$	$3\sigma_v$	
A_{1g}	1	1	1	1	1	1	1	1	1	1	1	1	
A_{2g}	1	1	1	1	-1	-1	1	1	1	1	-1	-1	
E_{1g}	2	1	-1	-2	0	0	2	1	-1	-2	0	0	
E_{2g}	2	-1	-1	2	0	0	2	-1	-1	2	0	0	
R	E	C_6	C_3	C_2	C_2'	C_2''	i	S_3	S_6	σ_h	σ_d	σ_v	
R^2	E	C_3	C_3	E	E	E	E	C_3	C_3	E	E	E	
$\chi(R)$ in E_{1g}	2	1	-1	-2	0	0	2	1	-1	-2	0	0	
$\chi^2(R)$ in E_{1g}	4	1	1	4	0	0	4	1	1	4	0	0	
$\chi(R^2)$ in E_{1g}	2	-1	-1	2	2	2	2	-1	-1	2	2	2	
χ (singlet)	3	0	0	3	1	1	3	0	0	3	1	1	$= A_{1g} + E_{2g}$
χ (triplet)	1	1	1	1	-1	-1	1	1	1	1	-1	-1	$= A_{2g}$

Therefore, $(e_{1g})^2 \Rightarrow {}^3A_{2g} + {}^1A_{1g} + {}^1E_{2g}$.

(ii)

D_{6h}	E	$2C_6$	$2C_3$	C_2	$3C_2'$	$3C_2''$	i	$2S_3$	$2S_6$	σ_h	$3\sigma_d$	$3\sigma_v$	
A_{1g}	1	1	1	1	1	1	1	1	1	1	1	1	
A_{2g}	1	1	1	1	−1	−1	1	1	1	1	−1	−1	
E_{2g}	2	−1	−1	2	0	0	2	−1	−1	2	0	0	
E_{2u}	2	−1	−1	2	0	0	−2	1	1	−2	0	0	
R	E	C_6	C_3	C_2	C_2'	C_2''	i	S_3	S_6	σ_h	σ_d	σ_v	
R^2	E	C_3	C_3	E	E	E	E	C_3	C_3	E	E	E	
$\chi(R)$ in E_{2u}	2	−1	−1	2	0	0	−2	1	1	−2	0	0	
$\chi^2(R)$ in E_{2u}	4	1	1	4	0	0	4	1	1	4	0	0	
$\chi(R^2)$ in E_{2u}	2	−1	−1	2	2	2	2	−1	−1	2	2	2	
χ (singlet)	3	0	0	3	1	1	3	0	0	3	1	1	$= A_{1g} + E_{2g}$
χ (triplet)	1	1	1	1	−1	−1	1	1	1	1	−1	−1	$= A_{2g}$

Therefore, $(e_{2u})^2 \Rightarrow {}^3A_{2g} + {}^1A_{1g} + {}^1E_{2g}$, the same three terms from $(e_{1g})^2$.

(iii)
$$
\begin{aligned}
&{}^3A_{2g} \times {}^3A_{2g} = {}^5A_{1g} + {}^3A_{1g} + {}^1A_{1g} \qquad {}^1A_{1g} \times {}^1E_{2g} = {}^1E_{2g} \\
&{}^3A_{2g} \times {}^1A_{1g} = {}^3A_{2g} \qquad\qquad\qquad\quad {}^1E_{2g} \times {}^3A_{2g} = {}^3E_{2g} \\
&{}^3A_{2g} \times {}^1E_{2g} = {}^3E_{2g} \qquad\qquad\qquad\quad {}^1E_{2g} \times {}^1A_{1g} = {}^1E_{2g} \\
&{}^1A_{1g} \times {}^3A_{2g} = {}^3A_{2g} \qquad\qquad\qquad\quad {}^1E_{2g} \times {}^1E_{2g} = {}^1A_{1g} + {}^1A_{2g} + {}^1E_{2g} \\
&{}^1A_{1g} \times {}^1A_{1g} = {}^1A_{1g}
\end{aligned}
$$

Altogether, there are 13 terms: ${}^5A_{1g}, {}^3A_{1g}, 3{}^1A_{1g}, 2{}^3A_{2g}, {}^1A_{2g}, 2{}^3E_{2g}, 3{}^1E_{2g}$.

A1.25 (i) Operation in the double group D_3': $E, R, C_3, C_3R, C_3^2, C_3^2R, 3C_2, 3C_2R$, a total of 12 operations.

(ii) The 12 operations can be divided into six classes: E, R, C_3 and C_3^2R, C_3^2 and C_3R, $3C_2, 3C_2R$.

A1.26

D_4'	E	R	C_4	C_4^3	C_2	$2C_2'$	$2C_2''$	
$(h = 16)$			C_4^3R	C_4R	C_2R	$2C_2'R$	$2C_2''R$	
$\chi(\alpha); J = 2\frac{1}{2}$	6	−6	$-\sqrt{2}$	$\sqrt{2}$	0	0	0	$= \Gamma_6 + 2\Gamma_7$

Details: $\chi(\alpha) = \dfrac{\sin\left(J + \frac{1}{2}\right)\alpha}{\sin\frac{\alpha}{2}} = \dfrac{\sin 3\alpha}{\sin\frac{\alpha}{2}}.$

$\alpha = 4\pi : \chi(4\pi) = \lim_{\alpha \to 4\pi} \dfrac{\sin 3\alpha}{\sin\frac{\alpha}{2}} = \lim_{\alpha \to 4\pi} \dfrac{3\cos 3\alpha}{\frac{1}{2}\cos\frac{\alpha}{2}} = 6.$

$\alpha = 2\pi : \chi(2\pi) = \lim_{\alpha \to 2\pi} \dfrac{\sin 3\alpha}{\sin\frac{\alpha}{2}} = \lim_{\alpha \to 2\pi} \dfrac{3\cos 3\alpha}{\frac{1}{2}\cos\frac{\alpha}{2}} = \dfrac{3 \cdot 1}{\frac{1}{2} \cdot (-1)} = -6.$

$\alpha = \dfrac{\pi}{2} : \chi\left(\dfrac{\pi}{2}\right) = \dfrac{\sin\frac{3\pi}{2}}{\sin\frac{\pi}{4}} = \dfrac{-1}{\left(\frac{1}{\sqrt{2}}\right)} = -\sqrt{2}.$

$\alpha = \dfrac{3\pi}{2} : \chi\left(\dfrac{3\pi}{2}\right) = \dfrac{\sin\frac{9\pi}{2}}{\sin\frac{3\pi}{4}} = \dfrac{1}{\left(\frac{1}{\sqrt{2}}\right)} = \sqrt{2}.$

$\alpha = \pi : \chi(\pi) = \dfrac{\sin 3\pi}{\sin\frac{\pi}{2}} = 0.$

A1.27 (i) Straightforward application of the given formula yields the following results:

J	E	R	$4C_3$ $4C_3^2 R$	$4C_3^2$ $4C_3 R$	$3C_2$ $3C_2 R$	$3C_4$ $3C_4^3 R$	$3C_4^3$ $3C_4 R$	$6C_2'$ $6C_2' R$	$\Gamma's$
$\frac{1}{2}$	2	-2	1	-1	0	$\sqrt{2}$	$-\sqrt{2}$	0	Γ_6
$1\frac{1}{2}$	4	-4	-1	1	0	0	0	0	Γ_8
$2\frac{1}{2}$	6	-6	0	0	0	$-\sqrt{2}$	$\sqrt{2}$	0	$\Gamma_7 + \Gamma_8$
$3\frac{1}{2}$	8	-8	1	-1	0	0	0	0	$\Gamma_6 + \Gamma_7 + \Gamma_8$
$4\frac{1}{2}$	10	-10	-1	1	0	$\sqrt{2}$	$-\sqrt{2}$	0	$\Gamma_6 + 2\Gamma_8$
$5\frac{1}{2}$	12	-12	0	0	0	0	0	0	$\Gamma_6 + \Gamma_7 + 2\Gamma_8$
$6\frac{1}{2}$	14	-14	1	-1	0	$-\sqrt{2}$	$\sqrt{2}$	0	$\Gamma_6 + 2\Gamma_7 + 2\Gamma_8$

As an example, take the case of $J = \frac{1}{2}$ and $\alpha = 2\pi$ (operation R).

$$\chi(2\pi) = \lim_{\alpha \to 2\pi} \dfrac{\sin\alpha}{\sin\frac{\alpha}{2}} = \lim_{\alpha \to 2\pi} \dfrac{\cos\alpha}{\frac{1}{2}\cos\frac{\alpha}{2}} = -2.$$

(ii) (a) In an O field, when spin-orbit interaction is ignored, the characters of the representations for ^4F $(L = 3)$ are determined below:

L	E	R	$4C_3$ $4C_3^2 R$	$4C_3^2$ $4C_3 R$	$3C_2$ $3C_2 R$	$3C_4$ $3C_4^3 R$	$3C_4^3$ $3C_4 R$	$6C_2'$ $6C_2' R$	Overall splitting
3	7	7	1	1	-1	-1	-1	-1	$= \Gamma_2 + \Gamma_4 + \Gamma_5$

Therefore, the ^4F term splits into $^4\Gamma_2$, $^4\Gamma_4$, and $^4\Gamma_5$ states. By considering spin-orbit coupling, these states are further split. From (i), when $S = 1\frac{1}{2}$, the spin state has Γ_8 symmetry. When this spin state is coupled with the orbital parts (Γ_2, Γ_4, and Γ_5), the resultant states are:

$$\Gamma_2 \times \Gamma_8 = \Gamma_8,$$

$$\Gamma_4 \times \Gamma_8 = \Gamma_6 + \Gamma_7 + 2\Gamma_8,$$

$$\Gamma_5 \times \Gamma_8 = \Gamma_6 + \Gamma_7 + 2\Gamma_8.$$

This means that $^4\Gamma_2$ is not split, but $^4\Gamma_4$ and $^4\Gamma_5$ are each split into four states. Note now that spin quantum number is no longer used to define the states.

(b)   When spin-orbit interaction is considered first, for ^4F, $J = 4\frac{1}{2}, 3\frac{1}{2}, 2\frac{1}{2}$, and $1\frac{1}{2}$. When these states are subjected to the effects of an O-field, from (i), all, except $J = 1\frac{1}{2}$, are further split:

$$J = 4\tfrac{1}{2} \Rightarrow \Gamma_6 + 2\Gamma_8,$$

$$J = 3\tfrac{1}{2} \Rightarrow \Gamma_6 + \Gamma_7 + \Gamma_8,$$

$$J = 2\tfrac{1}{2} \Rightarrow \Gamma_7 + \Gamma_8,$$

$$J = 1\tfrac{1}{2} \Rightarrow \Gamma_8.$$

Note that cases (a) and (b) lead to the same results, as should be the same case.

A1.28 (i)

	Splitting in octahedral crystal field	Further splitting with spin-orbit interaction "turned on"; $\Gamma(S = 1\frac{1}{2}) = \Gamma_8$ (Refer to the O' character table in previous problem.)	Overall splitting
^4G	Γ_1	$\Gamma_1 \times \Gamma_8 = \Gamma_8$	
	Γ_3	$\Gamma_3 \times \Gamma_8 = \Gamma_6 + \Gamma_7 + \Gamma_8$	$= 3\Gamma_6 + 3\Gamma_7 + 6\Gamma_8$
	Γ_4	$\Gamma_4 \times \Gamma_8 = \Gamma_6 + \Gamma_7 + 2\Gamma_8$	
	Γ_5	$\Gamma_5 \times \Gamma_8 = \Gamma_6 + \Gamma_7 + 2\Gamma_8$	

(ii)

	Splitting with spin-orbit interaction "turned on"; $L = 4, S = 1\frac{1}{2}$	Further splitting when placed in octahedral crystal field	Overall splitting
^4G	$J = 5\frac{1}{2}$	$\Gamma_6 + \Gamma_7 + 2\Gamma_8$	$= 3\Gamma_6 + 3\Gamma_7 + 6\Gamma_8$
	$J = 4\frac{1}{2}$	$\Gamma_6 + 2\Gamma_8$	
	$J = 3\frac{1}{2}$	$\Gamma_6 + \Gamma_7 + \Gamma_8$	
	$J = 2\frac{1}{2}$	$\Gamma_7 + \Gamma_8$	

A1.29 (i)

	Splitting in C_{4v} crystal field	Further splitting with spin-orbit interaction "turned on"; $\Gamma(S = \frac{1}{2}) = \Gamma_6$	Overall splitting
^2D	$^2A_1$ (or Γ_1)	$\Gamma_1 \times \Gamma_6 = \Gamma_6$	$= 2\Gamma_6 + 3\Gamma_7$
	$^2B_1$ (or Γ_3)	$\Gamma_3 \times \Gamma_6 = \Gamma_7$	
	$^2B_2$ (or Γ_4)	$\Gamma_4 \times \Gamma_6 = \Gamma_7$	
	^2E (or Γ_5)	$\Gamma_5 \times \Gamma_6 = \Gamma_6 + \Gamma_7$	

(ii)

	Splitting with spin-orbit interaction "turned on"; $L = 2, S = \frac{1}{2}$	Further splitting when placed in C_{4v} crystal field	Overall splitting
^2D	$J = 2\frac{1}{2}$	$\Gamma_6 + 2\Gamma_7$	$= 2\Gamma_6 + 3\Gamma_7$
	$J = 1\frac{1}{2}$	$\Gamma_6 + \Gamma_7$	

2 | Introductory Quantum Chemistry

PROBLEMS

2.1 Estimate the number of photons emitted per second by a sodium vapor lamp. Assume a wavelength of 589 nm and the light power is 3.5 W.

2.2 A red laser-pointer emits light of 650 nm and a green laser-pointer emits light of 532 nm. Estimate the number of photons emitted per second by each laser-pointer. Assume the light power is 1 mW for both laser-pointers.

2.3 Electron diffraction makes use of 40 keV (40,000 eV) electrons. Calculate their de Broglie wavelength.

2.4 Start with setting $\hbar$ ($= 1.05457 \times 10^{-34}$ J s), m_e ($= 9.109382 \times 10^{-31}$ kg), and a_0 ($= 5.29177 \times 10^{-11}$ m) to 1 a. u., express 1 a. u. for energy in terms of J.

2.5 The fine-structure constant, α, is given by: $\alpha = e^2/4\pi \varepsilon_0 \hbar c$.

   (i)   Determine α in atomic units. Note that $4\pi \varepsilon_0$ is set to 1 a. u.

   (ii) By direct substitution, determine α in SI units. Given: $e = 1.60217653 \times 10^{-19}$ C and $\varepsilon_0 = 8.85419 \times 10^{-12}$ J^{-1} C^2 m^{-1}.

REFERENCE: B. K. Teo and W.-K. Li, The scales of time, length, mass, energy, and other fundamental physical quantities in the atomic world and the use of atomic units in quantum mechanical calculations. *J. Chem. Educ.* **88**, 921–8 (2011).

2.6 Quantum mechanically, when a particle of mass m is confined to a one-dimensional box with length L and inside a bottomless well where there is no potential energy, i.e., $V = 0$, the ground state energy (E_1) of this particle is:

$$E_1 = \frac{h^2}{8mL^2} > 0.$$

Problems in Structural Inorganic Chemistry. Second edition. Wai-Kee Li, Hung Kay Lee, Dennis Kee Pui Ng, Yu-San Cheung, Kendrew Kin Wah Mak, and Thomas Chung Wai Mak. © Oxford University Press 2019. Published in 2019 by Oxford University Press. DOI: 10.1093/oso/9780198823902.001.0001

(i)   What is the uncertainty in position, Δx, for the particle?

(ii) What is the momentum p of the particle in its ground state?

(iii) Calculate $\Delta x \Delta p$ by approximating the uncertainty in momentum, Δp, to be twice the absolute value of the momentum you obtained in part (ii). Is your result in agreement with the uncertainty principle?

(iv) In part (iii) why should we approximate Δp to be twice the absolute value of the momentum?

2.7 (i) Apply the uncertainty principle, $\Delta p_x \Delta x \approx \frac{h}{4\pi} = 5.28 \times 10^{-35}$ J s, to complete the following table.

Particle	Mass (kg)	Speed (v) $(\mathrm{m\,s^{-1}})$	$\Delta v = 10\%$ of v $(\mathrm{m\,s^{-1}})$	Δx (m)
The Earth	5.98×10^{24}	2.97×10^3		
A 100-meter sprinter running at world-record speed	1.00×10^2	10.4		
Bullet	1.00×10^{-2}	1.00×10^3		
Dust	1.00×10^{-9}	1.00×10^1		
Pollen in Brownian motion	1.00×10^{-13}	1.00		
Xe atom at 298 K	2.18×10^{-25}	2.38×10^2		
Electron in a molecule	9.11×10^{-31}	1.00×10^3		

(ii) Based on the Δx values you have obtained, comment on from which case(s) the results are negligible, and which case(s) the results are intolerable. State clearly your reasoning.

2.8 For the particle-in-a-cube problem, the wavefunctions are:

$$\psi_{n_x\,n_y\,n_z}(x,\,y,\,z) = \left(\frac{2}{a}\right)^{3/2} \sin\left(\frac{n_x \pi x}{a}\right) \sin\left(\frac{n_y \pi y}{a}\right) \sin\left(\frac{n_z \pi z}{a}\right),$$

where a is the length of an edge of the cube and quantum numbers n_x, n_y, and n_z take on the values 1, 2, 3 .... The origin of the chosen coordinates is at one of the vertices of the cube and the positive x-, y-, and z-axes coincide with the edges as shown in the figure on the right. Consider the (2, 1, 1) state in a cube with dimensions $1\,\text{Å} \times 1\,\text{Å} \times 1\,\text{Å}$ and a volume element $\Delta\tau$ whose dimensions are $\Delta x = \Delta y = \Delta z = 0.001\,\text{Å}$.

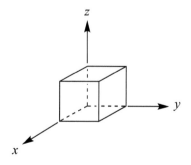

(i) Determine the probability of finding the particle in the aforementioned $\Delta\tau$ whose center is at $x = 0.2$ Å, $y = 0.3$ Å, $z = 0.5$ Å.

(ii) Determine the coordinates of the point(s) around which $\Delta\tau$ can be placed such that the probability of finding the particle is at a maximum. In addition, calculate this probability.

2.9 (i) Consider the state with quantum numbers n for a particle in a one-dimensional box with length a. Calculate the following average (or expectation) values: $<x>$, $<x^2>$, $<p_x>$, and $<p_x^2>$.

(ii) The "uncertainty" $\Delta\alpha$ of a dynamical variable α can be obtained from the relation: $(\Delta\alpha)^2 = <\alpha^2> - <\alpha>^2$. Use the results of part (i) to show that $\Delta x \Delta p_x > \dfrac{h}{4\pi}$, as required by the uncertainty principle.

2.10 Consider a particle with quantum number n moving in a one-dimensional box of length a.

(i)   Calculate the probability of finding the particle in the left quarter of the box.
(ii) For what value of n is this probability a maximum?
(iii) What is the limit of this probability for $n \to \infty$?
(iv) What principle is illustrated in (iii)?

Given: $\psi_n(x) = \left(\dfrac{2}{a}\right)^{1/2} \sin\left(\dfrac{n\pi x}{a}\right)$, $0 \le x \le a$.

2.11 When a particle is confined within a regular tetrahedral box, the energies and the wavefunctions of this particle cannot be solved exactly. As a result, we have to resort to approximations such as the variational theorem. Using this method, we can obtain a good approximation of the ground state energy of a regular tetrahedral box as $E_0 = 18.15\dfrac{\hbar^2}{mV^{2/3}}$, but the mathematics of this approximation are much too complicated to be presented here. The following formula has been proposed to determine the ground state energy of a regular polyhedron of volume V with n_v vertices:

$$E_0(n_v) = \left(12.82 + \frac{38.62}{n_v^{1.429}}\right)\frac{\hbar^2}{mV^{2/3}}.$$

For example, for a tetrahedron, $n_v = 4$, it gives $E_0(4) = 18.15\dfrac{\hbar^2}{mV^{2/3}}$. Apply this formula to determine the ground state energy E_0 (in $\dfrac{\hbar^2}{mV^{2/3}}$) of a particle in the following regular polyhedrons: an octahedron, a dodecahedron, an icosahedron, and a sphere. Hint: take $n_v = \infty$ for a sphere.

REFERENCE: W.-K. Li and S. M. Blinder, Variational solution for a particle in a regular tetrahedron, *Chem. Phys. Lett.* **496**, 339–40 (2010).

2.12 When a particle is trapped in an equilateral triangular box, the energy level diagram is considerably more complex than that when the box is a rectangle. For this two-dimensional triangular system, once again it takes a set of two quantum numbers to define a state. The energies of this system may be expressed as:

$$E_{p,q} = \left(p^2 + pq + q^2\right) \frac{2h^2}{3ma^2},$$

where m is the mass of the particle and a is the length of a side of the triangle. The two quantum numbers are related in a rather unconventional fashion:

$$q = 0, {}^1/_3, {}^2/_3, 1, \ldots, p = q + 1, q + 2, \ldots$$

So the ground state energy E_0 is

$$E_0 = E_{1,0} = \frac{2h^2}{3ma^2}.$$

The wavefunctions $\psi_{p,q}$ may be classified according to their symmetry properties. The symmetry point group of this system is C_{3v}, the wavefunctions may be classified to be A_1 (which is symmetric with respect to all symmetry operations), A_2 (symmetric with respect to the rotational axis but antisymmetric with respect to the vertical symmetry planes), or E (a two-dimensional representation).

For a level with energy $E_{p,0}$, the corresponding wavefunction $\psi_{p,0}$ has A_1 symmetry and the level is non-degenerate. For a level having energy $E_{p,q}$ with integral p and q (where $q > 0$), the level is doubly degenerate: one of the wavefunctions has A_1 symmetry, while the other one has A_2 symmetry. When p and q are non-integers $({}^1/_3, {}^2/_3, 1{}^1/_3, \ldots$ etc.), the doubly degenerate wavefunctions form an E set. The following table lists the quantum numbers, energies, and wavefunction symmetry for the first two states for a particle in an equilateral triangle.

Level	(p, q)	Symmetry	Energy (in E_0)
1	$(1, 0)$	A_1	1
2	$(1^1/_3, {}^1/_3)$	E	$2^1/_3$

Determine the quantum numbers, energies, and wavefunction symmetry for the next six states.

REFERENCE: W.-K. Li and S. M. Blinder, Particle in an equilateral triangle: Exact solution of a nonseparable problem. *J. Chem. Educ.* **64**, 130–2 (1987).

2.13 Degeneracy is an important concept in chemistry and one simple system to demonstrate this concept is the particle-in-a-square problem. When each side of the square is d, the energy of the particle with mass m is given by

$$E_{a,b} = \frac{\left(a^2 + b^2\right) h^2}{8md^2}, \qquad \text{where } a, b = 1, 2, \ldots$$

From this expression, it is clear that $E_{a,b} = E_{b,a}$. Hence, we have a degenerate energy level whenever $a \neq b$. However, there are also "unforeseen" or "accidental" degenerate situations. For example, $E_{1,7} = E_{7,1} = E_{5,5}$; also, $E_{1,8} = E_{8,1} = E_{7,4} = E_{4,7}$. So, when do these accidents arise? Mathematically, this question reduces to: given an integer k which is known to be a sum of two squares, how many (a, b) combinations are there that would satisfy $a^2 + b^2 = k$? In the cited reference, a general method is given to answer this question. Meanwhile, after mastering this general method, readers are urged to solve the following: What are the (a, b) combinations when $k = 585$ and when $k = 1105$?

REFERENCE: W.-K. Li, Degeneracy in the particle-in-a-square problem. *A. J. Phys.* **50**, 666 (1982).

2.14 (i) The electronic behavior of a one-electron atom A may be described by the particle-in-a-cube model. When two A atoms form diatomic molecule A_2, its stability is provided by a two-electron bond. In this question, we follow the free-electron model, i.e., the two electrons do not interact with each other. Applying the particle-in-a-box results, calculate the bond energy of A_2 in a.u. Then covert it to kJ mol^{-1} and compare it with the bond energy of hydrogen.

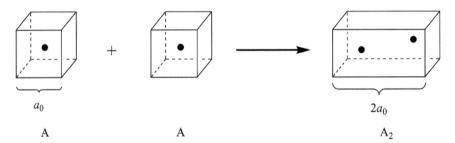

2.14 (ii) A third atom A is placed in touch of either end of the aforementioned A_2, forming a (linear) A_3 in the process. Again, use the particle-in-a-box results, calculate the bond energy (in a.u.) for the process $A_3 \rightarrow A_2 + A$.

REFERENCE: S. M. Blinder, *Introduction to Quantum Mechanics: in chemistry, materials science, and biology*, Amsterdam, Boston: Elsevier, 2004, pp. 43, pp. 293-4.

2.15 (i) In this calculation we will get an idea of the order of magnitude of nuclear energies. Assume that a nucleus can be represented as a cubic box of side 10^{-14} m. Calculate the lowest allowed energy of a hydrogen nucleus (proton mass $= 1.67 \times 10^{-27}$ kg). Express your result in MeV (1 MeV $= 10^6$ eV $= 1.602 \times 10^{-13}$ J). Also, calculate the "electronic energy" of an electron bounded in an atom, which can be represented as a cubic box of side 10^{-10} m.

(ii) Rutherford's famous experiment made use of a beam of 4.8 MeV α-particles (^{4}He nuclei) emitted by a sample of radium-226. Assume that the α-particle, before its emission, is in the lowest energy level of a particle in a cube (representing the radium nucleus). Calculate the dimension of the box.

2.16 The treatment of the particle-in-a-box problem in most physical chemistry textbooks does not take relativistic effect into consideration. This is entirely understandable as relativity is not a standard topic in a typical undergraduate chemistry curriculum. In this question we will examine the role relativity can play in this problem. Without going into the mathematical details, it can be shown that the electron is not relativistic unless the box length is equal to or shorter than the critical value $\frac{h}{2\pi mc}$, where m is the electronic mass and c is the speed of light. After putting in the numbers, the critical box length is 0.004 Å. When the box length is shorter than this value, the electron is relativistic.

When the free-electron model is used to describe the π electrons in conjugated molecules such as butadiene, give an estimate of the "box length" of this system. In addition, are the π electrons in this conjugated molecule relativistic?

REFERENCE: W.-K. Li and S. M. Blinder, Introducing Relativity into Quantum Chemistry. *J. Chem. Educ.* **88**, 71–3 (2011).

2.17 The potential energy (V) of this problem's quantum mechanical system is a combination of those of a particle-in-a-box and a harmonic oscillator. These two parts are clearly shown in the accompanying figure. On the left side, $V \rightarrow \infty$ as $x \leq 0$. On the right side, V for a harmonic oscillator is a function of both k and x: $V = \frac{1}{2}kx^2$, where k is the force constant of the oscillator.

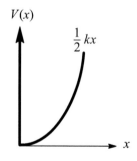

While the readers of this book are familiar with the solutions of the particle-in-a-box problem, they may be less conversant with those of the harmonic oscillator. First, the quantized energy values of this system may be expressed as

$$E_n = (n + {}^1\!/_2)h\nu_\mathrm{o}, \qquad n = 0, 1, 2, 3, \ldots$$

where ν_o is the classical frequency of the oscillator:

$$\nu_o = \frac{h}{2\pi}\sqrt{\frac{k}{\mu}},$$

where k is the aforementioned force constant, while μ is the reduced mass of the oscillator. So the allowed energy values (in units of $h\nu_o$) are $1/2$, $1^1/2$, $2^1/2$, $3^1/2$, etc.

The mathematical expressions for the corresponding wavefunctions are considerably more complex. For our purposes, their pictorial representations for $n = 0, 1, 2, 3, 4, 5$, as shown below, will suffice.

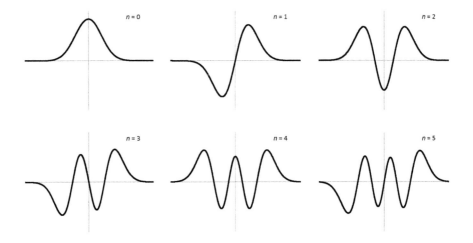

Write down the energy expression for our "half harmonic oscillator." In addition, describe the corresponding wavefunctions. Hint: To get the answers, you do NOT need to solve the Schrödinger Equation directly.

2.18 Suppose the electron in a hydrogen atom lies within a sphere of radius r from nucleus.

(i) What is Δr, the uncertainty in its position?

(ii) Use the relation $\Delta p \Delta r \geq \dfrac{h}{4\pi}$ from the Heisenberg uncertainty principle to deduce an estimate of ΔT, the uncertainty in kinetic energy.

(iii) Write down an expression for the total energy E by giving the kinetic energy T its minimum possible value (i.e., the lower limit of ΔT).

(iv) Derive an expression for the equilibrium distance r_e by minimizing E with respect to r.

(v) Derive an expression for the ground state energy E_0.

(vi) Comment on the results obtained in (iv) and (v).

2.19 This problem illustrates the application of the charge cloud model to the hydrogen atom. The hydrogen atom is pictured as a point charge nucleus located at the center of a spherical electron charge cloud of radius R. The charge density (charge/volume) is uniform at any point within the sphere and has the value $-\dfrac{e}{\frac{4}{3}\pi R^3}$. The energy levels in the hydrogen atom can now be calculated in the following steps. For convenience, use atomic units in all calculations.

    (i)   Kinetic energy (T)

          Derive an expression for T on the basis of de Broglie's relation and the following assumption concerning the wave nature of the charge cloud: "In a spherical charge cloud of radius R, standing waves of half-integral wavelength must fit a diameter of the sphere".

    (ii) Potential energy (V)

          Calculate the electric potential at the center of the charge cloud and hence the potential energy of interaction V between the charge cloud and the nucleus bearing charge $+e$.

    (iii) Total energy (E)

          Minimize the total energy E with respect to R to yield the equilibrium radius R_e. Finally, obtain an expression for the energy of the ground state of the hydrogen atom.

    (iv) Compare the results of this simple model with the known values for the hydrogen atom. Comment on the fact that the atom does not collapse under nuclear attraction.

REFERENCE: F. Rioux and P. Kroger, Charge cloud study of atomic and molecular structure. *Am. J. Phys.* **44**, 56–9 (1976).

2.20 Liu and co-workers used photodissociation mass spectrometry to study chloropropylene oxide (C_3H_5ClO), as reported in the cited reference. The following channel was observed:

$$C_3H_5ClO + h\nu \rightarrow C_3H_5O^+ + Cl + e^-, \qquad AE = 11.45\,\text{eV},$$

where AE is the experimental result measured for the energy change of the corresponding channel in their study.

    Based on the above result, determine $IE(C_3H_5ClO)$ and $D_0(C_3H_5O^+-Cl)$. Given: $IE(C_3H_5ClO) = 10.66\,\text{eV}$.

REFERENCES: F. Liu, C. Li, G. Wu, H. Gao, F. Qi, L. Sheng, Y. Zhang, S. Yu, S.-H. Chien, and W.-K. Li, Experimental and theoretical studies of the VUV photoionization of chloropropylene oxide, *J. Phys. Chem. A* **105**, 2973–9 (2001); A. S. Vorobev, I. I. Fuurlei, A. S. Sultanov, V. I. Khvostenko, G. V. Leplyanin, A. R. Derzhinskii, and G. A. Tolstikov, *Bull. Acad. Sci. USSR, Div. Chem. Sci.* 1388 (1989).

2.21 Zhao and co-workers used photodissociation mass spectrometry to study acetophenone $(C_6H_5COCH_3)$, as reported in the cited reference. The following channel was observed:

$$C_6H_5COCH_3 + h\nu(248\text{ nm}) \rightarrow C_6H_5CO + CH_3$$

From conservation of energy, $E_{int}(C_6H_5COCH_3) +$ energy of $h\nu(248\text{ nm})$

$$= \Delta E \text{ (energy change of the channel)} + \Delta E_{c.m.} + E_{int}(C_6H_5CO) + E_{int}(CH_3)$$

where E_{int} is the internal energy of the species and $\Delta E_{c.m.}$ is the total recoil kinetic energy of the fragments.

In their work, $E_{int}(C_6H_5COCH_3)$ was estimated to be in the range of 18 to 31 kJ mol^{-1} and we take $E_{int}(C_6H_5COCH_3) = (18 + 31)/2 = 25$ kJ mol^{-1} here. The onset of $\Delta E_{c.m.}$ was determined as 151 kJ mol^{-1}, corresponding to zero internal energy of the species, i.e., $E_{int}(C_6H_5CO) = E_{int}(CH_3) = 0$. Based on the above information, determine $D_0(C_6H_5CO-CH_3)$.

REFERENCE: H.-Q. Zhao, Y.-S. Cheung, C.-L. Liao, C.-X. Liao, C. Y. Ng, and W.-K. Li, A laser photofragmentation time-of-flight mass spectrometric study of acetophenone at 193 and 248 nm, *J. Chem. Phys.* **107**, 7230–41 (1997).

2.22 Zhao and co-workers used photodissociation mass spectrometry to study $HSCH_2CH_2SH$, as reported in the cited reference. The following channel was observed:

$$HSCH_2CH_2SH + h\nu(193\text{ nm}) \rightarrow HSCH_2CH_2 + SH$$

From conservation of energy, $E_{int}(HSCH_2CH_2SH) +$ energy of $h\nu(193\text{ nm})$

$$= \Delta E(\text{energy change of the channel}) + \Delta E_{c.m.} + E_{int}(HSCH_2CH_2) + E_{int}(SH)$$

where E_{int} is the internal energy of the species and $\Delta E_{c.m.}$ is the total recoil kinetic energy of the fragments.

In their work, $E_{int}(HSCH_2CH_2SH)$ was estimated to be 4 kJ mol^{-1} and the onset of $\Delta E_{c.m.}$ was determined as 314 kJ mol^{-1}, corresponding to zero internal energy of the species, i.e., $E_{int}(HSCH_2CH_2) = E_{int}(SH) = 0$. Based on the above information, determine $D_0(HSCH_2CH_2-SH)$.

REFERENCE: H.-Q. Zhao, Y.-S. Cheung, C.-X. Liao, C. Y. Ng, W.-K. Li, and S.-W. Chiu, 193 nm laser photofragmentation time-of-flight mass spectrometric study of HSCH$_2$CH$_2$SH, *J. Chem. Phys.* **104**, 130–8 (1996).

2.23 Ma and co-workers used photoionization mass spectrometry to study the CH_3SS radical, which was generated by the photodissociation of CH_3SSCH_3. The threshold for ionization of the radical was 143.0 nm. Determine the IE of the radical.

REFERENCE: Z.-X. Ma, C. L. Liao, C. Y. Ng, Y.-S. Cheung, W.-K. Li, and T. Baer, Experimental and theoretical studies of isomeric CH_3S_2 and $CH_3S_2^+$. *J. Chem. Phys.* **100**, 4870–5 (1994).

2.24 Liu and co-workers used photoionization mass spectrometry to study ethylene oxide (c-C_2H_4O), as reported in the cited reference. Two of the channels observed are shown below:

$$c\text{-}C_2H_4O + h\nu \rightarrow \quad c\text{-}C_2H_4O^+ + e^-, \qquad AE_1 = 10.51 \text{ eV},$$
$$\rightarrow \quad c\text{-}C_2H_3O^+ + H + e^-, \qquad AE_2 = 11.80 \text{ eV},$$

where the AEs are experimental results measured for the energy change of the corresponding channels in their study.

Based on the above results, determine $IE(c\text{-}C_2H_4O)$ and $D_0(c\text{-}C_2H_3O^+ - H)$.

REFERENCE: F. Liu, F. Qi, H. Gao, L. Sheng, Y. Zhang, S. Yu, K.-C. Lau, and W.-K. Li, A vacuum ultraviolet photoionization mass spectrometric study of ethylene oxide in the photon energy region of 10–40 eV, *J. Phys. Chem. A* **103**, 4155–61 (1999).

2.25 Qi and co-workers used photodissociation mass spectrometry to study ethylene sulfide (c-C_2H_4S), as reported in the cited reference. Two of the channels observed are shown below:

$$c\text{-}C_2H_4S + h\nu(193\,\text{nm}) \quad \rightarrow \quad C_2H_4\,({}^1A_g) + S\,({}^1D), \quad \Delta E_{\text{avai},1} = 270\,\text{kJ mol}^{-1},$$
$$\rightarrow \quad c\text{-}C_2H_3S + H, \qquad \Delta E_{\text{avai},2} = 180\,\text{kJ mol}^{-1},$$

where the ΔE_{avai}s are experimental available energy, i.e., ΔE (energy change of the channel) $+ \Delta E_{\text{avai}} = $ energy of $h\nu(193\,\text{nm})$.

Based on the above results, determine $D_0(c\text{-}C_2H_3S - H)$ and the average D_0 for the two C–S bonds in c-C_2H_4S.

REFERENCE: F. Qi, O. Sorkhabi, A. G. Suits, S.-H. Chien, and W.-K. Li, Photodissociation of ethylene sulfide at 193 nm: a photofragment translational spectroscopy study with VUV synchrotron radiation and ab initio calculations, *J. Am. Chem. Soc.* **123**, 148–61 (2001).

2.26 Li and co-workers used photoionization mass spectrometry to study CCl_4, as reported in the cited reference. Five channels were observed:

$$\begin{aligned}
CCl_4 + h\nu \quad &\rightarrow \quad CCl_3^+ + Cl + e^-, & \Delta E_1 &= 11.29 \text{ eV}, \\
&\rightarrow \quad CCl_2^+ + 2\,Cl + e^-, & \Delta E_2 &= 15.43 \text{ eV}, \\
&\rightarrow \quad CCl^+ + Cl + Cl_2 + e^-, & \Delta E_3 &= 15.60 \text{ eV}, \\
&\rightarrow \quad C^+ + 4\,Cl + e^-, & \Delta E_4 &= 24.60 \text{ eV}, \\
&\rightarrow \quad CCl_3 + Cl^+ + e^-, & \Delta E_5 &= 16.05 \text{ eV},
\end{aligned}$$

where the ΔEs are experimental results measured in their study.

Based on the above results, together with the well-known literature values of $IE(Cl) = 12.967$ eV and $D_0(Cl-Cl) = 2.52$ eV, determine bond dissociation energies $D_0(Cl_3C-Cl)$, $D_0(Cl_2C^+-Cl)$, $D_0(ClC^+-Cl)$, and $D_0(C^+-Cl)$, as well as $IE(CCl_3)$.

REFERENCE: Q. Li, Q. Ran, C. Chen, S. Yu, X. Ma, L. Sheng, Y. Zhang, and W.-K. Li, Experimental and theoretical study of the photoionization and dissociative photoionizations of carbon tetrachloride, *Int. J. Mass Spectrom. Ion Processes* **153**, 29–36 (1996).

2.27 Bond energy is an important concept in structural chemistry, and there are numerous (approximation) methods to determine a molecule's bond energies computationally. On the other hand, there are also various methods to do these determinations experimentally. One such procedure is photoionization mass spectrometry (PIMS), which was used by Qi and co-workers to study NH_3, as reported in the cited reference. To summarize: three channels were observed:

$$\begin{aligned}
NH_3 + h\nu \quad &\rightarrow \quad NH_3^+ + e^-, & \Delta E_1 &= 10.16 \text{ eV}, \\
&\rightarrow \quad NH_2^+ + H + e^-, & \Delta E_2 &= 15.75 \text{ eV}, \\
&\rightarrow \quad NH_2 + H^+ + e^-, & \Delta E_3 &= 18.57 \text{ eV}.
\end{aligned}$$

where the ΔEs are experimental results measured in their study.

Based on the above results, together with the well-known literature values of $IE(H) = 13.6$ eV, determine bond dissociation energies $D_0(H_2N-H)$, $D_0(H_2N^+-H)$, and $D_0(H_2N-H^+)$, as well as $IE(NH_2)$.

REFERENCE: F. Qi, L. Sheng, Y. Zhang, S. Yu, and W.-K. Li, Experimental and theoretical study of the dissociation energies $D_0(H_2N-H)$ and $D_0(H_2N^+-H)$ and other related quantities, *Chem. Phys. Lett.* **234**, 450–4 (1995).

2.28 Sheng and co-workers used photoionization mass spectrometry to study C_2H_3Cl, as reported in the cited reference. Five channels were observed:

$$\begin{aligned}
C_2H_3Cl + h\nu &\rightarrow C_2H_3Cl^+ + e^-, & \Delta E_1 &= 9.98 \text{ eV}, \\
&\rightarrow C_2H_2Cl^+ + H + e^-, & \Delta E_2 &= 14.90 \text{ eV}, \\
&\rightarrow C_2H_2Cl + H^+ + e^-, & \Delta E_3 &= 18.63 \text{ eV}, \\
&\rightarrow C_2H_3^+ + Cl + e^-, & \Delta E_4 &= 12.54 \text{ eV}, \\
&\rightarrow C_2H_3 + Cl^+ + e^-, & \Delta E_5 &= 19.30 \text{ eV},
\end{aligned}$$

where the ΔEs are experimental results measured in their study.

Based on the above results, together with the well-known literature values of IE(H) = 13.598 eV and IE(Cl) = 12.967 eV, determine bond dissociation energies $D_0(C_2H_2Cl-H)$, $D_0(C_2H_2Cl^+-H)$, $D_0(C_2H_2Cl-H^+)$, $D_0(C_2H_3-Cl)$, $D_0(C_2H_3^+-Cl)$, and $D_0(C_2H_3-Cl^+)$, as well as IE(C_2H_3Cl), IE(C_2H_2Cl), and IE(C_2H_3).

REFERENCE: L. Sheng, F. Qi, L. Tao, Y. Zhang, S. Yu, C.-K. Wong, and W.-K. Li, Experimental and theoretical studies of the photoionization and dissociative photoionizations of vinyl chloride, *Int. J. Mass Spectrom. Ion Processes* **148**, 179–89 (1995).

2.29 Sheng and co-workers used photoionization mass spectrometry to study CCl_2F_2, as reported in the cited reference. Eight channels were observed:

$$\begin{aligned}
CCl_2F_2 + h\nu &\rightarrow CCl_2F_2^+ + e^-, & AE_1 &= 11.84 \text{ eV}, \\
&\rightarrow CCl_2F^+ + F + e^-, & AE_2 &= 13.54 \text{ eV}, \\
&\rightarrow CClF_2^+ + Cl + e^-, & AE_3 &= 12.05 \text{ eV}, \\
&\rightarrow CClF^+ + Cl + F + e^-, & AE_4 &= 17.88 \text{ eV}, \\
&\rightarrow CF_2^+ + 2\,Cl + e^-, & AE_5 &= 17.08 \text{ eV}, \\
&\rightarrow CF^+ + 2\,Cl + F + e^-, & AE_6 &= 20.07 \text{ eV}, \\
&\rightarrow CF_2 + Cl + Cl^+ + e^-, & AE_7 &= 18.68 \text{ eV}, \\
&\rightarrow CCl^+ + F + FCl + e^-, & AE_8 &= 18.73 \text{ eV},
\end{aligned}$$

where the AEs are experimental results measured for the energy change of the corresponding channels in their study.

Based on the above results, together with the well-known literature values of IE(Cl) = 12.967 eV and $D_0(F-Cl)$ = 2.52 eV, determine bond dissociation energies $D_0(Cl_2FC^+-F)$, $D_0(ClF_2C^+-Cl)$, $D_0(ClFC^+-F)$, $D_0(ClFC^+-Cl)$, $D_0(F_2C^+-Cl)$, $D_0(FC^+-F)$, $D_0(FC^+-Cl)$, and $D_0(ClC^+-F)$, as well as IE(CF_2).

REFERENCE: L. Sheng, F. Qi, H. Gao, Y. Zhang, S. Yu, and W.-K. Li, Experimental and theoretical study of the photoionization and dissociative photoionizations of dichlorodifluoromethane, *Int. J. Mass Spectrom. Ion Processes* **161**, 151–9 (1997).

2.30 Wei and co-workers used photoionization mass spectrometry to study acetone $[CH_3C(O)CH_3]$ as reported in the cited reference. Listed below are some of the channels observed in their study:

$$
\begin{aligned}
CH_3C(O)CH_3 + h\nu \quad &\rightarrow \quad CH_3C(O)CH_3^+ + e^-, & AE_1 &= 9.69\,eV, \\
&\rightarrow \quad CH_3C(O)CH_2^+ + H + e^-, & AE_2 &= 13.10\,eV, \\
&\rightarrow \quad CH_3CO^+ + CH_3 + e^-, & AE_3 &= 10.49\,eV, \\
&\rightarrow \quad CH_2CO^+ + CH_4 + e^-, & AE_4 &= 10.53\,eV, \\
&\rightarrow \quad CH_2CO^+ + CH_3 + H + e^-, & AE_5 &= 14.97\,eV, \\
&\rightarrow \quad CH_3^+ + CO + CH_3 + e^-, & AE_6 &= 14.41\,eV,
\end{aligned}
$$

where the AEs are experimental results measured for the energy change of the corresponding channels in their study.

Based on the above results, determine bond dissociation energies $D_0[CH_3C(O)CH_2^+ - H]$, $D_0(CH_3CO^+ - CH_3)$, $D_0[CH_3 - C(O)CH_2^+]$, $D_0(H - CH_2CO^+)$, $D_0(CH_3^+ - CO)$, $D_0(CH_3 - H)$, as well as $IE[CH_3C(O)CH_3]$.

REFERENCE: L. Wei, B. Yang, R. Yang, C. Huang, J. Wang, X. Shan, L. Sheng, Y. Zhang, F. Qi, C.-S. Lam, and W.-K. Li, A vacuum ultraviolet photoionization mass spectrometric study of acetone, *J. Phys. Chem. A* **109**, 4231–41 (2005).

2.31 Cheung and co-workers determined the Gaussian-3 (G3) heats of formation for $(CH)_6$ isomers using both the atomization and isodesmic bond separation schemes. The G3 values of H_{298} of these isomers are tabulated below.

	Benzene (**1**)	Dewar benzene (**2**)	Prismane (**3**)	Benzvalene (**4**)	3,3'-Bicyclopropenyl (**5**)
H_{298} (hartree)	−232.04675	−231.92336	−231.86381	−231.93024	−231.85451

(i) The atomization scheme makes use of the following process:

$$(CH)_6 \rightarrow 6C + 6H$$

The G3 values of H_{298} and ΔH_{f298} of C and H are tabulated below.

	C	H
H_{298} (hartree)	−37.82536	−0.49864
ΔH_{f298} (kJ mol^{-1})	716.7	218.0

Calculate ΔH_{f298} for isomers **1** to **5**.

(ii) The isodesmic bond separation scheme makes use of the following process:

$$(CH)_6 + xCH_4 \rightarrow yC_2H_6 + zC_2H_4$$

where z = number of C=C double bond(s) in $(CH)_6$,
$y = 9 - 2z$,
$x = 2y + 2z - 6 = y + 3$

such that the number of each type of chemical bonds is conserved on both sides.

The G3 values of H_{298} and ΔH_{f298} of CH_4, C_2H_4, and C_2H_6 are tabulated below.

	CH_4	C_2H_6	C_2H_4
H_{298} (hartree)	−40.45380	−79.71890	−78.50341
ΔH_{f298} (kJ mol^{-1})	−74.5	−84.0	52.2

Calculate ΔH_{f298} for isomers **1** to **5**.

(iii) Compare your results obtained by the two schemes. Also comment on your results by comparing the results with the experimental results:

	Benzene (**1**)	Dewar benzene (**2**)	Prismane (**3**)	Benzvalene (**4**)	3,3′-Bicyclopropenyl (**5**)
ΔH_{f298} (kJ mol^{-1})	82.9 ± 0.3	364	–	363	–

REFERENCE: T.-S. Cheung, C.-K. Law, and W.-K. Li, Gaussian-3 heats of formation for (CH)$_6$ isomers, *J. Mol. Struct. (Theochem)* **572**, 243–7 (2001).

2.32 Cheng and co-workers determined the Gaussian-3 (G3) heats of formation for boron hydrides using both the atomization and isodesmic bond separation schemes. The G3 values of H_{298} for some boron hydrides are tabulated below.

	B_3H_7	B_4H_{10}	B_5H_{11}	B_6H_{10}	$B_{10}H_{14}$
H_{298} (hartree)	−78.59443	−105.21616	−130.64071	−154.91394	−256.69958

(i) The atomization scheme makes use of the following process:

$$B_xH_z \rightarrow xB + zH$$

The G3 values of H_{298} and ΔH_{f298} of B and H are tabulated below. Calculate ΔH_{f298} for the boron hydrides chosen.

	B	H
H_{298} (hartree)	-24.64034	-0.49864
ΔH_{f298} (kJ mol^{-1})	562.7	218.0

(ii) The isodesmic bond separation scheme makes use of the following process:

$$x\mathrm{BH}_3 \rightarrow \mathrm{B}_x\mathrm{H}_z + y\mathrm{H}_2$$

where $y = (3x - z)/2$.

The G3 values of H_{298} and ΔH_{f298} of BH$_3$ and H$_2$ are tabulated below.

	BH$_3$	H$_2$
H_{298} (hartree)	-26.56368	-1.16407
ΔH_{f298} (kJ mol^{-1})	100.0	0.0

Calculate ΔH_{f298} for the boron hydrides chosen.

REFERENCE: M.-F. Cheng, H.-O. Ho, C.-S. Lam, and W.-K. Li, Heats of formation for the boron hydrides: a Gaussian-3 study, *Chem. Phys. Lett.* **356**, 109–19 (2002).

2.33 The positronium atom, Pos, consisting of a positron and an electron, was introduced in Problem 3.2. The following table lists some properties (radius of 1s orbit, r_{1s}; atomic radius, r_{atom}; covalent radius, r_{cov}; energy of the 1s orbit, E_{1s}; ionization energy, IE; electron affinity, EA; Mulliken's electronegativity, χ'; and Pauling's electronegativity, χ) of the hydrogen atom. The radii are in Å units. The energies (including χ') are in eV units; χ is in $(\mathrm{eV})^{1/2}$ units.

	r_{1s}	r_{atom}	r_{cov}	E_{1s}	IE	EA	χ'	χ
H	0.529	0.529	0.371	-13.6	13.6	0.75	7.17	2.24
Pos								

Based on the results of its hydrogen analog, determine the corresponding values of Pos. Give brief justification for your answers.

Note: r_{1s}, r_{atom}, E_{1s}, and IE are Bohr atom results; χ_{cov} is half of the bond length found in H$_2$; EA is the experimental value; $\chi' = (\mathrm{IE} + \mathrm{EA})/2$; $\chi = 1.35(\chi')^{1/2} - 1.37$.

REFERENCE: B. K Teo and W.-K. Li, Exotic clusters formed by electrons and positrons: approximate calculations of binding energies and bonding capabilities of positronium atom and diprositronium and positronium hydride molecules based on their hydrogen analogs, *J. Clust. Sci.* **23**, 661–72 (2012).

SOLUTIONS

A2.1 The energy of a 589-nm photon is given by

$$E = h\nu = \frac{hc}{\lambda} = \frac{(6.6261 \times 10^{-34})\,(2.9979 \times 10^8)}{589 \times 10^{-9}} = 3.373 \times 10^{-19}\,\text{J}.$$

Since 3.5 W corresponds to 3.5 J s^{-1}, the lamp gives $3.5/(3.373 \times 10^{-19}) = 1.038 \times 10^{19}$ photons/sec.

A2.2 For the red laser-pointer:
Number of photons emitted per second

$$= (10^{-3}) \div \frac{(6.6261 \times 10^{-34})\,(2.9979 \times 10^8)}{650 \times 10^{-9}} = 3.27 \times 10^{15}.$$

For the green laser-pointer:
Number of photons emitted per second

$$= (10^{-3}) \div \frac{(6.6261 \times 10^{-34})\,(2.9979 \times 10^8)}{532 \times 10^{-9}} = 2.68 \times 10^{15}.$$

A2.3 Since $1\,\text{eV} = 1.602 \times 10^{-19}$ J, each electron has a kinetic energy of $(40 \times 10^3)(1.602 \times 10^{-19})$ J. According to the de Broglie relation, $\lambda = h/p$, the de Broglie wavelength of the electron is given by:

$$\lambda = \frac{h}{\sqrt{2mE}} = \frac{6.6261 \times 10^{-34}}{\sqrt{2(9.109 \times 10^{-31})(40 \times 10^3)(1.602 \times 10^{-19})}}$$

$$= 6.13 \times 10^{-12}\,\text{m}.$$

This gives sufficient resolution to study the geometric structure of molecules. [Since 40 keV electrons (with a non-relativistic speed of 1.19×10^8 m s^{-1}) travel at a significant fraction (0.397) of the speed of light, the relativistic energy-momentum relation must be used. The corrected de Broglie wavelength is actually 6.02×10^{-12} m.]

A2.4 To relate the following SI units: (J s) (as a whole), kg, m, and J, consider the kinetic energy of a particle: $E = \frac{1}{2}mv^2$, we have: $1\,\text{J} = 1\,\text{kg m}^2\,\text{s}^{-2} = 1\,\text{kg m}^2\frac{\text{J}^2}{(\text{J s})^2}$. Hence, $1\,\text{J} = 1\,(\text{Js})^2\,\text{kg}^{-1}\,\text{m}^{-2} = [(1.05457 \times 10^{-34})^{-1}\,\text{a. u.}]^2 \cdot [(9.109382 \times 10^{-31})^{-1}\,\text{a. u.}]^{-1} \cdot [(5.29177 \times 10^{-11})^{-1}\,\text{a. u.}]^{-2} = 2.29371 \times 10^{17}$ a. u. That is, 1 a. u. (for energy) $= (2.29371 \times 10^{17})^{-1} = 4.35974 \times 10^{-18}$ J.

A2.5 (i) $\alpha = \frac{(1)^2}{(1)(1)(137.036)} = 7.29735 \times 10^{-3}$ a. u

(ii) $\alpha = \frac{(1.60217653 \times 10^{-19}\ C)^2}{4\pi (8.85419 \times 10^{-12}\ J^{-1}\ C^2\ m^{-1})(1.05457 \times 10^{-34}\ J\ s)(2.99792458 \times 10^8\ m\ s^{-1})} =$

7.29735×10^{-3} (dimensionless)

A2.6 (i) $\Delta x = L.$

(ii) $E_1 = \frac{h^2}{8mL^2} = \frac{p^2}{2m}; \ p = \frac{h}{2L}.$

(iii) $\Delta x \Delta p = (L)\left(\frac{h}{L}\right) = h > \frac{h}{4\pi}.$

The result is in agreement with the uncertainty principle.

(iv) The particle moves in both directions and the momentum may be $+\frac{h}{2L}$ or $-\frac{h}{2L}$.

So the uncertainty in momentum is $\frac{h}{2L} - \left(-\frac{h}{2L}\right) = \frac{h}{L}.$

A2.7 (i)

Particle	Mass (kg)	Speed $(v)\ (m\ s^{-1})$	$\Delta v = 10\%$ of $v\ (m\ s^{-1})$	$\Delta x\ (m)$
The Earth	5.98×10^{24}	2.97×10^3	2.97×10^2	2.97×10^{-62}
A 100-meter sprinter running at world-record speed	1.00×10^2	10.4	1.04	5.07×10^{-37}
Bullet	1.00×10^{-2}	1.00×10^3	1.00×10^2	5.28×10^{-35}
Dust	1.00×10^{-9}	1.00×10^1	1.00	5.28×10^{-26}
Pollen in Brownian motion	1.00×10^{-13}	1.00	1.00×10^{-1}	5.28×10^{-21}
Xe atom at 298 K	2.18×10^{-25}	2.38×10^2	2.38×10^1	1.02×10^{-11}
Electron in a molecule	9.11×10^{-31}	1.00×10^3	1.00×10^2	5.79×10^{-7}

(ii) Only the Δx of an electron in a molecule is significant, as this Δx value is larger than the lengths of chemical bonds in a molecule. Recall that bond lengths are of the order of 10^{-10} m.

A2.8 (i) Probability $= \psi^2 \Delta \tau = [(2)^{3/2} \sin(0.4\pi) \sin(0.3\pi) \sin(0.5\pi)]^2 (10^{-3})^3 = 4.74 \times 10^{-9}.$

(ii) The points required are $(0.25\ \text{Å}, 0.5\ \text{Å}, 0.5\ \text{Å})$ and $(0.75\ \text{Å}, 0.5\ \text{Å}, 0.5\ \text{Å})$.

At both points, $\psi^2 = [(2)^{3/2}]^2 (1)(1)(1) = 8.$

Probability $= \psi^2 \Delta \tau = (8)(10^{-3})^3 = 8 \times 10^{-9}.$

A2.9   (i)
$$<x> = \frac{a}{2}; \quad <x^2> = a^2 \left[\frac{1}{3} - \left(\frac{1}{2n^2\pi^2} \right) \right];$$

$$<p_x> = 0; \quad <p_x^2> = 2mE = \left(\frac{n^2h^2}{4a} \right).$$

(ii)
$$\Delta x = a \left[\frac{1}{12} - \left(\frac{1}{2n^2\pi^2} \right) \right]^{1/2}, \quad \Delta p_x = \left(\frac{nh}{2a} \right).$$

$$\Delta x \Delta p_x = a \left[\frac{1}{12} - \left(\frac{1}{2n^2\pi^2} \right) \right]^{1/2} \left(\frac{nh}{2a} \right) = \left(\frac{n^2\pi^2 - 6}{3} \right)^{1/2} \left(\frac{h}{4\pi} \right)$$

$$\geq \left(\frac{1^2\pi^2 - 6}{3} \right)^{1/2} \left(\frac{h}{4\pi} \right) \quad (n_x \geq 1)$$

$$= 1.14 \left(\frac{h}{4\pi} \right) > \frac{h}{4\pi}.$$

A2.10   (i) $P = \int_0^{\frac{a}{4}} \left(\frac{2}{a} \right) \sin^2 \left(\frac{n\pi x}{a} \right) dx = \frac{1}{4} - \frac{1}{2n\pi} \sin \left(\frac{n\pi}{2} \right).$

(ii)
$$P = \begin{cases} \frac{1}{4} \text{ for } n \text{ even,} \\[2mm] \frac{1}{4} - \frac{1}{2n\pi} \text{ for } n = 1, 5, 9, \ldots, \\[2mm] \frac{1}{4} + \frac{1}{2n\pi} \text{ for } n = 3, 7, 11, \ldots. \end{cases}$$

Thus P is at a maximum for $n = 3$.

(iii) $P_{n\to\infty} = \frac{1}{4}.$

(iv)   Bohr's correspondence principle.

A2.11   The substitution into the mathematical expression is straightforward:

$$\text{Octahedron: } n_v = 6, \ E_0 = 15.80\frac{\hbar^2}{mV^{2/3}}.$$

$$\text{Cube: } n_v = 8, \ E_0 = 14.80\frac{\hbar^2}{mV^{2/3}}.^{\dagger}$$

$$\text{Icosahedron: } n_v = 12, \ E_0 = 13.92\frac{\hbar^2}{mV^{2/3}}.$$

$$\text{Dodecahedron: } n_v = 20, \ E_0 = 13.35\frac{\hbar^2}{mV^{2/3}}.$$

$$\text{Sphere: } n_v = \infty, \ E_0 = 12.82\frac{\hbar^2}{mV^{2/3}}.^{\dagger}$$

$\dagger$ Note: The ground state energy of a particle in a cube and in a sphere can be obtained exactly: $E_0(\text{cube}) = \frac{3h^2}{8mV^{2/3}} = \frac{12\pi^2\hbar^2}{8mV^{2/3}} = \frac{3\pi^2\hbar^2}{2mV^{2/3}} = 14.80\frac{\hbar^2}{mV^{2/3}}$ and $E_0(\text{sphere}) = \frac{\pi^2\hbar^2}{2mr^2} = 12.82\frac{\hbar^2}{mV^{2/3}}$, where r is the radius of the sphere.

A2.12

Level	(p, q)	Symmetry	Energy (in E_0)
1	$(1, 0)$	A_1	1
2	$(1^1/_3, {}^1/_3)$	E	$2^1/_3$
3	$(2, 0)$	A_1	4
4	$(1^2/_3, {}^2/_3)$	E	$4^1/_3$
5	$(2^1/_3, {}^1/_3)$	E	$6^1/_3$
6	$(2, 1)$	A_1, A_2	7
7	$(3, 0)$	A_1	9
8	$(2^2/_3, {}^2/_3)$	E	$9^1/_3$

Comments:

Since an equilateral triangle is a highly symmetric system, we expect degenerate states. For example, levels 25 and 26 with E symmetry and energy $30^1/_3$ units, have quantum number sets $(3^2/_3, 2^2/_3)$ and $(5^1/_3, {}^1/_3)$, respectively. Some of these degenerate states are listed below.

Level	(p, q)	Symmetry	Energy (in units of E_0)
...	...	...	...
25	$(3^2/_3, 2^2/_3)$	E	$30^1/_3$
26	$(5^1/_3, {}^1/_3)$	E	$30^1/_3$
...	...	...	...
38	$(4^1/_3, 3^1/_3)$	E	$44^1/_3$
39	$(5^2/_3, 1^2/_3)$	E	$44^1/_3$
...	...	...	...

A2.13 When $k = 585$, $(a, b) = (3, 24)$, $(24, 3)$, $(12, 21)$, and $(21, 12)$. When $k = 1105$, $(a, b) = (4, 33)$, $(33, 4)$, $(9, 32)$, $(32, 9)$, $(12, 31)$, $(31, 12)$, $(23, 24)$, and $(24, 23)$.

A2.14 (i) The energy level of A is given by $E(A) = \frac{h^2}{8m}\left(\frac{n_x^2}{a_0^2} + \frac{n_y^2}{a_0^2} + \frac{n_z^2}{a_0^2}\right) = \frac{3h^2}{8ma_0^2}$.

The energy level of A_2 is given by $E(A_2) = \frac{h^2}{8m}\left[\frac{n_x^2}{(2a_0)^2} + \frac{n_y^2}{a_0^2} + \frac{n_z^2}{a_0^2}\right] = \frac{9h^2}{32ma_0^2}$.

The bond energy of $A_2 = 2 \times \frac{3h^2}{8ma_0^2} - 2 \times \frac{9h^2}{32ma_0^2} = \frac{3h^2}{16ma_0^2} = 7.402$ a.u. $= 19434$ kJ mol^{-1} (*cf.* literature value of the bond energy of hydrogen is 432.1 kJ mol^{-1}).

(ii) The energy level of A_3 is given by $E(A_3) = \frac{h^2}{8m}\left[\frac{n_x^2}{(3a_0)^2} + \frac{n_y^2}{a_0^2} + \frac{n_z^2}{a_0^2}\right]$. The lowest level is $\frac{19h^2}{72ma_0^2}$, and the next level is $\frac{11h^2}{36ma_0^2}$.

The bond energy of the A—A_2 bond $= \left(\frac{3h^2}{8ma_0^2} + 2 \times \frac{9h^2}{32ma_0^2}\right) - \left(2 \times \frac{19h^2}{72ma_0^2}\right.$

$\left. + \frac{11h^2}{36ma_0^2}\right) = \frac{5h^2}{48ma_0^2} = 4.112$ a.u.

Note: This is a simple-minded attempt to mimic the two-electron bond in the hydrogen molecule with the particle-in-a-box model. Unfortunately, the quantitative result is not good at all. Additionally, as each electron in the diatomic molecule moves from one cube to two neighboring ones, the increase in volume leads to a lowering in kinetic energy, but leaving potential energy unchanged. This change may be taken as the origin of the "bonding" of the two cube atoms.

A2.15 (i) For proton, $M = 1.67 \times 10^{-27}$ kg in a cubic box with $a = 10^{-14}$ m, ground state energy is

$$E_{111} \approx \frac{h^2}{8Ma^2}\left(1^2 + 1^2 + 1^2\right) = 6.15 \text{ MeV}.$$

For the electron in an atom of cubic box:

$$E_{111} \approx 30.8 \text{ eV}, \ cf. \text{ IE of H atom} = 13.6 \text{ eV}.$$

Note: As expected, nuclear energies are several orders of magnitude higher than electronic energies.

$$E_{111} \approx \frac{h^2}{8Ma^2}\left(1^2 + 1^2 + 1^2\right) = 6.15 \text{ MeV}.$$

(ii) For α-particle, $M = 6.64 \times 10^{-27}$ kg, $E_{111} \approx 4.8$ MeV, hence, $a = 5.8 \times 10^{-15}$ m.

A2.16 The length d of the butadiene box may be estimated in the following manner. Typical C–C and C=C bond lengths are 1.54 and 1.35 Å, respectively. If we allow the π electrons to move a bit beyond the terminal carbon atoms, the length of the box may be rounded off to 4×1.40 Å $= 5.60$ Å $\approx 1400 \times 0.004$ Å. Since the calculated box length is about 1400 times $\frac{h}{2\pi mc}$, the π electrons in butadiene should not be relativistic.

A2.17 The Schrödinger Equation for the harmonic oscillator and our "half oscillator" are one and the same. But there is one important difference between their respective boundary conditions. For the "half oscillator," on account of the infinitely high potential wall at $x = 0$, the probability density of finding the particle at the wall must be zero, or $\psi(x = 0) = \psi(0) = 0$. In other words, the allowed wavefunctions must vanish at the origin, or must go through the origin. Upon examining the wavefunctions shown, it is clear only wavefunctions with $n = 1, 3, 5, \ldots$ are the ones we want, whereas $n = 0, 2, 4$ are not. Also, the energy expression now becomes:

$$E_n = (n + \tfrac{1}{2})h\nu_o, \qquad n = 1, 3, 5, \ldots$$

A2.18 (i) The electron-proton distance lies in the range of 0 to r; hence $\Delta r = r$.

(ii) $\Delta T = \dfrac{(\Delta p)^2}{2m} \geq \dfrac{\left(\frac{h}{4\pi \Delta r}\right)^2}{2m} = \dfrac{h^2}{32\pi^2 m r^2}.$

(iii) $T_{\min} = \dfrac{h^2}{32\pi^2 m r^2}; \; E = T + V = \dfrac{h^2}{32\pi^2 m r^2} - \dfrac{e^2}{4\pi \varepsilon_0 r}.$

(iv) $r_e = \dfrac{\varepsilon_0 h^2}{4\pi m e^2}.$

(v) $E_0 = -\dfrac{m e^4}{2\varepsilon_0^2 h^2}.$

(vi) The exact values for the radius and energy for the ground state hydrogen atom obtained from either the Bohr theory or the Schrödinger equation are $\dfrac{\varepsilon_0 h^2}{\pi m e^2}$ and $-\dfrac{m e^4}{8\varepsilon_0^2 h^2}$, respectively. These values are different from the corresponding results obtained in (iv) and (v) by a factor of 4.

A2.19 One way of writing out the atomic units is $|e| = m = \hbar = 1$.

(i) Relations $p = \dfrac{2\pi}{\lambda}$ and $\lambda\left(\dfrac{n}{2}\right) = 2R$ lead to $T = \dfrac{p^2}{2} = \dfrac{n^2\pi^2}{8R^2}.$

(ii) Electric potential (charge/distance) at the center due to spherical shell of radius r and thickness dr $= -\left(\dfrac{4}{3}\pi R^3\right)^{-1}(4\pi r^2 dr)\left(\dfrac{1}{r}\right) = -\dfrac{3}{R^3} r \, dr.$

$$V = -\dfrac{3}{R^3}\int_0^R r\,dr = -\dfrac{3}{2R}.$$

(iii) $E = \dfrac{n^2\pi^2}{8R^2} - \dfrac{3}{2R}.$

Minimization of E with respect to R yields: $R_e = \dfrac{n^2\pi^2}{6}$ and $E_H = -\dfrac{9}{2n^2\pi^2}.$

(iv) The energies obtained from the charge cloud model differ from the correct values by a factor of $9/\pi^2 (\approx 0.91)$. The atom does not collapse under nuclear attraction because, as the electron cloud shrinks in size, T (varying as $1/R^2$) dominates over V (varying as $-1/R$).

On the other hand, for the special case of $n = 1$, $R_e = \pi^2/6$, which is close to $<r_{1s}>$, or $\frac{3}{2}$ a.u.

A2.20 $D_0(C_3H_5O^+-Cl) = AE - IE(C_3H_5ClO) = 11.45 - 10.66 = 0.79$ eV.

A2.21 The energy of $h\nu(248 \text{ nm})$

$$= \frac{\left(6.6261 \times 10^{-34}\right)\left(2.9979 \times 10^8\right)}{248 \times 10^{-9}} \times (6.022 \times 10^{23})/1000 = 482 \text{ kJ mol}^{-1}.$$

$D_0(C_6H_5CO-CH_3) = 25 + 482 - 151 = 356 \text{ kJ mol}^{-1}.$

A2.22 The energy of $h\nu(193 \text{ nm})$

$$= \frac{\left(6.6261 \times 10^{-34}\right)\left(2.9979 \times 10^8\right)}{193 \times 10^{-9}} \times (6.022 \times 10^{23})/1000 = 620 \text{ kJ mol}^{-1}.$$

$D_0(HSCH_2CH_2-SH) = 4 + 620 - 314 = 310 \text{ kJ mol}^{-1}.$

A2.23 Wavelength of 143.0 nm $(= 143.0 \times 10^{-7} \text{ cm} = 1.430 \times 10^{-5} \text{ cm})$ corresponds to $1/(1.430 \times 10^{-5} \text{ cm}) = 69930 \text{ cm}^{-1} = 69930 \times (1.23985 \times 10^{-4} \text{ eV}) = 8.67$ eV. Therefore, IE(CH_3SS) = 8.67 eV.

Note: Although the appearance energy (AE) of the following dissociative photoionization channel was also measured:

$$CH_3SSCH_3 + h\nu \rightarrow CH_3S_2^+ + CH_3 + e^-, \quad AE = 11.07 \text{ eV},$$

the ion fragment was believed to be CH_2SSH^+, instead of CH_3SS^+. Hence, the AE cannot be combined with IE(CH_3SS) to obtain the bond energy $D_0(CH_3SS-CH_3)$.

A2.24 IE(c-C_2H_4O) = AE$_1$ = 10.51 eV.

$$D_0(c\text{-}C_2H_3O^+-H) = AE_2 - AE_1 = 11.80 - 10.51 = 1.29 \text{ eV}.$$

Note: Although 16 channels were observed and 28 fragments were studied computationally in the cited reference, most of them are not mentioned here because these channels and fragments did not originate from a simple bond breaking process.

A2.25 The energy of $h\nu(193 \text{ nm}) = 620 \text{ kJ mol}^{-1}$.
$\Delta E_1 = 620 - \Delta E_{\text{avai},1} = 350 \text{ kJ mol}^{-1}.$
$\Delta E_2 = 620 - \Delta E_{\text{avai},2} = 440 \text{ kJ mol}^{-1}.$
$D_0(c\text{-}C_2H_3S-H) = 350 \text{ kJ mol}^{-1}.$

Average D_0 for the two C–S bonds in $c\text{-}C_2H_3S = \Delta E_2/2 = 220 \, \text{kJ} \, \text{mol}^{-1}$.

Note: Although seven channels were observed and 11 fragments were studied computationally in the cited reference, most of them are not mentioned here because these channels and fragments did not originate from a simple single bond breaking process.

A2.26 The following diagram can be constructed:

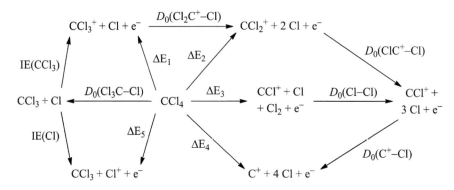

From the diagram:

$$D_0(Cl_3C-Cl) = \Delta E_5 - IE(Cl) = 3.08 \, \text{eV},$$
$$D_0(Cl_2C^+-Cl) = -\Delta E_1 + \Delta E_2 = 4.14 \, \text{eV},$$
$$D_0(ClC^+-Cl) = -\Delta E_2 + \Delta E_3 + D_0(Cl-Cl) = 2.69 \, \text{eV},$$
$$D_0(C^+-Cl) = -D_0(Cl-Cl) - \Delta E_3 + \Delta E_4 = 6.48 \, \text{eV},$$
$$IE(CCl_3) = IE(Cl) - \Delta E_5 + \Delta E_1 = 8.21 \, \text{eV}.$$

A2.27 The following diagram can be constructed:

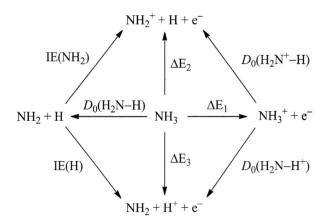

From the diagram:

$$D_0(H_2N-H) = \Delta E_3 - IE(H) = 4.97 \text{ eV},$$

$$D_0(H_2N^+-H) = -\Delta E_1 + \Delta E_2 = 5.59 \text{ eV},$$

$$D_0(H_2N-H^+) = -\Delta E_1 + \Delta E_3 = 8.41 \text{ eV},$$

$$IE(NH_2) = -D_0(H_2N-H) + \Delta E_2 = 10.78 \text{ eV}.$$

A2.28 Firstly, $IE(C_2H_3Cl) = \Delta E_1 = 9.98 \text{ eV}$.

Also, the following diagrams can be constructed:

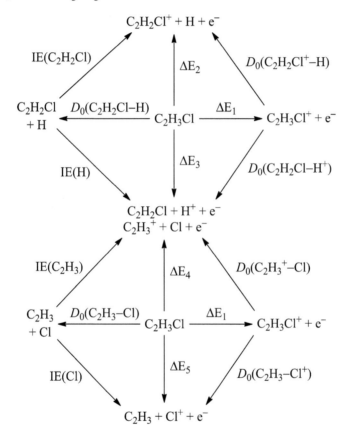

From the first diagram:

$$D_0(C_2H_2Cl-H) = \Delta E_3 - IE(H) = 5.03 \text{ eV},$$

$$D_0(C_2H_2Cl^+-H) = -\Delta E_1 + \Delta E_2 = 4.92 \text{ eV},$$

$$D_0(C_2H_2Cl-H^+) = -\Delta E_1 + \Delta E_3 = 8.65 \text{ eV},$$

$$IE(C_2H_2Cl) = -D_0(C_2H_2Cl-H) + \Delta E_2 = 9.87 \text{ eV}.$$

From the second diagram:

$$D_0(C_2H_3-Cl) = \Delta E_5 - IE(H) = 6.33 \text{ eV},$$

$$D_0(C_2H_3^+-Cl) = -\Delta E_1 + \Delta E_4 = 2.56 \text{ eV},$$

$$D_0(C_2H_3-Cl^+) = -\Delta E_1 + \Delta E_5 = 9.32 \text{ eV},$$

$$IE(C_2H_3) = -D_0(C_2H_3-Cl) + \Delta E_4 = 6.21 \text{ eV}.$$

A2.29 The following diagram can be constructed:

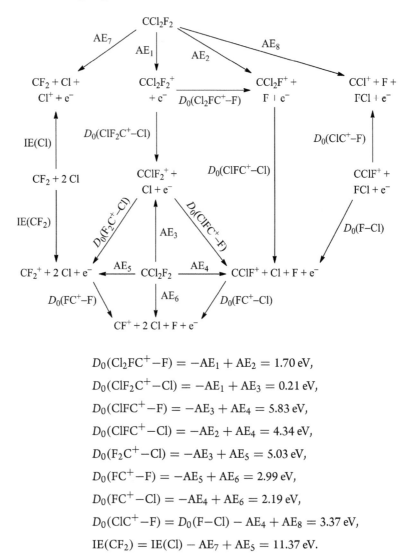

$$D_0(Cl_2FC^+-F) = -AE_1 + AE_2 = 1.70 \text{ eV},$$

$$D_0(ClF_2C^+-Cl) = -AE_1 + AE_3 = 0.21 \text{ eV},$$

$$D_0(ClFC^+-F) = -AE_3 + AE_4 = 5.83 \text{ eV},$$

$$D_0(ClFC^+-Cl) = -AE_2 + AE_4 = 4.34 \text{ eV},$$

$$D_0(F_2C^+-Cl) = -AE_3 + AE_5 = 5.03 \text{ eV},$$

$$D_0(FC^+-F) = -AE_5 + AE_6 = 2.99 \text{ eV},$$

$$D_0(FC^+-Cl) = -AE_4 + AE_6 = 2.19 \text{ eV},$$

$$D_0(ClC^+-F) = D_0(F-Cl) - AE_4 + AE_8 = 3.37 \text{ eV},$$

$$IE(CF_2) = IE(Cl) - AE_7 + AE_5 = 11.37 \text{ eV}.$$

A2.30 The following diagram can be constructed:

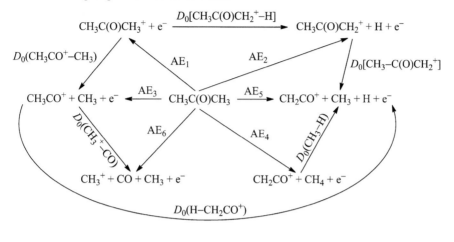

$$IE[CH_3C(O)CH_3] = AE_1 = 9.69 \text{ eV}.$$

$$D_0[CH_3C(O)CH_2^+ - H] = -AE_1 + AE_2 = 3.41 \text{ eV},$$

$$D_0(CH_3CO^+ - CH_3) = -AE_1 + AE_3 = 0.80 \text{ eV},$$

$$D_0[CH_3 - C(O)CH_2^+] = -AE_2 + AE_5 = 1.87 \text{ eV},$$

$$D_0(H - CH_2CO^+) = -AE_3 + AE_5 = 4.48 \text{ eV},$$

$$D_0(CH_3^+ - CO) = -AE_3 + AE_6 = 3.92 \text{ eV},$$

$$D_0(CH_3 - H) = -AE_4 + AE_5 = 4.44 \text{ eV}.$$

A2.31 (i) ΔH_{298} for atomization

$$= 6H_{298}(C) + 6H_{298}(H) - H_{298}[(CH)_6]$$
$$= 6\Delta H_{f298}(C) + 6\Delta H_{f298}(H) - \Delta H_{f298}[(CH)_6].$$

Hence, $\Delta H_{f298}[(CH)_6]$

$$= [6\Delta H_{f298}(C) + 6\Delta H_{f298}(H)] + H_{298}[(CH)_6] - \{[6H_{298}(C) + 6H_{298}(H)]\}.$$

The results are tabulated below.

	Benzene (**1**)	Dewar benzene (**2**)	Prismane (**3**)	Benzvalene (**4**)	3,3'- Bicyclopropenyl (**5**)
ΔH_{f298} (kJ mol^{-1})	87.4	411.4	567.7	393.3	592.2

(ii) ΔH_{298} for isodesmic bond separation.

$$= [yH_{298}(C_2H_6) + zH_{298}(C_2H_4)] - \{H_{298}[(CH)_6] + xH_{298}(CH_4)\}$$
$$= [y\Delta H_{f298}(C_2H_6) + z\Delta H_{f298}(C_2H_4)] - \{\Delta H_{f298}[(CH)_6] + x\Delta H_{f298}(CH_4)\}.$$

Hence, $\Delta H_{f298}[(CH)_6] = [y\Delta H_{f298}(C_2H_6) + z\Delta H_{f298}(C_2H_4) - x\Delta H_{f298}$
$(CH_4)] + H_{298}[(CH)_6] - \{yH_{298}(C_2H_6) + zH_{298}(C_2H_4) - xH_{298}(CH_4)\}.$
The results are tabulated below.

	Benzene (1)	Dewar benzene (2)	Prismane (3)	Benzvalene (4)	3,3'- Bicyclopropenyl (5)
ΔH_{f298} (kJ mol^{-1})	82.2	405.3	559.9	386.3	586.0

(iii) The differences between the results in the two schemes are between 5 and 8 kJ mol^{-1}. When compared with the experimental results, the errors in the atomization scheme are larger than those in the isodesmic bond separation due to the accumulation of errors in the bonds broken in the atomization scheme.

A2.32 (i) ΔH_{298} for atomization

$$= xH_{298}(B) + zH_{298}(H) - H_{298}(B_xH_z)$$
$$= x\Delta H_{f298}(B) + z\Delta H_{f298}(H) - \Delta H_{f298}(B_xH_z).$$

Hence, $\Delta H_{f298}(B_xH_z)$

$$= [x\Delta H_{f298}(B) + z\Delta H_{f298}(H)] + H_{298}(B_xH_z) - [xH_{298}(B) + zH_{298}(H)].$$

The results are tabulated below.

	B_3H_7	B_4H_{10}	B_5H_{11}	B_6H_{10}	$B_{10}H_{14}$
ΔH_{f298} (kJ mol^{-1})	108.3	50.4	81.4	80.7	-25.1

(ii) ΔH_{298} for isodesmic bond separation

$$= [yH_{298}(H_2) + H_{298}(B_xH_z)] - xH_{298}(BH_3)$$
$$= [y\Delta H_{f298}(H_2) + \Delta H_{f298}(B_xH_z)] - x\Delta H_{f298}(BH_3).$$

Hence, $\Delta H_{f298}(B_xH_z) = x\Delta H_{f298}(BH_3) - y\Delta H_{f298}(H_2) + [yH_{298}(H_2) + H_{298}(B_xH_z)] - xH_{298}(BH_3)$.

The results are tabulated below.

	B_3H_7	B_4H_{10}	B_5H_{11}	B_6H_{10}	$B_{10}H_{14}$
ΔH_{f298} (kJ mol^{-1})	122.9	70.5	105.0	106.0	14.5

A2.33

	r_{1s}	r_{atom}	r_{cov}	E_{1s}	IE	EA	χ'	χ
H	0.529	0.529	0.371	−13.6	13.6	0.75	7.17	2.24
Pos	1.058	1.058	0.742	−6.80	6.80	0.38	3.59	1.19

3 | Atomic Orbitals

PROBLEMS

3.1 In 1932, Urey noticed that a considerably over-exposed photograph of the visible spectrum of hydrogen showed weak satellite lines on the short wavelength side of the normal Balmer series. He made the (now) obvious conclusion that these satellite lines were due to an isotope of hydrogen, namely deuterium. If the H_α line and its satellite were observed at 6564.6 and 6562.8 Å, respectively, calculate the ratios M_1/m and M_2/M_1, where m, M_1, and M_2 are the masses of the electron, the proton, and the deuterium nucleus, respectively.

Given:

$$m = 9.109 \times 10^{-28} \text{ g,}$$

Atomic weight of H = 1.008,

$$N_0 \text{ (Avogadro's number)} = 6.022 \times 10^{23},$$

$$\tilde{v} = R_\infty \frac{Z^2}{1 + (m/M)} \left(\frac{1}{n_f^2} - \frac{1}{n_i^2} \right) \text{ for the } n_i \rightarrow n_f \text{ transition in a hydrogen-like}$$

system with nuclear mass M and charge $+Z$,

$$R_\infty = 109737.31 \text{ cm}^{-1}.$$

3.2 A positron is the antimatter counterpart of an electron: they have the same mass and the same spin, but opposite charges. In other words, a positron may be taken as a "positive electron".

A positronium atom (Pos) resembles a hydrogen atom. It has a positron, instead of a proton, as the nucleus, with an electron moving around it (or vice versa, as the two particles have the same mass). According to Bohr's model, the energy levels and the radii of the orbits for a hydrogen atom are:

$$E_n = -\frac{\mu e^4}{8\varepsilon_0^2 n^2 h^2} = -\frac{1}{2n^2} \text{ (a.u.),}$$

$$r_n = \frac{\varepsilon_0 n^2 h^2}{\pi \mu e^2} = n^2 \text{ (a.u.), } n = 1, 2, 3 \ldots .$$

54   *Problems in Structural Inorganic Chemistry.* Second edition. Wai-Kee Li, Hung Kay Lee, Dennis Kee Pui Ng, Yu-San Cheung, Kendrew Kin Wah Mak, and Thomas Chung Wai Mak. © Oxford University Press 2019. Published in 2019 by Oxford University Press. DOI: 10.1093/oso/9780198823902.001.0001

In these two expressions, the symbols have their usual meanings. One particularly relevant symbol for this question is μ, which denotes the reduced mass of the system.

Derive the corresponding expressions for Pos, and compare the results with those of a hydrogen atom.

3.3 A trial (Gaussian) wavefunction for the ground state of the hydrogen atom is
$$\psi = \left(\frac{2\alpha}{\pi}\right)^{3/4} e^{-\alpha r^2}.$$

(i) Show that ψ is normalized.

(ii) For the hydrogen atom, with wavefunctions independent of θ and ψ, the Hamiltonian operator (in atomic units) has the form:

$$\hat{H} = -\frac{1}{2}\nabla^2 - \frac{1}{r} = -\frac{1}{2}\left(\frac{d^2}{dr^2}\right) - \frac{1}{r}\left(\frac{d}{dr}\right) - \left(\frac{1}{r}\right).$$

It can be shown that $E(\alpha) = \int \psi \hat{H} \psi \, d\tau = \frac{3\alpha}{2} - \left(\frac{8\alpha}{\pi}\right)^{1/2}$. Apply the variational method to obtain the optimal α value and its corresponding minimum energy. Compare the calculated energy with the true value and give the percentage error.

(iii) Calculate the expectation value $<r>$ of ψ by evaluating the integral $\int \psi r \psi \, d\tau$.

(iv) Determine the most probable value of r by carrying out $\dfrac{d(r^2\psi^2)}{dr} = 0$.

For the mathematically inclined, it is a good exercise to confirm the $E(\alpha)$ expression in (ii).

3.4 In this question we treat the three-electron-atom Li with the unconventional configuration $1s^3$, i.e., $\psi_{trial} = (1, 2, 3) = e^{-\alpha(r_1+r_2+r_3)}$. The $\hat{H}$ operator of this atom has six terms:

$$\hat{H} = \sum_{i=1}^{3}\left(-\frac{1}{2}\nabla_i^2 - \frac{Z}{r_i}\right) + \frac{1}{r_{12}} + \frac{1}{r_{13}} + \frac{1}{r_{23}}.$$

$E(\alpha) = \int \psi_{trial}\hat{H}\psi_{trial}d\tau$ has five integrals in all. They are given below:

$$\int \psi\left(-\frac{1}{2}\nabla_i^2\right)\psi \, d\tau_i = \frac{1}{2}\alpha^2, \iint \psi\left(\frac{1}{r_{ij}}\right)\psi \, d\tau_i d\tau_j = -Z\alpha.$$

$$\iint \psi\left(\frac{1}{r_{12}}\right)\psi \, d\tau_1 d\tau_2 = \iint \psi\left(\frac{1}{r_{23}}\right)\psi \, d\tau_2 d\tau_3 = \iint \psi\left(\frac{1}{r_{13}}\right)\psi \, d\tau_1 d\tau_3 = \frac{5}{8}\alpha.$$

Calculate $E(\alpha)$ by the variational theorem. Compare with the experimental ground state energy (which can be taken as the "true" value), $E_0 = -7.478$ a.u. Also comment on the applicability of the variational theorem.

3.5 Using Slater's rules,

   (i)   Determine Z_{eff} for (a) a 3p electron of P; (b) a 4s and a 3d electron of Mn.

   (ii) Calculate the first and second ionization energies of Li. For reference, the experimental values are: $IE_1 = 5.39$, $IE_2 = 75.64$, and $IE_3 = 122.45$ eV.

  (iii) Estimate the first ionization energy and electron affinity (in eV) of Cl. Experimental values of IE_1 and EA for Cl are 12.967 and 3.61 eV, respectively.

3.6 The interionic distance in crystalline KCl is 3.14 Å. Use Slater's rules to determine the effective nuclear charge Z_{eff}, and hence deduce the ionic radii of K^+ and Cl^- by assuming that ionic radius is inversely proportional to effective nuclear charge estimated from Slater's rules. [For a discussion of univalent radii and crystal radii, see L. Pauling, *Nature of the Chemical Bond*, 3rd edn., Cornell University Press, Ithaca, 1960, pp. 511–9.]

3.7 The first and second ionization energies (denoted as IE_1 and IE_2) of helium are the energies required for the following processes to proceed:

$$He(g) \xrightarrow{\quad IE_1 \quad} He^+(g) + e^-,$$

$$He^+(g) \xrightarrow{\quad IE_2 \quad} He^{2+}(g) + e^-.$$

It is known that IE_1 for He is 24.6 eV. Determine (i) IE_2 for He, and (ii) the ground state energy for He. Express your answers in eV.

3.8 The following reaction might occur in the interior of a star:

$$He^{2+} + H \rightarrow He^+ + H^+.$$

Calculate the energy change (in eV). Assume all species are in their ground states.

3.9   (i) In the quantum mechanical treatment of Li^{2+}, the Θ and Φ equations and their solutions are the same as for the H atom, but the radial (R) equation is not. How does the radial equation for Li^{2+} differ from that for H?

     (ii) Show that the radial function, $R = 2\left(\frac{3}{a_0}\right)^{3/2} e^{-3r/a_0}$, for the 1s orbital of Li^{2+} satisfies the radial equation, $\frac{1}{R}\left(\frac{d}{dr}\right)\left(r^2\frac{dR}{dr}\right) + \left(\frac{8\pi^2\mu}{h^2}r^2\right)\left(E + \frac{e^2}{r}\right) = \ell(\ell + 1),$

and determine the energy eigenvalue. Compare this with the ground state energy E_H of the H atom.

(iii) Write down the 1s Slater orbital for the Li atom and compare it with the above Li^{2+} 1s orbital by means of an unscaled plot against r.

3.10 (i) Show that the wavefunction of the f_{xyz} orbital is proportional to $\sin^2 \theta \cos \theta \sin 2\phi$.

(ii) The wavefunction of this orbital has the form:

$\psi(f_{xyz}) = N \sin^2 \theta \cos \theta \sin 2\phi r^3 e^{-r/4}$. Determine normalization constant N.

(iii) Determine the directions, i.e., the θ and ϕ values, along which the electron is most likely to be found.

(iv) A function $f(x, y, z)$ is called "even" if $f(-x, -y, -z) = f(x, y, z)$. On the other hand, the function is "odd" if $f(-x, -y, -z) = -f(x, y, z)$. Determine whether the angular function of f_{xyz} orbital is an even or an odd function. Show your reasoning.

3.11 The angular wavefunctions of the p_z, d_{z^2} and f_{z^3} orbitals are:

$$p_z = \left(\frac{3}{4\pi}\right)^{1/2} \cos \theta,$$

$$d_{z^2} = \left(\frac{5}{16\pi}\right)^{1/2} (3 \cos^2 \theta - 1),$$

$$f_{z^3} = \left(\frac{7}{16\pi}\right)^{1/2} (5 \cos^3 \theta - 3 \cos \theta).$$

(i) Show that the three functions form an orthonormal set.

(ii) The shapes of the three orbitals are illustrated below. It is clear that all three functions have both positive and negative values. For each of the three orbitals, determine the position of the node(s) and hence the range(s) of angle θ which would give rise to a positive value and a negative value for the function.

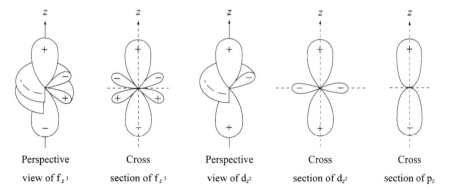

| Perspective view of f_{z^3} | Cross section of f_{z^3} | Perspective view of d_{z^2} | Cross section of d_{z^2} | Cross section of p_z |

3.12 The total wavefunctions for the $2p_x$ and $3d_{xz}$ orbitals (in atomic units) are:

$$\psi(2p_x) = (32\pi)^{-1/2}re^{-r/2}\sin\theta\cos\phi,$$
$$\psi(3d_{xz}) = (1/81)(2/\pi)^{1/2}r^2e^{-r/3}\sin\theta\cos\theta\cos\phi.$$

(i)   Show that $\psi(2p_x)$ is normalized.

(ii) Show that $\psi(2p_x)$ and $\psi(3d_{xz})$ are orthogonal to each other.

3.13 The wavefunctions of the 1s and 2s orbitals (in atomic units) are:

$$\psi(1s) = \frac{1}{\sqrt{\pi}}e^{-r}, \quad \psi(2s) = \frac{1}{\sqrt{32\pi}}(2-r)e^{-r/2}.$$

(i)   Show that $\psi(2s)$ is normalized.

(ii) Show that the two wavefunctions are orthogonal to each other.

3.14 (i)   For Li^{2+}, a hydrogenic ion with $Z = 3$, the ground state wavefunction (in a.u.) is

$$\psi_{1s}(Li^{2+}) = \pi^{-1/2}(Z)^{3/2}e^{-Zr} = \pi^{-1/2}(3)^{3/2}e^{-3r}.$$

Determine the most probable electron-nucleus separation for this ion. Express your answer in both a.u. and Å.

(ii) The most probable electron-nucleus separation for Li^{2+} is different from that for H. Suggest a reason for this difference.

3.15 In atomic units, the wavefunction of the hydrogen 1s orbital is: $\psi_{100} = \frac{1}{\sqrt{\pi}}e^{-r}$.

(i)   Calculate the probability of finding the electron in a spherical volume element $d\tau$ of radius $0.01\ a_0$ centered around a point which is $1\ a_0$ away from the nucleus. The volume of a sphere with radius b is $\frac{4}{3}\pi b^3$.

(ii) Calculate the probability of finding the electron in the volume element $d\tau$ bounded by $r = 0.995\ a_0$ and $1.005\ a_0$, $\theta = 0.199\ \pi$ and $0.201\ \pi$, and $\phi = 0.499\ \pi$ and $0.501\ \pi$.

(iii) Calculate the probability of finding the electron in a thin spherical shell of thickness 0.001 Å at a radius of $1\ a_0$ from the nucleus.

(iv) If we define a cumulative probability function $P_{n\ell}$ for an orbital with radial function $R_{n\ell}(r)$:

$$P_{n\ell}(\rho) = \int_0^\rho R_{n\ell}^2(r)r^2dr,$$

then this function gives the probability of finding the electron at a distance less than or equal to ρ from the nucleus.

(a) Show that P_{10} has the form: $P_{10}(\rho) = 1 - e^{-2\rho}(1 + 2\rho + 2\rho^2)$.

(b) Determine the probability of finding the electron inside a sphere centered at the nucleus with radius equal to 1 a_0.

(c) Determine the probability of finding the electron outside a sphere centered at the nucleus with radius equal to 2 a_0.

REFERENCE: T. C. W. Mak and W.-K. Li, Relative sizes of hydrogenic orbitals and the probability criterion. *J. Chem. Educ.* **52**, 90–1 (1975).

3.16 The wavefunction of the 2s orbital of the hydrogen atom (in atomic units) is:

$$\psi(2s) = \frac{1}{\sqrt{32\pi}}(2 - r)e^{-r/2}.$$

(i) Determine the positions of the node and maxima of the 2s radial probability distribution function. (The calculation is simplified by introducing the dimensionless variable $\rho = Zr/a_0$, i.e., measuring distances in units of a_0/Z.)

(ii) Calculate the probability of finding the electron within the nodal sphere.

3.17 This problem is concerned with graphical construction of the hydrogen-like $2p_z$ orbital. The wavefunction of the orbital is given by:

$$\psi(2p_z) = (32\pi)^{-1/2}(Z/a_0)^{5/2}re^{-(Zr/a_0)}\cos\theta = C\rho e^{-\rho/2}\cos\theta,$$

where C is a constant and $\rho = Zr/a_0$.

Let $P(\rho, \theta)$ (probability density) $= \psi^2(2p_z) = C^2\rho^2e^{-\rho}\cos^2\theta$. Note that $P(\rho, \theta)$ is (a) cylindrically symmetric with respect to the z-axis, i.e., it has no ϕ dependence, and (b) symmetric with respect to the xy-plane, i.e., the function remains unchanged upon changing θ to $180° - \theta$. Also, $P(\rho, \theta) = P(\rho, 0)\cos^2\theta$. P_{max}, the maximum value of $P(\rho, \theta)$, has the value $4C^2e^{-2}$ at $\rho = 2$, $\theta = 0$.

(i) Plot $P(\rho, 0)/P_{max}$ against ρ using the data given in Table 1. In the same graph, draw similar curves for $P(\rho, \theta)$ with $\theta = 15°, 30°, 45°$, and $60°$ by scaling the $\theta = 0°$ curve with the corresponding $\cos^2\theta$ values given in Table 2. A suitable scale is 2 cm $= 1$ ρ unit and 2 cm $= 0.1$ $P(\rho, 0)/P_{max}$.

Table 1 Variation of $P(\rho, 0)/P_{max} = (\rho^2 e^{2-\rho})/4$ with ρ

ρ	$P(\rho, 0)/P_{max}$	ρ	$P(\rho, 0)/P_{max}$	ρ	$P(\rho, 0)/P_{max}$
0	0	1.5	$9e^{0.5}/16 = 0.927$	4.5	$81/16e^{2.5} = 0.416$
0.25	$e^{1.75}/64 = 0.090$	2.0	1	5.0	$25/4e^{3} = 0.311$
0.5	$e^{1.5}/16 = 0.280$	2.5	$25/16e^{0.5} = 0.948$	6.0	$9/e^{4} = 0.165$
0.75	$9e^{1.25}/64 = 0.491$	3.0	$9/4e = 0.828$	7.0	$49/4e^{5} = 0.083$
1.0	$e/4 = 0.680$	3.5	$49/16e^{1.5} = 0.683$	8.0	$16/e^{6} = 0.040$
1.25	$25e^{0.75}/64 = 0.827$	4.0	$4/e^{2} = 0.541$	10.0	$25/e^{8} = 0.008$

Table 2 Variation of $\cos^2 \theta$ with θ

θ	0°	15°	30°	45°	60°	90°
$\cos^2 \theta$	1	0.933	0.75	0.5	0.25	0

(ii) From the above plot, choose a fixed value of $P(\rho, \theta)/P_{max}$, say 0.9, and read off the pairs of (ρ, θ) values of points where the horizontal line 0.9 intersects the curves. Use these data to plot the $P(\rho, \theta)/P_{max} = 0.9$ contour in a polar diagram.

(iii) Plot similar contours for $P(\rho, \theta)/P_{max} = 0.7, 0.5, 0.3,$ and 0.1 in the same diagram; bearing in mind the symmetry properties, sketch the complete $2p_z$ orbital.

Note: This problem shows how to plot an atomic orbital in a classical way. It is recognized that modern computer software makes this job much easier.

3.18 The wavefunction for hydrogenic $3d_{yz}$ atomic orbital (where a_0 is the Bohr radius) is given as:

$$\psi(3d_{yz}) = \left(\frac{1}{81}\sqrt{\frac{2}{\pi}}\right)\left(\frac{Z}{a_0}\right)^{7/2} r^2 e^{\frac{-Zr}{3a_0}} \sin \theta \cos \theta \sin \phi.$$

Also given below are the coordinates of two points:

$$A(x = \sqrt{3} a_0, y = 1 a_0, z = 1 a_0); \qquad B(x = -\sqrt{3} a_0, y = 1 a_0, z = 1 a_0).$$

Calculate the probability of finding the $3d_{yz}$ electron for an hydrogen atom in a sphere of volume $\Delta \tau = 1.00 \times 10^{-9} a_0^3$ centered at each of these two points.

3.19 The total wavefunction of the $3d_{x^2-y^2}$ orbital (in atomic units) is:

$$\psi(3d_{x^2-y^2}) = \left(\frac{1}{81\sqrt{2\pi}}\right) r^2 e^{-r/3} \sin^2 \theta \cos 2\phi.$$

Determine the maximum probability of locating a $3d_{x^2-y^2}$ electron in a volume element $\Delta\tau = 1.00 \times 10^{-9} a_0^3$ around the point (r, θ, ϕ). [Hint: The determination of θ and ϕ is straightforward. But the determination of r requires a maximization procedure.] So first you need to determine the coordinates r, θ, and ϕ of the point. You can then calculate the (maximum) probability. Note that there may be more than one point that would lead to the same maximum probability. You need to write down the coordinates of all these points.

3.20 The radial wavefunction for the hydrogenic 4f orbital is (in atomic units):

$$R_{4,3} = \left(\frac{1}{768\sqrt{35}}\right) r^3 e^{-r/4}.$$

(i) For any orbital, the average nucleus-electron distance, $<r>$, is given by:

$$<r> = \int_0^\infty r(R_{n,\ell})^2 d\tau = \int_0^\infty r(R_{n,\ell})^2 r^2 dr = \int_0^\infty r^3 (R_{n,\ell})^2 dr.$$

Determine $<r>$ (in a.u.) for the 4f orbital. Is this value equal to the most probable value of r?

(ii) Also, the average inverse nucleus-electron distance, $<r^{-1}>$, is given by

$$<r^{-1}> = \int_0^\infty r^{-1}(R_{n,\ell})^2 r^2 dr = \int_0^\infty r(R_{n,\ell})^2 dr.$$

Determine $<r^{-1}>$ (in a.u.) for the 4f orbital. (Note that $<r^{-1}>$ is not $<r>^{-1}$.) Does the virial theorem hold in this case?

3.21 Consider the transformation of an initial Cartesian coordinate system (x, y, z) to another (x', y', z') by means of three successive rotations over angles ϕ, θ, and ψ (the Eulerian angles) performed in the sequence depicted below.

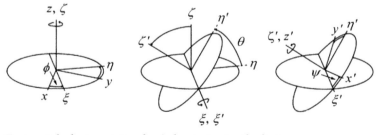

First, anticlockwise rotation by ϕ about z to give the $\xi\eta\zeta$-axes;

second, anticlockwise rotation by θ about ξ to give the $\xi'\eta'\zeta'$-axes;

third, anticlockwise rotation by ψ about ζ' to give the $x'y'z'$-axes.

(i) Express each of the three "new" p' atomic orbitals in terms of the original set of p orbitals.

(ii) Express the "new" $d_{z'^2}$ orbital in terms of the original d orbitals.

Note that the normalized angular parts of the d_{z^2}, $d_{x^2-y^2}$, and d_{xy} orbitals are

$$d_{z^2} = \sqrt{\frac{5}{16\pi}}(3\cos^2\theta - 1) = \sqrt{\frac{5}{16\pi}}\left(\frac{3z^2 - r^2}{r^2}\right);$$

$$d_{x^2-y^2} = \sqrt{\frac{15}{16\pi}}\left(\frac{x^2 - y^2}{r^2}\right);$$

$$d_{xy} = \sqrt{\frac{15}{16\pi}}\left(\frac{xy}{r^2}\right).$$

For a discussion on the definition of Eulerian angles, see H. Goldstein, *Classical Mechanics*, Addison-Wesley, Reading, 1950, pp. 107–9.

3.22 Let S_σ and S_δ represent the overlap integrals between d orbitals situated on two atoms of the same type as illustrated in (a) and (b), respectively. In terms of these, write down expressions for the overlap cases depicted in (c) and (d).

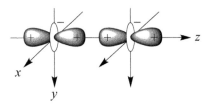

(a) End-on overlap of two d_{z^2} orbitals, overlap integral S_σ.

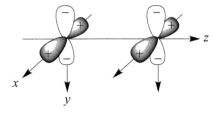

(b) Broadside-on overlap of two $d_{x^2-y^2}$ orbitals, overlap integral S_δ.

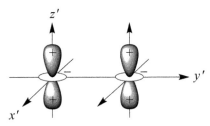

(c) Broadside-on (σ symbatic) overlap of two d_{z^2} orbitals.

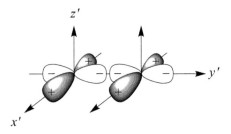

(d) End-on overlap of two $d_{x^2-y^2}$ orbitals.

REFERENCE: S. P. McGlynn, L. G. Vanquickenborne, M. Kinoshita, and D. G. Carroll, *Introduction to Applied Quantum Chemistry*, Holt, Rinehart, and Winston, New York, 1972, pp. 62–3.

3.23 Based on the idea that the two electrons in an helium atom experience different effective nuclear charges (α and β) at any given instant, the following approximate wavefunction has been proposed for the helium atom:

$$\psi = N(e^{-\alpha r_1}e^{-\beta r_2} + e^{-\beta r_1}e^{-\alpha r_2}),$$

where N is the normalization constant.

(i) Determine N.

(ii) Make an educated guess for α and β, without carrying out the actual optimization procedure.

REFERENCE: W.-K. Li, Two-parameter wave functions for the helium sequence. *J. Chem. Educ.* **64**, 128–9 (1987).

SOLUTIONS

A3.1 $M_1 \approx$ mass of proton $= (1.008 \text{ g})/(6.022 \times 10^{23}) = 1.674 \times 10^{-24}$ g,

$M_1/m = (1.674 \times 10^{-24} \text{ g})/(9.109 \times 10^{-28} \text{ g}) = 1838.$

$$\frac{\lambda_2}{\lambda_1} = \frac{1 + (m/M_2)}{1 + (m/M_1)}.$$

$m/M_2 = 1/3708,\ M_2/M_1 = 2.017 \approx 2.$

A3.2 For Pos, $\mu = \dfrac{m m_{e^+}}{m + m_{e^+}} = \dfrac{1}{2}$ a.u., while μ for a hydrogen atom is simply 1 a.u. Hence, for a Pos:

$$E_n = -\frac{1}{4n^2}\text{a.u.,} \quad r_n = 2n^2 \text{ a.u.,} \quad n = 1, 2, 3, \ldots.$$

In other words, the binding energy of a Pos is half that of a hydrogen atom, while the orbital radii of a Pos are twice as large as those for a hydrogen atom.

A3.3 (i)
$$\int |\psi|^2 d\tau = \left(\frac{2\alpha}{\pi}\right)^{3/2} \int_0^{2\pi} d\phi \int_0^{\pi} \sin\theta\, d\theta \int_0^{\infty} e^{-2ar^2} r^2 dr$$

$$= \left(\frac{2\alpha}{\pi}\right)^{3/2} (2\pi)(2) \left(\frac{1}{2^2(2\alpha)}\sqrt{\frac{\pi}{2\alpha}}\right) = 1.$$

(ii) $\dfrac{dE(\alpha)}{d\alpha} = 0 \Rightarrow \dfrac{3}{2} - \left(\dfrac{8}{\pi}\right)^{1/2}\left(\dfrac{1}{2}\right)\alpha^{-1/2} = 0 \Rightarrow$ or $\alpha = \dfrac{8}{9\pi} = 0.283.$

$$E_{\min} = \dfrac{3}{2}\cdot\dfrac{8}{9\pi} - \left(\dfrac{8}{\pi}\cdot\dfrac{8}{9\pi}\right)^{1/2} = -\dfrac{4}{3\pi} = -0.424 \text{ a.u.}$$

The true value E_0 is $-\frac{1}{2}$ a.u. The percentage error in the energy as calculated from a Gaussian-type trial function is 15%.

(iii) $<r> = \left(\dfrac{2\alpha}{\pi}\right)^{3/2}\displaystyle\int_0^{2\pi} d\phi \int_0^{\pi}\sin\theta\,d\theta \int_0^{\infty} e^{-2ar^2} r^3\,dr$

$= \left(\dfrac{2\alpha}{\pi}\right)^{3/2}(2\pi)(2)\left(\dfrac{1}{2(-2\alpha)^2}\right)$

$= \left(\dfrac{2}{\pi\alpha}\right)^{1/2} = \left(\dfrac{2}{\pi\left(\frac{8}{9\pi}\right)}\right)^{1/2} = 1.5$ a.u., equal to the exact value.

(iv) $\dfrac{d(r^2\psi^2)}{dr} = 0 \Rightarrow 2re^{-2\alpha r^2} + r^2 e^{-2\alpha r^2}(-4\alpha r) = 0$

or $r = \left(\dfrac{1}{2\alpha}\right)^{1/2} = \left(\dfrac{1}{2\left(\frac{8}{9\pi}\right)}\right)^{1/2} = 1.329$ a.u.; the exact value is 1 a.u.

A3.4 $E(\alpha) = \frac{3}{2}\alpha^2 - 3Z\alpha + \frac{15}{8}\alpha.$

Minimization leads to

$E' = 3\alpha - 3Z + \frac{15}{8} = 0$

$\alpha = 2.375,\ E = -8.4609$ a.u.

which is lower than the experimental value -7.478 a.u. This apparent violation of the variational theorem is due to the use of an unacceptable wavefunction (with three electrons housed in the same 1s orbital.)

A3.5 (i) (a) For P, $Z = 15$ and there are two /1s/, eight /2s,2p/, and five /3s,3p/ electrons. Hence, $Z_{\text{eff}}(3p) = 15 - 4(0.35) - 8(0.85) - 2(1.00) = 4.8$.

(b) For Mn, $Z = 25$ and there are two /1s/, eight /2s,2p/, eight /3s,3p/, five /3d/, and two /4s/ electrons. Hence,

$$Z_{\text{eff}}(4s) = 25 - 1(0.35) - 13(0.85) - 10(1.00) = 3.60.$$

$$Z_{\text{eff}}(3d) = 25 - 4(0.35) - 18(1.00) = 5.60.$$

(ii) $Z_{\text{eff}}(2s \text{ of Li}) = 3 - 2(0.85) = 1.30,\ n^*(2s \text{ of Li}) = 2,$
IE$_1$ of Li $= -E(2s \text{ of Li}) = (Z_{\text{eff}}/n^*)^2 \times 13.6$ eV $= 5.75$ eV, in good agreement with the experimental result.

$Z_{eff}(1s \text{ of } Li^+) = 3 - 1(0.3) = 2.70$, $n^*(1s \text{ of } Li^+) = 1$,

IE_2 of $Li = -E(1s \text{ of } Li^+) = 99.1$ eV, which is 31% larger than the experimental result. It is pointed that this calculation of IE_2 makes use of Koopmans' theorem, which states that the energy required to remove an electron from a closed-shell species can be *approximated* by the magnitude of the orbital energy of the electron removed. A better result can be obtained if we make use of the definition of IE_2, which is simply the difference between $E_{Li^{2+}}$ and E_{Li^+}:

$$Z_{eff}(1s \text{ of } Li^+) = 3 - 1(0.3) = 2.70; \; Z_{eff}(1s \text{ of } Li^{2+}) = 3.$$

$IE_2 = E_{Li^{2+}} - E_{Li^+} = -13.6[1(3/1)^2 - 2(2.70/1)^2]$ eV $= 75.9$ eV, which is only 3% larger than the experimental result.

(iii) Cl : $1s^2 2s^2 2p^6 3s^2 3p^5$, $Z_{eff}(3p) = 17 - 6(0.35) - 8(0.85) - 2(1.0) = 6.10$.

Cl^+ : $1s^2 2s^2 2p^6 3s^2 3p^4$, $Z_{eff}(3p) = 17 - 5(0.35) - 8(0.85) - 2(1.0) = 6.45$.

Cl^- : $1s^2 2s^2 2p^6 3s^2 3p^6$, $Z_{eff}(3p) = 17 - 7(0.35) - 8(0.85) - 2(1.0) = 5.75$.

n^* for 3p electron $= 3$.

$$IE_1 = E_{Cl^+} - E_{Cl} = -13.6[6(6.45/3)^2 - 7(6.10/3)^2] \text{ eV} = 16.4 \text{ eV}.$$
$$EA = E_{Cl} - E_{Cl^-} = -13.6[7(6.10/3)^2 - 8(5.75/3)^2] \text{ eV} = 6.09 \text{ eV}.$$

Both values are significantly larger than the corresponding experimental result.

A3.6 K^+ : Z_{eff} (with respect to Cl^-) $= 19 - 8(0.35) - 8(0.85) - 2(1.0) = 7.40$.
Cl^- : Z_{eff} (with respect to K^+) $= 17 - 8(0.35) - 8(0.85) - 2(1.0) = 5.40$.
Assuming $r = C/Z_{eff}$, we have

$$r_{K^+}/r_{Cl^-} = 5.40/7.40, \; r_{K^+} + r_{Cl^-} = 3.14.$$

Solving, $r_{Cl^-} = 1.82$ Å, $r_{K^+} = 1.32$ Å.

A3.7 $IE_1 = E(He^+) - E(He)$,
$IE_2 = E(He^{2+}) - E(He^+)$.

Summing up, $IE_1 + IE_2 = -E(He)$, and
$IE_2 = \left(\frac{2^2}{1^2}\right) \times 13.6$ eV $= 54.4$ eV.
$E(He) = -(IE_1 + IE_2) = -(24.6 + 54.4) = -79.0$ eV.

A3.8 He^{2+} and H^+ are bare nuclei so their (electronic) energies equal zero. He^+ and H are hydrogen-like so their 1s energies equal $-\frac{Z^2}{2}$. Thus $\Delta E = -\frac{4}{2} + \frac{1}{2} = -\frac{3}{2}$ a.u. $= -40.8$ eV.

A3.9   (i)   The potential energy term is changed from e^2/r to $3e^2/r$ and the reduced mass μ is also different (to a very slight extent).

(ii) $E(\text{1s of Li}^{2+}) = -9e^2/2a_0 = 9E_H$.

(iii)   For a 1s electron of the Li atom,

$$Z_{\text{eff}} = 2.7, \psi(\text{1s}) = 2\,(4\pi)^{-1/2}\left(\frac{2.7}{a_0}\right)^{3/2} e^{-2.7r/a_0}.$$

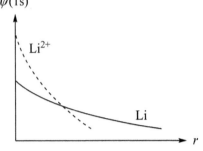

$\psi(\text{1s})$

Li^{2+}

Li

r

A3.10   (i) $\psi\,(\text{f}_{xyz}) \propto x \cdot y \cdot z = (r\sin\theta\,\cos\phi)(r\sin\theta\,\sin\phi)(r\cos\theta)$
$$= \tfrac{1}{2}r^3\sin^2\theta\,\cos\theta\,\sin 2\phi.$$

Hence, $\psi\,(\text{f}_{xyz})$ is proportional to $\sin^2\theta\,\cos\theta\,\sin 2\phi$.

(ii) $$\int_0^{2\pi} \sin^2 2\phi\,d\phi = \pi.$$

$$\int_0^\pi \sin^4\theta\,\cos^2\theta\,\sin\theta\,d\theta = \frac{16}{105}.$$

$$\int_0^\infty r^8 e^{-r/2}dr = \frac{8!}{(1/2)^9} = 2^{16}\times 3 \times 105.$$

$$N = \left(\pi \times \frac{16}{105}\times 2^{16}\times 3 \times 105\right)^{-1/2} = \frac{1}{1024\sqrt{3\pi}}.$$

(iii)   Since the angular function is proportional to $\sin^2\theta\,\cos\theta\,\sin 2\phi$, the maxima of electron density are determined by setting $\dfrac{d}{d\theta}[(\sin^2\theta\,\cos\theta)^2] = \dfrac{d}{d\phi}[(\sin 2\phi)^2] = 0.$

$$\frac{d}{d\theta}[(\sin^2\theta\,\cos\theta)^2] = 0 \Rightarrow \tan^2\theta = 2 \Rightarrow \theta = 54.74°, 125.26°.$$

$$\frac{d}{d\phi}[(\sin 2\phi)^2] = 0 \Rightarrow \cos 2\phi = 0 \Rightarrow \phi = 45°, 135°, 225°, 315°.$$

There are eight directions, along which the electron is most likely to be found with identical probability density.

The same results can be more conveniently obtained pictorially by noting that the eight lobes of the f_{xyz} orbital point toward the corners of a cube with the origin as its center. The cube is orientated in such a way that x-, y-, and z-axes pass through the centers of the six faces as shown below. The directions of the lobes are all the combinations of $(\pm 1, \pm 1, \pm 1)$. Thus, $\theta = \tan^{-1}\left(\dfrac{\sqrt{2}}{1}\right) = 54.74°$ (for the four lobes above the xy-plane) and $180° - 54.74° = 125.26°$ (for the four lobes under the xy-plane); $\phi = 45°, 135°, 225°$, and $315°$.

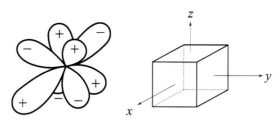

(iv) Since $\psi(f_{xyz}) \propto xyz$, it can be expressed in the form of $xyz \cdot f(r)$. The transformation $(x, y, z) \to (-x, -y, -z)$ does not change r. Hence,
$$\psi[f_{xyz}(-x, -y, -z)] = (-x)(-y)(-z) \cdot f(r) = -xyz \cdot f(r) = -\psi[f_{xyz}(x, y, z)]$$
and it is an odd function. This is illustrated pictorially above.

Remark: It can easily be shown that s and d orbitals are even functions, while p and f orbitals are odd.

A3.11 (i) $\displaystyle\int_0^\pi \cos^n\theta \sin\theta\,d\theta = -\frac{1}{n+1}[\cos^{n+1}(\pi) - \cos^{n+1}(0)] = \frac{2}{n+1}$ for even n,

$$= 0 \text{ for odd } n.$$

Note that $d\tau = \sin\theta\,d\theta\,d\phi$ for the angular functions.

$$I_{11} = \int |\psi(p_z)|^2 d\tau = \frac{3}{4\pi}\int_0^{2\pi} d\phi \int_0^\pi \cos^2\theta \sin\theta\,d\theta = \frac{3}{4\pi}(2\pi)\left(\frac{2}{3}\right) = 1.$$

$$I_{22} = \int |\psi(d_{z^2})|^2 d\tau = \frac{5}{16\pi}\int_0^{2\pi} d\phi \int_0^\pi (3\cos^2\theta - 1)^2 \sin\theta\,d\theta = \ldots = 1.$$

$$I_{33} = \int |\psi(f_{z^3})|^2 d\tau = \frac{7}{16\pi}\int_0^{2\pi} d\phi \int_0^\pi (5\cos^3\theta - 3\cos\theta)^2 \sin\theta\,d\theta = \ldots = 1.$$

$$I_{12} = \int \psi(p_z)\psi(d_{z^2})\,d\tau =$$

$$\left(\frac{3}{4\pi}\right)^{1/2}\left(\frac{5}{16\pi}\right)^{1/2}\int_0^{2\pi} d\phi \int_0^\pi \cos\theta(3\cos^2\theta - 1)\sin\theta\,d\theta = 0.$$

$$I_{23} = \int \psi(d_{z^2})\psi(f_{z^3})d\tau = \left(\frac{5}{16\pi}\right)^{1/2}\left(\frac{7}{16\pi}\right)^{1/2} \times$$

$$\int_0^{2\pi} d\phi \int_0^{\pi} (3\cos^2\theta - 1)(5\cos^3\theta - 3\cos\theta)\sin\theta\, d\theta = 0.$$

$$I_{13} = \int \psi(p_z)\psi(f_{z^3})d\tau = \left(\frac{3}{4\pi}\right)^{1/2}\left(\frac{7}{16\pi}\right)^{1/2} \times$$

$$\int_0^{2\pi} d\phi \int_0^{\pi} \cos\theta(5\cos^3\theta - 3\cos\theta)\sin\theta\, d\theta = 0.$$

(ii) Orbital p_z is positive when $0° \leq \theta < 90°$ and negative when $90° < \theta \leq 180°$. The node is at $\theta = 90°$.

For d_{z^2}, when $3\cos^2\theta - 1 = 0$, we have nodes at $\theta = \cos^{-1}\left(\pm\frac{1}{\sqrt{3}}\right) = 54.7°$ or $125.3°$.

So, d_{z^2} is positive when $0° \leq \theta < 54.7°$ and $125.3° < \theta \leq 180°$; it is negative when $54.7° < \theta < 125.3°$.

For f_{z^3}, when $5\cos^3\theta - 3\cos\theta = 0$, we have nodes at $\theta = \cos^{-1}\left(\pm\sqrt{\frac{3}{5}}\right)$ or $\cos^{-1}(0)$, resulting in $\theta = 39.2°$ or $140.8°$ or $90°$.

So, f_{z^3} is positive for the ranges $0° \leq \theta < 39.2°$ and $90° < \theta \leq 140.8°$; and negative for the ranges $39.2° < \theta < 90°$ and $140.8° < \theta \leq 180°$.

A3.12 (i)
$$I_1 = \int |\psi(2p_x)|^2 d\tau = \frac{1}{32\pi} \int_0^{\infty} r^2 e^{-r} r^2 dr \int_0^{\pi} \sin^2\theta \sin\theta\, d\theta \int_0^{2\pi} \cos^2\phi\, d\phi.$$

$$I_r = \int_0^{\infty} r^4 e^{-r} dr = \frac{4!}{1^5} = 24.$$

$$I_\theta = \int_0^{\pi} \sin^3\theta\, d\theta = \frac{1}{3}\left[-\cos\theta(\sin^2\theta + 2)\right]_0^{\pi} = \frac{4}{3}.$$

$$I_\phi = \int_0^{2\pi} \cos^2\phi\, d\phi = \frac{1}{2}\left[\phi + \sin\phi\cos\phi\right]_0^{2\pi} = \pi.$$

$$I_1 = \frac{1}{32\pi} \times 24 \times \frac{4}{3} \times \pi = 1.$$

(ii) $$I_2 = \int \psi(2p_x)\psi(3d_{xz})d\tau$$

$$= (32\pi)^{-1/2}(1/81)(2/\pi)^{1/2} \int_0^{\infty} r^3 e^{-5r/6} r^2 dr \int_0^{\pi} \sin^2\theta\cos\theta\sin\theta\, d\theta \int_0^{2\pi} \cos^2\phi\, d\phi.$$

$$I_r = \int_0^{\infty} r^5 e^{-5r/6} dr = \frac{5!}{(5/6)^6}.$$

$$I_\theta = \int_0^\pi \sin^2\theta\cos\theta\sin\theta\,d\theta = 0.$$

$$I_\phi = \int_0^{2\pi} \cos^2\phi\,d\phi = \pi.$$

$$I_2 = 0.$$

So, it is not necessary to calculate I_r and I_ϕ in this case.

A3.13 (i) $\quad I_1 = \int |\psi(2s)|^2 d\tau = \dfrac{1}{32\pi}\int_0^\infty (2-r)^2 e^{-r} r^2 dr \int_0^\pi \sin\theta\,d\theta \int_0^{2\pi} d\phi.$

$$I_r = \int_0^\infty (2-r)^2 e^{-r} r^2 dr = 8.$$

$$I_\theta = \int_0^\pi \sin\theta\,d\theta = 2.$$

$$I_\phi = \int_0^{2\pi} d\phi = 2\pi.$$

$$I_1 = \frac{1}{32\pi} \times 8 \times 2 \times 2\pi = 1.$$

So, $\psi(2s)$ is normalized.

(ii) $\quad I_2 = \int \psi(1s)\psi(2s)d\tau$

$$= \left(\frac{1}{\sqrt{\pi}}\right)\left(\frac{1}{\sqrt{32\pi}}\right)\int_0^\infty (2-r)e^{-3r/2} r^2 dr \int_0^\pi \sin\theta\,d\theta \int_0^{2\pi} d\phi.$$

$$I_r = \int_0^\infty (2-r)e^{-3r/2} r^2 dr = 0.$$

$$I_\theta = \int_0^\pi \sin\theta\,d\theta = 2.$$

$$I_\phi = \int_0^{2\pi} d\phi = 2\pi.$$

$I_2 = 0$; so $\psi(1s)$ and $\psi(2s)$ are orthogonal to each other.

A3.14 (i) $\quad D(r) = \text{constant} \times r^2 e^{-6r}.$

$$\frac{dD(r)}{dr} = 0 \Rightarrow 2re^{-6r} + r^2 e^{-6r}(-6) = 0.$$

$r_{max} = a_0/3$ for the 1s orbital of Li^{2+}, i.e., $r_{max} = \frac{1}{3}(\text{a.u.}) = 0.176\,\text{Å}.$

(ii) For the 1s orbital of H, $r_{max} = 1a_0$, larger than that for the 1s orbital of Li^{2+}. This is because Z for Li^{2+} is 3 and Z for H is 1. The larger attraction brings in the electron closer.

A3.15 All answers are in atomic units.

(i) $d\tau = (4/3)\pi(0.01)^3 a_0^3$.

$$P = |\psi|^2 d\tau = \pi^{-1}e^{-2r}(\text{with } r = 1a_0)d\tau = \pi^{-1}e^{-2}(4/3)\pi(0.01)^3$$

$$= 1.80 \times 10^{-7}.$$

(ii) $d\tau = r^2 dr \sin\theta \, d\theta \, d\phi = (1a_0)^2(0.01a_0)\sin(0.2\pi)(0.002\pi)(0.002\pi)$

$$= 7.39 \times 10^{-8}\pi a_0^3.$$

$$P = |\psi|^2 d\tau = \pi^{-1}e^{-2r}(\text{with } r = 1a_0)d\tau = \pi^{-1}e^{-2}(7.39 \times 10^{-8}\,\pi)$$

$$= 1.00 \times 10^{-8}.$$

(iii) $0.001\,\text{Å} = (0.001/0.529)a_0$.

$$d\tau = 4\pi r^2 dr = 4\pi(1a_0)^2(0.001/0.529)a_0 = 4\pi(0.001/0.529)a_0^3.$$

$$P = |\psi|^2 d\tau = \pi^{-1}e^{-2r}(\text{with } r = 1a_0)d\tau = \pi^{-1}e^{-2}[4\pi(0.001/0.529)]$$

$$= 1.02 \times 10^{-3}.$$

(iv) (a) $P_{10} = 4\int_0^\rho e^{-2r}r^2 dr = 4\left[\left(-\dfrac{1}{2}\rho^2 - \dfrac{1}{2}\rho - \dfrac{1}{4}\right)e^{-2\rho}\right]_0^\rho$

$$= 1 - (2\rho^2 + 2\rho + 1)e^{-2\rho}.$$

(b) $P = [1 - (2\rho^2 + 2\rho + 1)e^{-2\rho}]_{\rho=1} = 1 - 5e^{-2} = 0.323.$

(c) $P = 1 - [1 - (2\rho^2 + 2\rho + 1)e^{-2\rho}]_{\rho=2} = 13e^{-4} = 0.238.$

A3.16 (i) $D(\text{2s radial probability distribution function}) = |\psi(2s)|^2(4\pi r^2)$
$$= \frac{1}{8}r^2(2-r)^2 e^{-r}.$$

For $\dfrac{dD}{dr} = 0$, we have $r = 2, 0.764, 5.236$ a.u. By considering either $\dfrac{d^2D}{dr^2}$ or the nodal property of $\psi(2s)$, it can be readily shown that the node lies at $r = 2$ a.u. and the maxima at $r = 0.764$ and 5.236 a.u., the latter being the principal maximum.

(ii) $P = \int D dr = \int_0^2 \dfrac{1}{8}r^2(2-r)^2 e^{-r} dr = 1 - 7e^{-2} = 0.0527.$

A3.17

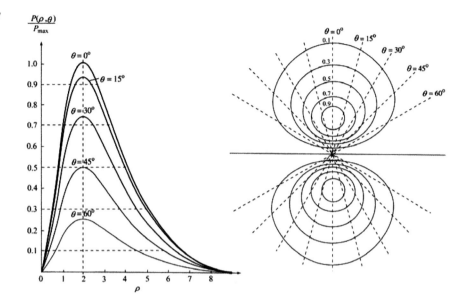

$$\frac{P(\rho,\theta)}{P_{max}}$$

A3.18 At A : $r = \sqrt{5}a_0, \theta = \cos^{-1}\left(\frac{z}{r}\right) = \cos^{-1}\left(\frac{1}{\sqrt{5}}\right) = 63.43°, \phi = \tan^{-1}\left(\frac{1}{\sqrt{3}}\right) = 30°.$

$$\psi(3d_{yz}) = \left(\frac{1}{81}\sqrt{\frac{2}{\pi}}\right)(a_0^{-7/2})(5a_0^2)e^{\frac{-\sqrt{5}}{3}}\left(0.894 \times 0.447 \times \frac{1}{2}\right)$$

$$= 4.684 \times 10^{-3}(a_0^{-3/2}).$$

$$P = |4.684 \times 10^{-3}(a_0^{-3/2})|^2 \times (1.00 \times 10^{-9}a_0^3) = 2.194 \times 10^{-14}.$$

At B : $r = \sqrt{5}a_0, \theta = \cos^{-1}\left(\frac{z}{r}\right) = \cos^{-1}\left(\frac{1}{\sqrt{5}}\right) = 63.43°,$

$$\phi = \tan^{-1}\left(\frac{1}{-\sqrt{3}}\right) = -30°.$$

$P = 2.194 \times 10^{-14}$ also.

A3.19 $P = |\psi(3d_{x^2-y^2})|^2 d\tau = \left(\frac{1}{81\sqrt{2\pi}}\right)^2 r^4 e^{-2r/3} \sin^4\theta \cos^2 2\phi \propto (r^4 e^{-2r/3})(\sin^4\theta) \times$

$(\cos^2 2\phi)$. The maximum values for P are obtained for $r = 6$ a.u.; $\theta = 90°$; $\phi = 0°,$
$90°, 180°$, and $270°$. So there are four points that have the maximum probability density:

$$P_{max} = \left(\frac{1}{81\sqrt{2\pi}}\right)^2 (6^4)(e^{-4})(1 \times 10^{-9}) = 5.76 \times 10^{-13}.$$

A3.20 (i) $<r> = \left(\dfrac{1}{768\sqrt{35}}\right)^2 \int_0^\infty r^9 e^{-r/2} dr = \left(\dfrac{1}{768\sqrt{35}}\right)^2 (9!)(2^{10}) = 18$ a.u.

The probability density of finding the electron at a distance r regardless of direction is: $D_{4,3} = 4\pi r^2 |R_{4,3}|^2 \propto r^8 e^{-r/2}$.

To obtain r_{max}, the most probable value of r, set $\dfrac{dD_{4,3}}{d\rho} = 0$ for maximum value of

$D_{4,3}$: $\dfrac{dD_{4,3}}{d\rho} = 0 \Rightarrow 8r^7 e^{-r/2} - \dfrac{1}{2}r^8 e^{-r/2} = 0 \Rightarrow r_{max} = 16$, which is different from $<r>$.

(ii) $<r^{-1}> = \left(\dfrac{1}{768\sqrt{35}}\right)^2 \int_0^\infty r^7 e^{-r/2} dr = \left(\dfrac{1}{768\sqrt{35}}\right)^2 (7!)(2^8) = {}^1\!/_{16}$ a.u.

Average potential energy $<PE> = <-r^{-1}> = -{}^1\!/_{16}$ a.u.

Total energy of a 4f electron $= -1/(2n^2) = -\dfrac{1}{32}$ a.u.

Average kinetic energy $<KE> = -\dfrac{1}{32} - \left(-\dfrac{1}{16}\right) = \dfrac{1}{32}$ a.u.

$<PE> : <KE> = -2 : 1$.

So the virial theorem holds in this case.

A3.21 The two sets of axes are related by the following transformations:

$$\begin{bmatrix} x' \\ y' \\ z' \end{bmatrix} = \begin{bmatrix} \cos\psi\cos\phi - \cos\theta\sin\phi\sin\psi & \cos\psi\sin\phi + \cos\theta\cos\phi\sin\psi & \sin\psi\sin\theta \\ -\sin\psi\cos\phi - \cos\theta\sin\phi\cos\psi & -\sin\psi\sin\phi + \cos\theta\cos\phi\cos\psi & \cos\psi\sin\theta \\ \sin\theta\sin\phi & -\sin\theta\cos\phi & \cos\theta \end{bmatrix} \begin{bmatrix} x \\ y \\ z \end{bmatrix}$$

$$\begin{bmatrix} x \\ y \\ z \end{bmatrix} = \begin{bmatrix} \cos\psi\cos\phi - \cos\theta\sin\phi\sin\psi & -\sin\psi\cos\phi - \cos\theta\sin\phi\cos\psi & \sin\theta\sin\phi \\ \cos\psi\sin\phi + \cos\theta\cos\phi\sin\psi & -\sin\psi\sin\phi + \cos\theta\cos\phi\cos\psi & -\sin\theta\cos\phi \\ \sin\theta\sin\psi & \sin\theta\cos\psi & \cos\theta \end{bmatrix} \begin{bmatrix} x' \\ y' \\ z' \end{bmatrix}$$

(i) A p orbital transforms in the same way as the corresponding Cartesian axis. Expressions for the p$'$ orbitals can be readily written down by inspection of the first matrix:

$$p_x{}' = (\cos\psi\cos\phi - \cos\theta\sin\phi\sin\psi)p_x$$
$$+ (\cos\psi\sin\phi + \cos\theta\cos\phi\sin\psi)p_y + (\sin\psi\sin\theta)p_z;$$
$$p_y{}' = (-\sin\psi\cos\phi - \cos\theta\sin\phi\cos\psi)p_x$$
$$+ (-\sin\psi\sin\phi + \cos\theta\cos\phi\cos\psi)p_y + (\cos\psi\sin\theta)p_z;$$
$$p_z{}' = (\sin\theta\sin\phi)p_x + (-\sin\theta\cos\phi)p_y + (\cos\theta)p_z.$$

(ii) $z' = (\sin\theta\sin\phi)x - (\sin\theta\cos\phi)y + (\cos\theta)z, r' = r.$

$$\frac{3z'^2 - r'^2}{r'^2} = \frac{3x^2 - r^2}{r^2}\sin^2\theta\sin^2\phi + \frac{3y^2 - r^2}{r^2}\sin^2\theta\cos^2\phi + \frac{3z^2 - r^2}{r^2}\cos^2\theta$$
$$- 6\left(\frac{xy}{r^2}\right)\sin^2\theta\sin\phi\cos\phi + 6\left(\frac{xz}{r^2}\right)\sin\theta\cos\theta\sin\phi$$
$$- 6\left(\frac{yz}{r^2}\right)\sin\theta\cos\theta\cos\phi.$$

The function $\dfrac{3x^2 - r^2}{r^2}$ corresponds to a d_{x^2} orbital, exactly like the d_{z^2} orbital but directed along the x-axis. Such an orbital can be expressed as a linear combination of the conventional d_{z^2} and $d_{x^2-y^2}$ orbitals:

$$\frac{3x^2 - r^2}{r^2} = \frac{3}{2}\left(\frac{x^2 - y^2}{r^2}\right) - \frac{1}{2}\left(\frac{3z^2 - r^2}{r^2}\right).$$

Similarly, $\dfrac{3y^2 - r^2}{r^2} = -\dfrac{3}{2}\left(\dfrac{x^2 - y^2}{r^2}\right) - \dfrac{1}{2}\left(\dfrac{3z^2 - r^2}{r^2}\right).$

Taking into account the different normalizing factors of various d orbitals, combination of the three preceding equations yields

$$d_{z^2} = (\cos^2\theta - \frac{1}{2}\sin^2\theta)d_{z^2} + \sqrt{\frac{3}{4}}[\sin^2\theta(\sin^2\phi - \cos^2\phi)]d_{x^2-y^2}$$
$$- \sqrt{3}(\sin^2\theta\sin\phi\cos\phi)d_{xy} + \sqrt{3}(\sin\theta\cos\theta\sin\phi)d_{xz}$$
$$- \sqrt{3}(\sin\theta\cos\theta\cos\phi)d_{yz}.$$

A3.22 Re-orientation (a) → (c) of the pair of d_{z^2} orbitals may be achieved by the following transformation ($\phi = 0$, $\theta = 90°$, $\psi = 0$; see the previous problem):

$$\begin{bmatrix} x' \\ y' \\ z' \end{bmatrix} = \begin{bmatrix} 1 & 0 & 0 \\ 0 & 0 & 1 \\ 0 & -1 & 0 \end{bmatrix} \begin{bmatrix} x \\ y \\ z \end{bmatrix},$$

from which it is readily deduced that $d_{z^2} = -\frac{1}{2}d_{z^2} - \sqrt{\frac{3}{4}}d_{x^2-y^2}$. Since, by symmetry, the overlap of d_{z^2} and $d_{x^2-y^2}$ on the two atomic centers vanishes, the overlap integral in case (c) is $\frac{1}{4}S_\sigma + \frac{3}{4}S_\delta$. In other words, the broadside-on (σ symbatic) overlap of two d_{z^2} orbitals has 25% σ and 75% δ character.

The same transformation applies to the reorientation (b) → (d). The overlap integral for the end-on overlap of two $d_{x^2-y^2}$ orbitals is $\frac{3}{4}S_\sigma + \frac{1}{4}S_\delta$.

A3.23 (i)

$$\int |\psi|^2 d\tau = N^2 \int \int (e^{-\alpha r_1} e^{-\beta r_2} + e^{-\beta r_1} e^{-\alpha r_2})^2 d\tau_1 d\tau_2$$

$$= N^2 \left[\int e^{-2\alpha r_1} d\tau_1 \int e^{-2\beta r_2} d\tau_2 + 2 \int (e^{-(\alpha+\beta)r_1}) d\tau_1 \int (e^{-(\alpha+\beta)r_2}) d\tau_2 \right.$$

$$\left. + \int e^{-2\beta r_1} d\tau_1 \int e^{-2\alpha r_2} d\tau_2 \right] = 1.$$

$$I_1 = \int e^{-2\alpha r_1} d\tau_1 = \int e^{-2\alpha r_2} d\tau_2 = \int_0^\infty e^{-2\alpha r_2} (4\pi r_2^2) dr_2 = (4\pi) \frac{2!}{(2\alpha)^3} = \frac{\pi}{\alpha^3}.$$

$$I_2 = \int e^{-2\beta r_1} d\tau_1 = \int e^{-2\beta r_2} d\tau_2 = \int_0^\infty e^{-2\beta r_2} (4\pi r_2^2) dr_2 = (4\pi) \frac{2!}{(2\beta)^3} = \frac{\pi}{\beta^3}.$$

$$I_3 = \int e^{-(\alpha+\beta)r_1} d\tau_1 = \int e^{-(\alpha+\beta)r_2} d\tau_2 - \int_0^\infty e^{-(\alpha+\beta)r_2} (4\pi r_2^2) dr_2$$

$$= (4\pi) \frac{2!}{(\alpha+\beta)^3} = \frac{8\pi}{(\alpha+\beta)^3}.$$

Hence, we have $N^2(I_1 \cdot I_2 + 2I_3 \cdot I_3 + I_2 \cdot I_1) = 1$,

$$N^2 \left[\frac{\pi}{\alpha^3} \cdot \frac{\pi}{\beta^3} + 2 \cdot \frac{8\pi}{(\alpha+\beta)^3} \cdot \frac{8\pi}{(\alpha+\beta)^3} + \frac{\pi}{\beta^3} \cdot \frac{\pi}{\alpha^3} \right] = 1,$$

$$N = \sqrt{\frac{\alpha^3 \beta^3 (\alpha+\beta)^6}{2\pi^2 [64\alpha^3\beta^3 + (\alpha+\beta)^6]}}.$$

(ii) Since, at any instant, one electron is closer to the nucleus than the other, the "inner" electron should experience an effective nuclear charge close to 2, while that for the "outer" electron is approximately 1. As it turns out, for He, the two charges are 2.1832 and 1.1885. For details, refer to the reference cited.

Hybrid Orbitals | 4

PROBLEMS

4.1 In the figure on the right, equivalent hybrid orbitals h_1 and h_2 are shown to lie on the xy-plane. In addition, hybrid h_1 forms an angle of $45°$ with the $-x$-axis and the angle formed by hybrids h_1 and h_2 is called θ.

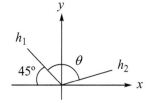

    (i)   Determine the (normalized) wavelfunction of hybrid orbital h_1 when it is (a) an sp hybrid orbital, (b) an sp^2 hybrid orbital, and (c) an sp^3 hybrid orbital.

    (ii) When both h_1 and h_2 are sp^2 hybrids, determine the (normalized) wavefunction of h_2.

    (iii) When both h_1 and h_2 are sp^3 hybrids, determine the (normalized) wavefunction of h_2.

4.2 Four hybrid orbitals are formed by one s and three p orbitals. As shown on the right, three of the hybrids: h_2, h_3, and h_4, are equivalent to each other, while the remaining one, h_1, is unique. Furthermore, as can be seen in the figure, h_1 lies on the x-axis and h_2 lies on the xy-plane.

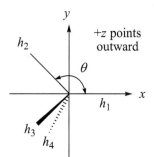

    (i)   It is known that h_1 is an sp^2 hybrid. What is the wavefunction for h_1?

    (ii) Write down the wavefunctions for the remaining three hybrid orbitals.

    (iii) What are the hybridization indices for h_2, h_3, and h_4?

    (iv) Determine θ_{12} and θ_{23}, where θ_{ij} denotes the angle formed by hybrids h_i and h_j.

Problems in Structural Inorganic Chemistry. Second edition. Wai-Kee Li, Hung Kay Lee, Dennis Kee Pui Ng, Yu-San Cheung, Kendrew Kin Wah Mak, and Thomas Chung Wai Mak. © Oxford University Press 2019. Published in 2019 by Oxford University Press. DOI: 10.1093/oso/9780198823902.001.0001

4.3 Diborane, B_2H_6, has the structure shown below (left). Note that terminal hydrogens H^a, H^b, H^e, and H^f lie in the xz-plane, while bridging hydrogens H^c and H^d are in the xy-plane.

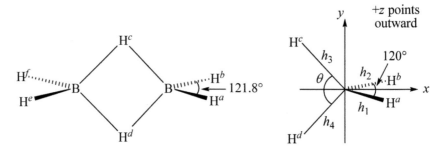

(i) Now we consider the boron atom bonded to hydrogens H^a, H^b, H^c, and H^d. Since $\angle H^aBH^b$ is about $120°$, we may assume that hybrids h_1 and h_2 (pointing to H^a and H^b, respectively, shown above on the right) are sp^2 hybrids. (Note again that h_1 and h_2 lie in the xz-plane.) Write down the wavefunctions of h_1 and h_2 with justification.

(ii) The remaining two (equivalent) hybrids, h_3 and h_4, lie in the xy-plane, pointing to H^c and H^d, respectively. Write down the wavefunctions for these two hybrids. Note that hybrids h_1 and h_3 are *not* equivalent to each other; and the 2s and three 2p orbitals of B are "used up" in forming these four hybrids.

(iii) Determine the hybridization index of h_3 and h_4. In addition, what is the angle θ formed by h_3 and h_4?

4.4 Consider two arbitrary and inequivalent hybrid orbitals, h_i and h_j, each formed by one s orbital and three p orbitals. The hybridization indices for h_i and h_j are n_i and n_j, respectively.

(i) By applying the orthogonal relationship, deduce the equation relating the angle θ_{ij} (between h_i and h_j), n_i and n_j.

(ii) Verify the result of (i) by considering the specific case of two sp^3 hybrids.

4.5 In a certain system, one s orbital and three p orbitals undergo hybridization to form four inequivalent hybrids, h_1 to h_4, as expressed below.

$$\begin{bmatrix} h_1 \\ h_2 \\ h_3 \\ h_4 \end{bmatrix} = \begin{bmatrix} 0.2794 & 0.5726 & 0.6871 & -0.3492 \\ 0.5587 & -0.2863 & -0.3436 & -0.6984 \\ 0.6397 & 0.4405 & -0.3671 & 0.5118 \\ 0.4478 & -0.6294 & 0.5245 & 0.3582 \end{bmatrix} \begin{bmatrix} s \\ p_x \\ p_y \\ p_z \end{bmatrix}$$

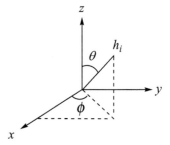

(i)   Verify that the hybrids form an orthonormal set.
(ii) Determine the hybridization indices, and angles ϕ and θ for the hybrids (ϕ and θ are defined in the figure above).
(iii) Determine all the inter-hybrid angles.

4.6   For each of the following: (i) a (closed-shell singlet) CH_2-like system, (ii) an NH_3-like system, and (iii) a H_2O-like system, derive an expression relating the hybridization indices (ratio of p and s orbital populations) n_b and n_ℓ of the bond and lone pair hybrids, respectively, where only ns and np orbitals of the central atom are assumed to take part in hybridization.

4.7   At a normal C–H bond length, we have the following overlap integrals between the carbon valence orbitals and the hydrogen 1s orbital:

$$\int \psi(C_{2s})\psi(H_{1s})d\tau = 0.57 \quad \text{and} \quad \int \psi(C_{2p_z})\psi(H_{1s})d\tau = 0.46,$$

with the z-axis lying along the C–H bond. [The numerical values of the integrals are taken from J. N. Murrell, S. F. A. Kettle, and J. M. Tedder, *Valence Theory*, 2nd edn., Wiley, New York, 1970, p. 65.]

(i)   Determine the overlap integral between the hydrogen 1s orbital and carbon sp, sp^2, and sp^3 hybrid orbitals which are directed along the z-axis.
(ii) The C–H stretching vibrational frequencies for $HC\equiv CH$, $CH_2=CH_2$, and CH_4 are 3423–3476, 2989–3106, and 2916–3019 cm^{-1}, respectively. These frequencies may be taken as a measure of the C–H bond strength in these molecules. Comment on whether the overlap integrals you have determined in (i) correlate well with the given spectral data.

4.8   The overlap integral between a 1s and a $2p_\sigma$ orbital on two different nuclei is given by:

$$S = \left(R + R^2 + \frac{R^3}{3} \right) e^{-R}$$

where R (in a.u.) is the distance between the nuclei.
Determine the value of R which gives the maximum overlap. (It may be of interest that the internuclear distance in HF equals 0.916 Å.)

4.9   The $XeOF_4$ molecule belongs to point group C_{4v}.

(i)   Work out the character of the representation Γ_{hyb} based on a set of six hybrid orbitals, five of which point to the O and F atoms and the remaining one to the lone electron pair.

(ii)   Reduce Γ_{hyb} to the irreducible representations of C_{4v}.

(iii) Derive a plausible hybridization scheme.

4.10   The IF_7 molecule has a pentagonal bipyramidal structure which belongs to point group D_{5h}. Deduce Γ_{hyb}, the representation based on the seven hybrid orbitals of the central iodine atom that are directed toward the ligand fluorine atoms. Hence propose a plausible hybridization scheme for description of the bonding in this molecule.

4.11   The two most important eight-coordinate geometries for metal complexes are the square (or Archimedian) antiprism (D_{4d}) and the triangulated dodecahedron (D_{2d}). As shown in the following figure, both polyhedra can be derived by distortion of the cube (O_h) in different ways.

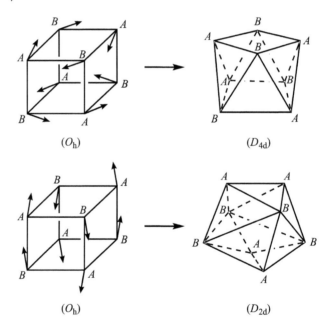

For each possible eight-coordinate geometry, D_{4d}, D_{2d}, and O_h,

(i)   Work out the character of the representation Γ_{hyb} based on the set of eight hybrid orbitals emanating from the central metal atom.

(ii)   Reduce Γ_{hyb} to the irreducible representations of the molecular point group.

(iii) Derive the plausible hybridization scheme(s) and comment on your choice(s).

4.12 In an octahedral AX_6 molecule, which orbitals of the central atom are involved in hybridization? Derive the linear combinations for the hybrid orbitals. Adopt the system of coordinates shown on the right.

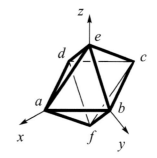

4.13 A hypothetical ion MX_6^{n-} has a trigonal antiprismatic structure (D_{3d} symmetry). Deduce the possible hybridization (σ only) schemes for the central metal ion. In addition, choose one of the schemes to derive the explicit linear combinations for the hybrid orbitals. Adopt the system of coordinates shown on the right.

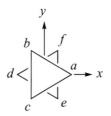

$+z$ points outward

SOLUTIONS

A4.1 (i) $h_1 = \dfrac{1}{\sqrt{1+n}}s + \sqrt{\dfrac{n}{1+n}}\left(-\dfrac{1}{\sqrt{2}}p_x + \dfrac{1}{\sqrt{2}}p_y\right)$, where n is the hybridization index. When h_1 is an sp hybrid, $h_1 = \dfrac{1}{\sqrt{2}}s - \dfrac{1}{2}p_x + \dfrac{1}{2}p_y$;

When h_1 is an sp^2 hybrid, $h_1 = \dfrac{1}{\sqrt{3}}s - \dfrac{1}{\sqrt{3}}p_x + \dfrac{1}{\sqrt{3}}p_y$;

When h_1 is an sp^3 hybrid, $h_1 = \dfrac{1}{2}s - \sqrt{\dfrac{3}{8}}p_x + \sqrt{\dfrac{3}{8}}p_y$.

(ii) h_2 can be written as:

$$h_2 = \frac{1}{\sqrt{3}}s + ap_x + bp_y, \quad \text{with } a^2 + b^2 = \frac{2}{3}. \tag{1}$$

Orthogonal relationship for h_1 and h_2 : $\dfrac{1}{3} - \dfrac{1}{\sqrt{3}}a + \dfrac{1}{\sqrt{3}}b = 0$,

$$\text{or } a = \frac{1}{\sqrt{3}} + b. \tag{2}$$

Substituting (2) into (1) : $\left(\dfrac{1}{\sqrt{3}} + b\right)^2 + b^2 = \dfrac{2}{3}$, or $b = 0.2113$.

$a = \dfrac{1}{\sqrt{3}} + 0.2113 = 0.7887.$

$h_2 = 0.5774s + 0.7887p_x + 0.2113p_y.$

Alternative method:

The angle between two sp^2 hybrids is $120°$. Hence, the angle between the x-axis and h_2 is $15°$.

$$h_2 = \frac{1}{\sqrt{3}}s + \sqrt{\frac{2}{3}}[(\cos 15°)p_x + (\sin 15°)p_y] = 0.5774s + 0.7887p_x + 0.2113p_y.$$

(iii) The angle between two sp^3 hybrids is $\cos^{-1}(-\frac{1}{3}) \approx 109.47°$. Hence, the angle between the x-axis and h_2 is $25.53°$.

$$h_2 = \frac{1}{2}s + \frac{\sqrt{3}}{2}[p_x \cos(25.53°) + p_y \sin(25.53°)] = 0.5s + 0.7815p_x + 0.3732p_y.$$

A4.2 (i) $h_1 = \frac{1}{\sqrt{3}}s + \sqrt{\frac{2}{3}}p_x.$

(ii) Combining the symmetry conditions given in the question and the result in (i):

$$\begin{bmatrix} h_1 \\ h_2 \\ h_3 \\ h_4 \end{bmatrix} = \begin{bmatrix} \frac{1}{\sqrt{3}} & \sqrt{\frac{2}{3}} & 0 & 0 \\ a & -b & c & 0 \\ a & -b & -c(\cos 60°) & d \\ a & -b & -c(\cos 60°) & -d \end{bmatrix} \begin{bmatrix} s \\ p_x \\ p_y \\ p_z \end{bmatrix}$$

Since the sum of squares of the coefficients in each column is equal to 1, we have:
$$a = \frac{\sqrt{2}}{3}, b = \frac{1}{3}, c = \sqrt{\frac{2}{3}}, d = \frac{1}{\sqrt{2}}. \text{ So,}$$

$$\begin{bmatrix} h_1 \\ h_2 \\ h_3 \\ h_4 \end{bmatrix} = \begin{bmatrix} \frac{1}{\sqrt{3}} & \sqrt{\frac{2}{3}} & 0 & 0 \\ \frac{\sqrt{2}}{3} & -\frac{1}{3} & \sqrt{\frac{2}{3}} & 0 \\ \frac{\sqrt{2}}{3} & -\frac{1}{3} & -\frac{1}{\sqrt{6}} & \frac{1}{\sqrt{2}} \\ \frac{\sqrt{2}}{3} & -\frac{1}{3} & -\frac{1}{\sqrt{6}} & -\frac{1}{\sqrt{2}} \end{bmatrix} \begin{bmatrix} s \\ p_x \\ p_y \\ p_z \end{bmatrix}.$$

(iii) $n_2(= n_3 = n_4) = \dfrac{\frac{1}{9} + \frac{2}{3}}{\frac{2}{9}} = 3\frac{1}{2}.$

(iv)

θ_{12} is given by: $\tan(180° - \theta_{12}) = \dfrac{\sqrt{\frac{2}{3}}}{\frac{1}{3}} = 2.449; \theta_{12} = 112.2°.$

$\theta_{23} = \theta_{34};$ and θ_{34} is given by: $\tan(\theta_{34}/2) = \dfrac{\frac{1}{\sqrt{2}}}{\sqrt{\frac{1}{9} + \frac{1}{6}}} = 1.342;$

$$\theta_{23} = \theta_{34} = 106.6°.$$

A4.3   (i)   Hybrid h_1 makes an angle of $60°$ with the x-axis and lies in the first quadrant of the xz-plane. So, $h_1 = \dfrac{1}{\sqrt{3}}s + \sqrt{\dfrac{2}{3}}(\dfrac{1}{2}p_x + \dfrac{\sqrt{3}}{2}p_z) = \dfrac{1}{\sqrt{3}}s + \dfrac{1}{\sqrt{6}}p_x + \dfrac{1}{\sqrt{2}}p_z$.

Similarly, $h_2 = \dfrac{1}{\sqrt{3}}s + \dfrac{1}{\sqrt{6}}p_x - \dfrac{1}{\sqrt{2}}p_z$.

(ii)   To obtain h_3 and h_4, we first write down the following coefficient matrix, in which hybrids h_1 and h_2 have been determined in (i),

$$
\begin{bmatrix} h_1 \\ h_2 \\ h_3 \\ h_4 \end{bmatrix} =
\begin{bmatrix}
\frac{1}{\sqrt{3}} & \frac{1}{\sqrt{6}} & 0 & \frac{1}{\sqrt{2}} \\
\frac{1}{\sqrt{3}} & \frac{1}{\sqrt{6}} & 0 & -\frac{1}{\sqrt{2}} \\
a & -b & c & 0 \\
a & -b & -c & 0
\end{bmatrix}
\begin{bmatrix} s \\ p_x \\ p_y \\ p_z \end{bmatrix}
$$

Since the sum of squares of the coefficients in each column is equal to 1, we have: $a = \dfrac{1}{\sqrt{6}}, b = \dfrac{1}{\sqrt{3}}, c = \dfrac{1}{\sqrt{2}}$. So,

$$
\begin{bmatrix} h_1 \\ h_2 \\ h_3 \\ h_4 \end{bmatrix} =
\begin{bmatrix}
\frac{1}{\sqrt{3}} & \frac{1}{\sqrt{6}} & 0 & \frac{1}{\sqrt{2}} \\
\frac{1}{\sqrt{3}} & \frac{1}{\sqrt{6}} & 0 & -\frac{1}{\sqrt{2}} \\
\frac{1}{\sqrt{6}} & -\frac{1}{\sqrt{3}} & \frac{1}{\sqrt{2}} & 0 \\
\frac{1}{\sqrt{6}} & -\frac{1}{\sqrt{3}} & -\frac{1}{\sqrt{2}} & 0
\end{bmatrix}
\begin{bmatrix} s \\ p_x \\ p_y \\ p_z \end{bmatrix}
$$

(iii) $n_3 = n_4 = \dfrac{\frac{1}{3} + \frac{1}{2} + 0}{\frac{1}{6}} = 5;\ \theta = 2\tan^{-1}\left(\dfrac{1/\sqrt{2}}{1/\sqrt{3}}\right) = 101.5°$. This result is fairly close to the experimental value of $97°$. It is stressed that these results are based on the assumption that all the hybrids point straight toward the corresponding hydrogens.

A4.4   (i)   The two hybrids can be reoriented, without changing their angle, such that one hybrid lies along the positive x-axis and the other in the second quadrant of xy-plane as shown on the right. Let θ_{ij} be the angle between them. Orbitals h_i and h_j can be expressed as:

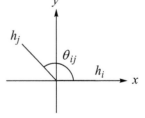

$$ h_i = \dfrac{1}{\sqrt{1 + n_i}}(s + \sqrt{n_i}\ p_x), $$

$$ h_j = \dfrac{1}{\sqrt{1 + n_j}}[s + \sqrt{n_j}\ (p_x \cos\theta_{ij} + p_y \sin\theta_{ij})]. $$

Orthogonal relationship:

$$\frac{1}{\sqrt{1+n_i}}\frac{1}{\sqrt{1+n_j}}(1+\sqrt{n_i}\sqrt{n_j}\cos\theta_{ij}) = 0,$$

$$\cos\theta_{ij} = -\frac{1}{\sqrt{n_in_j}}.$$

(ii) For sp^3 hybrids, $n_i = n_j = 3$, $\theta_{ij} = \cos^{-1}\left(-\frac{1}{3}\right) \approx 109.47°$ (equal to the bond angle in a tetrahedral molecule such as CH_4, which adopts an ideal sp^3 hybridization scheme).

A4.5 (i) For example, $\int|h_1|^2 d\tau = (0.2794)^2+(0.5726)^2+(0.6871)^2+(-0.3492)^2 = 1$. Also, $\int h_1h_2 d\tau = (0.2794)(0.5587)+(0.5726)(-0.2863)+(0.6871)(-0.3436) + (-0.3492)(-0.6984) = 0$.

(ii)

Hybridization index	ϕ	θ
h_1	11.8	50.2°	111.3°
h_2	2.2	230.2°	147.4°
h_3	1.4	320.2°	48.3°
h_4	4.0	140.2°	66.4°

(iii) Define θ_{ij} to be the angle formed by h_i and h_j. By using $\cos\theta_{ij} = -\dfrac{1}{\sqrt{n_in_j}}$, we get $\theta_{12} = 101.3°$, $\theta_{13} = 104.0°$, $\theta_{14} = 98.4°$, $\theta_{23} = 124.1°$, $\theta_{24} = 109.7°$, and $\theta_{34} = 114.6°$.

A4.6 (i) Adopt the system of coordinates shown on the right. By symmetry, it is obvious that

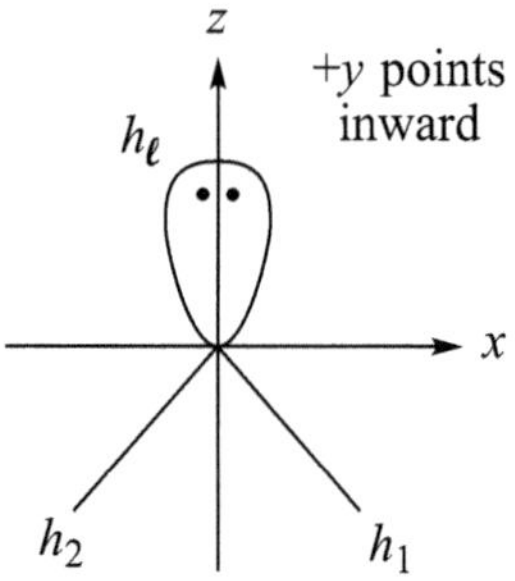

$$\begin{bmatrix} h_\ell \\ h_1 \\ h_2 \end{bmatrix} = \begin{bmatrix} a & 0 & d \\ b & c & -e \\ b & -c & -e \end{bmatrix}\begin{bmatrix} s \\ p_x \\ p_z \end{bmatrix}$$

Since the sum of squares of the coefficients in each column and each row is equal to 1, we have: $b = \sqrt{\dfrac{1-a^2}{2}}, c = \dfrac{1}{\sqrt{2}}, d = \sqrt{1-a^2}, e = \sqrt{\dfrac{1-d^2}{2}} = \dfrac{a}{\sqrt{2}}$. Now the coefficient matrix becomes

Solutions

$$\begin{bmatrix} h_\ell \\ h_1 \\ h_2 \end{bmatrix} = \begin{bmatrix} a & 0 & \sqrt{1-a^2} \\ \sqrt{\frac{1-a^2}{2}} & \frac{1}{\sqrt{2}} & -\frac{a}{\sqrt{2}} \\ \sqrt{\frac{1-a^2}{2}} & -\frac{1}{\sqrt{2}} & -\frac{a}{\sqrt{2}} \end{bmatrix} \begin{bmatrix} s \\ p_x \\ p_z \end{bmatrix}$$

The hybridization indices are $n_\ell = \dfrac{1-a^2}{a^2}$ and $n_b = \dfrac{\frac{1}{2} + a^2/2}{(1-a^2)/2} = \dfrac{1+a^2}{1-a^2}$.

Eliminating a^2 in these expressions yields $n_\ell = \dfrac{2}{n_b - 1}$.

(ii) Adopt the system of coordinates shown on the right, in which the lone pair points to the $+z$ direction. By symmetry, it is obvious that

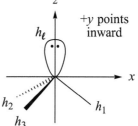

$$\begin{bmatrix} h_\ell \\ h_1 \\ h_2 \\ h_3 \end{bmatrix} = \begin{bmatrix} a & 0 & 0 & e \\ b & c & 0 & -f \\ b & -c(\cos 60^\circ) & d & -f \\ b & -c(\cos 60^\circ) & -d & -f \end{bmatrix} \begin{bmatrix} s \\ p_x \\ p_y \\ p_z \end{bmatrix}$$

Since the sum of squares of the coefficients in each column and each row is equal to 1, we have: $b = \sqrt{\dfrac{1-a^2}{3}}, c = \sqrt{\dfrac{2}{3}}, d = \dfrac{1}{\sqrt{2}}, e = \sqrt{1-a^2}, f = \sqrt{\dfrac{1-e^2}{3}} = \dfrac{a}{\sqrt{3}}$. Now the coefficient matrix becomes

$$\begin{bmatrix} h_\ell \\ h_1 \\ h_2 \\ h_3 \end{bmatrix} = \begin{bmatrix} a & 0 & 0 & \sqrt{1-a^2} \\ \sqrt{\frac{1-a^2}{3}} & \sqrt{\frac{2}{3}} & 0 & -\frac{a}{\sqrt{3}} \\ \sqrt{\frac{1-a^2}{3}} & -\frac{1}{\sqrt{6}} & \frac{1}{\sqrt{2}} & -\frac{a}{\sqrt{3}} \\ \sqrt{\frac{1-a^2}{3}} & -\frac{1}{\sqrt{6}} & -\frac{1}{\sqrt{2}} & -\frac{a}{\sqrt{3}} \end{bmatrix} \begin{bmatrix} s \\ p_x \\ p_y \\ p_z \end{bmatrix}$$

The hybridization indices are $n_\ell = \dfrac{1-a^2}{a^2}$ and $n_b = \dfrac{\frac{2}{3} + \frac{a^2}{3}}{\frac{(1-a^2)}{3}} = \dfrac{2+a^2}{1-a^2}$.

Eliminating a^2 in these expressions yields $n_\ell = \dfrac{3}{n_b - 2}$.

(iii) Adopt the system of coordinates shown on the right, in which the lone pairs and bond pairs lie on the xz- and yz-planes, respectively. By symmetry, it is obvious that

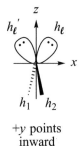

+y points
inward

$$
\begin{bmatrix} h_\ell \\ h'_\ell \\ h_1 \\ h_2 \end{bmatrix} = \begin{bmatrix} a & c & 0 & e \\ a & -c & 0 & e \\ b & 0 & d & -f \\ b & 0 & -d & -f \end{bmatrix} \begin{bmatrix} s \\ p_x \\ p_y \\ p_z \end{bmatrix}
$$

Since the sum of squares of the coefficients in each column and each row is equal to 1, we have: $b = \sqrt{\frac{1-2a^2}{2}}, c = \frac{1}{\sqrt{2}}, d = \frac{1}{\sqrt{2}}, e = \sqrt{1 - a^2 - c^2} = \sqrt{\frac{1-2a^2}{2}}$,

$f = \sqrt{\frac{1 - 2e^2}{2}} = a$. Now the coefficient matrix becomes

$$
\begin{bmatrix} h_\ell \\ h'_\ell \\ h_1 \\ h_2 \end{bmatrix} = \begin{bmatrix} a & \frac{1}{\sqrt{2}} & 0 & \sqrt{\frac{1-2a^2}{2}} \\ a & -\frac{1}{\sqrt{2}} & 0 & \sqrt{\frac{1-2a^2}{2}} \\ \sqrt{\frac{1-2a^2}{2}} & 0 & \frac{1}{\sqrt{2}} & -a \\ \sqrt{\frac{1-2a^2}{2}} & 0 & -\frac{1}{\sqrt{2}} & -a \end{bmatrix} \begin{bmatrix} s \\ p_x \\ p_y \\ p_z \end{bmatrix}
$$

The hybridization indices are $n_\ell = \dfrac{\frac{1}{2} + (1 - 2a^2)/2}{a^2} = \dfrac{1 - a^2}{a^2}$ and $n_b = \dfrac{\frac{1}{2} + a^2}{(1 - 2a^2)/2} = \dfrac{1 + 2a^2}{1 - 2a^2}$. Eliminating a^2 in these expressions yields $n_\ell = \dfrac{n_b + 3}{n_b - 1}$.

Alternative method:

This problem can be solved by a more mathematical approach.
Orbital h_i can be expressed as: $h_i = a_i s + b_i p_x + c_i p_y + d_i p_z$.
Since $a_i^2 + b_i^2 + c_i^2 + d_i^2 = 1$, the hybridization index, n_i, is given by:

$$
n_i = (b_i^2 + c_i^2 + d_i^2)/a_i^2 = (1 - a_i^2)/a_i^2.
$$

We have: $a_i^2 = (1 + n_i)^{-1}$.

(i) For a (closed-shell singlet) CH_2-like system, an sp^2 hybridization scheme is employed and hence $a_1^2 + a_2^2 + a_3^2 = 1$. There is one hybrid for a lone pair and

84

two hybrids for bond pairs. So, $(1 + n_\ell)^{-1} + (1 + n_b)^{-1} + (1 + n_b)^{-1} = 1$, or

$$n_\ell = \frac{2}{n_b - 1}.$$

(ii) For an NH_3-like system, the sp^3 hybridization scheme is employed and hence $a_1^2 + a_2^2 + a_3^2 + a_4^2 = 1$. There is one hybrid for a lone pair and three hybrids for bond pairs. So, $(1 + n_\ell)^{-1} + (1 + n_b)^{-1} + (1 + n_b)^{-1} + (1 + n_b)^{-1} = 1$, or $n_\ell = \dfrac{3}{n_b - 2}$.

(iii) For an H_2O-like system, an sp^3 hybridization scheme is employed and hence $a_1^2 + a_2^2 + a_3^2 + a_4^2 = 1$. There are two hybrids for lone pairs and two hybrids for bond pairs. So, $(1 + n_\ell)^{-1} + (1 + n_\ell)^{-1} + (1 + n_b)^{-1} + (1 + n_b)^{-1} = 1$, or

$$n_\ell = \frac{n_b + 3}{n_b - 1}.$$

A4.7 (i) The wavefunctions of the carbon sp^n hybrid orbitals which are directed along the z-axis are given by: $h = \dfrac{1}{\sqrt{1 + n}}[\psi(C_{2s}) + \sqrt{n}\,\psi(C_{2p_z})]$.

The overlap is $\int h\psi(H_{1s})d\tau = \dfrac{1}{\sqrt{1 + n}}\Big[\int \psi(C_{2s})\psi(H_{1s})d\tau$

$$+ \sqrt{n}\int \psi(C_{2p_z})\psi(H_{1s})d\tau\Big]$$

$$= \frac{1}{\sqrt{1 + n}}(0.57 + 0.46\sqrt{n}).$$

For sp hybrid orbitals, $n = 1$, overlap $= 0.73$.
For sp^2 hybrid orbitals, $n = 2$, overlap $= 0.71$.
For sp^3 hybrid orbitals, $n = 3$, overlap $= 0.68$.

(ii) The overlap integrals between the carbon valence orbitals and the hydrogen 1s orbital decrease in the order of $HC{\equiv}CH$, $CH_2{=}CH_2$, and CH_4 (with the hybridization index for the carbon atom equal to 1, 2, and 3, respectively). Assuming that the C–H bond strength increases with the overlap integral, we can conclude that the overlap integrals determined in (i) correlate well with the given spectral data.

A4.8 Setting $\frac{dS}{dR} = 0$, maximum overlap occurs at $R = 2.1038$ a.u. or $= 1.115$ Å.

A4.9 (i)

C_{4v}	E	$2C_4$	C_4^2	$2\sigma_v$	$2\sigma_d$
Γ_{hyb}	6	2	2	4	2

(ii)
$$\Gamma_{hyb} = \quad 3A_1 \quad + \quad B_1 \quad + \quad E.$$
$$\text{s, p}_z\text{, d}_{z^2} \qquad \text{d}_{x^2-y^2} \qquad (\text{p}_x\text{, p}_y)\text{, (d}_{xz}\text{, d}_{yz})$$

(iii) Possible hybridization schemes include sp^3d^2 and spd^4. For $XeOF_4$, the former scheme is more plausible.

A4.10

D_{5h}	E	$2C_5$	$2C_5^2$	$5C_2$	σ_h	$2S_5$	$2S_5^3$	$5\sigma_v$
Γ_{hyb}	7	2	2	1	5	0	0	3

$$\Gamma_{hyb} = \quad 2A_1' \quad + \quad A_2'' \quad + \quad E_1' \quad + \quad E_2'.$$
$$\phantom{\Gamma_{hyb} =}\quad s,\, d_{z^2} \qquad\quad p_z \qquad\quad (p_x,\, p_y) \qquad (d_{x^2-y^2},\, d_{xy})$$

The hybridization scheme is sp^3d^3. The configuration $5s^1 5p^3 5d^3$ for this hybridization scheme may be obtained upon excitation from the ground configuration $5s^2 5p^5 5d^0$ of iodine.

A4.11 (i) to (iii)

D_{4d} geometry:

D_{4d}	E	$2S_8$	$2C_4$	$2S_8^3$	C_2	$4C_2'$	$4\sigma_d$
Γ_{hyb}	8	0	0	0	0	0	2

$$\Gamma_{hyb} = \quad A_1 \quad + \quad B_2 \quad + \quad E_1 \quad + \quad E_2 \quad + \quad E_3.$$
$$\phantom{\Gamma_{hyb} =}\quad s,\, d_{z^2} \qquad p_z \qquad (p_x,\, p_y) \qquad (d_{xy},\, d_{x^2-y^2}) \qquad (d_{xz},\, d_{yz})$$

According to group theory, the possible hybridization schemes are d^4sp^3 and d^5p^3. We need to know the electronic configuration of the central metal atom before we can comment on the most likely hybridization scheme.

D_{2d} geometry:

D_{2d}	E	$2S_4$	C_2	$2C_2'$	$2\sigma_d$
Γ_{hyb}	8	0	0	0	4

$$\Gamma_{hyb} = \quad 2A_1 \quad + \quad 2B_2 \quad + \quad 2E.$$
$$\phantom{\Gamma_{hyb} =}\quad s,\, d_{z^2} \qquad p_z,\, d_{xy} \qquad (p_x,\, p_y),\, (d_{xz},\, d_{yz})$$

The only possible hybridization scheme is d^4sp^3.

O_h geometry:

O_h	E	$8C_3$	$6C_2$	$6C_4$	$3C_4^2$	i	$6S_4$	$8S_6$	$3\sigma_h$	$6\sigma_d$
Γ_{hyb}	8	2	0	0	0	0	0	0	0	4
$\Gamma(f_{xyz})$	1	1	−1	−1	1	−1	1	−1	−1	1

$$\Gamma_{hyb} = \quad A_{1g} \quad + \quad A_{2u} \quad + \quad T_{1u} \quad + \quad T_{2g}.$$
$$\phantom{\Gamma_{hyb} =}\quad s \qquad\qquad\qquad\qquad (p_x,\, p_y,\, p_z) \qquad (d_{xy},\, d_{yz},\, d_{xz})$$

The hybrid orbitals cannot be constructed using only s, p, and d atomic orbitals since none of them transforms as A_{2u}. Noting that f_{xyz} has A_{2u} symmetry, we can write the hybridization scheme as d^3fsp^3.

A4.12

O_h	E	$8C_3$	$6C_2$	$6C_4$	$3C_4^2$	i	$6S_4$	$8S_6$	$3\sigma_h$	$6\sigma_d$
Γ_{hyb}	6	0	0	2	2	0	0	0	4	2

$$\Gamma_{hyb} = \underset{s}{A_{1g}} \quad + \quad \underset{(d_{z^2},\, d_{x^2-y^2})}{E_g} \quad + \quad \underset{(p_x,\, p_y,\, p_z)}{T_{1u}}.$$

The hybrids have the form:

$$
\begin{bmatrix} h_a \\ h_b \\ h_c \\ h_d \\ h_e \\ h_f \end{bmatrix}
=
\begin{bmatrix}
(6)^{-1/2} & -(12)^{-1/2} & \frac{1}{2} & (2)^{-1/2} & 0 & 0 \\
(6)^{-1/2} & -(12)^{-1/2} & -\frac{1}{2} & 0 & (2)^{-1/2} & 0 \\
(6)^{-1/2} & -(12)^{-1/2} & \frac{1}{2} & -(2)^{-1/2} & 0 & 0 \\
(6)^{-1/2} & -(12)^{-1/2} & -\frac{1}{2} & 0 & -(2)^{-1/2} & 0 \\
(6)^{-1/2} & (3)^{-1/2} & 0 & 0 & 0 & (2)^{-1/2} \\
(6)^{-1/2} & (3)^{-1/2} & 0 & 0 & 0 & -(2)^{-1/2}
\end{bmatrix}
\begin{bmatrix} s \\ d_{z^2} \\ d_{x^2-y^2} \\ p_x \\ p_y \\ p_z \end{bmatrix}
$$

A4.13

D_{3d}	E	$2C_3$	$3C_2$	i	$2S_6$	$3\sigma_d$
Γ_{hyb}	6	0	0	0	0	2

$$\Gamma_{hyb} = \underset{s,\, d_{z^2}}{A_{1g}} \quad + \quad \underset{(d_{xy},\, d_{x^2-y^2}),\, (d_{xz},\, d_{yz})}{E_g} \quad + \quad \underset{p_z}{A_{2u}} \quad + \quad \underset{(p_x,\, p_y)}{E_u}.$$

The hybrids are either d^2sp^3 or d^3p^3. If the coordinate system shown in the question is adopted and s, $d_{x^2-y^2}$, d_{xy}, p_z, p_x, and p_y orbitals are used to form hybrids, we will have

$$
\begin{bmatrix} h_a \\ h_b \\ h_c \\ h_d \\ h_e \\ h_f \end{bmatrix}
=
\begin{bmatrix}
(6)^{-1/2} & (3)^{-1/2} & 0 & (6)^{-1/2} & (3)^{-1/2} & 0 \\
(6)^{-1/2} & -(12)^{-1/2} & -\frac{1}{2} & (6)^{-1/2} & -(12)^{-1/2} & \frac{1}{2} \\
(6)^{-1/2} & -(12)^{-1/2} & \frac{1}{2} & (6)^{-1/2} & -(12)^{-1/2} & -\frac{1}{2} \\
(6)^{-1/2} & (3)^{-1/2} & 0 & -(6)^{-1/2} & -(3)^{-1/2} & 0 \\
(6)^{-1/2} & -(12)^{-1/2} & -\frac{1}{2} & -(6)^{-1/2} & (12)^{-1/2} & -\frac{1}{2} \\
(6)^{-1/2} & -(12)^{-1/2} & \frac{1}{2} & -(6)^{-1/2} & (12)^{-1/2} & \frac{1}{2}
\end{bmatrix}
\begin{bmatrix} s \\ d_{x^2-y^2} \\ d_{xy} \\ p_z \\ p_x \\ p_y \end{bmatrix}
$$

5 | Molecular Symmetry

PROBLEMS

5.1 A student sets up the coordinates for NH_3 rather unconventionally, as shown on the right. In this coordinate system, the x-axis is the three-fold axis. One of the σ-planes, σ_b, makes an angle of $45°$ with the y-axis. By considering the transformation of a general point in space, deduce the six 3×3 matrices which form a set of reducible representations for the C_{3v} point group.

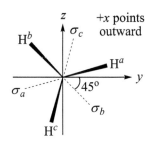

5.2 There are eight equivalent lobes for the f_{xyz} orbital. They point toward the corners of a cube with the origin as its center and have alternating signs, as shown in Fig. A below. The cube is oriented in such a way that the x-, y-, and z-axes pass through the centers of the six faces, as shown in Fig. B below. Rotation of f_{xyz} orbital about z-axis by $-45°$ results in $f_{z(x^2-y^2)}$ orbital. The top views of f_{xyz} and $f_{z(x^2-y^2)}$ orbitals are shown below in Figs C and D, respectively.

Assign these two orbitals to their proper irreducible representation in point groups C_{2v} and D_{4h}.

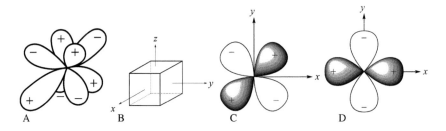

 Problems in Structural Inorganic Chemistry. Second edition. Wai-Kee Li, Hung Kay Lee, Dennis Kee Pui Ng, Yu-San Cheung, Kendrew Kin Wah Mak, and Thomas Chung Wai Mak. © Oxford University Press 2019. Published in 2019 by Oxford University Press. DOI: 10.1093/oso/9780198823902.001.0001

5.3 The square pyramidal $[InCl_5]^{2-}$ anion has C_{4v} symmetry. Of the four vertical mirror planes, one σ_v is in the xz-plane, the other σ_v in the yz-plane, and the two σ_ds lie in between them.

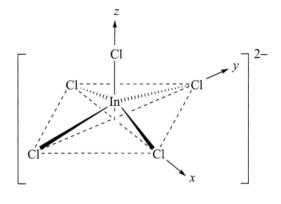

(i) With reference to the coordinate system based on the orthonormal base vectors $\hat{e}_1$, $\hat{e}_2$, and $\hat{e}_3$ (pointing in the x-, y-, and z-directions, respectively), write down all eight matrices that represent the symmetry operations of the C_{4v} group.

(ii) Deduce the character of Γ_e, the representation based on $\{\hat{e}_1, \hat{e}_2, \hat{e}_3\}$. Reduce Γ_e to the irreducible representations of C_{4v}.

REFERENCE: D. S. Brown, F. W. B. Einstein, and D. G. Tuck, Tetragonal-pyramidal indium(III) species. Crystal structure of tetraethylammonium pentachloroindate(III). *Inorg. Chem.* **8**, 14–8 (1969).

5.4 Point groups S_n, have only two symmetry elements: E and S_n. Explain why n must be an even integer and $n \geq 4$.

5.5 *Trans*-N_2F_2 has C_{2h} symmetry and its coordinates are taken as those shown below. The character table of the C_{2h} point group is shown below.

C_{2h}	E	C_2	i	σ_h
A_g	1	1	1	1
B_g	1	−1	1	−1
A_u	1	1	−1	−1
B_u	1	−1	−1	1

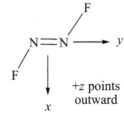

(i) To which representations do the Cartesian coordinates x, y, and z belong?

(ii) To which representations do the binary products x^2, y^2, z^2, xy, xz, and yz belong?

5.6 In a reference book, the character table of the D_{3d} point group has the following form:

D_{3d}	E	$2C_3$	$3C_2$	i	$2S_6$	$3\sigma_d$
A_{1g}	1	1	1	1	1	1
A_{2g}	1	1	−1	1	1	−1
E_g	2	−1	0	2	−1	0
A_{1u}	1	1	1	−1	−1	−1
A_{2u}	1	1	−1	−1	−1	1
E_u	2	−1	0	−2	1	0

(i) To which representations do the Cartesian coordinates x, y, and z belong?

(ii) Taking the z-axis as the C_3 axis, to which representations do the binary products x^2, y^2, z^2, xy, xz, and yz belong?

5.7 (i) Borane dianions $(BH)_n^{2-}$ adopt structures with high symmetry. For instance, *closo*-dodecaborane(12) dianion, $(BH)_{12}^{2-}$, has the structure of an icosahedron, a polyhedron with 12 vertices and 20 faces. The symmetry point group for this structure is I_h, which has 120 symmetry operations: E, $12C_5$, $12C_5^2$, $20C_3$, $15C_2$, i, $12S_{10}$, $12S_{10}^3$, $20S_6$, 15σ. The structure of $(BH)_{12}^{2-}$ is shown below (left), where each vertex is occupied by a BH group.

When one or more BH group(s) in $(BH)_{12}^{2-}$ is (are) substituted by C (which is isoelectronic to BH), carborane dianions $C_m(BH)_{12-m}^{2-}$ are obtained. The only possible mono- and di-substituted *closo*-dodecaborane(12) dianions are displayed below [(a) to (d), where C atoms are represented by black dots]. Determine the symmetry point group of each carborane dianion shown.

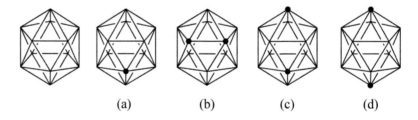

(a)   (b)   (c)   (d)

(ii) Fullerene C_{20} has the structure of a dodecahedron, a polyhedron with 20 vertices and 12 faces. The symmetry point group for this structure is also I_h. The structure of C_{20} is shown below (left), where each vertex is occupied by a C atom.

When one or more C atom(s) of C_{20} is (are) replaced by Si, substituted fullerenes Si_mC_{20-m} are obtained. The only possible mono- and di-substituted C_{20} with the general formula Si_mC_{20-m} ($m = 1$ or 2) are displayed below [(a) to (f), where Si atoms are represented by black dots]. Determine the symmetry point group of each substituted fullerene shown.

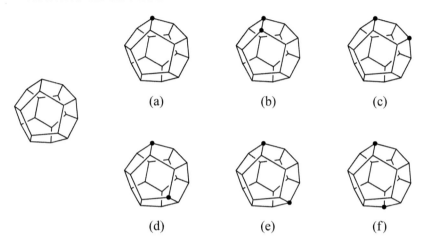

(a) (b) (c)

(d) (e) (f)

5.8 Determine the point group of each of the following silicate anions:

   (i) $[SiO_4]^{4-}$

  (ii) $[O_3Si-O-SiO_3]^{6-}$ with linear Si–O–Si linkage. Consider both staggered and eclipsed conformation.

 (iii) $[Si_3O_9]^{6-}$

 (iv) $[Si_4O_{12}]^{8-}$

  (v) $[Si_6O_{18}]^{12-}$

The structures for the last three species are shown below. The $(SiO)_n$ ($n = 3, 4, 6$) rings are planar.

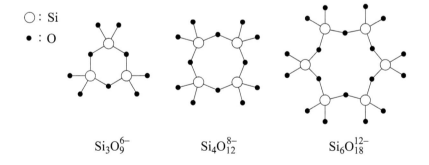

$\bigcirc$: Si
$\bullet$: O

$Si_3O_9^{6-}$ $Si_4O_{12}^{8-}$ $Si_6O_{18}^{12-}$

5.9 Assign point group symbols for the following cyclic S_n molecules ($n = 6, 7, 8, 10, 11, 12, 13, 18$ in α and β forms, 20):

(i) S_6; (ii) S_7; (iii) S_8; (iv) S_{10}; (v) S_{11}; (vi) S_{12}; (vii) S_{13}; (viii) α–S_{18}; (ix) β–S_{18}; (x) S_{20}.

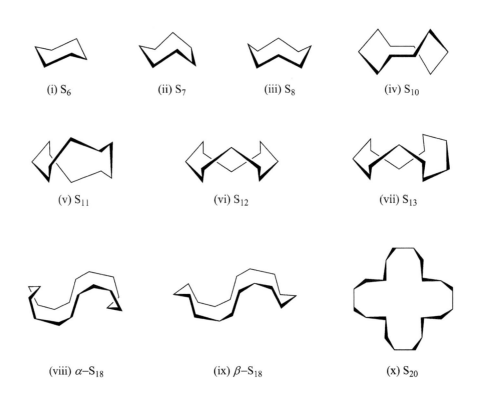

(i) S_6 (ii) S_7 (iii) S_8 (iv) S_{10}

(v) S_{11} (vi) S_{12} (vii) S_{13}

(viii) α–S_{18} (ix) β–S_{18} (x) S_{20}

5.10 (i) Give the point groups and chiral nomenclature for the isomeric forms of tris(ethylenediamine)cobalt(III) illustrated below:

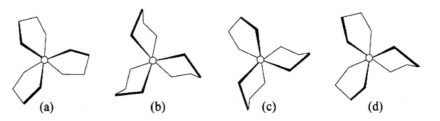

(a)   (b)   (c)   (d)

(ii) List the remaining possible configurations of $[\mathrm{Co(en)_3}]^{3+}$.

A note on the nomenclature of chiral metal complex:

First we view the optically active tris(chelate) complexes down the three-fold rotational axis. If the helix viewed in this way is left-handed, it is called a "Λ-isomer". Its mirror image is the "Δ-isomer". The naming is illustrated in the figure on the right.

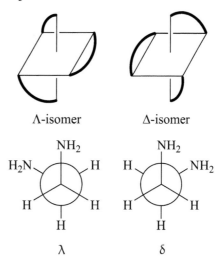

Besides the asymmetry due to the tris(chelate) structure of octahedral complexes, asymmetry may also be generated in the ligands such as ethylenediamine. The enantiomeric gauche conformations may be "left-handed" (represented by "λ") or "right-handed" (represented by "δ"), which are shown in the figure above. Thus, one of the possible complete notations showing the configuration of tris(ethylenediamine) cobalt(III) is Λ-M$\lambda\lambda\lambda$, where M denotes metal.

5.11 There are five possible conformations for cyclohexane-1,4-dione: a chair form, two boat forms, a symmetric twist form, and a twist form of lower symmetry generally referred to as the twist-boat form. Sketch all five distinct conformations and identify their point groups.

REFERENCE: C.-S. Tse, D. Y. Chang, K.-Y. Law, and T. C. W. Mak, The crystal and molecular structure of trans-2,5-di-p-bromobenzyl-2, 5-diethoxycarbonylcyclo-hexane 1,4-dione. Acta Cryst. **B32**, 1216–9 (1976).

5.12 The structure of ethane, C_2H_6, is shown on the right. Consider the fluoro-substituted ethanes with the general formula $C_2H_{6-n}F_n$, $n = 1, 2, \ldots, 6$. Draw members of this series that have the following symmetries: (i) C_1; (ii) C_2; (iii) C_s; (iv) C_{2h}; (v) C_{3v}; (vi) D_{3d}.

5.13 Cyclohexane molecule (C_6H_{12}, shown on the right) has D_{3d} symmetry. Consider the chloro-substituted cyclohexanes with the general formula $C_6H_mCl_n$, with $n = 1, 2, \ldots$ and $m + n = 12$. Write the structural formulas for members from the $C_6H_mCl_n$ series that have the following symmetries: (i) C_1; (ii) C_2; (iii) C_s; (iv) C_{2h}; (v) C_{3v}.

5.14 Elemental sulfur, S_8, has the "crown" structure shown below (left). Viewed from the "top", this structure becomes an octagon. This octagon is also shown below (right), where the "+" sign denotes atoms that are above the paper plane and the "−" sign indicates atoms below. All eight atoms are labeled in the figures.

    (i)   What is the symmetry point group of S_8? Also, if one of the sulfur atoms in S_8 is replaced by (the isoelectronic) oxygen, the hypothetical compound S_7O is formed. What is the point group of S_7O?

    (ii) When four of the sulfur atoms in S_8 are replaced by oxygen atoms, another hypothetical compound, S_4O_4, is formed. Clearly there are several isomers for S_4O_4. Write down which sulfur atoms are to be replaced to obtain the isomers which have the following symmetries: (a) C_1; (b) C_2; (c) C_s; (d) C_{4v}; (e) D_{2d}.

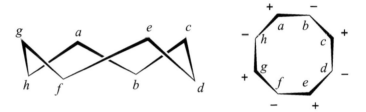

5.15 Cyclooctatetraene, C_8H_8, has the "tub" structure shown below (left). This molecule belongs to the D_{2d} group, whose symmetry operations are easily visualized as in the top view displayed below (right). Here "+" signs indicate atoms that are above the plane of the paper and the "−" signs indicate atoms below.

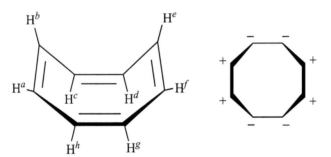

Consider the halogen-substituted cyclooctatetraenes with general formula $C_8H_{8-n}X_n$, $n = 0, 1, 2, \ldots, 8$. Write down which H atoms are to be replaced to obtain the isomers which have the following symmetries: (i) C_1; (ii) C_s; (iii) C_2; (iv) C_{2v}; (v) S_4; (vi) D_2.

REFERENCE: W.-K. Li, Identification of molecular point groups. *J. Chem. Educ.* **70**, 485–7 (1993).

5.16 Shown on the right is the structure of metal cluster $Mo_6Cl_8^{4+}$, with the six Mo atoms forming an octahedron and the eight Cl atoms placed above the eight faces of the octahedron. This cluster has O_h symmetry. Note that Cl atoms a, b, c, d, e, f, g, and h are above faces 135, 145, 146, 136, 235, 245, 246, and 236, respectively.

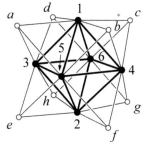

(i) If the following Mo atom(s) is (are) replaced by the isoelectronic W and/or Cr atom(s), determine the symmetry point group of the substituted clusters: (a) $1 = W$; (b) $1 = 2 = W$; (c) $3 = 5 = W$; (d) $1 = 2 = 3 = W$; (e) $1 = 3 = 5 = W$; (f) $1 = W$ and $2 = Cr$; (g) $3 = W$ and $5 = Cr$; (h) $1 = 5 = W$ and $2 = 4 = Cr$; (i) $1 = 2 = W$ and $3 = 4 = Cr$.

(ii) Replace one or more Cl atoms with F atoms so that the resulting cluster will belong to the following point groups: (a) C_{2v}; (b) C_s; (c) C_{4v}; (d) D_{2h}; (e) C_{3v}; (f) D_{3d}; (g) T_d; (h) C_2. You only need to give the labels of the Cl atoms (a to h) replaced.

5.17 Shown on the right is the structure of metal cluster $Ta_6Cl_{12}^{2+}$, with the six Ta atoms forming an octahedron and the 12 Cl atoms placed above the 12 edges of the octahedron. This cluster has O_h symmetry. Replace one or more Cl atoms with F atoms so that the resultant cluster will belong to the following point groups: (i) C_{4v}; (ii) D_{4h}; (iii) C_s; (iv) C_1; (v) C_{2v}; (vi) C_2; (vii) C_{2h}; (viii) C_{3v}; (ix) C_3; (x) D_{3d}; (xi) D_3; (xii) D_{2h}; (xiii) D_{2d}; (xiv) D_2. You only need to give the labels of the Cl atoms (a to l) replaced.

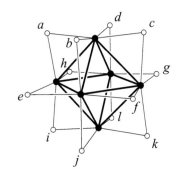

5.18 The structure of P_4O_{10} is shown on the right. In this highly symmetrical and pretty structure, the four P atoms form a perfect tetrahedron. Each P atom is linked to a terminal oxygen (with a double bond); these O atoms are labeled a to d in the figure. The remaining six oxygens, with labels e to j, are bridging atoms, located above the edges of the P_4 tetrahedron. So P_4O_{10} has T_d symmetry.

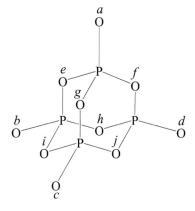

Replace one or more O atoms in P_4O_{10} with S atoms so that the resulting molecules will have the symmetries given below. Also, you need to observe the condition imposed. You only need to give the label(s) of the O atoms replaced. Do not draw the structures of your answers.

(i) D_{2d} [replacing only bridging O atom(s)];
(ii) D_{2d} [replacing both bridging and terminal O atom(s)].
(iii) C_{3v} [replacing only terminal O atom(s)];
(iv) C_{3v} [replacing only bridging O atom(s)];
(v) C_{3v} [replacing both terminal and bridging O atom(s), but each P atom is bonded to only either a terminal S atom or bridging S atom(s), and not both];
(vi) C_{2v} [replacing only terminal O atom(s)];
(vii) C_{2v} [replacing only bridging O atom(s)];
(viii) C_{2v} [replacing both terminal and bridging O atom(s), but each P atom is bonded to only either terminal S atom or bridging S atom(s), and not both];
(ix) C_s [replacing only bridging O atom(s)];
(x) C_s [replacing both bridging and terminal O atom(s)];
(xi) C_2 [replacing only bridging O atom(s)];
(xii) C_2 [replacing both bridging and terminal O atom(s)];
(xiii) C_1 [replacing ONE terminal O atom and bridging O atom(s)];
(xiv) C_1 [replacing TWO terminal O atoms and bridging O atom(s)].

5.19 Adamantane, $C_{10}H_{16}$, has the structure shown below. This molecule has T_d symmetry and two types of carbon atoms. There are four methine carbons (each of which is bonded to one hydrogen with label a, h, k, or n) which occupy the corners of a tetrahedron. In addition, there are six methylene carbons (each of which is bonded to two hydrogens) which are located above the edges of the aforementioned tetrahedron. All 16 hydrogens are labeled in the figure.

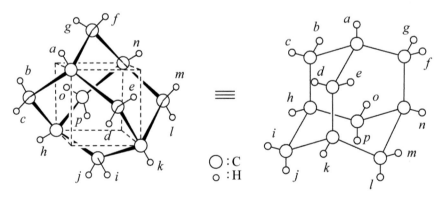

$\bigcirc : C$
$\circ : H$

Consider the chloro-substituted adamantanes with the general formula $C_{10}H_{16-n}Cl_n$, $n = 0, 1, 2, ..., 16$. Draw members of this series that have the following symmetries: (i) C_1; (ii) C_2; (iii) C_3; (iv) C_s; (v) C_{2v}; (vi) C_{3v}; (vii) D_{2d}. Provide only one answer for each part. In the answers, only list the label(s) of the hydrogen atom(s) to be replaced.

REFERENCE: W.-K. Li, Identification of molecular point groups. *J. Chem. Educ.* **70**, 485–7 (1993).

5.20 In a classic paper on optically active coordination compounds, Mills and Quibell reported the synthesis of a Pt complex containing one molecule of *meso*-stilbenediamine ($H_2N–CHPh–CHPh–NH_2$) and one molecule of isobutylenediamine ($H_2N–CH_2–CMe_2–NH_2$). Also, the complex was successfully resolved into its enantiomers. Determine whether the coordination around the metal ion is tetrahedral or square planar. Also illustrate these modes pictorially.

REFERENCE: W. H. Mills and T. H. H. Quibell, The configuration of the valencies of 4-covalent platinum: the optical resolution of *meso*-stilbenediaminoisobutylenediaminoplatinous salts. *J. Chem. Soc.* 839–46 (1935).

5.21 Consider the series of octahedral complexes with the general formula $[MA_mB_n(XX)_p(XY)_q]$, where $m, n, p, q = 0, 1, 2, ..., m + n + 2p + 2q = 6$, A and B are monodentate ligands such as Cl and F, XX is a bidentate ligand with two identical donating groups, and XY is a bidentate ligand with two different donating groups. An example of XX is $H_2NCH_2CH_2NH_2$, and an example of XY is $(CH_3)_2NCH_2CH_2NH_2$. Write the structural formulas for members from this series that have the symmetry requested below and that meet the following requirements. (i) C_1, with at least one monodentate ligand; (ii) C_1, with no monodentate ligand; (iii) C_2, with at least one monodentate ligand; (iv) C_2, with no monodentate ligand; (v) C_3; (vi) C_s; (vii) C_{2v}; (viii) C_{2h}; (ix) D_{2h}; (x) C_{3v}; (xi) D_3; (xii) C_{4v}; (xiii) D_{4h}.

REFERENCE: W.-K. Li, Identification of molecular point groups. *J. Chem. Educ.* **70**, 485–7 (1993).

5.22 Draw all geometrical isomers of $[Co(trien)Cl_2]^+$, where "trien" is the tetradentate ligand triethylenetetramine which may be conveniently represented by "N–N–N–N" in your drawings. Also, ignoring the conformation of chelate rings, identify the symmetry point group of each isomer and predict whether it is optically active or not.

5.23 Enumerate all the possible symmetry groups for hypothetical molecules with the formula X_2Y_2, with a sketch of each structure. These will include (i) linear structures, (ii) planar nonlinear structures, and (iii) nonplanar structures. Whenever possible, give a known example for each structure sketched.

SOLUTIONS

A5.1

$$E = \begin{bmatrix} 1 & 0 & 0 \\ 0 & 1 & 0 \\ 0 & 0 & 1 \end{bmatrix}, \quad C_3 = \begin{bmatrix} 1 & 0 & 0 \\ 0 & -\frac{1}{2} & -\frac{\sqrt{3}}{2} \\ 0 & \frac{\sqrt{3}}{2} & -\frac{1}{2} \end{bmatrix},$$

$$C_3^{-1} = \begin{bmatrix} 1 & 0 & 0 \\ 0 & -\frac{1}{2} & \frac{\sqrt{3}}{2} \\ 0 & -\frac{\sqrt{3}}{2} & -\frac{1}{2} \end{bmatrix}, \quad \sigma_b = \begin{bmatrix} 1 & 0 & 0 \\ 0 & 0 & -1 \\ 0 & -1 & 0 \end{bmatrix},$$

$$\sigma_a = C_3 \times \sigma_b = \begin{bmatrix} 1 & 0 & 0 \\ 0 & \frac{\sqrt{3}}{2} & \frac{1}{2} \\ 0 & \frac{1}{2} & -\frac{\sqrt{3}}{2} \end{bmatrix}, \quad \sigma_c = \sigma_b \times C_3 = \begin{bmatrix} 1 & 0 & 0 \\ 0 & -\frac{\sqrt{3}}{2} & \frac{1}{2} \\ 0 & \frac{1}{2} & \frac{\sqrt{3}}{2} \end{bmatrix}.$$

A5.2

C_{2v}	E	$C_2(z)$	$\sigma_v(xz)$	$\sigma'_v(yz)$	
$\Gamma(f_{xyz})$	1	1	-1	-1	$= A_2$
$\Gamma(f_{z(x^2-y^2)})$	1	1	1	1	$= A_1$

D_{4h}	E	$2C_4(z)$	C_2	$2C'_2$	$2C''_2$	i	$2S_4$	σ_h	$2\sigma_v$	$2\sigma_d$
$\Gamma(f_{xyz})$	1	-1	1	1	-1	-1	1	-1	-1	1
$\Gamma(f_{z(x^2-y^2)})$	1	-1	1	-1	1	-1	1	-1	1	-1

Thus, $\Gamma(f_{xyz})$ transforms as B_{1u}; $\Gamma(f_{z(x^2-y^2)})$ transforms as B_{2u}.

A5.3 (i) For the symmetry operations σ_{va}, σ_{vb}, σ_{da}, and σ_{db}, $\beta = 0, 90, 45$, and $-45°$, respectively.

$$E: \begin{bmatrix} 1 & 0 & 0 \\ 0 & 1 & 0 \\ 0 & 0 & 1 \end{bmatrix} \qquad C_4: \begin{bmatrix} 0 & -1 & 0 \\ 1 & 0 & 0 \\ 0 & 0 & 1 \end{bmatrix}$$

$$C_4^3: \begin{bmatrix} 0 & 1 & 0 \\ -1 & 0 & 0 \\ 0 & 0 & 1 \end{bmatrix} \qquad C_2: \begin{bmatrix} -1 & 0 & 0 \\ 0 & -1 & 0 \\ 0 & 0 & 1 \end{bmatrix}$$

$$\sigma_{va}: \begin{bmatrix} 1 & 0 & 0 \\ 0 & -1 & 0 \\ 0 & 0 & 1 \end{bmatrix} \qquad \sigma_{vb}: \begin{bmatrix} -1 & 0 & 0 \\ 0 & 1 & 0 \\ 0 & 0 & 1 \end{bmatrix}$$

$$\sigma_{da}: \begin{bmatrix} 0 & 1 & 0 \\ 1 & 0 & 0 \\ 0 & 0 & 1 \end{bmatrix} \qquad \sigma_{db}: \begin{bmatrix} 0 & -1 & 0 \\ -1 & 0 & 0 \\ 0 & 0 & 1 \end{bmatrix}$$

(ii)

C_{4v}	E	$2C_4$	C_4^2	$2\sigma_v$	$2\sigma_d$	
Γ_e	3	1	-1	1	1	$= A_1 + E$.

A5.4 If n is an odd number, S_n^n generates an additional operation σ perpendicular to the S_n axis and hence group S_n becomes C_{nh}. So n cannot be odd. In addition, $S_2 = i$ and hence the point groups S_2 and C_i are equivalent. To sum up: $S_1 = C_s$, $S_2 = C_i$, and $S_3 = C_{3h}$. Hence n must be equal to or larger than 4.

A5.5

C_{2h}	E	C_2	i	σ_h		
A_g	1	1	1	1		x^2, y^2, z^2, xy
B_g	1	-1	1	-1		xz, yz
A_u	1	1	-1	-1	z	
B_u	1	-1	-1	1	x, y	

Reasoning:

(i) Functions x, y, z must have u symmetry. It is easily seen that z is symmetric with respect to C_2, while both x and y are antisymmetric with respect to C_2.

(ii) All the binary products must have g symmetry. Also: $\Gamma(z^2) = \Gamma(z) \times \Gamma(z) = A_u \times A_u = A_g$; $\Gamma(x^2) = \Gamma(y^2) = \Gamma(xy) = B_u \times B_u = A_g$; $\Gamma(xz) = \Gamma(yz) = A_u \times B_u = B_g$.

A5.6

D_{3d}	E	$2C_3$	$3C_2$	i	$2S_6$	$3\sigma_d$		
A_{1g}	1	1	1	1	1	1		$x^2 + y^2, z^2$
A_{2g}	1	1	-1	1	1	-1		
E_g	2	-1	0	2	-1	0		$(x^2 - y^2, xy); (xz, yz)$
A_{1u}	1	1	1	-1	-1	-1		
A_{2u}	1	1	-1	-1	-1	1	z	
E_u	2	-1	0	-2	1	0	(x, y)	

Reasoning:

(i) It is straightforward to obtain $\Gamma(z)$ by applying all the operations on z. On the other hand, each of x and y is neither symmetric nor antisymmetric with respect to C_3, so they must form a degenerate set. Since they have u symmetry, they form an E_u set.

(ii) $\Gamma(z^2) = \Gamma(z) \times \Gamma(z) = A_{2u} \times A_{2u} = A_{1g}$.
Since $x^2 + y^2 = r^2 - z^2$ and $\Gamma(r^2) = \Gamma(z^2) = A_{1g}$, $\Gamma(x^2 + y^2)$ is also A_{1g}.
$\Gamma(xz, yz) = \Gamma(x, y) \times \Gamma(z) = E_u \times A_{2u} = E_g$.

None of the remaining $x^2 - y^2$ and xy is symmetric or antisymmetric with respect to C_3, so they also form a degenerate set. Since they have g symmetry, they form another E_g set.

A5.7 (i) (a) C_{5v}; (b) C_{2v}; (c) C_{2v}; (d) D_{5d}.
(ii) (a) C_{3v}; (b) C_{2v}; (c) C_s; (d) C_2; (e) C_{2v}; (f) D_{3d}.

A5.8 (i) T_d. (ii) D_{3d} for staggered conformation and D_{3h} for eclipsed. (iii) D_{3h}. (iv) D_{4h}. (v) D_{6h}.

A5.9 (i) D_{3d}; (ii) C_s; (iii) D_{4d}; (iv) D_2; (v) C_2; (vi) D_{3d}; (vii) C_2; (viii) C_{2h}; (ix) C_{2h}; (x) D_4.

A5.10 (i) (a) Δ-M$\delta\delta\delta$, D_3;
(b) Δ-M$\lambda\lambda\lambda$ (more stable than Δ-M$\delta\delta\delta$ form by about 7.5 kJ mol^{-1}), D_3;
(c) Λ-M$\delta\delta\lambda$, C_2;
(d) Δ-M$\lambda\delta\delta$, C_2.
(ii) Λ-M$\lambda\lambda\lambda$, Λ-M$\delta\delta\delta$, Δ-M$\lambda\lambda\delta$, and Λ-M$\delta\lambda\lambda$, which are enantiomers of (a), (b), (c), and (d), respectively.

A5.11

   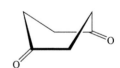

Chair (C_{2h}) Boat (C_{2v}) Boat (C_2)

Symmetric twist (D_2) Twist-boat (C_2)

A5.12 (i) (ii) (iii)

C_1 C_2 C_s

(iv) 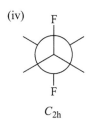 (v) 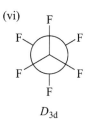 (vi)

C_{2h} C_{3v} D_{3d}

A5.13 (i) 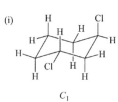 (ii)

C_1 C_2

(iii) 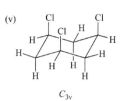 (iv) (v)

C_s C_{2h} C_{3v}

A5.14 (i) S_8: D_{4d}; S_7O: C_s.

     (ii) (a) *abdf* or *abce*; (b) *abcd* or *abdg*; (c) *abde* or *abcf*; (d) *aceg*; (e) *abef*.

A5.15 (i) *a*; (ii) *ab*; (iii) *ah*; (iv) *abef*; (v) *aceg*; (vi) *adeh*.

A5.16 (i) (a) C_{4v}; (b) D_{4h}; (c) C_{2v}; (d) C_{2v}; (e) C_{3v}; (f) C_{4v}; (g) C_s; (h) C_1; (i) D_{2h}.

     (ii) (a) *ab, ac*; (b) *abc, abg*; (c) *abcd*; (d) *abgh*; (e) *a*; (f) *ag*; (g) *acfh*; (h) *abdf*.

A5.17 (i) *abcd*; (ii) *efgh*; (iii) *ab*; (iv) *ace*; (v) *e*; (vi) *aj*; (vii) *abkl*; (viii) *abe*; (ix) *abcehj*;

     (x) *abegkl*; (xi) *chj*; (xii) *ak*; (xiii) *acjl*; (xiv) *acegjl*.

A5.18 (i) *ej*; (ii) *abcdej*; (iii) *a*; (iv) *efg*; (v) *ahij*; (vi) *ab*; (vii) *e*; (viii) *abj*; (ix) *ef*; (x) *aef*; (xi) *efi*; (xii) *abfi*; (xiii) *aeh*; (xiv) *abfj*.

A5.19 (i) *af*; (ii) *bm*; (iii) *bdf*; (iv) *b*; (v) *de*; (vi) *a*; (vii) *gfij*.

A5.20

Square planar (optically active, as found by Mills and Quibell)

Tetrahedral (may show slight optical activity since the chelate rings are non-planar)

A5.21 There are other possible answers, in addition to the ones given below.

(i)

C_1

(ii)

C_1

(iii)

C_2

(iv)

C_2

(v)

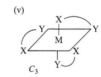

C_3

(vi)

C_s

(vii)

C_{2v}

(viii)

C_{2h}

(ix)

D_{2h}

(x)

C_{3v}

(xi)

D_3

(xii)

C_{4v}

(xiii)

D_{4h}

A5.22

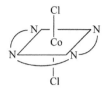
C_{2v}

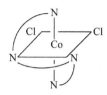
C_2 (optically active)

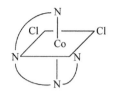
C_1 (optically active)

A5.23 Linear structures:

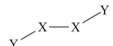

Y—X—X—Y

$D_{\infty h}$

e.g., C_2H_2, C_2F_2

Y—Y—X—X

$C_{\infty v}$

Y—X—Y—X

$C_{\infty v}$

Planar nonlinear structures:

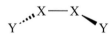

C_{2h}

e.g., *trans*-N_2H_2, *trans*-N_2F_2

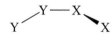

C_{2v}

e.g., *cis*-N_2H_2, *cis*-N_2F_2

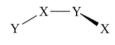

C_s

C_s

D_{2h}

C_{2v}

Nonplanar structures:

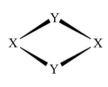

C_2

e.g., H_2O_2, F_2O_2

C_1

C_1

C_{2v}

C_s

6 | Molecular Geometry and Bonding

PROBLEMS

6.1 (i) (a) Consider a "distorted tetrahedral" AX_2Y_2 molecule of C_{2v} symmetry. Derive a relation between $\angle X-A-Y(\gamma_1)$, $\angle X-A-X(\alpha_1)$, and $\angle Y-A-Y(\beta_1)$. Given that, in propane, $CH_3CH_2^*CH_3$, $\angle H^*-C-H^* = 106.7°$ and $\angle C-C-C = 112.4°$, use the relation you have derived to calculate $\angle H^*-C-C$.

(b) Suppose four hybrid orbitals are formed by one s and three p orbitals at the central atom A. Two equivalent hybrid orbitals with hybridization indices n_x bond to atoms X, and the other two equivalent hybrid orbitals with hybridization indices n_y bond to atoms Y. Derive a relation between n_x and n_y. Compare your result with that in Problem 4.6(iii). (The equation obtained in Problem 4.4 should be useful.)

(ii) Consider a disphenoidal molecule such as Cl_2TeMe_2. Derive a relation between $\angle Cl-Te-Me$ (γ_2), $\angle Me-Te-Me$ (α_2), and $\angle Cl-Te-Cl$ (β_2). Note that α_2 is the obtuse angle of $\angle Me-Te-Me$, as illustrated on the right.

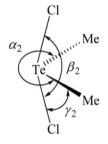

(iii) Consider a square pyramidal molecule such as $XeOF_4$. Derive a relation between $\angle F-Xe-F$ (α_3) and $\angle O-Xe-F$ (β_3).

(iv) (a) Consider a C_{3v} molecule AX_3Y. Derive a relation between $\angle X-C-X$ (α) and $\angle X-A-Y$ (β). Additionally, in chloromethane, it is known that $\angle H-C-Cl = 108.0°$, calculate angle $\angle H-C-H$.

(b) Suppose four hybrid orbitals are formed by one s and three p orbitals at the central atom A. One hybrid orbital with hybridization indices n_Y bonds to atom Y and the other three equivalent hybrid orbitals with hybridization index n_X bond to atoms X. Derive a relation between n_Y and n_X. Compare your result with that in Problem 4.6(ii). (The equation obtained in Problem 4.4 should be useful.)

104 *Problems in Structural Inorganic Chemistry.* Second edition. Wai-Kee Li, Hung Kay Lee, Dennis Kee Pui Ng, Yu-San Cheung, Kendrew Kin Wah Mak, and Thomas Chung Wai Mak. © Oxford University Press 2019. Published in 2019 by Oxford University Press. DOI: 10.1093/oso/9780198823902.001.0001

REFERENCE: W.-K. Li and T. C. W. Mak, Bond angle relationships in some AX_nY_m molecules. *J. Chem. Educ.* **51**, 571 (1974).

6.2 Metallocenes are compounds with a metal atom sandwiched between two π-bonded hydrocarbon ligands. Shown below are two examples: ruthenocene $Ru(C_5H_5)_2$ (left) and ferrocene $Fe(C_5H_5)_2$ (right). In each structure shown, the carbon atoms in the two rings are labeled from 1 to 10, while points A and B are at the center of the upper and lower rings, respectively. It is important to note that the two rings are in an eclipsed conformation in $Ru(C_5H_5)_2$, and they are in a staggered conformation in $Fe(C_5H_5)_2$.

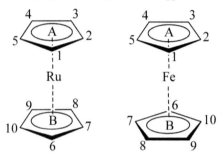

(i) For $Ru(C_5H_5)_2$, determine the values of the following seven dihedral angles: $\angle C^3-A-B-C^8$, $\angle C^3-A-B-C^7$, $\angle C^3-A-B-C^9$, $\angle C^3-A-B-C^6$, $\angle C^6-C^1-A-C^3$, $\angle C^9-C^6-B-C^1$, and $\angle B-A-C^2-C^7$.

(ii) For $Fe(C_5H_5)_2$, determine the values of the following five dihedral angles: $\angle C^1-A-B-C^6$, $\angle C^1-A-B-C^9$, $\angle C^4-A-B-C^3$, $\angle C^1-A-B-C^{10}$, and $\angle B-A-C^2-C^3$.

6.3 (i) For dichloromethane, H_2CCl_2, determine the dihedral angle $\angle H^3-C-^*-Cl^2$ and $\angle H^3-C-^*-H^4$, where the line $C-^*$ is the bisector of $\angle H-C-H$, as shown below.

(ii) Referring to the ethane molecule (staggered conformation) shown below, determine the dihedral angles $\angle H^3-C^1-C^2-H^7$, $\angle H^3-C^1-C^2-H^4$, $\angle H^4-C^1-C^2-H^8$, and $\angle H^4-C^1-C^2-H^7$.

(iii) Prismane, $(CH)_6$, is an isomer of benzene. It has a trigonal prismatic structure (shown below) where each apex is occupied by a CH group. Determine the dihedral angles $\angle 1-2-5-6$, $\angle 1-2-3-6$, and $\angle 1-4-6-5$.

(iv) Phosphorus pentafluoride, PF_5, has a trigonal bipyramidal structure, as shown below. Note that atoms F^3, F^4, and F^5 form an equilateral triangle, and atom P is at the center of this triangle. Also, straight line F^1-P-F^2 is perpendicular to the triangle. Determine the values of dihedral angles $\angle F^1-P-F^3-F^2$, $\angle F^3-P-F^1-F^5$, $\angle F^3-P-F^5-F^4$, $\angle F^4-F^5-F^3-P$, and $\angle F^4-P-F^3-F^1$.

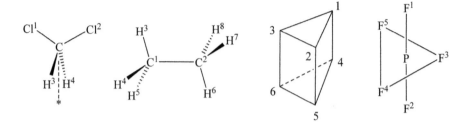

6.4 (i) Consider a (polygonal) pyramid illustrated on the right. The base $BCDE\ldots$ is an n-sided regular polygon. The projection of the vertex at the top (labeled as A) onto the base is the center of the base (labeled as A'). In addition, vertex A forms an equilateral triangle with each edge of the base, i.e., triangles ABC, $ACD,\ldots$ are all equilateral. Determine the dihedral angle between triangle ACD and the base $BCDEF\ldots$. Express this dihedral angle in terms of n.

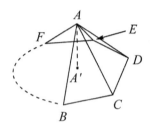

(ii) The B_4 skeleton of $(BCl)_4$, the B_6 skeleton of $(BCl)_6^{2-}$, and the B_{12} skeleton of $(BH)_{12}^{2-}$ form a perfect tetrahedron, octahedron, and icosahedron, respectively. Determine the dihedral angles $\angle G-H-I-J$, $\angle L-M-N-O$, and $\angle R-S-T-U$ in each of the polyhedra, as labeled below.

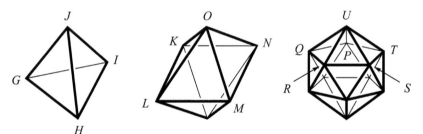

(iii) The C_8 skeleton of $(CH)_8$ is a perfect cube. Determine dihedral angle $\angle V-W-X-Y$, as shown in the figure on the right.

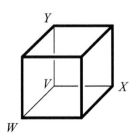

6.5 The dihedral angle $\angle A-B-C-D$ (ω) for four atoms A, B, C, and D, is defined as the angle between the projections of $\overrightarrow{BA}$ and $\overrightarrow{CD}$ onto $\overrightarrow{BC}$. A positive value of ω means that the $ABCD$ sequence forms a right-hand screw. It can be shown that:

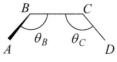

$$\cos \omega = \frac{(\overrightarrow{AB} \times \overrightarrow{BC}) \cdot (\overrightarrow{BC} \times \overrightarrow{CD})}{AB(BC)^2 CD \sin \theta_B \sin \theta_C}.$$

Using this given formula, derive a relationship between bond angle θ ($\angle A-B-C = \angle B-C-D$) and the dihedral angle ω ($\angle A-B-C-D$) in the chair form of cyclohexane.

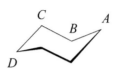

6.6 Consider the carbon skeleton of non-planar cyclobutane. Derive a relation between

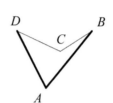

(i) the bond angle θ ($\angle A-B-C$) and the dihedral angle ω ($\angle B-A-D-C$);

(ii) the bond angle θ and the angle δ between planes sharing a common diagonal ($\angle D-A-C-B$).

6.7 A regular helical chain is characterized by three parameters: r, the radius of the enveloping cylinder; ϕ, the rotation angle in projections; and z, the difference in height between neighboring atoms. Derive expressions for

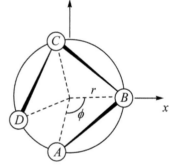

(i) the bond length $\ell = AB$,

(ii) the bond angle $\theta = \angle A-B-C$, and

(iii) the dihedral angle $\omega = \angle A-B-C-D$ in terms of the chosen parameters.

Also calculate their values for fibrous sulfur, which has a 10_3 helical structure [i.e., $\phi = (3/10) \times 360° = 108°$] with $r = 0.95$ Å and $z = 1.38$ Å. Hint: Make use of the coordinate system shown above, in which the axis of the enveloping cylinder is taken as the z-axis and B lies on the positive x-axis. The coordinates are $(r \cos(n\phi), r \sin(n\phi), nz)$, where $n = -1, 0, 1,$ and 2 for A, B, C, and D, respectively.

REFERENCE: M. D. Lind and S. Geller, Structure of pressure-induced fibrous sulfur. *J. Chem. Phys.* **51**, 348–53 (1969).

6.8 For a molecular compound $C_5H_2O_5$, write plausible VB structural formulas if the molecular skeleton is (i) aliphatic (either *syn* or *anti* form is acceptable) and (ii) monocyclic (i.e. containing a single ring). In case (b), discuss the structure and bonding of the fully deprotonated dianion $C_5O_5^{2-}$.

REFERENCES: C.-K. Lam, M.-F. Cheng, C.-L. Li, J.-P. Zhang, X.-M. Chen, W.-K. Li, and T. C. W. Mak, Stabilization of D_{5h} and C_{2v} valence tautomers of the croconate dianion. *Chem Commun.* 448–9 (2004); C.-K. Lam and T. C. W. Mak, Generation and stabilization of D_{6h} and C_{2v} valence tautomeric structures of the rhodizonate dianion in hydrogen-bonded host lattices. *Angew. Chem. Int. Ed.* **40**, 3453–5 (2001).

6.9 Based on the VSEPR theory, determine the structure and the symmetry point group of the following species.

    (i) $(CN)_2$ (ii)   HNC (hydrogen isocyanide)

   (iii) HNCO (isocyanic acid)   (iv) H_2CN_2 (diazomethane)

    (v) HN_3 (vi) NO_2F (N is the central atom)

  (vii) $SeCl_3^+$ (viii) Hg_2Cl_2

   (ix) $(NH_2)_2CS$ (x) $B_3N_3H_6$ (this is a ring compound)

6.10 Based on the VSEPR theory, determine the structure and the symmetry point group of the following species.

    (i) $O(SF_5)_2$ (ii) $HClO_4$ (iii) F_3SCCF_3 (iv) $SF_2(CH_3)_2$

    (v) OIF_5 (vi) SPF_3 (P is the central atom)   (vii) SO_3

  (viii) SCl_2O_2 (ix) $S_2O_3^{2-}$

    (x) $Sb(SCH_2CO_2)_2^-$ (Hint: This anion is formed by Sb^- with two

$$\cdot O-\overset{\displaystyle O}{\overset{\displaystyle \|}{C}}-CH_2-S\cdot \text{ "fragments")}$$

   (xi) $SbCl_3 \cdot C_6H_5NH_2$ (Sb is the central atom)

  (xii) TiX_4 and ZnX_4^{2-} (where X is a halogen)

6.11 The compound HXeOXeH was prepared in 2008. Write down a plausible electron-dot structure for this novel molecule. Then apply the VSEPR theory to predict the structure of this molecule, assuming that it is planar.

REFERENCE: L. Khriachtchev, K. Isokoski, A. Cohen, M. Räsänen, and R. B. Gerber, A small neutral molecule with two noble-gas atoms: HXeOXeH. *J. Am. Chem. Soc.* **130**, 6114–8 (2008).

6.12 Ions $TeBr_6^{2-}$ and $SbBr_6^{3-}$ have octahedral structures even though there is a lone pair on the central atom in both cases. Comment.

REFERENCE: R. J. Gillespie, The electron-pair repulsion model for molecular geometry. *J. Chem. Educ.* **47**, 18–23 (1970).

6.13   (i) Predict the structure of P_4.

      (ii) Two structures have been suggested for S_8: crown (shown on the right) and cubic (*not* shown). Which is more plausible?

6.14 Consider the bond angles in difluorides CF_2, SiF_2, and GeF_2. Predict which molecule, among the three, will have the largest bond angle and which will have the smallest bond angle. Repeat this process for the germanium halides GeF_2, $GeCl_2$, and $GeBr_2$. Justify your predictions in terms of the VSEPR theory.

6.15 (i) In high school, our chemistry teacher told us: "Most (inorganic) chlorides are soluble in water; mercurous chloride is one of the few chlorides that is not." Explain:

      (a) Why are most chlorides soluble in water?

      (b) Why is mercurous chloride insoluble in water?

      (c) What is the structure of mercurous chloride, according to the VSEPR theory?

    (ii) Another lesson from our high school chemistry teacher: "Silver chloride is also insoluble in water, but it is soluble in ammonia water." What is the silver-containing species that is soluble in ammonia water? What is the structure of this species, as predicted by the VSEPR model?

6.16 It is well known that, in aqueous solutions, HCl dissociates into $H^+(aq)$ (or H_3O^+) and $Cl^-(aq)$. On the other hand, in the gas phase, HCl simply exists as a diatomic molecule. Write down the three most obvious resonance structures for gaseous HCl, and then arrange them in order of importance.

6.17 Polynitrogen compound $N_5^+AsF_6^-$ was synthesized in 1999 and calculations showed that the most stable N_5^+ isomer has a V-shaped structure as shown.

Discuss the structure and bonding of this cation using the hybridization theory. In particular, your discussion should include the following: (i) sketch the important resonance structures; (ii) the hybridization schemes for N^a, N^b, and N^c; and (iii) the bond orders for N^a-N^b and N^b-N^c. Finally, comment on whether or not the calculated structural parameters

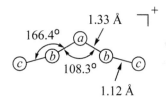

are consistent with the VSEPR theory and determine the structure of anion AsF_6^- using this theory.

REFERENCE: K. O. Christe, W. W. Wilson, J. A. Sheehy, and J. A. Boatz, N_5^+: A novel homoleptic polynitrogen ion as a high energy density material. *Angew. Chem. Int. Ed.* **38**, 2004–9 (1999).

6.18 Theoretical calculation showed that nitrosyl azide has a planar non-linear chain-like structure: $N^a-N^b-N^c-N^d-O$ (the nitrogen atoms are labeled for easy reference). Discuss the bonding of this molecule using the hybridization theory. In particular, your discussion should include the following points:

(i)   the important resonance structures;

(ii)   the hybridization schemes at atoms N^b, N^c, and N^d;

(iii)   the bond orders for N^a-N^b, N^b-N^c, N^c-N^d, and N^d-O;

(iv)   the approximate bond angles of $\angle N^a-N^b-N^c$, $\angle N^b-N^c-N^d$, and $\angle N^c-N^d-O$.

REFERENCE: J. M. Galbraith and H. F. Schaefer III, The nitrosyl azide potential energy hypersurface: a high-energy-density boom or bust? *J. Am. Chem. Soc.*, **118**, 4860–70 (1996).

6.19   The interesting compound $S_3N_2O_2$ has a planar structure with a two-fold symmetry axis as shown. Propose a bonding scheme which accounts for the measured bond distances. Give reasonable estimates of the bond angles α, β, and γ.

REFERENCE: G. MacLean, J. Passmore, P. S. White, A. Banister, and J. A. Durrant, The redetermination of the crystal structure of trisulphur dinitrogen dioxide. *Can. J. Chem.* **59**, 187–90 (1981).

6.20   Determine the bond order of all the Cl–O bonds for chlorine perchlorate, $ClOClO_3$.

REFERENCE: C. J. Schack and D. Pilipovich, Chlorine perchlorate. *Inorg. Chem.* **9**, 1387–90 (1970).

6.21   Electron diffraction experiments on dichlorine heptoxide, Cl_2O_7, determined values 1.405 ± 0.002 and $1.709 \pm 0.004\,\text{Å}$ for two types of Cl–O bonds. Also there is a Cl–O–Cl angle of $118.6 \pm 0.7°$. Sketch the structure of Cl_2O_7 and discuss the bonding on the basis of the valence bond theory.

REFERENCE: B. Beagley, Electron diffraction study of gaseous chlorine heptoxide. *Trans. Faraday Soc.* **61**, 1821–30 (1965).

6.22   In $S_4O_6^{2-}$, there are two different S–S bond lengths and only one S–O bond distance. Deduce the structure of $S_4O_6^{2-}$. What is the bond order of the S–O bonds and why?

6.23   Molecule F_4SCH_2 has the structure as shown. Note that the four atoms of S, C, and two hydrogens are coplanar to each other. Determine whether the CH_2 plane is coplanar with or perpendicular to the equatorial plane of the trigonal bipyramid.

REFERENCE: H. Bock, J. E. Boggs, G. Kleemann, D. Lentz, H. Oberhammer, E. M. Peters, K. Seppelt, A. Simon, and B. Solouki, Structure and reactions of methylenesulfur tetrafluoride. *Angew. Chem. Int. Ed.* **18**, 944–5 (1979).

6.24   (i)   In the solid phase, PCl_5 exists as ionic compound $[PCl_4^+][PCl_6^-]$. What are the structures of these two ions?

(ii) In the solid phase, PBr$_5$ exists as ionic compound $[PBr_4^+]Br^-$. Why does solid PBr$_5$ not exist as $[PBr_4^+][PBr_6^-]$, as in the case of solid PCl$_5$?

6.25 The bond angles, $\angle C-N-C$, in N(CH$_3$)$_3$ and N(CF$_3$)$_3$ are 110.9° and 117.9°, respectively. In addition, the C–N bond lengths in N(CH$_3$)$_3$ and N(CF$_3$)$_3$ are 1.458 and 1.426 Å, respectively. Explain why the carbon and nitrogen atoms in N(CF$_3$)$_3$ are almost coplanar, while these same atoms in N(CH$_3$)$_3$ form a pyramid.

6.26 Discuss the structure and bonding of the F$_3$CNO (trifluoronitrosomethane) and F$_3$CN$_3$ (azidotrifluoromethane) molecules on the basis of the valence bond theory.

REFERENCE: K. O. Christie and C. J. Schack, Properties of azidotrifluoromethane. *Inorg. Chem.* **20**, 2566–70 (1981).

6.27 Methyl isothiocyanate, MeNCS, and silyl isothiocyanate, H$_3$SiNCS, have different molecular geometries. Comment.

6.28 Write Lewis-type formulas for the linear triatomic molecules UO$_2^{2+}$ (uranyl dication) and CUO (carbon uranium oxide). Note that the uranium atom is in the +VI oxidation state, i.e., it can use up to six valence electrons to form conventional two-electron bonds.

REFERENCE: H.-S. Hu, Y.-H. Qiu, X.-G. Xiong, W. H. Eugen Schwarz, and J. Li, On the maximum bond multiplicity of carbon: unusual C≡U quadruple bonding in molecular CUO. *Chem. Sci.* **3**, 2786–96 (2012).

6.29 Most five-coordinate compounds adopt either the trigonal bipyramidal (TBP) or square pyramidal (SP) structure. Comment on the following structural data.

(i) When the central atom is a main group element, the axial bonds are longer than the equatorial bonds in the TBP structure; the reverse is the case for the SP structure.

(ii) When the central atom is a transition metal, e.g., in TBP $[CuCl_5]^{3-}$, the axial Cu–Cl bonds (2.296 Å) is shorter than the equatorial ones (2.391 Å). On the other hand, in SP $[Ni(CN)_5]^{3-}$, the axial Ni–C bond (2.17 Å) is longer than the equatorial ones (1.87 Å).

6.30 In bromine pentafluoride, BrF$_5$, the bromine atom is below the basal plane of the square pyramid, but in $[Ni(CN)_5]^{3-}$, the nickel atom is above the basal plane. Comment. (Note that $[Ni(CN)_5]^{3-}$ can exist in both C_{4v} and D_{3h} forms; the present discussion is concerned with the former.)

6.31 Apply the VSEPR theory to determine the structures of the following xenon compounds. Illustrate the structures with clearly drawn figures and with written descriptions (pyramidal, tetrahedral, etc.). In addition, determine the symmetry point group of each of the structures you have deduced.

    (i)   Xenon oxides: XeO, XeO_2, XeO_3, XeO_4.

    (ii) Xenon fluorides: XeF_2, XeF_4, XeF_6, XeF_8.

    (iii) Xenon oxyfluorides: $XeOF_2$, XeO_2F_2, XeO_3F_2.

6.32 Comment on the geometrical shapes of Group 2 dihalide molecules tabulated below, where L and B denote linear and bent molecules, respectively.

Halogen \ Metal	Be	Mg	Ca	Sr	Ba	Hg
F	L	L	B	B	B	L
Cl	L	L	L	B	B	L
Br	L	L	L	L	B	L
I	L	L	L	L	B	L

REFERENCE: C. A. Coulson, d-Electron in chemical bonding. *Proceedings of Welch Conferences on Chemical Research,* **16**, 61–98 (1972).

6.33 Bond angles remain approximately constant across each of the following two series:

    $SiH_2(92°)$, $PH_2(92°)$, $H_2S(92°)$;

    $CH_2(^1A_1)(102°)$, $NH_2(103°)$, $H_2O(105°)$.

This fact is not anticipated by the VSEPR theory. Rationalize this with the aid of the given Walsh diagram for AH_2 molecules. Note that orbital $1b_1$ is essentially the p orbital perpendicular to the molecular plane.

REFERENCE: N. C. Baird, Molecular geometry predictions using simple MO theory. AX_n systems ($n = 2 - 7$). *J. Chem. Educ.* **55**, 412–7 (1978).

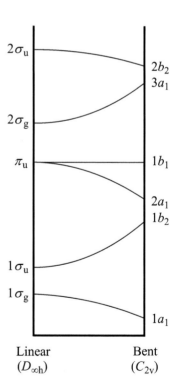

Linear $(D_{\infty h})$ Bent (C_{2v})

6.34 The energy of a diatomic molecule can be approximated by the Morse function:

$$E(R) = D\left[1 - e^{-\beta(R-R_e)}\right]^2$$

where R_e is the equilibrium internuclear separation while D and β are constants.

(i) Find the bond energy D_e.

(ii) Sketch the Morse function, labelling D_e and R_e.

(iii) Expand the Morse function up to terms quadratic in $(R - R_e)$. Show that this approximates a harmonic oscillator potential and identify the force constant k.

6.35 Consider species ClO_2^+, ClO_2, and ClO_2^-.

(i) Arrange them in the order of increasing O–Cl bond length.

(ii) Arrange them in the order of increasing bond angle.

6.36 Complete the following table by matching the species $(CH)_5N$ (pyridine), NCO^- (the cyanate anion), CH_3NH_2 (methylamine), CH_3CN (methyl cyanide or acetonitrile), and CH_3CHNCH_3 (N—methylethanimine) with the corresponding C–N bond distances in the last column.

Species	C–N bond order	C–N bond length (Å)
		1.469
		1.337
		1.280*
		1.171
		1.158

*This bond length is obtained using high-level *ab initio* calculations, while the remaining bond lengths are experimental values found in the literature.

6.37 Complete the following table by matching the species hydroxylamine (H_2NOH), nitrosonium ion (NO^+), nitronium ion (NO_2^+), nitrite ion (NO_2^-), and nitrate ion (NO_3^-) with the corresponding N–O bond distances in the last column.

Species	N–O bond order	N–O bond length (Å)
		1.062
		1.15
		1.236
		1.25
		1.47

6.38 Complete the following table by matching the species CO, CO_2, CO_3^{2-}, H_3COCH_3 (dimethylether), and $H_3CCO_2^-$ (acetate ion) with the corresponding C–O bond distances in the last column.

Species	C–O bond order	C–O bond length (Å)
		1.128
		1.163
		1.250
		1.290
		1.416

6.39 Complete the following table by matching the species SO_4^{2-}, SO_3^{2-}, S_2O (linked in the non-linear form of S–S–O), and $S_2O_7^{2-}$ (linked in the non-linear form $[O_3S-O-SO_3]^{2-}$) with the corresponding S O bond distances in the last column. Note that there are two different S–O distances (terminal and bridging) in $S_2O_7^{2-}$.

Species	S–O bond order	S–O bond length (Å)
		1.645
		1.51
		1.49
		1.437
		1.43

6.40 Complete the following table by matching the species P_4O_{10}, PO_4^{3-}, HPO_3^{2-}, and $H_2PO_2^-$ with the corresponding P–O bond distances in the last column.

Species	P–O bond order	P–O bond length (Å)
		1.40
		1.49
		1.51
		1.52
		1.65

Structural notes on the species:

(a) The structure of P_4O_{10} is shown on the right. Note that the four P atoms form a tetrahedron and there are six bridging P—O—P bonds along the edges of the tetrahedron. Also, there are four terminal P–O bonds. So P_4O_{10} will appear twice in the table: once for bridging P—O—P bonds and once for terminal P–O bonds.

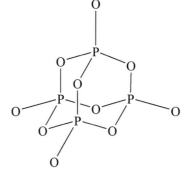

(b) Anion PO_4^{3-} has a tetrahedral structure.

(c) For HPO_3^{2-}, P is bonded to one H and three O atoms in a tetrahedral arrangement.

(d) For $H_2PO_2^-$, P is bonded to two H atoms and two O atoms in a tetrahedral manner.

6.41 Complete the following table by matching the given S–N bond distances with the species:

(i) NS_2^+ (where N is the central atom),

(ii) NSF_3 (where S is the central atom),

(iii) S_2N_2 (with a square planar structure having only one type of sulfur–nitrogen bond),

(iv) $S_4N_4^{2+}$ (with a planar eight-membered ring, as shown below), and

(v) $S_4(NEt)_2$ (with a chair-shaped six-membered ring, also shown below).

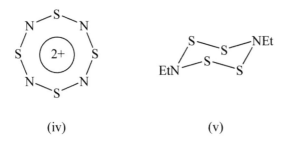

(iv)   (v)

Species	S–N bond order	S–N bond length (Å)
		1.71
		1.65
		1.55
		1.46
		1.42

6.42 Cyclopropenyl anion, with four π electrons, is the smallest "antiaromatic" species. Since this anion is not known experimentally, its structure and bonding have generated a fair amount of interest among the theoretical chemists.

From the theoretical calculations, we have the following four structures for the cyclopropenyl anion:

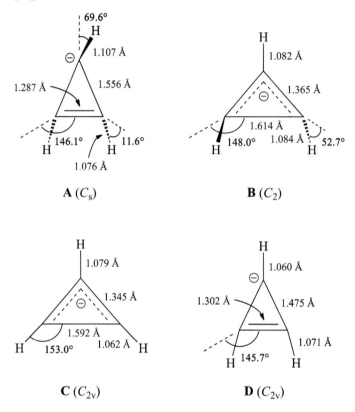

A (C_s)

B (C_2)

C (C_{2v})

D (C_{2v})

Upon calculating the vibrational frequencies of these structures, it is found that **A**(C_s symmetry), **B**(C_2), **C**(C_{2v}), and **D**(C_{2v}) have zero, one, two, and two imaginary frequencies, respectively.

(i) Comment on these structures and the relationship among them.

(ii) The most symmetrical (equilateral triangular) structure for $(CH)_3^-$ with D_{3h} symmetry is not included above. Comment on the number of imaginary frequencies for this structure.

REFERENCE: W.-K. Li, The cyclopropenyl anion: an *ab initio* molecular orbital study. *J. Chem. Research (S)* 220–1 (1987).

6.43 This problem is related to the 1,2-hydrogen shift for the ethenyl anion (**A**, shown below, with C_s symmetry, structural parameters are theoretically calculated numbers).

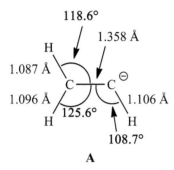

A

One possible 1,2-hydrogen shift channel is the migration of the *syn*-hydrogen on the molecular plane:

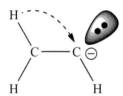

The transition state is expected to be a bridging structure. Indeed, a stationary point with such a structure is found. The structure is planar with C_{2v} symmetry as shown below:

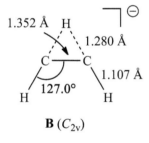

B (C_{2v})

According to kinetics theory, the transition state of a reaction has one and only one imaginary vibrational frequency. Upon a vibrational frequency analysis, it is found that **B** has two imaginary frequencies. In order to find the true transition state, which has only one imaginary frequency, subsequent calculation was carried out by starting with **B** with all symmetry constraints removed (i.e., under C_1 symmetry). A transition state (**C**, shown below) with only one imaginary frequency was found. It is the true transition state for the 1,2-hydrogen shift of the *syn*-hydrogen.

$$H^c$$

1.117 Å   1.292 Å

H^a ···· C^a——C^b ····· H^b

112.5°   1.420 Å

$\angle H^a–C^a–C^b–H^c = 103.8°$

C (C_s)

Further theoretical calculation found another similar transition state, also with only one imaginary frequency but having C_2 symmetry:

$$H^c$$

1.101 Å   1.308 Å

H^a ···· C^a——C^b —— H^b

108.6°   1.414 Å

$\angle H^a–C^a–C^b–H^c = 87.6°$

D (C_2)

From the theoretical calculations, the energy order was found to be **B** > **C** > **D**.

(i)   Why is **C** lower than **B** in energy?

(ii) Why is **D** lower than **C** in energy?

(iii) Why, in the theoretical calculation, does **B** collapse to **C** instead of **D**, even though **D** is lower than **C** in energy? (Note that all symmetry constraints were removed and hence symmetry was not the issue.)

(iv) What equilibrium structures are connected by transition state **D**?

(v)   Another hydrogen shift process for **A** is the 1,1-hydrogen shift. Without breaking the C–H bond, the shift may be achieved by swinging of the C–H bond and C–C bond rotation:

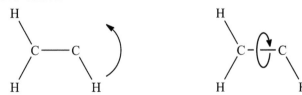

Which one do you think is more favorable? Hence, sketch the transition state for the process.

REFERENCE: W.-K. Li, R. H. Nobes, and L. Radom, Structures and rearrangement processes for the prototype alkyl, alkenyl and alkynyl anions: A theoretical study of the ethyl, ethenyl and ethynyl anions, *J. Mol. Struct. (Theochem)* **149**, 67–79 (1987).

6.44 Electron-deficient compound B_4Cl_4 has a very pretty structure: a tetrahedral B_4 skeleton with four B–Cl terminal bonds, giving rise to T_d symmetry for the molecule. Apply VSEPR theory to describe the bonding of this molecule and deduce the bond order of each B–B linkage.

6.45 In deriving the $2n+2$ rules for carborane clusters, we assume that there are n cluster atoms at the vertices of a regular deltahedron. The four valence orbitals on each vertex atom may be divided into one external sp hybrid orbital (pointing away from the polyhedron), two equivalent tangential p orbitals, and one internal sp hybrid orbital (pointing toward the center of the polyhedron). This question concerns only the internal sp hybrids of the vertex atoms.

It is asserted that the combination of n internal sp hybrids generates one strong bonding molecular orbital and $n - 1$ nonbonding (or antibonding) orbitals. For the special case of a trigonal bipyramid, the five vertex atoms are labeled in the figure. Linearly combine the five internal sp hybrids to form five molecular orbitals and show that only one of the five molecular orbitals is strongly bonding in nature.

6.46 By considering the $\nu(CO)$ spectra and applying Wade's rules, suggest a plausible structure (with special comment on the location of the H atom) for each of the following transition-metal hydride clusters:

(i) The neutral $HFeCo_3(CO)_9[P(OMe)_3]_3$ complex, which has carbonyl absorptions (in cm^{-1}) at 2040(s), 2009(m), 1990(s), 1963(w), 1833(m), and 1821(m).

(ii) The $[HCo_6(CO)_{15}]^-$ anion in its $[(Ph_3P)_2N]^+$ salt, which shows the following infrared bands (in cm^{-1}) in CH_2Cl_2: 2060(w), 2005(s), 1975(ms), and 1830(ms).

6.47 Shown on the right is a substituted borane cluster, $(BH)_{10}XY$, with all vertices other than X and Y being BH groups. Apply Wade's $2n + 2$ rules to predict which of the following four clusters would be most likely to exist.

Cluster	A	B	C	D
X	Bi	S	P	P
Y	S	S	CH	Bi

6.48 Make use of the skeletal electron counting rules to predict the core structures of the following cluster compounds. Sketch the molecular skeleton in each case and determine the symmetry point group of each structure. Fill in the table below.

(i) Tl_5^{7-}
(ii) $TlSn_9^{3-}$
(iii) $C_2B_7H_9$
(iv) $C_2B_8H_{10}$
(v) $Fe_3As_2(CO)_9$
(vi) $Co_2(CO)_6(CR)_2$
(vii) $(CO)_3Fe(C_2B_3H_7)$
(viii) $Fe_4BH_2(CO)_{12}H$
(ix) $(C_6H_6)Fe(Et_2C_2B_4H_4)$
(x) $(C_5H_5)_2Co_2(C_2B_3H_5)$
(xi) $[Rh_{12}(CO)_{27}Sb]^{3-}$ in which Sb is an interstitial atom
(xii) $[Os_4(CO)_{12}N]^-$ in which N is an interstitial atom
(xiii) $[Ru_6(CO)_{14}(C_6H_5Me)C]$ in which C is an interstitial atom
(xiv) $[Rh_9(CO)_{21}P]^{2-}$ in which P is an interstitial atom
(xv) $[Os_{10}C(CO)_{24}]^{2-}$ in which C is an interstitial atom

Cluster compound	No. of skeletal bonding electron pairs	Reference polyhedron
(i)	. . .	. . .
. . .	. . .	. . .

6.49 Nickel and carbon monoxide form a variety of carbonyl compounds with the general formula $Ni_m(CO)_n$. Derive the structure of the following compounds by employing Wade's $2n + 2$ rules and determine the symmetry point group of each of your structures. (i) $Ni_5(CO)_{11}$ (ii) $Ni_6(CO)_{14}$ (iii) $Ni_7(CO)_{17}$ (iv) $Ni_8(CO)_{18}$ (v) $Ni_9(CO)_{19}$

6.50 Metalloborane $[(\eta^5\text{-}C_5H_5)Co]_3(P\phi)(B\phi)$, where ϕ stands for phenyl group, was synthesized and characterized in 1988 by T. P. Fehlner and co-workers. Answer the following questions regarding the structure of this complex by employing the $2n+2$ rules developed by Wade.

(i) What structure would be predicted if both P and B atoms are considered as vertex atoms?

(ii) What structure would be predicted if Pϕ and Bϕ are considered as bridging groups?

(iii) The same structure should be obtained from (i) and (ii). Illustrate this structure pictorially.

6.51 Compound $Fe_4(C_5H_5)_4(CO)_4$ is a dark-green solid. Its infrared spectrum shows one single CO stretching band at $1640\ cm^{-1}$. Its proton NMR spectrum exhibits one single line even at very low temperature. Apply the $2n + 2$ rules to predict the structure of the compound. Your answer needs to be in accord with these spectral data. In other words, for this molecule, you also need to show how the CO and C_5H_5 ligands are bonded to the metals.

REFERENCE: R. B. King, Organometallic chemistry of the transition metals. XVI. Polynuclear cyclopentadienylmetal carbonyls of iron and cobalt. *Inorg. Chem.* **5**, 2227–30 (1966).

6.52 Transition metal complexes may be broadly classified into the following three categories:

(i) Those with electronic configurations which are completely unrelated to the 18-electron rule. Typical examples include many first-row transition metal complexes.

(ii) Those complexes which follow the 18-electron rule in the sense that they never have more than 18 valence electrons. Examples include many second- and third-row transition metal complexes.

(iii) Complexes which conform rigorously to the 18-electron rule. Examples include many of the metal carbonyls.

With the aid of the energy level diagram shown on the right (or its variant), give a satisfactory explanation for these three types of behavior on the basis of a simple molecular orbital treatment of the metal–ligand bonding.

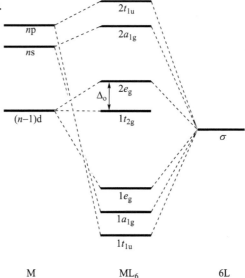

SOLUTIONS

A6.1 (i) (a) Place the two A–X bonds in the xz-plane and the two A–Y bonds in the xy-plane with the x-axis bisecting both α_1 and β_1, as shown in the figure on the right. We have:

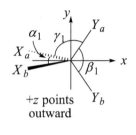

+z points outward

$$\overrightarrow{AX_a} = \left(-AX\cos\left(\frac{\alpha_1}{2}\right), 0, -AX\sin\left(\frac{\alpha_1}{2}\right)\right)$$

$$\overrightarrow{AY_a} = \left(AY\cos\left(\frac{\beta_1}{2}\right), AY\sin\left(\frac{\beta_1}{2}\right), 0\right)$$

The dot product of vectors $\overrightarrow{AX_a}$ and $\overrightarrow{AY_a}$ is given by:

$$\overrightarrow{AX_a} \cdot \overrightarrow{AY_a} = AX \cdot AY\cos\gamma_1 = \left(-AX \cdot AY\cos\left(\frac{\alpha_1}{2}\right)\cos\left(\frac{\beta_1}{2}\right) + 0 + 0\right).$$

Hence, $\cos\gamma_1 = -\cos\left(\frac{\alpha_1}{2}\right)\cos\left(\frac{\beta_1}{2}\right)$.

In $CH_3CH_2^*CH_3$, given $\angle H^* - C - H^* = 106.7°$ and $\angle C - C - C = 112.4°$, we can readily obtain $\angle H^* - C - C = 109.4°$.

(b) Using the equation obtained in Problem 4.4 for γ_1, α_1, and β_1:

$$\cos\gamma_1 = -(n_x n_y)^{-1/2},$$

$$\cos\alpha_1 = -(n_x n_x)^{-1/2} \Rightarrow 2\cos^2\left(\frac{\alpha_1}{2}\right) - 1 = -n_x^{-1}$$

$$\Rightarrow \cos\left(\frac{\alpha_1}{2}\right) = \sqrt{\frac{n_x - 1}{2n_x}},$$

$$\cos\beta_1 = -(n_y n_y)^{-1/2} \Rightarrow \cos\left(\frac{\beta_1}{2}\right) = \sqrt{\frac{n_y - 1}{2n_y}}.$$

Substituting these results into the last equation in (a), $-(n_x n_y)^{-1/2} = -\sqrt{\frac{n_x - 1}{2n_x}}\sqrt{\frac{n_y - 1}{2n_y}}$, leading to $(n_x - 1)(n_y - 1) = 4$. The result is actually identical to that obtained in Problem 4.6(iii).

(ii) When α_1 of AX_2Y_2 in (i) is increased beyond $180°$, the molecule becomes disphenoidal. Hence, the equation in (i) can be used directly: $\cos\gamma_2 =$

$- \cos \left(\dfrac{\alpha_2}{2} \right) \cos \left(\dfrac{\beta_2}{2} \right)$. A more convenient equation can be obtained by intro-

ducing the acute bond angle of $\angle \text{Cl}-\text{Te}-\text{Cl}$, $\alpha_2' (= 360° - \alpha_2)$. Then the equa-

tion becomes: $\cos \gamma_2 = -\cos \left(\dfrac{360° - \alpha_2'}{2} \right) \cos \left(\dfrac{\beta_2}{2} \right) = \cos \left(\dfrac{\alpha_2'}{2} \right) \cos \left(\dfrac{\beta_2}{2} \right)$.

(iii) The XeF_4 moiety of square pyramidal XeOF_4 molecule is a special case of disphenoidal molecule with
$\alpha_2' = \beta_2$. Referring to the figure on the right, $2\beta_3 + \beta_2 = 360°$. Putting $\gamma_2 = \alpha_3$ and $\alpha_2' = \beta_2 = 360° - 2\beta_3$ into the equation in (ii), we have: $\cos \alpha_3 = \cos \left(\dfrac{360° - 2\beta_3}{2} \right) \times \cos \left(\dfrac{360° - 2\beta_3}{2} \right) = \cos^2 \beta_3$.

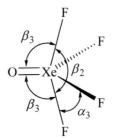

(iv) (a) For an AX_3Y molecule of C_{3v} symmetry we can choose the coordinate
system shown in the figures below. We have:

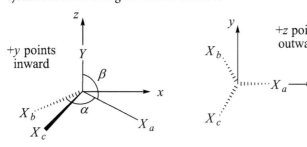

$$\overrightarrow{AX_a} = (AX \sin \beta,\ 0,\ AX \cos \beta),$$

$$\overrightarrow{AX_b} = \left(-\dfrac{1}{2} AX \sin \beta,\ \dfrac{\sqrt{3}}{2} AX \sin \beta,\ AX \cos \beta \right).$$

The dot product of vectors $\overrightarrow{AX_a}$ and $\overrightarrow{AX_b}$ is given by:

$$\overrightarrow{AX_a} \cdot \overrightarrow{AX_b} = (AX)^2 \cos \alpha = (AX)^2 \left(-\dfrac{1}{2} \sin^2 \beta + 0 + \cos^2 \beta \right).$$

Hence, $1 + 2 \cos \alpha = 3 \cos^2 \beta$.

In H_3CCl, given $\angle \text{H}-\text{C}-\text{Cl} = 108.0°$, $\angle \text{H}-\text{C}-\text{H}$ is $110.9°$.

(b) Using the equation obtained in Problem 4.4 for α and β:

$$\cos \alpha = -(n_X n_X)^{-1/2} = -n_X^{-1},$$

$$\cos \beta = -(n_X n_Y)^{-1/2}.$$

Substituting these results into the equation obtained in (a),
$1 - 2n_X^{-1} = 3(n_X n_Y)^{-1}$, leading to $n_X n_Y = 2n_Y + 3$. The result is actually identical to that obtained in Problem 4.6(ii) with $n_X = n_b$ and $n_Y = n_\ell$.

A6.2 (i) For $Ru(C_5H_5)_2$:

$\angle C^3-A-B-C^8 = 0°$, $\angle C^3-A-B-C^7 = 360°/5 = 72°$,

$\angle C^3-A-B-C^9 = -360°/5 = -72°$,

$\angle C^3-A-B-C^6 = (360°/5) \times 2 = 144°$, $\angle C^6-C^1-A-C^3 = -90°$,

$\angle C^9-C^6-B-C^1 = 90°$, $\angle B-A-C^2-C^7 = 0°$.

(ii) For $Fe(C_5H_5)_2$:

$\angle C^1-A-B-C^6 = 180°$, $\angle C^1-A-B-C^9 = -360°/10 = -36°$,

$\angle C^4-A-B-C^3 = 360°/5 = 72°$,

$\angle C^1-A-B-C^{10} = -(36° + 72°) = -108°$, $\angle B-A-C^2-C^3 = 90°$.

A6.3 (i) $\angle H^3-C-*-Cl^2 = -90°$, $\angle H^3-C-*-H^4 = 180$.

(ii) $\angle H^3-C^1-C^2-H^7 = 60°$, $\angle H^3-C^1-C^2-H^4 = 120°$,

$\angle H^4-C^1-C^2-H^8 = 180°$, $\angle H^4-C^1-C^2-H^7 = -60°$.

(iii) $\angle 1-2-5-6 = -60°$, $\angle 1-2-3-6 = 90°$, $\angle 1-4-6-5 = -90°$.

(iv) $\angle F^1-P-F^3-F^2 = 180°$, $\angle F^3-P-F^1-F^5 = 120°$, $\angle F^3-P-F^5-F^4 = 180°$,

$\angle F^4-F^5-F^3-P = 0°$, $\angle F^4-P-F^3-F^1 = -90°$.

A6.4 (i) Consider triangles $A'CD$ and ACD in the figure below. Let C' be the midpoint of CD. We have $\angle A'-C'-C = \angle A-C'-C = \angle A-A'-C' = 90°$. For triangle $A'C'C$, $CC'/A'C' = \tan(\angle C-A'-C') = \tan(\angle C-A'-D/2) = \tan[(360°/n)/2] = \tan(180°/n)$, or $A'C' = CC'/\tan(180°/n)$. For triangle $AC'C$, $AC'/CC' = \tan(60°) = 3^{1/2}$, or $AC' = 3^{1/2}CC'$. For triangle $AA'C'$, $\cos(\angle A'-C'-A) = A'C'/AC' = [3^{1/2}\tan(180°/n)]^{-1}$. The dihedral angle between triangle ACD and polygon $BCDEF \ldots = \angle A'C'A = \cos^{-1}\{[3^{1/2}\tan(180°/n)]^{-1}\}$.

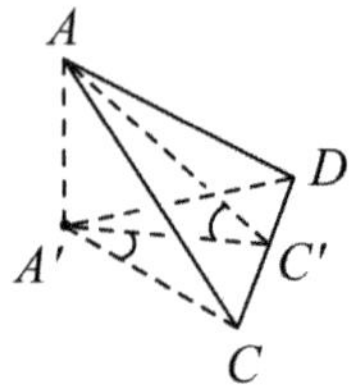

(ii) Note that pyramids $GHIJ$, $KLMNO$, and $PQRSTU$ are special cases of the pyramid described in (i) with $n = 3, 4$, and 5, respectively.

$\angle G-H-I-J =$ dihedral angle between triangle HIJ and the base (triangle GHI) $= \cos^{-1}\{[3^{1/2}\tan(180°/3)]^{-1}\} = \cos^{-1}(1/3) = 70.53°$.

$\angle L - M - N - O$ = dihedral angle between triangle MNO and the base (square $KLMN$) = $\cos^{-1}\{[3^{1/2}\tan(180°/4)]^{-1}\} = \cos^{-1}(3^{-1/2}) = 54.74°$.

$\angle R - S - T - U$ = dihedral angle between triangle STU and the base (pentagon $PQRST$) = $\cos^{-1}\{[3^{1/2}\tan(180°/5)]^{-1}\} = 37.38°$.

By inspection, it is clear that the signs of these dihedral angles are all positive.

(iii) Referring to the labeling in the figure on the right, let V' denote the center of the square $XVWY'$, which is also the mid-point of VY'. We have $VV' = VY'/2 = (2^{1/2}VW)/2 = 2^{-1/2}VW = 2^{-1/2}VY$. The dihedral angle between triangles VWX and WXY is simply $\tan(\angle Y - V' - V) = VY/VV' = 2^{1/2}$. By definition, this is simply $\angle V - W - X - Y$, or $\angle V - W - X - Y = \tan^{-1}(2^{1/2}) = 54.74°$. Again, by inspection, the sign of $\angle V - W - X - Y$ is positive. (It is interesting to note that the answer is the same as that of $\angle L - M - N - O$ in Part (ii).)

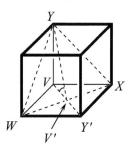

A6.5 Place the cyclohexane ring in a Cartesian frame as shown below. (The xy-plane is between and equidistant from the triangles ACE and BDF.)

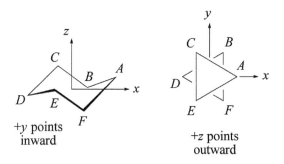

+y points inward

+z points outward

The coordinates of the carbon atoms A to F are $(x, 0, z)$, $\left(\frac{1}{2}x, \frac{\sqrt{3}}{2}x, -z\right)$, $\left(-\frac{1}{2}x, \frac{\sqrt{3}}{2}x, z\right)$, $(-x, 0, -z)$, $\left(-\frac{1}{2}x, -\frac{\sqrt{3}}{2}x, z\right)$, and $\left(\frac{1}{2}x, -\frac{\sqrt{3}}{2}x, -z\right)$, respectively. The C–C bond length may be taken as unity without loss of generality. Considering the distance between B and C, we have $x^2 + 4z^2 = 1$, i.e., $2z = \sqrt{1 - x^2}$.

The three vectors, $\vec{AB}$, $\vec{BC}$, and $\vec{CD}$, can be expressed in terms of x as:

$$\vec{AB} = \left(-\frac{1}{2}x, \frac{\sqrt{3}}{2}x, -\sqrt{1-x^2}\right).$$

$$\vec{BC} = \left(-x, 0, \sqrt{1-x^2}\right).$$

$$\vec{CD} = \left(-\frac{1}{2}x, -\frac{\sqrt{3}}{2}x, -\sqrt{1-x^2}\right).$$

$$\vec{AB} \times \vec{BC} = \begin{vmatrix} \hat{i} & \hat{j} & \hat{k} \\ -\frac{1}{2}x & \frac{\sqrt{3}}{2}x & -\sqrt{1-x^2} \\ -x & 0 & \sqrt{1-x^2} \end{vmatrix} = \left(\frac{\sqrt{3}}{2}x\sqrt{1-x^2}, \frac{3}{2}x\sqrt{1-x^2}, \frac{\sqrt{3}}{2}x^2\right).$$

$$\vec{BC} \times \vec{CD} = \begin{vmatrix} \hat{i} & \hat{j} & \hat{k} \\ -x & 0 & \sqrt{1-x^2} \\ -\frac{1}{2}x & -\frac{\sqrt{3}}{2}x & -\sqrt{1-x^2} \end{vmatrix} = \left(\frac{\sqrt{3}}{2}x\sqrt{1-x^2}, -\frac{3}{2}x\sqrt{1-x^2}, \frac{\sqrt{3}}{2}x^2\right).$$

$$(\vec{AB} \times \vec{BC}) \cdot (\vec{BC} \times \vec{CD}) = \left[\frac{3}{4}x^2(1-x^2) - \frac{9}{4}x^2(1-x^2) + \frac{3}{4}x^4\right]$$

$$= \frac{3}{4}x^2(3x^2 - 2). \tag{6.1}$$

$$\cos\theta = \cos\theta_B = \cos\theta_C = \vec{BA} \cdot \vec{BC} = 1 - \frac{3}{2}x^2 \Rightarrow x^2 = \frac{2}{3}(1 - \cos\theta).$$

Substituting into Eq. (6.1),

$$(\vec{AB} \times \vec{BC}) \cdot (\vec{BC} \times \vec{CD}) = \frac{3}{4}\left[\frac{2}{3}(1 - \cos\theta)\right](-2\cos\theta) = \cos\theta(\cos\theta - 1).$$

Substituting into the given formula, $\cos\omega = \dfrac{\cos\theta(\cos\theta - 1)}{\sin^2\theta} = -\dfrac{\cos\theta}{1 + \cos\theta}$.

By inspection, the $ABCD$ sequence in the present case forms a left-hand screw. Hence, the negative of the principal arcosine value should be taken for ω, i.e.,

$$\omega = -\cos^{-1}\left(-\frac{\cos\theta}{1 + \cos\theta}\right).$$

In general, for even-membered rings of symmetry $D_{(n/2)d}$, L. Pauling has shown that

$$\cos \omega = \frac{1 - \cos \theta - 2 \cos \left(\frac{360°}{n} \right)}{1 + \cos \theta}.$$

REFERENCE: L. Pauling, On the stability of the S_8 molecule and the structure of fibrous sulfur. *Proc. Nat. Acad. Sci. U.S.A.* **35**, 495–9 (1949).

A6.6 (i) Non-planar cyclobutane has D_{2d} symmetry and is a special case of the general formula given in A6.5 with $n = 4$. Hence, $\cos \omega = \dfrac{1 - \cos \theta}{1 + \cos \theta}$.

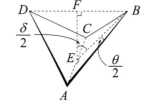

(ii) Refer to the figure on the previous page: A, B, C, and D are the positions of carbon atoms, E and F are the mid-points of AC and BD, respectively.

Since $BE = DE$ and $AB = BC$, both BDE and ABC are isosceles triangles. So both $\angle AEB$ and $\angle BFE$ are right angles.

For triangle AEB: $\dfrac{AE}{BE} = \tan \left(\dfrac{\theta}{2} \right)$.

For triangle BFE: $\dfrac{BF}{BE} = \sin \left(\dfrac{\delta}{2} \right)$.

Since $BD = AC$, $AE = BF$. Hence, we have $\tan \left(\dfrac{\theta}{2} \right) = \sin \left(\dfrac{\delta}{2} \right)$.

A6.7 The coordinates of A, B, C, and D are $(r \cos \phi, -r \sin \phi, -z)$, $(r,0,0)$, $(r \cos \phi, r \sin \phi, z)$, $(r \cos 2\phi, r \sin 2\phi, 2z)$, respectively.

(i) $\ell = AB = \sqrt{(r \cos \phi - r)^2 + (-r \sin \phi - 0)^2 + (-z - 0)^2}$

$= \sqrt{2r^2(1 - \cos \phi) + z^2}$.

(ii) $\overrightarrow{AB} = (r - r \cos \phi, 0 - (-r \sin \phi), 0 - (-z)) = (r(1 - \cos \phi), r \sin \phi, z)$.

$\overrightarrow{BC} = (r \cos \phi - r, r \sin \phi - 0, z - 0) = (r(\cos \phi - 1), r \sin \phi, z)$.

$$\ell^2 \cos\theta = \vec{BA} \cdot \vec{BC} = -\vec{AB} \cdot \vec{BC} = -[r^2(1 - \cos\phi)(\cos\phi - 1) + r^2 \sin^2\phi + z^2]$$

$$= -[2r^2 \cos\phi(1 - \cos\phi) + z^2].$$

$$\cos\theta = \frac{-(2r^2 \cos\phi(1 - \cos\phi) + z^2)}{2r^2(1 - \cos\phi) + z^2} = -\frac{2\cos\phi(1 - \cos\phi) + \left(\frac{z}{r}\right)^2}{2(1 - \cos\phi) + \left(\frac{z}{r}\right)^2}.$$

(iii) $\vec{CD} = (r\cos 2\phi - r\cos\phi, r\sin 2\phi - r\sin\phi, 2z - z)$

$$= (r(\cos 2\phi - \cos\phi), r(\sin 2\phi - \sin\phi), z)$$

$$= (r(2\cos\phi + 1)(\cos\phi - 1), r\sin\phi(2\cos\phi - 1), z).$$

$$\vec{AB} \times \vec{BC} = \begin{vmatrix} \hat{i} & \hat{j} & \hat{k} \\ r(1 - \cos\phi) & r\sin\phi & z \\ r(\cos\phi - 1) & r\sin\phi & z \end{vmatrix}$$

$$= (0, 2rz(\cos\phi - 1), 2r^2\sin\phi(1 - \cos\phi)).$$

$$\vec{BC} \times \vec{CD} = \begin{vmatrix} \hat{i} & \hat{j} & \hat{k} \\ r(\cos\phi - 1) & r\sin\phi & z \\ r(2\cos\phi + 1)(\cos\phi - 1) & r\sin\phi(2\cos\phi - 1) & z \end{vmatrix}$$

$$= (2rz\sin\phi(1 - \cos\phi), 2rz\cos\phi(\cos\phi - 1), 2r^2\sin\phi(1 - \cos\phi)).$$

$$(\vec{AB} \times \vec{BC}) \cdot (\vec{BC} \times \vec{CD}) = 0 + 4r^2z^2\cos\phi(\cos\phi - 1)^2 + 4r^4\sin^2\phi(\cos\phi - 1)^2$$

$$= 4r^4(1 - \cos\phi)^2\left[\left(\frac{z}{r}\right)^2\cos\phi + \sin^2\phi\right].$$

From (i), $\ell^2 = 2r^2(1 - \cos\phi) + z^2$.

From (ii), $\ell^2 \cos\theta = -2r^2\cos\phi(1 - \cos\phi) - z^2$.

Hence, $\ell^4 \sin^2\theta = \ell^4(1 - \cos^2\phi) = \ell^4(1 - \cos\phi)(1 + \cos\phi)$

$$= (\ell^2 - \ell^2\cos\phi)(\ell^2 + \ell^2\cos\phi)$$

$$= 2[r^2(1 - \cos^2\phi) + z^2] \cdot 2r^2(1 - \cos\phi)^2$$

$$= 4r^4(1 - \cos\phi)^2\left[\sin^2\phi + \left(\frac{z}{r}\right)^2\right].$$

$$\cos\omega = \frac{(\vec{AB} \times \vec{BC}) \cdot (\vec{BC} \times \vec{CD})}{AB(BC)^2CD \sin\theta_B \sin\theta_C} = \frac{(\vec{AB} \times \vec{BC}) \cdot (\vec{BC} \times \vec{CD})}{\ell^4 \sin^2\theta}$$

$$= \frac{\sin^2\phi + \left(\frac{z}{r}\right)^2\cos\phi}{\sin^2\phi + \left(\frac{z}{r}\right)^2}.$$

For fibrous sulfur, $\ell = 2.07$ Å, $\theta = 106°$, and $\omega = 85°$ (not $95°$ as reported).

A6.8 (i)

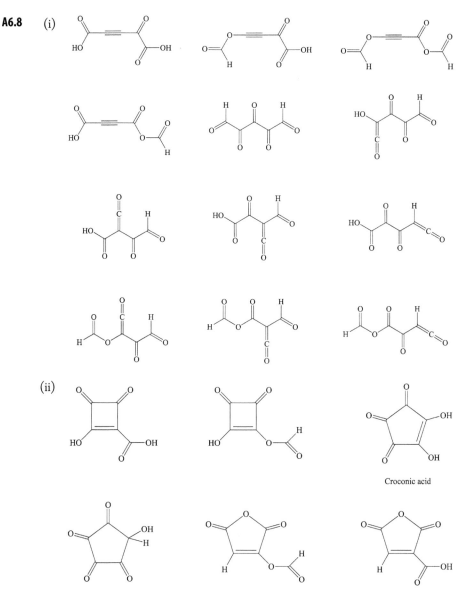

(ii)

Croconic acid

The croconate dianion $C_2O_5^{2-}$ has a D_{5h} structure in which the two negative charges are equally distributed on all five exocyclic oxygen atoms:

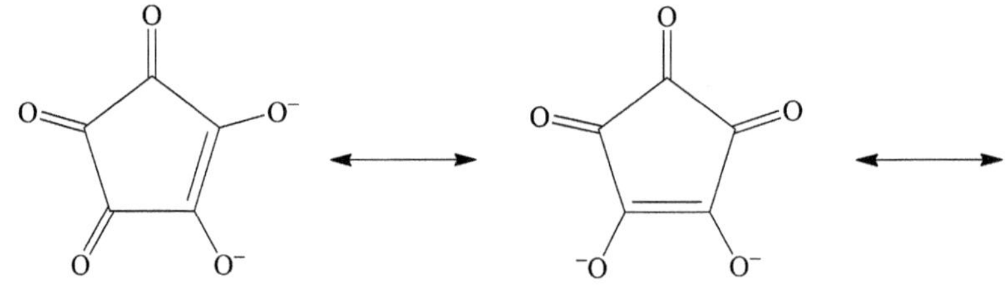

etc.

or

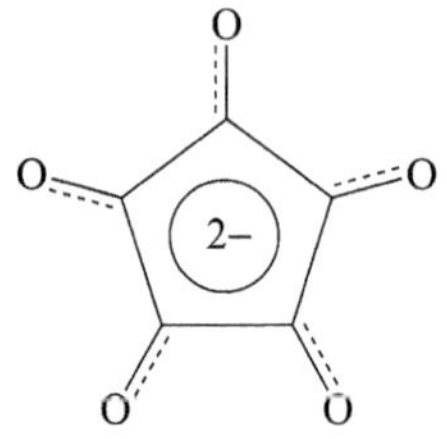

Croconate

which is a member of the non-benzenoid aromatic oxocarbon dianions $C_nO_n^{2-}$ ($n = 3$ for deltate; $n = 4$ for squarate; $n = 5$ for croconate; and $n = 6$ for rhodizonate). In the crystalline state, the charge-localized (enolate C_{2v}) and charge-delocalized (non-benzenoid aromatic D_{nh}) valence tautomeric forms of the croconate and rhodizonate dianions can be stabilized by donor hydrogen bonding with designed urea derivatives.

A6.9

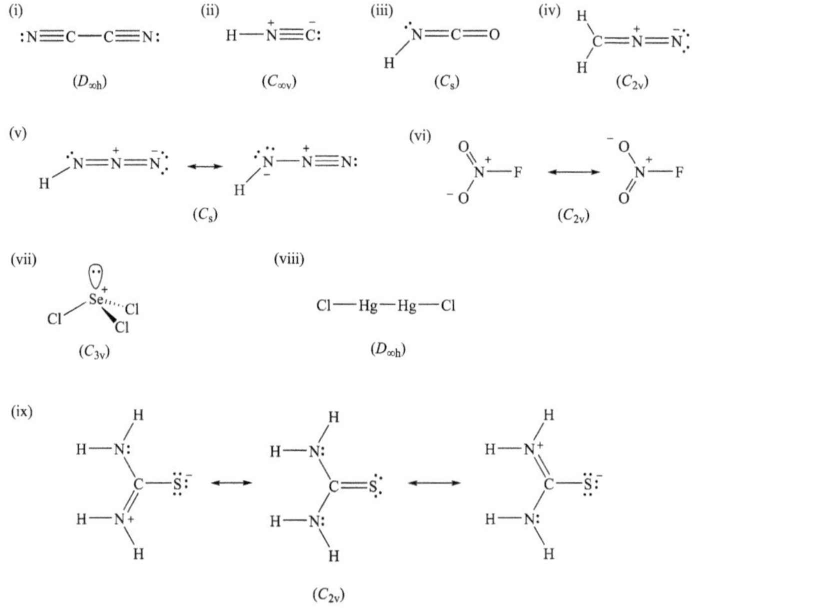

(i)

:N≡C—C≡N:

($D_{\infty h}$)

(ii)

H—N≡C:

($C_{\infty v}$)

(iii)

N=C=O

(C_s)

(iv)

C=N=N:

(C_{2v})

(v)

N=N=N: ⟷ N—N≡N:

(C_s)

(vi)

(C_{2v})

(vii)

(C_{3v})

(viii)

Cl—Hg—Hg—Cl

($D_{\infty h}$)

(ix)

(C_{2v})

(x)

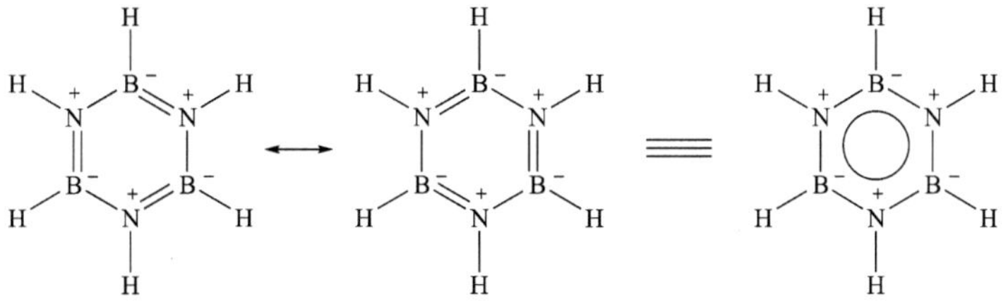

(D_{3h})

A6.10

(i)

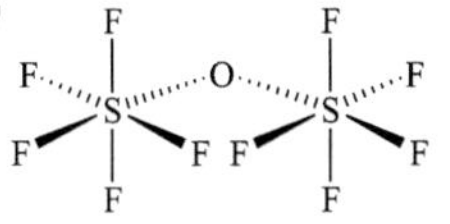

$(C_{2v};$ octahedral arrangement around each S atom)

(ii)

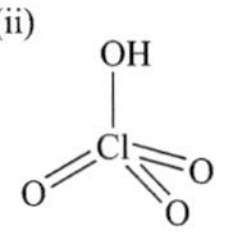

$(C_{3v};$ idealized; may be C_s or even $C_1)$

(iii)

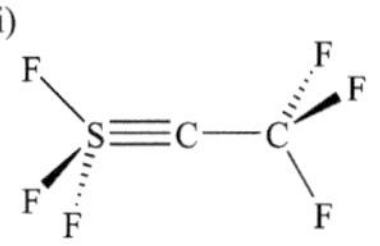

(C_{3v})

(iv)

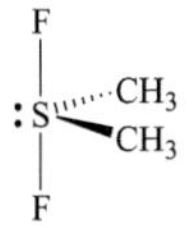

$(C_{2v};$ disphenoidal)

(v)

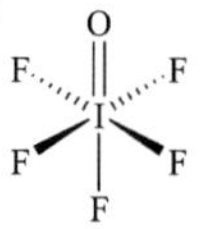

(C_{4v})

(vi)

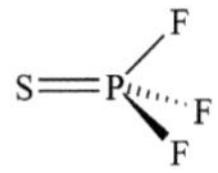

(C_{3v})

(vii)

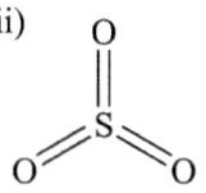

(D_{3h})

(viii)

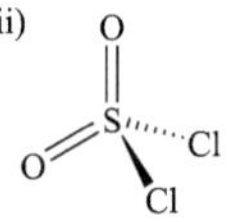

(C_{2v})

(ix)

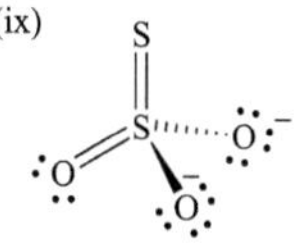

, etc.

(C_{3v})

(x)

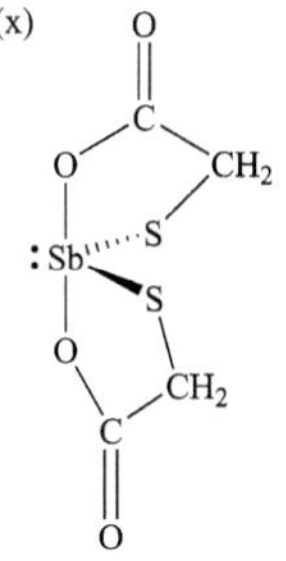

(C_2)

(xi)

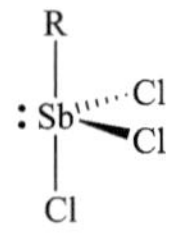

$R = C_6H_6NH_2$
$(C_s;$ idealized)

(xii)

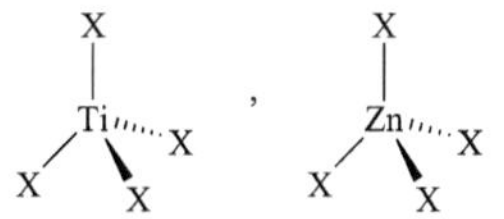

(T_d)

A6.11 Once we know the connectivity of the atoms in the molecule, writing out the electron-dot structure becomes straightforward:

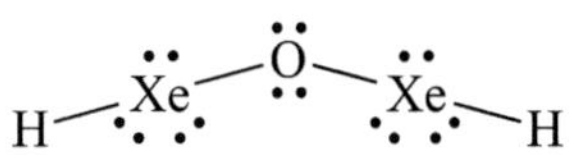

It is clear the molecule has C_{2v} symmetry.

A6.12 When there are many large ligands surrounding a central atom, ligand–ligand repulsions, usually ignored in VSEPR treatments, have to be taken into account. When all the ligands are touching each other, the lone pair is forced to go inside the valence shell, occupying a spherical s-type orbital. Take $TeBr_6^{2-}$ as an example. The Br–Br separation is 3.81 Å, which is slightly smaller than the van der Waals distance of 3.9 Å, indicating that the Br atoms are touching each other with no room left for the lone pair. Also, the forced-in lone pair would decrease the electronegativity of Te and lead to a longer bond length. The observed bond length of 2.75 Å in $TeBr_6^{2-}$, which is longer than the value of 2.51 Å predicted from the sum of covalent radii, is consistent with this picture.

A6.13 (i) P_4 has the structure of tetrahedron which is consistent with the bonding picture of P atom forming three bonds and bearing a lone pair.

(ii) S_8 has the crown structure which is consistent with each S atom forming two bonds and bearing two lone pairs.

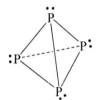

A6.14 $\angle F-C-F(104.8°) > \angle F-Si-F(100.8°) > \angle F-Ge-F(97.2°)$.
Bond pair–bond pair repulsion decreases as the central atom becomes larger, if the ligand is more electronegative than the central atom.
$\angle F-Ge-F(97.2°) < \angle Cl-Ge-Cl(100.4°) < \angle Br-Ge-Br(101.4°)$.
Bond pair–bond pair repulsion increases as the electronegativity of the ligand decreases.

A6.15 (i)   Most chlorides are soluble in water because they are ionic. Mercurous chloride, Hg_2Cl_2, is not soluble because it is a covalent compound. It has a linear structure: Cl–Hg–Hg–Cl, in accord with the VSEPR theory.

(ii)   The silver complex is $Ag(NH_3)_2^+$, with the linear structure $H_3N-Ag^+-NH_3$.

A6.16 :$\overset{\cdot\cdot}{\underset{\cdot\cdot}{Cl}}{}^+$:H^- :$\overset{\cdot\cdot}{\underset{\cdot\cdot}{Cl}}$—H   :$\overset{\cdot\cdot}{\underset{\cdot\cdot}{Cl}}$: H^+

Importance →

A6.17 (i)

(ii)   Hybridization schemes: sp^2 at N^a, sp at N^b and N^c.

(iii)   Bond orders: $N^a-N^b = 1\frac{1}{3}$; $N^b-N^c = 2\frac{2}{3}$.

$\angle N^a-N^b-N^c$ is about 14° away from theory, and $\angle N^b-N^a-N^b$ is about 12° away from theory. Finally, AsF_6^- has an octahedral structure.

A6.18 (i)

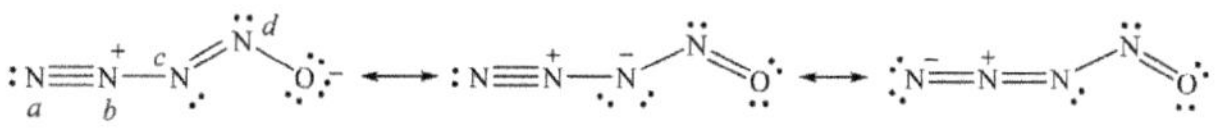

(ii) Hybridization schemes: sp at N^b, sp^2 at N^c and N^d.

(iii) Bond orders: $N^a-N^b = 2\frac{2}{3}$, $N^b-N^c = 1\frac{1}{3}$, $N^c-N^d = 1\frac{1}{3}$, and $N^d-O = 1\frac{2}{3}$.

(iv) Bond angles: $\angle N^a-N^b-N^c \approx 180°$, $\angle N^b-N^c-N^d \approx 120°$, and $\angle N^c-N^d-O \approx 120°$. The bond lengths and bond angle of this molecule are shown below for reference.

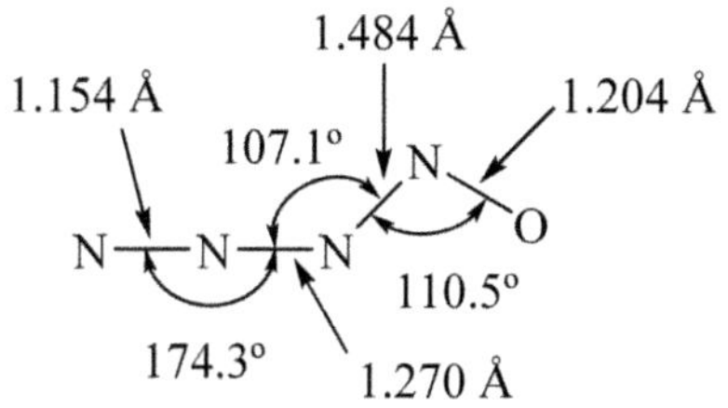

A6.19 The bond length variation is consistent with the Lewis structural formula on the right. The central S atom is sp^3 hybridized with two lone pairs; the N and penultimate S atoms are each sp^2 hybridized with one lone pair. The pen-

ultimate S atoms make use of their p and d orbitals in π bonding. Consideration of electron pair repulsions alone leads to the following expectation: $\alpha < 109.5° < \beta < \gamma \approx 120°$. The experimental (X-ray diffraction) values are: $\alpha = 97.3°$, $\beta = 124.3°$, and $\gamma = 118.8°$.

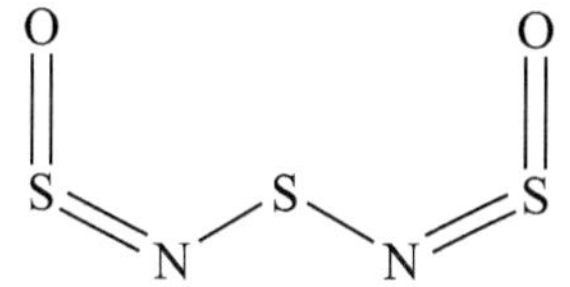

A6.20 The structure is shown on the right. The 'bridging' Cl–O bonds are single bonds and the 'terminal' Cl–O bonds are double bonds.

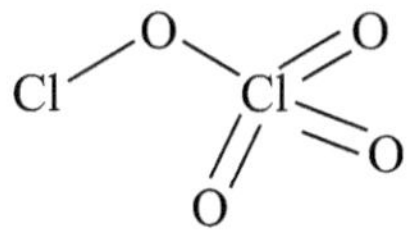

A6.21 As a first approximation, the Cl atoms and the central O atom may be considered to be sp^3 hybridized. The two lone pairs of the central O atom might be expected to make the Cl–O–Cl angle less than the ideal tetrahedral angle of $\approx 109.5°$. Actually in this case the angle is slightly opened up because of steric repulsion of the two bulky ClO_3 groups.

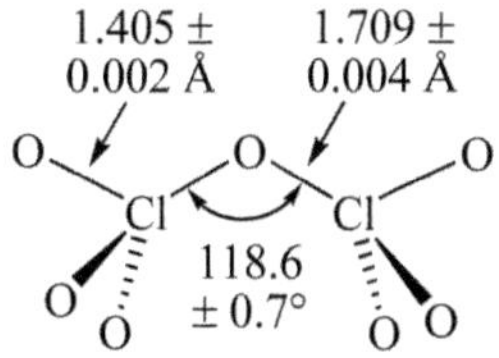

The large difference in the Cl–O bond distances indicates multiple bonding between Cl and terminal O atoms. This is possible since the Cl atoms can make use of their 3d orbitals for π-overlap with the p orbitals of terminal O atoms. The bridging Cl–O bonds may be considered as single bonds, while the remaining Cl–O bonds are double bonds. As a reference: the Cl–O bond length in Cl_2O is 1.70 Å, while that in $FClO_3$ is 1.404 Å. So the bridging Cl–O bonds are single bonds and the terminal Cl–O bonds are double bonds.

A6.22 There are several equivalent resonance structures for $S_4O_6^{2-}$, one of which is shown on the right. The S–O bond order is $1\frac{2}{3}$.

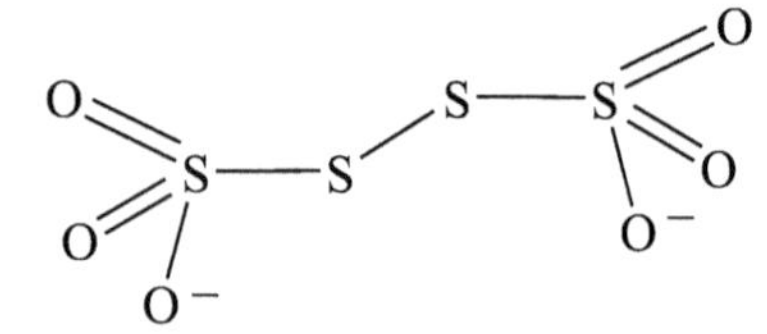

A6.23 The two hydrogens of the methylene groups should be co-planar with the two axial Fs of the SF_4 moiety. Structure (I) may be represented by linking a tetrahedron to an octahedron through two bent bonds.

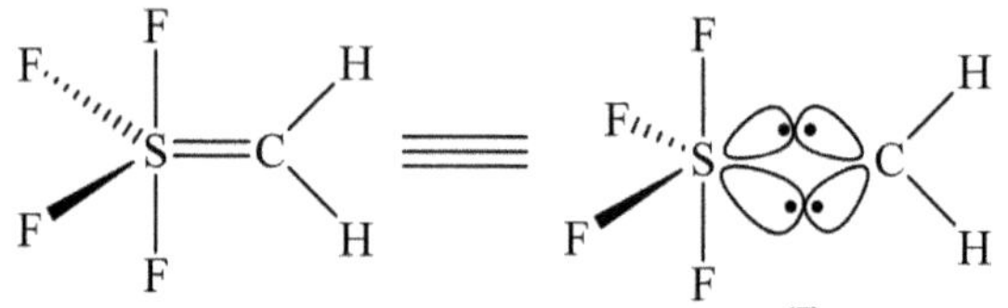

A6.24 (i) The geometries of PCl_4^+ and PCl_6^- are tetrahedral and octahedral, respectively.
(ii) To form the PBr_6^- ion, six large Br atoms need to be arranged around a relatively small P atom. This is not favorable sterically.

A6.25 Resonance structures such as the one shown are possible for $N(CF_3)_3$ but not for $N(CH_3)_3$. In these structures, there is C=N double bond character and sp^2 hybridization character at C and N atoms, resulting in larger (close to 120°) C–N–C angles.

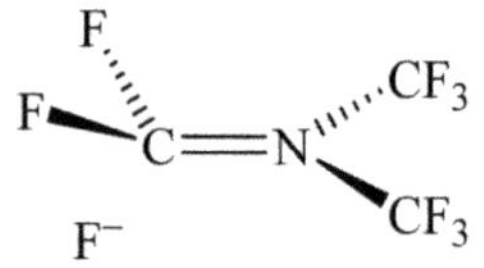

A6.26 The F_3CNO molecule is expected to possess a structure with C_s symmetry in which the CF_3 and NO groups are eclipsed since the nitrogen lone pair tends to avoid the fluorine atoms.

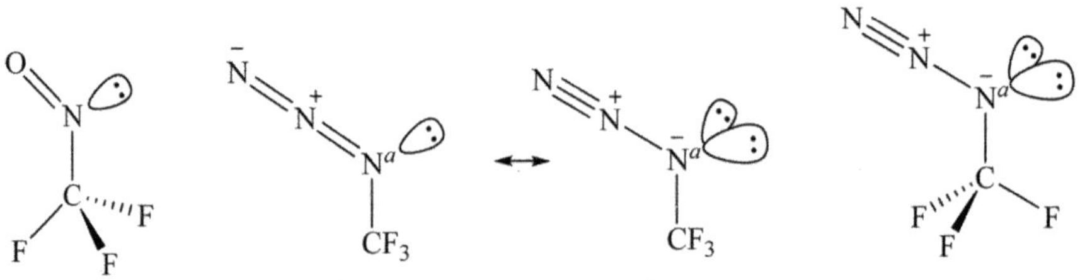

The F_3CN_3 molecule is expected to have a planar CN_3 backbone and an approximately linear N_3 group. It has two resonance structures (structures I and II) with different numbers of lone pairs localized on N^a (regardless of the CF_3 conformation relative to the N_3 group). Since the C atom is "electron deficient" (caused by the electron-withdrawing inductive effect of the F atoms), structure II is favored as the formal negative charge on N^a compensates for the electron deficiency. Hence, the CF_3 group should be staggered (structure III) relative to the N_3 group to reduce "repulsion" between fluorine atoms and the lone pairs on N^a.

A6.27 MeNCS and H_3SiNCS have the following structures:

$$\underset{Me}{\nwarrow}\overset{-}{N}=C=S \qquad\qquad H_3\overset{-}{Si}=\overset{+}{N}=C=S$$

$$(C_s) \qquad\qquad\qquad\qquad (C_{3v})$$

In the silyl compound, the silicon atom is pentavalent and forms a double bond with the nitrogen atom. Here N is sp hybridized and the π bond between N and Si is of the $p_\pi - d_\pi$ variety.

A6.28 For UO_2^{2+}, $:\!\ddot{O}\!=\!\overset{2+}{U}\!=\!\ddot{O}\!:$ with two positive charges residing on the central U atom.

For CUO, $:\!\overset{-}{C}\!\equiv\!\overset{+}{U}\!\equiv\!\ddot{O}\!: \longleftrightarrow C\!\equiv\!U\!=\!\ddot{O}\!:$, since both carbon and oxygen must satisfy

$$(a) \qquad\qquad\qquad (b)$$

the octet rule.

Resonance structure (b) shows an unusual $C \equiv U$ quadruple bond, which contradicts the generally accepted belief that the 2s and 2p orbitals of carbon can only form single, double, and triple bonds (one σ and two π) with other atoms. For theoretical justification for a quadruple bond in CUO, see the cited reference. Briefly, in addition to forming two C-U π-bonds with the C $2p_x$ and $2p_y$ orbitals, the U 6d orbitals interact more strongly with C 2s than C $2p_z$, whereas the U 5f orbitals interact with C $2p_z$ nearly as well as C 2s. This difference accounts for the formation of two weak σ bonds between C and U in molecular CUO.

A6.29 The situation for (i) is well known and may be explained in terms of either sp^3d hybridization or the VSEPR theory.

For $[Ni(CN)_5]^{3-}$ in (ii), the highest d orbital, $d_{x^2-y^2}$, is empty, thus allowing the basal ligands to move in closer to the metal. For $[CuCl_5]^{3-}$, the highest d orbital, d_{z^2}, is only half filled, leading to shorter axial bonds.

A6.30 The lone pair electrons below the basal plane in BrF_5 repel the in-plane ligands upward. On the other hand, no such effect is found in $[Ni(CN)_5]^{3-}$. In addition, ligand repulsions will force the basal ligands downward.

A6.31

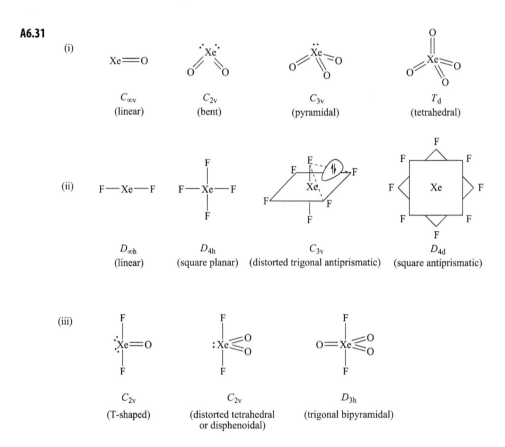

(i)

$C_{\infty v}$ (linear) C_{2v} (bent) C_{3v} (pyramidal) T_d (tetrahedral)

(ii)

$D_{\infty h}$ (linear) D_{4h} (square planar) C_{3v} (distorted trigonal antiprismatic) D_{4d} (square antiprismatic)

(iii)

C_{2v} (T-shaped) C_{2v} (distorted tetrahedral or disphenoidal) D_{3h} (trigonal bipyramidal)

A6.32 From the information given, it may be noted that non-linearity is favored by a heavy metal (in Group 2) atom and/or an electronegative ligand.

When $HgCl_2$ and $BaCl_2$ are compared, it is seen that both Hg and Ba have a $6s^2$ ground configuration. However, when one of the 6s electrons is promoted to yield the lowest excited configuration, $6s^1 6p^1$ and $6s^1 5d^1$ are obtained for Hg and Ba, respectively. For HgX_2, sp hybridization yields a linear geometry; for BaX_2, $sd_{x^2-y^2}$ hybridization leads to a bent shape with bond angles around $90°$. Significant d orbital participation thus results in a non-linear molecule.

Since F is the most electronegative element, it would draw electronic charge away from the metal, leaving a positive charge on the latter. This charge causes contraction of the d orbitals, making them more efficient in metal-ligand overlap. Hence, electronegative ligands favor non-linearity.

It is instructive to compare the arguments given here with those of the following reference: R. J. Gillespie, The electron-pair repulsion model for molecular geometry. *J. Chem. Educ.* **47**, 18–23 (1970).

A6.33 Across each of the series, as far as electronic configurations are concerned, the only difference is in the population of the $1b_1$ orbital:

$$SiH_2, \ CH_2: \ \ldots (2a_1)^2(1b_1)^0; \ PH_2, \ NH_2: \ \ldots (2a_1)^2(1b_1)^1;$$
$$H_2S, \ H_2O: \ \ldots (2a_1)^2(1b_1)^2.$$

Since the energy of the $1b_1$ orbital is independent of the molecular geometry (or bond angle), any change in its population does not significantly alter the bond angle.

A6.34 (i) Minimum value of $E(R)$ can be found by setting $E'(R) = 0$. It is easy to see from the formula itself that $E(R)$ will have a minimum value of 0 when $R = R_e$. As $R \to \infty$, $E(R)$ approaches D. Thus $D_e = D$.

(ii)

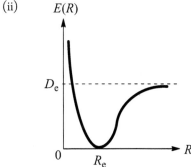

(iii) The expansion for the exponential:

$$e^x = \sum_{n=0}^{\infty} \frac{x^n}{n!} = 1 + x + \frac{x^2}{2} + \frac{x^3}{6} + \ldots$$

Expanding the Morse function up to terms quadratic in $(R - R_e)$ gives:

$$E(R) = 0 + D\beta^2(R - R_e)^2 = D\beta^2(R - R_e)^2$$

This has the form of a harmonic oscillator potential $V(x) = \frac{1}{2}kx^2$ with $x = R - R_e$ and $k = 2D\beta^2$.

A6.35

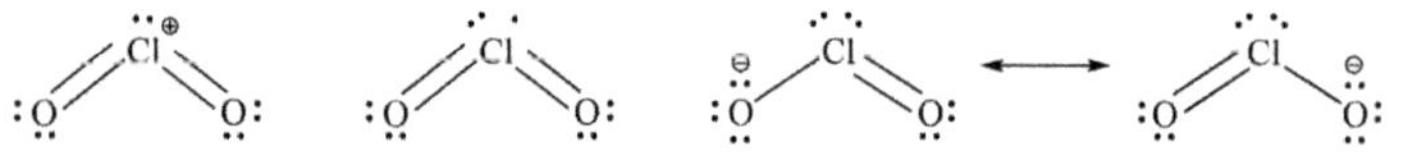

∠OClO	122°	118°		110°
O–Cl bond length	1.31 Å	1.47 Å		1.56 Å
No. of lone pairs on Cl	1	$1\frac{1}{2}$		2

A6.36

Species	C–N bond order	C–N bond length (Å)
$CH_3C\equiv N$	3	1.158
	$2\frac{1}{2}$	1.171
$CH_3CH=NCH_3$	2	1.280
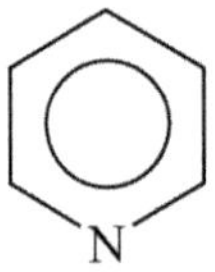	$1\frac{1}{2}$	1.337
CH_3-NH_2	1	1.469

A6.37

Species	N–O bond order	N–O bond length (Å)
$NO^+(:N\equiv\overset{+}{O}:)$	3	1.062
NO_2^+ $(\ddot{O}=\overset{+}{N}=\ddot{O})$	2	1.15
NO_2^- 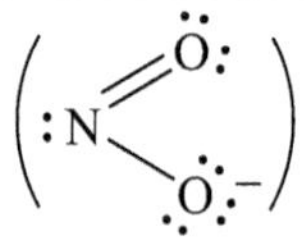 , etc.	$1\frac{1}{2}$	1.236
NO_3^- 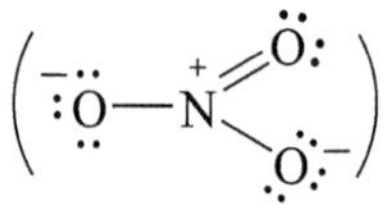 , etc.	$1\frac{1}{3}$	1.25
H_2N-OH	1	1.47

A6.38

Species	C–O bond order	C–O bond length (Å)
CO 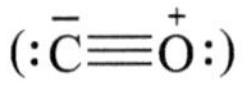 ($:\bar{C}{\equiv}\overset{+}{O}:$)	3	1.128
CO_2 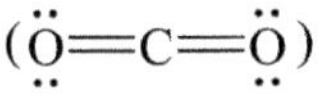 ($\ddot{O}{=}C{=}\ddot{O}$)	2	1.163
$CH_3CO_2^-$ 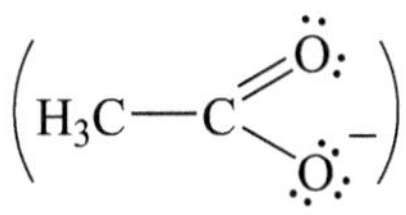 $\left(H_3C{-}C\diagup^{\ddot{O}:}_{\diagdown\ddot{O}:^-} \right)$, etc.	$1\frac{1}{2}$	1.250
CO_3^{2-} 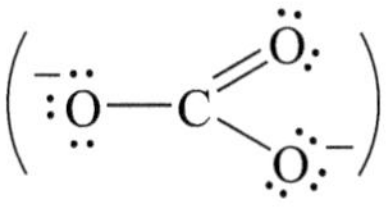 $\left(:\overset{-}{\ddot{O}}{-}C\diagup^{\ddot{O}:}_{\diagdown\ddot{O}:^-} \right)$	$1\frac{1}{3}$	1.290
$(CH_3)_2O$ 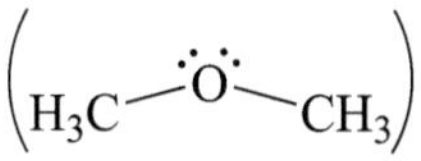 $\left(H_3C\diagdown^{:\ddot{O}:}\diagup CH_3 \right)$	1	1.416

A6.39

Species	S–O bond order	S–O bond length (Å)
Bridging S–O in $S_2O_7^{2-}$ ¶	1	1.645
SO_3^{2-} 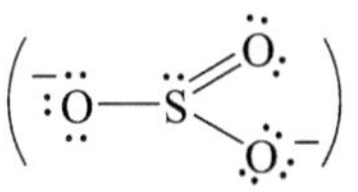 $\left(:\overset{-}{\ddot{O}}{-}\overset{}{S}\diagup^{\ddot{O}:}_{\diagdown\ddot{O}:^-} \right)$, etc.	$1\frac{1}{3}$	1.51
SO_4^{2-} 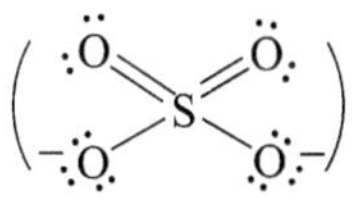 $\left(\overset{:\ddot{O}}{_-\!:\ddot{O}}{\diagdown}S{\diagup}\overset{\ddot{O}:}{\ddot{O}:^-} \right)$, etc.	$1\frac{1}{2}$	1.49
Terminal S–O in $S_2O_7^{2-}$ ¶	$1\frac{2}{3}$	1.437
S_2O 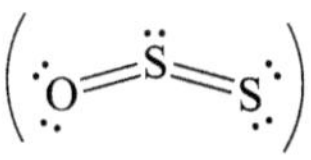 $\left(:\ddot{O}{=}\ddot{S}{=}\ddot{S}: \right)$	2	1.43

¶
$$\overset{-}{:}\overset{\ddot{O}}{}\diagdown S\diagup^{:\ddot{O}:}\diagdown \overset{\overset{-}{:\ddot{O}:}}{S}\diagup \text{, etc.}$$

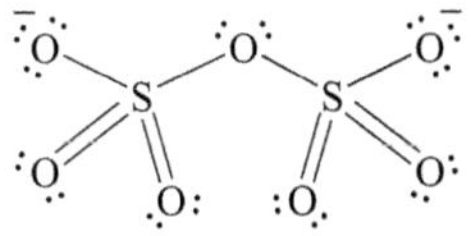

A6.40

Species	P–O bond order	P–O bond length (Å)
Terminal P=O in $P_4O_{10}^{\text{¶}}$	2	1.40
$H_2PO_2^-$ $\left(\begin{array}{c}H\\[-2pt]\diagdown\\[-6pt]H\diagup P \diagup\!\!\diagdown \begin{array}{c}\ddot{O}:\\ \ddot{O}:^-\end{array}\end{array}\right)$	$1\frac{1}{2}$	1.49
HPO_3^{2-} $\left(H-P\diagup\!\!\!\diagdown\begin{array}{c}\ddot{O}:\\ \ddot{O}:^-\\ \ddot{O}:^-\end{array}\right)$	$1\frac{1}{3}$	1.51
PO_4^{3-} $\left(:\!O\!=\!P\diagup\!\!\!\diagdown\begin{array}{c}\ddot{O}:^-\\ \ddot{O}:^-\\ \ddot{O}:^-\end{array}\right)$	$1\frac{1}{4}$	1.52
Bridging P–O in $P_4O_{10}^{\text{¶}}$	1	1.65

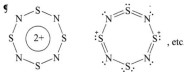

¶

A6.41

Species	S–N bond order	S–N bond length (Å)
$S_4(NEt)_2$ $\left(\begin{array}{c}S\!-\!S\!-\!NEt\\ EtN\!-\!S\!-\!S\end{array}\right)$	1	1.71
S_2N_2 $\left(\begin{array}{c}:\ddot{S}\!-\!\ddot{N}:^-\\ \mid\quad\mid_+\\ .\ddot{N}\!=\!\ddot{S}.\end{array}, \text{etc.}\right)$	$1\frac{1}{4}$	1.65
$S_4N_4^{2+\,\text{¶}}$	$1\frac{1}{2}$	1.55
NS_2^+ ($:\ddot{S}\!=\!\overset{+}{N}\!=\!\ddot{S}:$)	2	1.46
NSF_3 $\left(:N\!\equiv\!S\diagup\!\!\!\diagdown\begin{array}{c}F\\ F\\ F\end{array}\right)$	3	1.42

Note: Interestingly, both S_2N_2 and $S_4N_4^{2+}$, with two and 10 π electrons, respectively, are aromatic.

A6.42 (i) Structures **C** and **D**, having two imaginary vibrational frequencies, do not have any chemical relevance and immediate relationship with **A** and **B**. Meanwhile, **A**, with no imaginary frequencies, represents the equilibrium structure of the anion. As has been suggested, the anion is "doing all that is possible to avoid cyclic conjugation of the four π electrons." The remaining structure, **B**, with a single imaginary frequency, should be a transition state of a process involving **A**. Studying the structures of **A** and **B** more closely, it is not difficult to conclude that **B** is the transition state of the pseudorotation of **A**, as shown below.

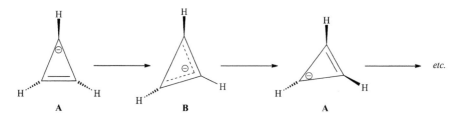

(ii) As the equilateral triangular structure has a higher symmetry (D_{3h}) than **C** or **D**, it will have at least, or very likely more than, two imaginary frequencies. Therefore it is of no interest to us.

A6.43 The TSs of ethenyl anion (**B**, **C**, and **D**) may be treated as an adduct of ethylene dianion ($C_2H_2^{2-}$) and H^+.

(i) One of the factors determining the stability of the adduct is the interaction between $C_2H_2^{2-}$ and H^+. As shown below, in **B**, the two lone pairs point away from the center of **B**, hence the bridging H^+ cannot accept electron density from them efficiently.

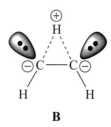

B

In **C**, H^+ is located above the C=C bond and is able to accept electron density from the π bond. Therefore, the interaction between $C_2H_2^{2-}$ and H^+ is stronger in **C** than in **B**.

(ii) In addition, the relative stability of the individual species, $C_2H_2^{2-}$ and H^+ also affects the relative stability of the whole adduct. In both **C** and **D**, H^+ accepts electron density from the π bond. However, the configuration of the $C_2H_2^{2-}$ moiety is different: *cis* in **C** and *trans* in **D**. Being of the same size, the two lone pairs have larger mutual repulsion in **C** than **D**. On the whole, therefore, **D** is lower than **C** in energy.

(iii) Without any symmetry constraints, **B** can collapse into **C** by conrotatory rotation of the two ethylenic hydrogen atoms about the C=C bond and into **D** by disrotatory rotation:

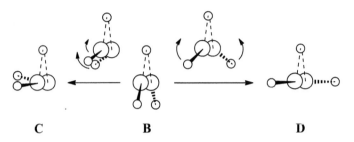

C	**B**	**D**

Both processes end up with H^+ sitting above the C=C bond to interact with the π bond of the $C_2H_2^{2-}$ moiety in the final products (**C** and **D**). However, during the disrotatory rotation for **B** → **D**, the planarity of the $C_2H_2^{2-}$ is destroyed, so is the π bond. While the lone pairs repulsion is reduced, the stability so gained cannot compensate the stability loss due to π bond breaking. Therefore, though **D** is thermodynamically more stable than **C**, it is kinetically unfavorable in the "collapse" process of **B**.

(iv) Species **D** is the transition state for the 1,2-hydrogen shift of the *anti*-hydrogen in **A**.

(v) Swinging of the C–H bond should be more favorable because it does not break the C=C double bond. The transition state for the process is expected to be

$$\begin{array}{c} H \\ \backslash \\ C = C - H \\ / \quad \ominus \\ H \end{array}$$

A6.44 After forming four regular 2c–2e B–Cl bonds, there are 12 boron atomic orbitals and eight electrons remaining for the bonding of the B_4 skeleton. Clearly four 3c–2e bonds can be formed with these orbitals and electrons, and the four bonding electron pairs are localized on the four faces of the tetrahedron. These four pairs of bonding electrons are equally shared by the six B–B linkages. Hence each linkage has $^4/_6$ electron pair and its bond order is $^2/_3$. Additionally, it is of interest to note that there are four electron pairs around each boron atom and they are arranged in an approximately tetrahedral manner, in accordance with the VSEPR theory. As these four electron pairs are not equivalent to each other (one is a 2c–2e pair while the other three are 3c–2e), their spatial arrangement is not exactly tetrahedral.

A6.45 The five molecular orbitals formed are:

$$1a_1' = p(a + b) + q(c + d + e),$$
$$2a_1' = p'(a + b) - q'(c + d + e),$$
$$a_2'' = (2)^{-1/2}(a - b),$$
$$e' = [(6)^{-1/2}(2c - d - e); (2)^{-1/2}(d - e)].$$

Among them, only $1a_1'$ is bonding. The remaining ones are either nonbonding or antibonding.

A6.46 (i) $V = 8 + 3 \times 9 = 35; L = 9 \times 2 + 3 \times 2 = 24; m = 1;$

$t = 12 \times (1 + 3) = 48; q = 0;$

$C = 35 + 24 + 1 - 48 - 0 = 12 = 2n + 4$ with $n = 4$.

The transition metal skeleton of the cluster has a trigonal pyramidal structure which is a *nido* derivative of a trigonal bipyramid (a trigonal bipyramid minus one apex).

The infrared spectrum indicates the presence of both terminal and bridging carbonyl groups. They constitute three sets of carbonyl groups. One set consists of three terminal carbonyl groups attaching to Fe. Another set also consists of three terminal carbonyl groups, but each of them attaches to one Co atom. The last set consists of three carbonyl groups which bridge the three Co atoms.

Each Co atom also bonds to one $P(OMe)_3$ ligand. The structure is analogous to that of $Co_4(CO)_{12}$.

The H atom may be expected to lie above or below the Co_3 basal plane. In reality it is located outside the cluster, thus capping the Co_3 face. The structure is shown on the right (with ligands omitted for clarity).

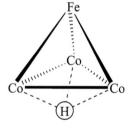

(ii) $V = 6 \times 9 = 54; L = 15 \times 2 = 30; m = 1; t = 12 \times 6 = 72; q = -1;$

$C = 54 + 30 + 1 - 72 - (-1) = 14 = 2n + 2$ with $n = 6$.

The transition metal skeleton of the cluster has a *closo* octahedral structure. The H atom might be expected to situate at the cluster center, playing the role of a six-coordinate "interstitial" hydride ligand. The structure is shown on the right with ligands omitted for clarity. The infrared spectrum indicates the presence of both terminal and bridging carbonyl groups. Cobalts 5 and 6 have one terminal CO group each, while cobalts 1, 2, 3, and 4 have two terminal CO groups each. Edges 3–5, 4–5, 3–6, and 4–6 are asymmetrically bridged by a CO group and edge 1–2 is symmetrically bridged by a CO group.

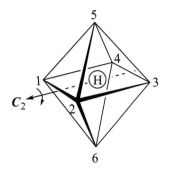

Idealized symmetry C_{2v}

A6.47 The clusters have the general formula $(BH)_{10}XY$ and an icosahedral structure with $n = 12$. To arrive at this structure, we need $2n + 2 = 26$ skeletal electrons, i.e., X and Y need to contribute six electrons to the framework. This happens when X is P and Y is Bi or CH.

A6.48

Cluster compound	No. of skeletal bonding electron pairs	Reference polyhedron
(i)	6	*Closo*, trigonal bipyramid
(ii)	11	*Closo*, bicapped square antiprism
(iii)	8	*Hypho*, isosahedron
(iv)	11	*Closo*, bicapped square antiprism
(v)	6	*Closo*, trigonal bipyramid
(vi)	6	*Nido*, trigonal bipyramid
(vii)	8	*Nido*, pentagonal bipyramid
(viii)	6	*Closo*, trigonal bipyramid
(ix)	8	*Closo*, pentagonal bipyramid
(x)	8	*Closo*, pentagonal bipyramid
(xi)	13	*Closo*, isosahedron
(xii)	7	*Arachno*, octahedron
(xiii)	7	*Closo*, octahedron
(xiv)	11	*Nido*, bicapped square antiprism
(xv)	7	Tetracapped octahedron

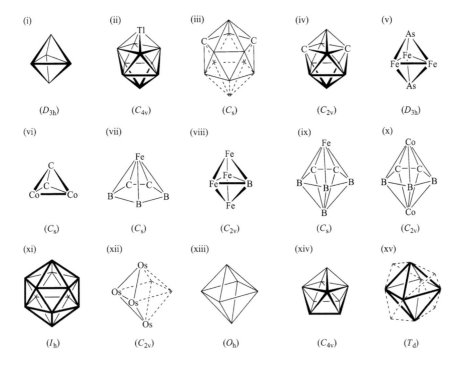

(i)	(ii)	(iii)	(iv)	(v)
(D_{3h})	(C_{4v})	(C_s)	(C_{2v})	(D_{3h})
(vi)	(vii)	(viii)	(ix)	(x)
(C_s)	(C_s)	(C_{2v})	(C_s)	(C_{2v})
(xi)	(xii)	(xiii)	(xiv)	(xv)
(I_h)	(C_{2v})	(O_h)	(C_{4v})	(T_d)

A6.49 (i) $Ni_5(CO)_{11}$: $n = 5$, $C = 10 \times 5 + 2 \times 11 - 12 \times 5 = 12 = 2n + 2$; trigonal bipyramid; D_{3h} symmetry.

(ii) $Ni_6(CO)_{14}$: $n = 6$, $C = 10 \times 6 + 2 \times 14 - 12 \times 6 = 16 = 2n + 4$; *nido* derivative of a pentagonal bipyramid, or a pentagonal pyramid; C_{5v} symmetry.

(iii) $Ni_7(CO)_{17}$: $n = 7$, $C = 10 \times 7 + 2 \times 17 - 12 \times 7 = 20 = 2n + 6$; *arachno* derivative of a tricapped trigonal prism, or a monocapped trigonal prism; C_{2v} symmetry.

(iv) $Ni_8(CO)_{18}$: $n = 8$, $C = 10 \times 8 + 2 \times 18 - 12 \times 8 = 20 = 2n + 4$; *nido* derivative of a tricapped trigonal prism, or a bicapped trigonal prism; C_{2v} symmetry.

(v) $Ni_9(CO)_{19}$: $n = 9$, $C = 10 \times 9 + 2 \times 19 - 12 \times 9 = 20 = 2n + 2$; tricapped trigonal prism; D_{3h} symmetry.

A6.50 (i) With $n = 5$, $C = 2 \times 3 + 4 + 2 = 12 = 2n + 2$. So the structure has the *closo* geometry of a trigonal bipyramid.

(ii) With $n = 3$, $C = 9 \times 3 + 3 \times 5 + 4 + 2 - 36$
$$= 12 = 2n + 6.$$
So the three metal atoms are arranged in a triangular manner. Such a triangle may be considered to be an *arachno* structure derived from a trigonal bipyramid.

(iii) The complex should have the structure shown in the accompanying figure.

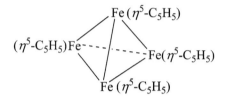

A6.51 $C = 32 + 20 + 8 - 48 = 12 = 2n + 4$ with $n = 4$. The transition metal skeleton of the cluster is a *nido* derivative of a trigonal bipyramid (a trigonal bipyramid minus one apex). From the given spectroscopic data, the skeleton is a tetrahedral structure. Note that the CO groups are η^3 ligands, each of which caps a Fe_3 face.

A6.52 Referring to the energy level diagram shown, for a complex to conform the 18-electron rule, the $1t_{2g}$ orbitals are completely filled.

For group (i) complexes, the $1t_{2g}$ orbitals are essentially nonbonding and Δ_o is small, i.e., the $2e_g$ orbitals are only slightly antibonding. Hence, the $2e_g$ orbitals are available for occupation without much loss in energy. As a result, there is no restriction on the number of d electrons.

For group (ii) complexes, the $1t_{2g}$ orbitals are still nonbonding, while the $2e_g$ orbitals are strongly antibonding, as Δ_o is considerably larger now. Hence, there is a tendency not to fill the $2e_g$ orbitals, while there is still no restriction on the number of electrons that occupy the $1t_{2g}$ orbitals.

For group (iii) complexes, the $1t_{2g}$ orbitals become strongly bonding due to back donation. Thus, while it is still imperative not to have any electrons in the antibonding $2e_g$ orbitals, it is equally important to have six electrons in the $1t_{2g}$ orbitals, since removal of these electrons would destabilize the complex by loss of bond energy.

Crystal Field Theory | **7**

PROBLEMS

7.1 (i)   For a tetrahedral complex, if the z-axis of the central atom is chosen as one of the molecular three-fold axes, how does the set of d orbitals split?

(ii) What is the d orbital splitting pattern in a flattened (along the z-axis) tetrahedral crystal field? In an elongated (same axis) tetrahedral field?

(iii) What is the d orbital splitting pattern for an ML_8 complex with the ligands at the vertices of a cube? How would the magnitude of the splitting(s) compare with that in an analogous tetrahedral ML_4 complex? Comment also on the case where the eight ligands are arranged in a square antiprismatic manner.

7.2 The crystal field splitting diagram for a square pyramidal structure is given below.

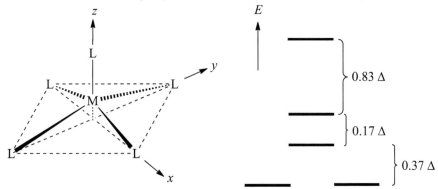

(i)   Assign with justification each energy level to an appropriate 3d orbital.

(ii) Determine the position of barycenter.

(iii) A given square pyramidal RuL_5 complex containing Ru^{2+} has a magnetic moment of 2.90 BM. Calculate the crystal field stabilization energy (CFSE) of this complex.

7.3 (i)   The hydrated Cu(II) complex $[Cu(H_2O)_5]^{2+}$ was found to assume a trigonal bipyramidal structure. According to the following crystal field splitting diagram for

Problems in Structural Inorganic Chemistry. Second edition. Wai-Kee Li, Hung Kay Lee,
Dennis Kee Pui Ng, Yu-San Cheung, Kendrew Kin Wah Mak, and Thomas Chung Wai Mak.
© Oxford University Press 2019. Published in 2019 by Oxford University Press.
DOI: 10.1093/oso/9780198823902.001.0001

a trigonal bipyramidal crystal field, calculate the crystal field stabilization energy (CFSE) of this complex.

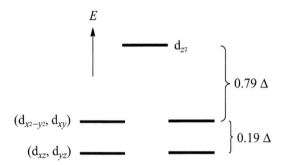

(ii) What is the other common structure for complexes with five-fold coordination around the metal center? Predict the structural behavior of $[Cu(H_2O)_5]^{2+}$ in aqueous solution.

REFERENCE: V. Shivaiah and S. K. Das, Fivefold coordination of a Cu^{II}–aqua ion: a supramolecular sandwich consisting of two crown ether molecules and a trigonal-bipyramidal $[Cu(H_2O)_5]^{2+}$ complex. *Angew. Chem. Int. Ed.* **45**, 245–8 (2006).

7.4 (i) Calculate the change in crystal field stabilization energy (CFSE) for a high-spin d^3 metal ion when the field changes from octahedral to trigonal bipyramidal.

(ii) Potassium hexaiodomanganate(X) (X is the oxidation state of Mn to be determined) has a magnetic moment of 3.82 BM. Predict with explanation whether it would be labile or inert. Assume that a trigonal bipyramidal transition state is involved.

7.5 An unlabeled crystal field splitting diagram for a linear crystal field is given below.

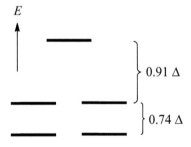

(i) Assign, with justification, each energy level to an appropriate 3d orbital. Assume the molecular axis is the same as the z-axis.

(ii) Calculate the crystal field stabilization energy (CFSE) of a high-spin d^4 configuration in a linear field.

7.6 $[ArMn(^tBu)]$ ($^tBu = tert$-butyl) is the first example of a heteroleptic two-coordinate manganese aryl/alkyl complex. X-ray diffraction analysis has revealed that it also has a quasi-linear structure. The complex follows the Curie-Weiss law with an effective magnetic moment of 5.73 BM. Suggest the electronic configuration of this complex.

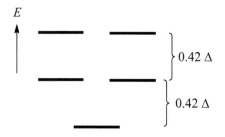

$$Ar =$$

$^iPr = iso$-propyl

REFERENCE: C. Ni, J. C. Fettinger, G. J. Long, and P. P. Power, Terphenyl substituted derivatives of manganese(II): distorted geometries and resistance to elimination. *Dalton Trans.* **39**, 10664–70 (2010).

7.7 (i) Draw a labeled crystal field splitting diagram for 3d orbitals in a cubic crystal field.

(ii) Square antiprism is another possible structure for eight-coordination which can be considered as a distortion of the cube by a rotation of one of the square planes by 45° around the perpendicular axis. Given the following crystal field splitting pattern, assign, with justification, each energy level to an appropriate d orbital.

E

$\Big\} 0.42 \, \Delta$

$\Big\} 0.42 \, \Delta$

(iii) Both $[Nb(ox)_4]^{4-}$ (ox = oxalato) and the ZrF_8 unit in $Na_7Zr_6F_{31}$ adopt a square antiprismatic structure. Calculate the crystal field stabilization energy (CFSE) of these complexes.

(iv) The tetradentate ligand L_{N4} (shown below) reacts with $[Fe(H_2O)_6](BF_4)_2$ in a 2:1 molar ratio in acetonitrile to afford $[Fe(L_{N4})_2](BF_4)_2$, which displays a distorted square antiprismatic structure. The $[Fe(L_{N4})_2]^{2+}$ complex in this compound has a high-spin configuration. Determine the total spin angular momentum quantum number (S) and predict its spin-only magnetic moment.

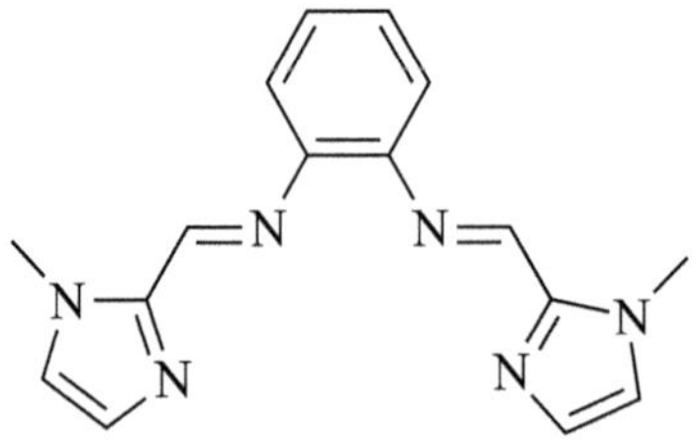

L_{N4}

REFERENCES: F. A. Cotton, Michael P. Diebold, and W. J. Roth, Variable stereochemistry of the eight-coordinate tetrakis(oxalato)niobate(IV), $Nb(C_2O_4)_4^{4-}$. *Inorg. Chem.* **26**, 2889–93 (1987); J. H. Burns, R. D. Ellison, and H. A. Levy, The crystal structure of $Na_7Zr_6F_{31}$. *Acta Cryst.* **B24**, 230–7 (1968); A. K. Patra, K. S. Dube, G. C. Papaefthymiou, J. Conradie, A. Ghosh, and T. C. Harrop, Stable eight-coordinate iron(III/II) complexes. *Inorg. Chem.* **49**, 2032–4 (2010).

7.8 Köhn et al. reported the following reaction and electronic absorption data:

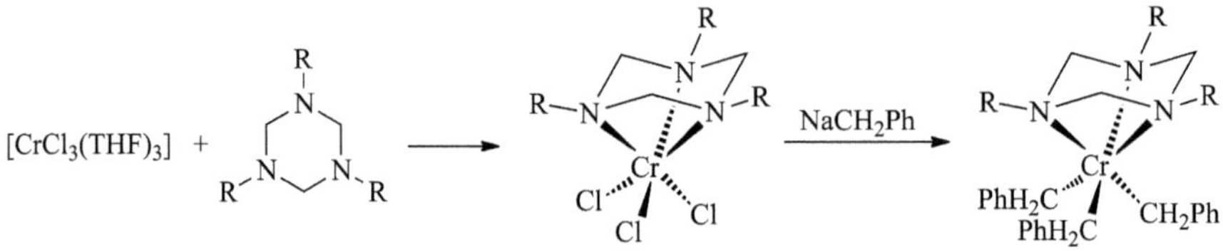

1a R = CH_2Ph
1b R = c-C_6H_{11}

2a R = CH_2Ph
2b R = c-C_6H_{11}

Complex	$\tilde{\upsilon}_1$ (cm^{-1})	$\tilde{\upsilon}_2$ (cm^{-1})	$\tilde{\upsilon}_3$ (cm^{-1})
1a	13989	19650	29917 (shoulder)
1b	13651	19145	27400 (shoulder)
2a	15822	22401	----
2b	15815	20786	----

(i) With the aid of the diagram in Problem 7.36, assign all the absorption bands in the table.

(ii) The following graph shows the variation of the ratio $\tilde{\upsilon}_2/\tilde{\upsilon}_1$ with Dq/B' for A_{2g} ground state ions. Determine the 10 Dq and B' for each of these complexes, and then calculate $\tilde{\upsilon}_3$ from the corresponding values of $\tilde{\upsilon}_1$, $\tilde{\upsilon}_2$, and B'.

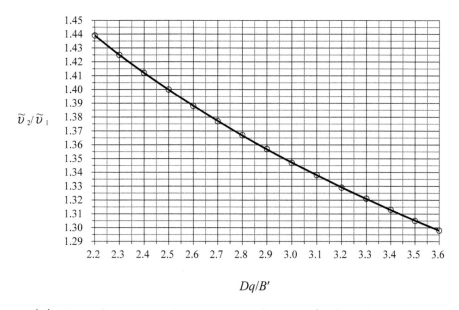

$\tilde{v}_2/\tilde{v}_1$

Dq/B'

(iii) Does the 10 Dq value increase or decrease after benzylation? Suggest an explanation.

REFERENCES: R. D. Köhn, G. Kociok-Köhn, and M. Haufe, η^3-1,3,5-triazacyclohexane complexes of tribenzylchromium(III). *Organometal. Chem.* **501**, 303–7 (1995); A. B. P. Lever, Electronic spectra of some transition metal complexes: derivation of Dq and B. *J. Chem. Educ.* **45**, 711–2 (1968).

7.9 (i) The electronic absorption spectra of three Cu(II) complexes, namely $[\mathrm{Cu(H_2O)_6}]^{2+}$, *cis*-$[\mathrm{Cu(en)_2(H_2O)_2}]^{2+}$ (en = $\mathrm{H_2NCH_2CH_2NH_2}$) and $[\mathrm{Cu(edta)}]^{2-}$ $[\mathrm{edta} = (^{-}\mathrm{O_2CCH_2})_2\mathrm{NCH_2CH_2N(CH_2CO_2^-)_2}]$ are given below. Assign the electronic transition for these absorption bands. Assume all these complexes have an octahedral structure.

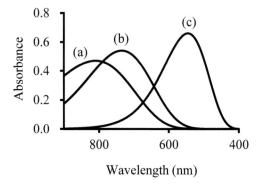

(ii) Based on the absorption positions, assign these spectra to the three complexes.

(iii) Rationalize the low molar absorptivities of these absorptions, in particular for spectrum (a).

(iv) Solution spectra of these complexes are very broad. However, the solid state spectra are much sharper. For $[Cu(H_2O)_6]^{2+}$, the solid state spectrum shows a splitting of the absorption band. Explain these observations.

REFERENCE: A. T. Baker, The ligand field spectra of copper(II) complexes. *J. Chem. Educ.* **75**, 98–9 (1998).

7.10 If an octahedral Fe(II) complex has a large paramagnetic susceptibility, what is the ground-state label according to the Tanabe-Sugano diagram? What electronic states are involved in the spin-allowed electronic transition(s)?

7.11 (i) Draw all the possible geometric isomers of the oxo-Mo(V) octahedral complex MoOCl(L). (The structure of L is shown below.) Also write down their molecular symmetries and indicate whether these isomers are optically active.

(ii) The vanadium(V) complex VO(L') was synthesized to study the insulin-mimetic effect observed for vanadium. (The structure of L' is shown below.) It was found by X-ray diffraction analysis that the vanadium center adopts a trigonal bipyramidal geometry and the amine nitrogen is *trans* to the oxo group. Draw the structure of this complex and predict whether it is optically active.

L   L'

REFERENCES: K. R. Barnard, M. Bruck, S. Huber, C. Grittini, J. H. Enemark, R. W. Gable, and A. G. Wedd, Mononuclear and binuclear molybdenum(V) complexes of the ligand N, N'-dimethyl-N, N'-bis(2-mercaptophenyl)ethylenediamine: geometric isomers. *Inorg. Chem.* **36**, 637–49 (1997); C. R. Cornman, T. C. Stauffer, and P. D. Boyle, Oxidation of a vanadium(V)–dithiolate complex to a vanadium(V)–η^2,η^2-disulfenate complex. *J. Am. Chem. Soc.* **119**, 986–7 (1997).

7.12 (i) The complex $[NiF_6]^{2-}$ is diamagnetic despite the fact that F^- is one of the weakest ligands as far as crystal field splitting is concerned. Comment.

(ii) The Co^{2+} ion readily forms a tetrahedral chloride complex but Ni^{2+} does not. Comment.

7.13 The lattice energies of bivalent first transition series metal chlorides, MCl_2, where M^{2+} is surrounded by six octahedrally disposed Cl^- ligands, are given below (in 10^3 kJ mol^{-1}):

$CaCl_2$	$ScCl_2$	$TiCl_2$	VCl_2	$CrCl_2$	$MnCl_2$	$FeCl_2$	$CoCl_2$	$NiCl_2$	$CuCl_2$	$ZnCl_2$
2.23	—	2.48	2.55	2.57	2.50	2.61	2.69	2.75	2.79	2.71

Estimate the crystal field strengths of the chloride ions on V^{2+} and Co^{2+}.

7.14 A spinel is a metal oxide, M_3O_4, which has a nearly close-packed array of oxide ions. The three metal ions occupy one tetrahedral and two octahedral sites. The formula may then be written as $A[BC]O_4$, where the ions in octahedral sites are enclosed in brackets. Furthermore, a spinel in which the octahedral sites are occupied by M(III) ions is known as a normal spinel. If one of the octahedral sites is occupied by an M(II) ion, it is known as an inverse spinel.

For Mn_2FeO_4, calculate the total crystal field stabilization energy (CFSE) in units of Δ_o (or Dq) for all permutations of ions and oxidation states. (As indicated earlier, only M(II) and M(III) states need to be considered.) Other assumptions are: $\Delta_t = \frac{4}{9}\Delta_o$; $(\Delta_o)_{Fe} = (\Delta_o)_{Mn}$; $(\Delta_o)_{2+} = (\Delta_o)_{3+}$; and both iron and manganese oxides are high-spin cases. Deduce the correct structure on the basis of the calculated CFSE. Is the deduced spinel normal or inverse?

7.15 If $[CoF_6]^{3-}$ is a high-spin complex, what do you predict for $[CoBr_6]^{3-}$? In fact, $[CoF_6]^{3-}$ is the only known binary Co^{3+}/halogen complex. Comment on the "non-existence" of $[CoX_6]^{3-}$, $X = Cl^-$, Br^-, and I^-.

7.16 (i) The complex $[Cr(dipy)_3]^{2+}$, where dipy = 2, 2'-dipyridyl, is known to possess a distorted octahedral structure and magnetic moment of ≈ 5 BM. What is the electron distribution of Cr^{2+} in this complex that is consistent with the observations?

(ii) What is the spin-only magnetic moment of the Co^{3+} ion? Would the magnetic moment of the gaseous ion be close to the spin-only value? Comment.

(iii) In high-spin Fe^{3+} octahedral complexes, the magnetic moments are always very close to the spin-only value of 5.9 BM. On the other hand, low-spin Fe^{3+} complexes have magnetic moments of about 2.3 BM at room temperature, which differ significantly from the spin-only value of 1.7 BM. Furthermore, these (low-spin) moments drop to about 1.9 BM at liquid nitrogen temperature. Comment.

7.17 Comment on possible distortion of the following octahedral complexes. The observed magnetic moments are given in BM.

(i) $[Ni(NH_3)_6]^{2+}$, 2.89;   (ii)   trans-$[Ni(CN)_4(H_2O)_2]^{2-}$, diamagnetic;

(iii) $[Co(H_2O)_6]^{2+}$, 4.2;   (iv) $[Co(CN)_6]^{4-}$, 1.7;

(v) $[FeF_6]^{3-}$, 5.9;   (vi) $[Fe(CN)_6]^{3-}$, 1.7.

7.18 (i)   Under what circumstances would a square planar complex of Co^{2+} exhibit a spin-only magnetic moment μ of 2.5 BM in (aqueous) solution at 25°C? In addition, predict whether μ will increase or decrease with increasing temperature.

(ii) How would μ change if a large amount of pyridine is added to the above solution? (Hint: When a square planar complex is in solution, two solvent molecules would enter the "vacant" ligand sites, forming a tetragonal complex.)

7.19 For tetrahedral complexes of the first transition series metals, which electronic configurations would lead to Jahn–Teller distortion? Upon such distortion, how do the splittings in t_2 and e levels compare?

7.20 The trans isomer $Cu(en)_2Cl_2$, where "en" stands for $H_2NCH_2CH_2NH_2$, is more stable than the cis isomer. Rationalize this in terms of (i) the classical ionic model, and (ii) the crystal field theory. Which of the above arguments is applicable to the low-spin complex $[Co(en)_2Cl_2]^+$?

7.21 (i)   The complex ion $[CoF_6]^{3-}$ is probably tetragonally distorted. Will the distortion be large compared to that of $[CuX_6]^{4-}$ ions and will it exhibit axial elongation or axial compression?

(ii) In crystalline $CrCl_2$, two ligands are much farther from the metal ion (2.90 Å) than the other four (2.39 Å). Comment.

(iii) The internuclear distances found in $CuCl_2 \cdot H_2O$ are: two Cu–O at 2.01 Å, two Cu–Cl at 2.31 Å, and two Cu–Cl at 2.98 Å. Comment.

7.22 In each of the following pairs of complexes select the one with the larger value of Δ_o and give the basis for your choice:

(i) $[Fe(H_2O)_6]^{2+}$ and $[Fe(H_2O)_6]^{3+}$

(ii) $[CoCl_6]^{4-}$ and $[CoCl_4]^{2-}$

(iii) $[CoCl_6]^{3-}$ and $[CoF_6]^{3-}$

(iv) $[Fe(CN)_6]^{4-}$ and $[Os(CN)_6]^{4-}$

7.23 In a square planar crystal field, the d orbital splitting pattern is

$$d_{x^2-y^2} > d_{xy} > d_{z^2} > (d_{xz}, d_{yz}).$$

Let Δ_1 be the energy difference between the $d_{x^2-y^2}$ and d_{xy} levels. According to Gray and Ballhausen, the Δ_1 values for $[Pt(CN)_4]^{2-}$, $[PtCl_4]^{2-}$, $[PtBr_4]^{2-}$, and $[PdCl_4]^{2-}$ are (in units of 10^3 cm^{-1}) 30.0, 23.5, 22.2, and 19.2, respectively. Comment on the following trends for these values:

(i) $[Pt(CN)_4]^{2-} > [PtCl_4]^{2-} > [PtBr_4]^{2-}$
(ii) $[PtCl_4]^{2-} > [PdCl_4]^{2-}$

7.24 An abbreviated Tanabe-Sugano diagram for a d^7 metal ion is given on the right.

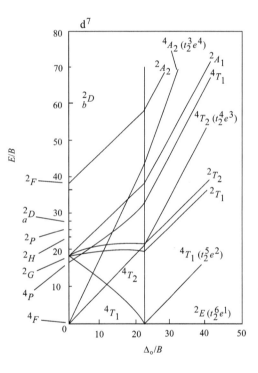

(i)   What are the ground states of weak-field and strong-field O_h complexes for this configuration?

(ii) For the weak-field complex, list the expected spin-allowed transitions. Which one should be the weakest? Why?

(iii) For the strong-field complex, list the expected spin-allowed transitions.

(iv) What is the ground state of a d^3 ion in an octahedral field?

(v)   What is the ground state of a d^7 ion in a tetrahedral field?

7.25 Illustrate schematically the d-orbital splitting patterns for the following metal complexes:

(i) ML_2, linear, $D_{\infty h}$ symmetry (ligands on the z-axis)
(ii) ML_3, equilateral triangular, D_{3h} symmetry (ligands on the xy-plane)
(iii) ML_8, cubic, O_h symmetry (the origin is placed at the center of the cube)
(iv) ML_8, square antiprismatic, D_{4d} symmetry
(v) ML_{12}, icosahedral, I_h symmetry

7.26 (i) Sketch and explain briefly the structure for the complex $[Re(S_2C_2Ph_2)_3]$.

(ii) What is the d orbital splitting pattern expected for this complex? Give brief explanations for your answer.

7.27 Both CrO_4^{2-} and MnO_4^- are d^0 complexes and yet are very colorful. What type of electronic transitions is responsible for the colors of the complexes? Explain.

7.28 The molar extinction coefficients (in $L\,mol^{-1}\,cm^{-1}$) of the strongest visible absorption bands of the following complexes are $[Mn(H_2O)_6]^{2+}$: 0.035; $[MnBr_4]^{2-}$: 4.0; $[Co(H_2O)_6]^{2+}$: 10; $[CoCl_4]^{2-}$: 600. Comment.

7.29 In each of the following pairs of electronic transitions, state which would be expected to be more intense and explain your choice:

(i) $^1A_{1g} \rightarrow {}^1T_{2g}$ in $[Co(NH_3)_6]^{3+}$ or $^1A_1 \rightarrow {}^1T_2$ in $[Co(en)_3]^{3+}$

(ii) $^4A_2 \rightarrow {}^4E$ or $^4A_2 \rightarrow {}^2E$ in $[V(C_2O_4)_3]^{4-}$

(iii) $^3A_2 \rightarrow {}^3E$ or $^3A_2 \rightarrow {}^3A_2$ in $[Pd(en)_3]^{2+}$

(iv) the most intense d–d band in $[NiCl_4]^{2-}$ or the most intense d–d band in $[MnCl_4]^{2-}$

7.30 (i) Most high-spin Mn^{2+} complexes are very pale in color, whereas high-spin Fe^{3+} complexes, with the same d^5 configuration, exhibit a range of different colors, some weak and some quite intense. Comment.

(ii) Between complexes $[Co(NH_3)_5Cl]^{2+}$ and $[Co(NH_3)_5Br]^{2+}$, which one should have a lower-energy charge-transfer band?

7.31 The absorption spectra of $[Mn(H_2O)_6]^{2+}$ and $[MnBr_4]^{2-}$ ions are given below together with the partial energy level diagram for the Mn^{2+} ion.

(i) Which spectrum belongs to which ion? Justify your choice.

(ii) Briefly explain the weakness of the bands.

(iii) There is great variation in the widths of the bands, with one being extremely narrow. Comment.

(iv) What is the color of each of the complexes?

(v) Assign the various bands to the proper state transitions.

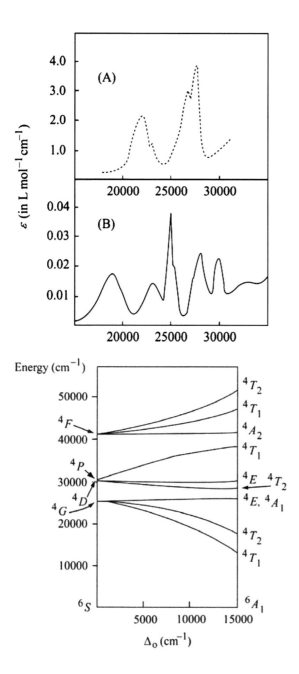

7.32 The visible absorption spectra of four cobalt complexes are shown in the following figure. Assign, with justification, the spectral curves A, B, C, and D to complexes $[Co(NH_3)_6]^{2+}$, $[Co(NH_3)_6]^{3+}$, $[Co(H_2O)_6]^{2+}$, and $[CoCl_4]^{2-}$.

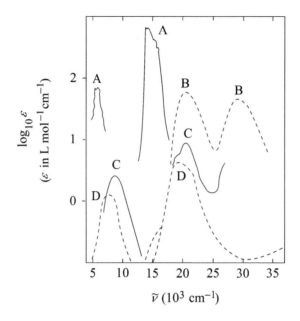

REFERENCE: D. Sutton, *Electronic Spectra of Transition Metal Complexes*, McGraw-Hill, New York, 1968, p. 16.

7.33 The visible absorption spectrum of $[Cr(H_2O)_6]^{3+}$ shows three bands at 17400, 24700, and 37000 cm^{-1}.

(i) Decide, with brief but clear explanation, which of the following energy level (or Orgel) diagrams can be used to interpret the observed spectral data below.

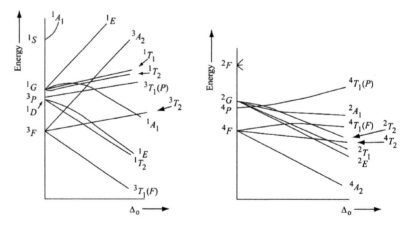

(ii) Assign the observed bands to transitions from the ground state to the excited states.

(iii)   What is the color of the $[Cr(H_2O)_6]^{3+}$ ion?

(iv)   Predict a value for the magnetic moment of the $[Cr(H_2O)_6]^{3+}$ ion.

7.34   The absorption spectra of *cis*- and *trans*-$[Co(en)_2F_2]^+$ are given below. Note that the first band splits to a very slight extent for the *cis* form (the dotted curve shows where the low frequency side of the band of the *cis* isomer would be if the band were symmetrical). However, the same splitting has the magnitude of a few thousand wavenumbers for the *trans* isomer.

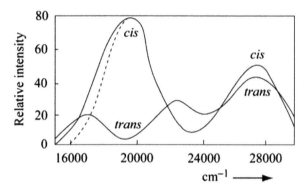

(i)   Why do the *cis* bands have higher intensities than the *trans* bands?

(ii)   What causes the splitting in the first band and why is the splitting for the *trans* isomer greater than that for the *cis* isomer?

REFERENCE: F. A. Cotton and G. Wilkinson, *Advanced Inorganic Chemistry*, 5th edn., Wiley, New York, 1988, p. 734.

7.35   The absorption spectra of $[Ni(H_2O)_6]^{2+}$ and $[Ni(en)_3]^{2+}$ are shown in the figure below, along with a "color map" and an abbreviated d^8 Orgel diagram.

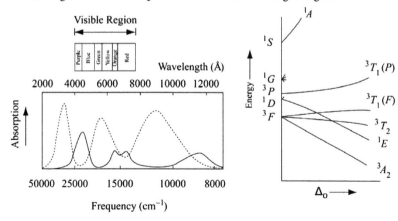

(i)   Which spectrum belongs to $[Ni(H_2O)_6]^{2+}$?

(ii) What is the color of each of the complexes?

(iii) With the aid of the given Orgel diagram, assign the electronic transitions.

(iv) What causes the middle band in the "solid-curve" spectrum to split?

REFERENCE: F. A. Cotton, Ligand field theory. *J. Chem. Educ.* **41**, 466–76 (1964).

7.36   The crystal field splitting of terms ^3F and ^3P for a d^8 ion in an O_h field illustrating the nephelauxetic effect and the mixing of T_{1g} terms is shown below.

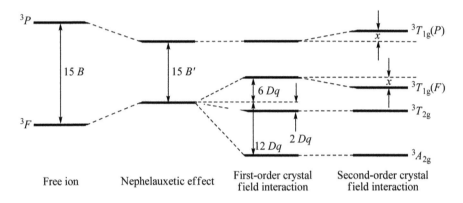

(i)   The following absorption bands are found in the spectrum of $[Ni(H_2O)_6]^{2+}$: 8500, 15400, and 26200 cm^{-1}. Determine the values of B' and Dq for this ion.

(ii) Two absorption bands at 23900 and 17500 cm^{-1} are found in the spectrum of $[Cr(ox)_3]^{3-}$ (ox^{2-} = oxalate), while the third expected band is masked by the charge-transfer transition and, hence, not observed. Given that B for Cr^{3+} is 918 cm^{-1} and the nephelauxetic parameter β for this complex is 0.7, determine the values of B' and Dq for this ion.

Note on nephelauxetic effect:

Nephelauxetic means cloud-expanding in Greek. The nephelauxetic effect refers to a decrease in the electronic interaction parameter B which occurs when a transition metal ion is placed in a ligand field. If B' denotes the aforementioned parameter in a complex, the nephelauxetic parameter β is simply the ratio B'/B, which is smaller than 1.

7.37   (i)   For $[Ti(H_2O)_6]^{3+}$, a value of Δ_o can be determined directly from λ_{max} in the electronic absorption spectrum. Why is this not possible for most other octahedral ions?

(ii) The isoelectronic ions VO_4^{3-}, CrO_4^{2-}, and MnO_4^-, all have intense charge-transfer transitions. The wavelengths of these transitions increase in this series. Explain this trend.

7.38 Why do the electronic absorption spectra of the lanthanide ions show sharp bands differing from the broad bands in the spectra of the 3d transition elements?

7.39 It is generally accepted that vibronic coupling gives intensity to some symmetry-forbidden transitions. Consider the transitions $^1A_{1g} \rightarrow {}^1T_{1g}$ and $^1A_{1g} \rightarrow {}^1T_{2g}$ for complex $[Co(NH_3)_6]^{3+}$. Given that the symmetries of the normal modes of an octahedral ML_6 molecule are A_{1g}, E_g, $2T_{1u}$, T_{2u}, and T_{2g}, which vibrations are responsible for the occurrence of the observed transitions?

7.40 The crystal field energy level scheme for $V(C_2O_4)_3^{3-}$ (a d^2 system; D_3 symmetry) is shown on the right. In the crystal spectrum of this complex ion, the following bands are found: (x, y)-polarized: 16500 and 23500 cm^{-1}; z-polarized: 16350 cm^{-1}. Make an assignment of these observed bands.

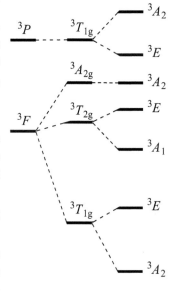

7.41 A schematic molecular orbital energy level diagram for metallocenes $M(C_5H_5)_2$ (with D_{5d} symmetry) is given below. (In this diagram C_5H_5 is abbreviated as "cp".)

(i) What is the ground electronic configuration and ground state of ferrocene, $Fe(C_5H_5)_2$?

(ii) The electronic spectrum of $Fe(C_5H_5)_2$ in the range from 18000 to 32000 cm^{-1} exhibits the following bands (with molar extinction coefficient ε given in parentheses): 18900 (7), 20900 (< 1), 21800 (36), 22400 (< 1), 24000 (72), and 30800 (49). Assign, with justification, each observed band to an electronic transition, indicating both the orbitals and states involved. Should you have doubts for a certain observed band, list all possible assignments for this band and explain why a clear assignment cannot be made.

Hints: (1) All observed bands are for d—d transitions. (2) Consider both spin-allowed and spin-forbidden transitions. (3) Useful(?) direct product in the D_{5d} point group: $E_{1g} \times E_{2g} = E_{1g} + E_{2g}$.

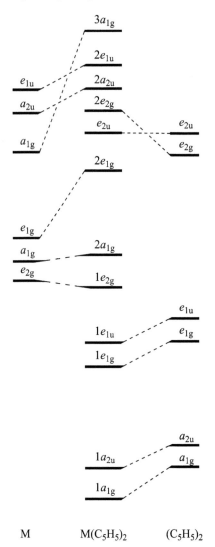

M M(C$_5$H$_5$)$_2$ (C$_5$H$_5$)$_2$

SOLUTIONS

A7.1 (i) Obviously, the splitting pattern (i.e., a set of three-fold degenerate orbitals above a set of two-fold degenerate orbitals by Δ_t) is independent of the coordinate system chosen. However, the components of the upper set are now the d_{z^2}, d_{xz}, and d_{yz} orbitals, while the remaining two d orbitals make up the lower set.

(ii) When the tetrahedron is flattened, the splitting pattern becomes $d_{xy} > (d_{xz}, d_{yz}) \gg d_{x^2-y^2} > d_{z^2}$. In the elongated case, it is $(d_{xz}, d_{yz}) > d_{xy} \gg d_{z^2} > d_{x^2-y^2}$.

(iii) The d orbital splitting pattern for cubic ML_8 should be identical to that for ML_4 (tetrahedral); however, the magnitude of the splitting is now doubled. When the eight ligands are arranged in a square antiprismatic manner, the system has D_{4d} symmetry. Upon inspecting the D_{4d} character table and the way that the d orbital lobes are arranged, it is fairly obvious that the energy ordering should be $(d_{xz}, d_{yz}) > (d_{x^2-y^2}, d_{xy}) > d_{z^2}$. However, the magnitude of the splittings cannot be determined from simple qualitative considerations.

A7.2 (i) The lobes of the $3d_{x^2-y^2}$ and $3d_{z^2}$ orbitals point directly at ligands and hence these two orbitals are destabilized. Since the lobes of $3d_{x^2-y^2}$ orbitals point at four ligands, these are more destabilized than the $3d_{z^2}$ which points at only one ligand. On the other hand, the lobes of $3d_{xy}$, $3d_{yz}$, and $3d_{xz}$ orbitals point in between the ligands and are stabilized. Among these three orbitals, the $3d_{yz}$ and $3d_{xz}$ orbitals have components along the z-axis, where there is only one ligand. Hence, these two orbitals are more stable than the $3d_{xy}$ orbital. It is also worth noting that, the $3d_{yz}$ and $3d_{xz}$ orbitals are interchangeable by the four-fold z-axis, and hence they form a pair of degenerate states (at the bottom of the energy level diagram).

(ii) Let the barycenter be x above the $3d_{xy}$ level (as shown below). Since $E_{\text{stabilization}} = E_{\text{destabilization}}$, we have

$$2(0.37\,\Delta + x) + x = (0.17\,\Delta - x) + \left[(0.17\,\Delta - x) + 0.83\,\Delta\right]$$

Solving, $x = 0.086\,\Delta$.

Therefore, the barycenter sits about midway between the $3d_{z^2}$ and $3d_{xy}$ orbitals.

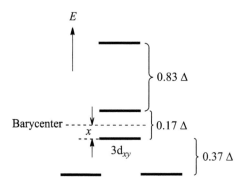

(iii)   The complex RuL_5 has $\mu = 2.90$ BM. According to the spin-only formula, $\mu_s = \sqrt{n(n+2)}$, we have: $\sqrt{n(n+2)} = 2.90$. Solving: $n = 2.2$. Hence, the complex has two unpaired electrons. Since Ru^{2+} ion consists of six d electrons, the electronic energy level is:

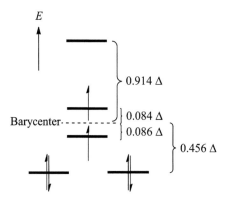

Hence, $CFSE = 4(0.456\,\Delta) + 0.086\,\Delta - 0.084\Delta = 1.826\Delta$.

A7.3   (i)   Let the barycenter be x above the $(d_{x^2-y^2}, d_{xy})$ level (as shown below).

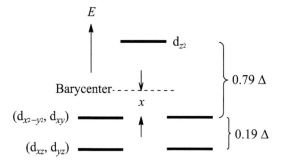

Since $E_{stabilization} = E_{destabilization}$, we have

$$2(0.19\,\Delta + x) + 2x = 0.79\,\Delta - x$$

Solving, $x = 0.082\,\Delta$.

Since Cu^{2+} ion consists of nine d electrons, CFSE $= 4(0.19\,\Delta + 0.082\,\Delta) + 4(0.082\,\Delta) + (0.79\,\Delta - 0.082\,\Delta) = 0.708\,\Delta$.

(ii) Square pyramid.

In terms of CFSE, the energy difference between the trigonal bipyramidal and square pyramidal structures is small. Hence, it is expected that the solvated complex $[Cu(H_2O)_5]^{2+}$ undergoes frequent transformations between these two structures. The transformations can be achieved by a Berry twist mechanism as shown below:

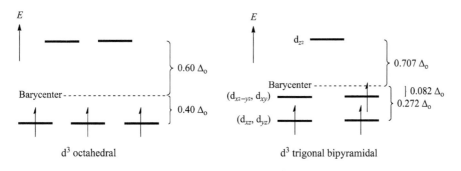

Trigonal bipyramid   Square pyramid

A7.4 (i) The energy level diagrams for octahedral and trigonal bipyramidal structures of a d^3 metal complex are shown below. Note that the one for the trigonal bipyramidal structure is obtained in the previous problem.

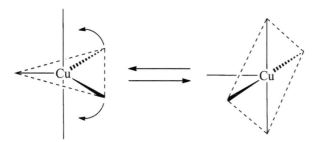

d³ octahedral   d³ trigonal bipyramidal

CFSE for octahedral structure $= 3(0.40)\,\Delta_o = 1.2\,\Delta_o$.

CFSE for trigonal bipyramidal structure $= 2(0.272) + 1(0.082) = 0.626\,\Delta_o$.

The change in CFSE from octahedral structure to trigonal bipyramidal structure
$$= CFSE_t - CFSE_o = 0.626\,\Delta_o - 1.2\,\Delta_o = -0.574\,\Delta_o.$$

(ii) The complex has $\mu = 2.90$ BM. According to the spin-only formula, $\mu_s = \sqrt{n(n+2)}$, we have: $\sqrt{n(n+2)} = 3.82$. Solving: $n = 2.95$. Hence, the complex consists of three unpaired electrons, i.e., $[MnF_6]^{2-}$. So the result in (i) can be used. It indicates a large loss of CFSE upon a d^3 octahedral complex when losing a ligand to yield a five-coordinated trigonal bipyramidal transition state. The loss will be added directly onto the E_a of the rate determining step and will result in a very slow reaction. So the complex is kinetically inert.

A7.5 (i) The lobes of the $3d_{z^2}$ orbital point directly at the two ligands and hence this orbital is destabilized. The lobes of the $3d_{yz}$ and $3d_{xz}$ orbitals do not point at the ligands but the directions of the lobes consist of z-components, these two orbitals are therefore destabilized also, but to a much lesser extent. The lobes of the $3d_{x^2-y^2}$ and $3d_{xy}$ orbitals lie on the xy-plane and hence these two orbitals are stabilized.

(ii) Let the barycenter be x above the $(d_{x^2-y^2}, d_{xy})$ level (as shown below).

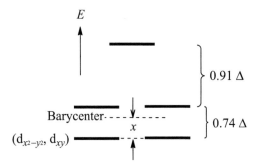

Since $E_{stabilization} = E_{destabilization}$, we have

$$2x = 2(0.74\,\Delta - x) + 1(0.74\,\Delta - x) + 0.91\,\Delta.$$

Solving, $x = 0.63\,\Delta$.

$$CFSE = 2(0.63\,\Delta) - 2(0.11\,\Delta) = 1.04\,\Delta.$$

A7.6 The complex has $\mu = 5.73$ BM. According to the spin-only formula, $\mu_s = \sqrt{n(n+2)}$, we have: $\sqrt{n(n+2)} = 5.73$. Solving: $n = 4.82$. Hence, the complex consists of five unpaired electrons and it should have a high-spin d^5 configuration.

A7.7 (i) The crystal field splitting diagram for a cubic structure is similar to that for a tetrahedral field except that the splitting is larger.

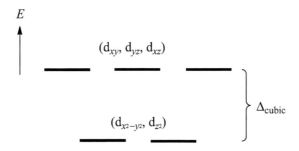

(ii) Assume a square plane is rotated by 45° around the z-axis. The lobes of the $3d_{z^2}$ orbital point through the centers of the two square planes and do not point directly at the ligands. Hence, the $3d_{z^2}$ orbital is most stable. The lobes of the $3d_{x^2-y^2}$ and $3d_{xy}$ lie between the square planes but point to the same directions of the ligands of one of the square planes. Therefore, these two orbitals are less stable than the $3d_{z^2}$ orbital. Finally, two of the lobes of each of the $3d_{xz}$ and $3d_{yz}$ orbitals point to the ligands. Hence, these two orbitals are least stable.

(iii) Let the barycenter be x above the $(d_{x^2-y^2}, d_{xy})$ level (as shown below).

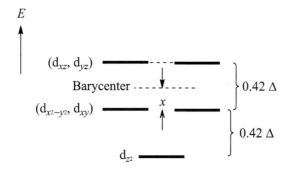

Since $E_{stabilization} = E_{destabilization}$, we have

$$(0.42\,\Delta + x) + 2x = 2(0.42\,\Delta - x)$$

Solving, $x = 0.08\,\Delta$.

$[Nb(ox)_4]^{4-}$ is a d^1 system, CFSE $= (0.42\,\Delta + 0.08\,\Delta) = 0.50\,\Delta$.

In the ZrF_8 unit, Zr is in $+IV$ oxidation state, which is a d^0 system. CFSE $= 0$.

(iv) There are four unpaired electrons, so $S = 2$ and $\mu_s = \sqrt{4(4+2)} = 4.90\,BM$.

A7.8 (i) For d^3 octahedral complexes:

$$\tilde{v}_1 : {}^4A_{2g} \rightarrow {}^4T_{2g}, \ \tilde{v}_2 : {}^4A_{2g} \rightarrow {}^4T_{1g}(F), \ \tilde{v}_3 : {}^4A_{2g} \rightarrow {}^4T_{1g}(P).$$

(ii) Refer to the diagram in Problem 6.36, first of all, $\tilde{v}_1 = 10\ Dq$. With the given graph, the following table is constructed.

Complex	$10\ Dq$ (cm^{-1})	Dq (cm^{-1})	$\tilde{v}_2/\tilde{v}_1$	Dq/B'	B' (cm^{-1})
1a	13989	1398.9	1.405	2.46	569
1b	13651	1365.1	1.402	2.49	548
2a	15822	1582.2	1.416	2.36	670
2b	15815	1581.5	1.314	3.39	467

Furthermore,

$\tilde{v}_2 = 18\ Dq - x$ (x arises from the mixing of two ${}^4T_{1g}$ terms). $\tilde{v}_3 - 12\ Dq + 15\ B' + x$

$\therefore \tilde{v}_3 = 3\tilde{v}_1 - \tilde{v}_2 + 15\ B' = 30852\ cm^{-1}$ for (**1a**) and $30028\ cm^{-1}$ for (**1b**),

which are in good agreement with the experimental values.

(iii) The $10\ Dq$ value is increased by benzylation. The Cl^- anion is a π-donor ligand, which reduces the Δ and hence it is a weak field ligand. The CH_2PH^- group is a strong σ-donor, which creates a strong ligand field.

A7.9 (i) ${}^2E_g \rightarrow {}^2T_{2g}$.

(ii) (a): $[Cu(H_2O)_6]^{2+}$, (b): $[Cu(edta)]^{2-}$, (c): cis-$[Cu(en)_2(H_2O)_2]^{2+}$.

According to the spectrochemical series, the N-donor ligands are usually stronger than the O-donor ligands. The value of Δ_o should therefore increase from an "O_6" donor set to "N_2O_4" (edta) to "N_4O_2".

(iii) All the d-d transitions for octahedral complexes are Laporte forbidden, in particular for $[Cu(H_2O)_6]^{2+}$, which is highly symmetric and contains a center of inversion. For the other two complexes, since there is no center of inversion, this selection rule is partially relaxed.

(iv) The electronic absorption band is broad because the M–L bonds are constantly vibrating. Such ligand motions will be exaggerated through molecular collisions in solution. In the solid state, the absorption band will usually become sharpened due to the restricted M–L vibration. Cu^{2+} is a d^9 ion which is subjected to the Jahn-Teller effect. This will split the degenerate energy level and give additional electronic transitions.

A7.10 A d^6 octahedral Fe(II) complex can be either high-spin ($t_{2g}^4 e_g^2$, $S = 2$) or low-spin (t_{2g}^6, $S = 0$). If the complex has a large paramagnetic susceptibility, it must be high-spin, since the low-spin complex would be diamagnetic. According to the d^6 Tanabe-Sugano diagram, the ground term for the high-spin case (i.e. to the left of the discontinuity) is $^5T_{2g}$. The only other quintet term is ^5E_g, so the only spin-allowed transition is $^5T_{2g} \rightarrow {}^5E_g$.

A7.11 (i)

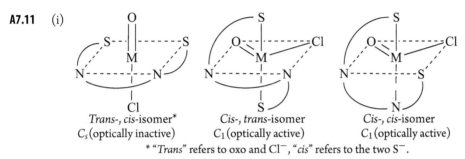

	Trans-, cis-isomer*	Cis-, trans-isomer	Cis-, cis-isomer
	C_s (optically inactive)	C_1 (optically active)	C_1 (optically active)

* "*Trans*" refers to oxo and Cl^-, "*cis*" refers to the two S^-.

(ii) There is an internal mirror for this structure (passing through V, oxo group, N, O, and the benzene ring). So this compound is optically inactive.

A7.12 (i) Crystal field splitting increases as the charge on the metal ion becomes higher; thus $[Ni(IV)F_6]^{2-}$ has a spin-paired configuration.

(ii) The $[CFSE(O_h) - CFSE(T_d)]$ value for Ni^{2+} (d^8) is much larger than that for Co^{2+} (d^7), where CFSE denotes crystal field stabilization energy. Hence, the tetrahedral complex of Co^{2+} is relatively more stable.

A7.13 For V^{2+}, crystal field stabilization energy (CFSE) is $12\,Dq$, which, from the graph shown, corresponds to $170\,kJ\,mol^{-1}$. Hence, $\Delta_o \approx 140\,kJ\,mol^{-1}$. Similar treatment yields a Δ_o value of $92\,kJ\,mol^{-1}$ for Co^{2+}.

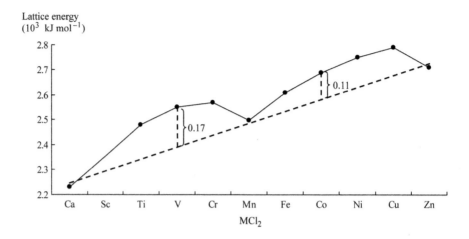

Lattice energy $(10^3 \ kJ \ mol^{-1})$

A7.14   CFSE for each metal ion:

$$Mn^{2+} \text{ and } Fe^{3+}(d^5), \qquad CFSE(O_h) = CFSE(T_d) = 0.$$

$$Mn^{3+}(d^4), \qquad CFSE(O_h) = 6 \, Dq, \, CFSE(T_d) = \frac{16}{9} \, Dq.$$

$$Fe^{2+}(d^6), \qquad CFSE(O_h) = 4 \, Dq, \, CFSE(T_d) = \frac{8}{3} \, Dq.$$

Normal spinels:

$$Fe^{2+}[Mn^{3+}Mn^{3+}]O_4, \, CFSE = 14.7 \, Dq.$$
$$Mn^{2+}[Mn^{3+}Fe^{3+}]O_4, \, CFSE = 6 \, Dq.$$

Inverse spinels:

$$Mn^{3+}[Fe^{2+}Mn^{3+}]O_4, \, CFSE = 11.8 \, Dq.$$
$$Mn^{3+}[Mn^{2+}Fe^{3+}]O_4, \, CFSE = 1.8 \, Dq.$$
$$Fe^{3+}[Mn^{2+}Mn^{3+}]O_4, \, CFSE = 6 \, Dq.$$

Hence, based on CFSE, the normal spinel $Fe^{2+}[Mn^{3+}Mn^{3+}]O_4$ is the expected structure for Mn_2FeO_4.

A7.15   Since the ligand field of Br^- is weaker than F^-, according to the spectrochemical series, $[CoBr_6]^{3-}$ should also be expected to be a high-spin complex. For $[CoX_6]^{3-}$, $X = Cl^-, Br^-,$ or I^-, the high-spin complexes give rise to crystal field stabilization energies

(CFSE) of 4 Dq only. The repulsion of six large ligands causes the complexes to be unstable and transform to the tetrahedral complexes $[CoX_4]^-$ with a CFSE of $6 Dq$ (T_d) or 2.7 Dq (O_h). Although the CFSE is somewhat smaller, the ligand–ligand repulsion in the tetrahedral complexes is considerably reduced. Hence, they are more stable than the corresponding octahedral complexes.

A7.16 (i) The high-spin configuration $t_{2g}^3 e_g^1$ will give rise to a spin-only magnetic moment $[\sqrt{4S(S + 1)}]$ of 4.90 BM. Jahn–Teller distortion is also involved in this case.

(ii) The ground term for Co^{3+} ($3d^6$) is ^5D and the spin-only magnetic moment $[\sqrt{4S(S + 1)}]$ is 4.90 BM. In the gas phase, the magnetic moment of Co^{3+} will not be close to this value since there will be a contribution from the orbital motion.

(iii) High-spin Fe^{3+} octahedral complexes have a $^6A_{1g}$ ground state, which has no angular momentum to contribute to the magnetic moment. However, the ground state of low-spin Fe^{3+} complexes is $^2T_{2g}$, which usually makes an orbital contribution to the magnetic moment. Furthermore, such contribution is effectively quenched at low temperatures.

A7.17 (i) An octahedral complex with no distortion; a d^8 complex with O_h symmetry would have a spin-only magnetic moment of 2.83 BM.

(ii) A tetragonal complex with significant axial elongation; the distortion is not due to the Jahn–Teller effect, but due to the different field strengths of CN^- and H_2O.

(iii) A tetragonal complex with slight axial elongation due to Jahn–Teller effect, since the orbital degeneracy is in the t_{2g} orbitals; the spin-only magnetic moment is 3.87 BM.

(iv) A low-spin complex with tetragonal distortion due to the Jahn–Teller effect; the nature of the distortion cannot be predicted on symmetry grounds. (Axial elongation is likely, which is observed for high-spin Cr^{2+} and Mn^{3+} complexes.)

(v) An octahedral high-spin complex with no distortion.

(vi) A low-spin tetragonal complex with slight axial elongation due to the Jahn–Teller effect.

A7.18 (i) The spin-only magnetic moments for high-spin and low-spin square planar complexes of Co^{2+} (d^7) are 3.87 and 1.73 BM, respectively. The observed value of 2.5 BM indicates that there is a mixture of both states in the sample, the low-spin case being the more dominant ($64\% \times 1.73 + 36\% \times 3.87 = 2.5$). Hence, it may be assumed that the low-spin state is slightly lower in energy. With the distribution given above, it is predicted that μ will increase with increasing temperature.

(ii) Since pyridine is a stronger base than water, some coordination of pyridine along the four-fold axis is expected, which in turn leads to a decrease in tetragonality. In other words, the high-spin state is stabilized, leading to an increase in μ.

A7.19 Tetrahedral complexes with the following configurations are subjected to Jahn–Teller distortion: $d^1(e^1)$, $d^3(e^2t_2^1)$, $d^4(e^2t_2^2)$, $d^6(e^3t_2^3)$, $d^8(e^4t_2^4)$, and $d^9(e^4t_2^5)$. Note that there are very few low-spin tetrahedral complexes of the first transition series metals and they are not considered here. Upon Jahn–Teller distortion, the splitting in t_2 level should be larger than that for e orbitals.

A7.20 In the classical ionic theory, the *trans* isomer is predicted to be more stable since such a configuration minimizes the Cl^--Cl^- repulsion. In the crystal field theory, Jahn–Teller distortion will cause ring strain for the *cis* isomer. For $[Co(en)_2Cl_2]^+$, with a "t_{2g}^6" configuration, no distortion results. So the classical model is invoked: the *trans* isomer is more stable due to the reduced Cl^--Cl^- interaction.

A7.21 (i) Distortion in $[CoF_6]^{3-}$ is small compared to that of $[CuX_6]^{4-}$, since the former is caused by degeneracy in the t_{2g} orbitals and the latter in the e_g orbitals. (Recall that the e_g orbitals have lobes pointing directly at the ligands.) Also, the distortion in $[CoF_6]^{3-}$ is axial compression.

(ii) The $Cr^{2+}(d^4)$ ion in this crystal has a high-spin $(t_{2g}^3e_g^1)$ configuration, which is subject to Jahn–Teller distortion in an octahedral environment. The distortion in this case is axial elongation.

(iii) The Cu–O bonds are shorter than the Cu–Cl bonds, mainly because oxygen has a smaller covalent radius than chlorine. Furthermore, in the electrostatic model, a ligand generating a stronger field is expected to lead to a shorter metal–ligand distance than a weaker ligand. Since H_2O is a stronger ligand than Cl^-, the Cu–O distances are shorter than those for Cu–Cl. In addition, Jahn–Teller distortion leads to two different Cu–Cl lengths.

A7.22 (i) $[Fe(H_2O)_6]^{3+}$; there is a higher charge on the metal.

(ii) $[CoCl_6]^{4-}$; $\Delta_o > \Delta_t$.

(iii) $[CoF_6]^{3-}$; F^- is stronger than Cl^- as a ligand.

(iv) $[Os(CN)_6]^{4-}$; splitting for third transition series metal is greater than that for the first series.

A7.23 It is important to note that, within the crystal field theory, Δ_1 (for square planar MX_4) is exactly the same as Δ_o (for octahedral MX_6). Hence, the trend for Δ_1 is that for Δ_o, as found here.

(i) $\Delta_o(CN^-) > \Delta_o(Cl^-) > \Delta_o(Br^-)$, according to the spectrochemical series.

(ii) Δ_o (third transition series metals) $> \Delta_o$ (second transition series metals).

A7.24 (i) Weak-field or high-spin complex: $^4T_{1g}$; strong-field or low-spin complex: ^2E_g.

(ii) Spin-allowed transitions: $^4T_{1g} \rightarrow {}^4T_{2g}$, $^4T_{1g} \rightarrow {}^4A_{2g}$, and $^4T_{1g} \rightarrow {}^4T_{1g}$. Among these, the second one should be the weakest, since it is a two-electron transition $(t_{2g}^5 e_g^2 \rightarrow t_{2g}^3 e_g^4)$.

(iii) $^2E_g \rightarrow {}^2T_{1g}, {}^2T_{2g}, {}^2A_{1g}$, and $^2A_{2g}$.

(iv) $^4A_{2g}$.

(v) $^4A_2$.

A7.25 (i) $\delta_g(x^2 - y^2, xy) < \pi_g(xz, yz) < \sigma_g^+(z^2)$.

(ii) $e''(xz, yz) < a_1'(z^2) < e'(x^2 - y^2, xy)$.

(iii) $e_g(z^2, x^2 - y^2) < t_{2g}(xy, xz, yz)$.

(iv) $a_1(z^2) < e_2(x^2 - y^2, xy) < e_3(xz, yz)$.

(v) The five d orbitals do not split in an icosahedral crystal field. Together they have the symmetry of the H_g representation of the I_h group.

A7.26 (i) $Re(S_2C_2Ph_2)_3$ is a trigonal–prismatic complex with D_{3h} symmetry as shown. The central Re ion with formal charge of $+6$ is surrounded by three dithiolene ligands

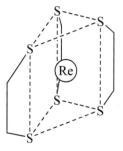

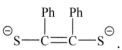

(ii) $e'(x^2 - y^2, xy) < a_1'(z^2) < e''(xz, yz)$.

A7.27 In both splitting patterns: CrO_4^- and MnO_4^-, the (symmetry-allowed) $L \rightarrow M$ charge transfers are responsible for the colors of the complexes.

A7.28

Complex	Molar extinction coefficient	Comment
$[Mn(H_2O)_6]^{2+}$	0.035	With a center of symmetry; spin-forbidden
$[MnBr_4]^{2-}$	4.0	No center of symmetry; spin-forbidden
$[Co(H_2O)_6]^{2+}$	10.0	With a center of symmetry; spin-allowed
$[CoCl_4]^{2-}$	600	No center of symmetry; spin-allowed

A7.29 (i) Transition $^1A_1 \rightarrow {}^1T_2$ in $[\text{Co(en)}_3]^{3+}$; Laporte selection rule is not operative in this instance.

(ii) Transition $^4A_2 \rightarrow {}^4E$, since it is spin-allowed; $^4A_2 \rightarrow {}^2E$ is spin-forbidden.

(iii) Transition $^3A_2 \rightarrow {}^3E$, since $^3A_2 \rightarrow {}^3A_2$ is symmetry-forbidden ($A_2 \times A_2 = A_1$) in D_3 complexes; the former is symmetry-allowed.

(iv) The most intense d–d band in $[\text{NiCl}_4]^{2-}$, since in $[\text{MnCl}_4]^{2-}$ the Mn^{2+} has a high-spin d^5 ground configuration and all transitions are spin-forbidden.

A7.30 (i) Complexes of Fe^{3+}, which is more highly charged than Mn^{2+}, have lower-energy charge-transfer bands. These bands sometimes appear in the near ultraviolet region, thus obscuring the weak spin-forbidden d–d transitions.

(ii) The charge-transfer bands for these complexes are of the $L(np) \rightarrow M(3d)$ type. Since Cl is more electronegative than Br, $[\text{Co(NH}_3)_5\text{Br}]^{2+}$ should have a lower-energy charge-transfer band.

A7.31 (i) (A): $[\text{MnBr}_4]^{2-}$.   (B): $[\text{Mn(H}_2\text{O})_6]^{2+}$.

Tetrahedral complexes, being non-centrosymmetric, have more intense d–d transition bands.

(ii) The ground state of the d^5 system is ^6S, which is non-degenerate, and leads to no splitting in a crystal field. Also, this is the only sextet term arising from the configuration, the rest being either doublets or quartets. Hence, all d–d transitions in a high-spin d^5 complex, regardless of its symmetry, are spin-forbidden and low in intensity. This is why many of these complexes, especially the octahedral ones, are very pale in color (see Problem 7.29). In addition, since $[\text{MnBr}_4]^{2-}$ lacks an inversion center, its spectral intensity is higher than that of $[\text{Mn(H}_2\text{O})_6]^{2+}$.

(iii) Theoretical considerations show that the widths of the d–d transition bands should be proportional to the slope of the upper state relative to that of the lower state. In the present case, where the ground state energy is independent of the crystal field strength, this means that the band widths should be proportional to the slopes of the lines for the respective upper states. Thus, the narrowest bands are those at 25000 and 29750 cm^{-1}, which correspond to the transitions to upper states with zero slope. The widths of the other lines are also seen to be greater in proportion to the slopes of the upper state energy lines.

(iv) Complexes A and B are yellow–green and pale pink (less intense), respectively.

(v)   Band assignments:

A:	21600 cm^{-1}	$^6A_1$	$\rightarrow$ $^4T_1(G)$
	22160 cm^{-1}		$\rightarrow$ $^4T_2(G)$
	23040 cm^{-1} (shoulder)		$\rightarrow$ $^4A_1, ^4E(G)$
	26750 cm^{-1}		$\rightarrow$ $^4T_2(D)$
	27690 cm^{-1}		$\rightarrow$ $^4E(D)$
B:	18870 cm^{-1}	$^6A_{1g}$	$\rightarrow$ $^4T_{1g}(G)$
	23120 cm^{-1}		$\rightarrow$ $^4T_{2g}(G)$
	25000 cm^{-1}		$\rightarrow$ $^4A_{1g}, ^4E_g(G)$
	27980 cm^{-1}		$\rightarrow$ $^4T_{2g}(D)$
	29750 cm^{-1}		$\rightarrow$ $^4E_g(D)$
	32960 cm^{-1}		$\rightarrow$ $^4T_{1g}(D)$

[Readers are not expected to obtain four significant figures for the bands from the given spectra.]

A7.32   A : $[CoCl_4]^{2-}$, tetrahedral complexes have more intense d–d transition due to partial breakdown of the Laporte selection rule.

B: $[Co(NH_3)_6]^{3+}$,   C: $[Co(NH_3)_6]^{2+}$,   D: $[Co(H_2O)_6]^{2+}$.

NH$_3$ is a stronger ligand than H$_2$O; Co^{3+} causes a larger ligand field splitting compared with Co^{2+}.

A7.33   (i)   For Cr^{3+} (d^3), the ground term is ^4F. Therefore the diagram on the right should be applicable.

(ii)   Using the spin selection rule, the following assignments can be made: 17400 cm^{-1} ($^4A_{2g} \rightarrow {}^4T_{2g}$), 24700 cm^{-1} ($^4A_{2g} \rightarrow {}^4T_{1g}(F)$), and 37000 cm^{-1} ($^4A_{2g} \rightarrow {}^4T_{1g}(P)$).

(iii)   The color of the $[Cr(H_2O)_6]^{3+}$ ion is green.

(iv)   For the first transition series ions the spin-only formula should be used. The expected magnetic moment is $\sqrt{n(n+2)} = 3.87$ BM.

A7.34   (i)   The *trans* isomer has a center of symmetry, which preserves the g (even symmetry) character of the energy levels. According to Laporte's rule, transitions between such levels are forbidden.

(ii)   The splitting of the bands, or the energy levels, is caused by the difference in field strengths of ligands F$^-$ and en, yielding a tetragonal structure with axial elongation for the *trans* isomer. On the other hand, the tetragonality of the *cis* isomer is not as marked; hence, the splitting of the band is not as pronounced.

A7.35 (i) Based on the crystal field strengths of the ligands $[\Delta_o(\text{en}) > \Delta_o(\text{H}_2\text{O})]$ and the intensities of the bands ($[\text{Ni}(\text{H}_2\text{O})_6]^{2+}$ has an inversion center), it is obvious that the "dotted-curve" spectrum belongs to $[\text{Ni}(\text{en})_3]^{2+}$.

(ii) The color of $[\text{Ni}(\text{H}_2\text{O})_6]^{2+}$ is green; $[\text{Ni}(\text{en})_3]^{2+}$ is blue.

(iii) In ascending order of energy: $^3A_{2g} \rightarrow {}^3T_{2g}$, $^3A_{2g} \rightarrow {}^3T_{1g}(F)$, and $^3A_{2g} \rightarrow {}^3T_{1g}(P)$, assuming idealized O_h symmetry for both complexes.

(iv) The splitting is due to the spin-orbit interaction between the $^3T_{2g}$ and ^1E_g states, which happens to lie closely in energy to the Δ_o value for the aquo complex.

A7.36 (i) If the energy of the ground state $^3A_{2g}$ is taken as zero and the transitions from this state to states $^3T_{2g}$, $^3T_{1g}(F)$, and $^3T_{1g}(P)$ have energies $\tilde{\nu}_1$, $\tilde{\nu}_2$, and $\tilde{\nu}_3$, respectively, then

$$\tilde{\nu}_1 = 10\,Dq, \tag{1}$$

$$\tilde{\nu}_2 = 18\,Dq - x, \tag{2}$$

$$\tilde{\nu}_3 = 12\,Dq + 15\,B' + x. \tag{3}$$

Thus $1\,Dq = \tilde{\nu}_1/10 = 850\,\text{cm}^{-1}$.

Appropriate combinations of Equations (1), (2), and (3) yield

$$15\,B' = \tilde{\nu}_3 + \tilde{\nu}_2 - 3\tilde{\nu}_1,$$

which leads to $B' = 1060\,\text{cm}^{-1}$.

(ii) The ion Cr^{3+} has a d^3 configuration and a ^4F ground term. Under the influence of a crystal field with O_h symmetry, this ion has a splitting pattern analogous to that of Ni^{2+}. Since $\tilde{\nu}_3$ has the highest energy, it is the band which is masked. However, for Dq, the calculation is similar to that in (i):

$$1\,Dq = \tilde{\nu}_1/10 = 1750\,\text{cm}^{-1}.$$

For B', it is known that,
$$B'/B = \beta = 0.7,$$

hence $B' = \beta B = 0.7 \times 918 = 640\,\text{cm}^{-1}$.

A7.37 (i) $[\text{Ti}(\text{H}_2\text{O})_6]^{3+}$ has a d^1 metal center and there is only one d-d transition from t_{2g} orbital to e_g orbital. These Δ_o can be calculated from the $\lambda_{\max}$ of this band. For most other octahedral ions, there are electron-electron repulsions which complicate the situation. Hence, Δ_o usually cannot be determined directly.

(ii) The charge-transfer transitions are from ligand to metal (LMCT transitions). Along the series, the oxidation state of the metal changes from +V to +VII, so the energy levels of the 3d metal orbitals are lower across the series. This reduces the gap between the ligand and the metal energy levels. Hence, MnO_4^- has the longest-wavelength LMCT transition.

A7.38 The basic reason for the difference is that the electrons responsible for the electronic transitions of lanthanide ions are 4f electrons, and the 4f orbitals are very effectively shielded from external influence by the overlying 5s and 5p subshells. Hence, the states arising from the various $4f^n$ configurations are only slightly split (of the order of 10^2 cm^{-1}) by external fields. Thus, when electronic transitions occur from one state to another of an $4f^n$ configuration, the absorption bands (f–f transitions) are extremely sharp and the broadening effect of ligand vibration is minimized.

A7.39 In the group O_h, the coordinates x, y, and z jointly form a basis for the T_{1u} representation. For the $^1A_{1g} \rightarrow {}^1T_{1g}$ transition,

$$A_{1g} \times T_{1u} \times T_{1g} = A_{1u} + E_u + T_{1u} + T_{2u}.$$

Thus, while the pure electronic transition $^1A_{1g} \rightarrow {}^1T_{1g}$ is forbidden, it will occur if there is a simultaneous excitation of a vibration of A_{1u}, E_u, T_{1u}, or T_{2u} symmetry. Similarly, for the $^1A_{1g} \rightarrow {}^1T_{2g}$ transition,

$$A_{1g} \times T_{1u} \times T_{2g} = A_{2u} + E_u + T_{1u} + T_{2u}.$$

Hence, this transition will have non-zero intensity if it is accompanied by the excitation of a A_{2u}, E_u, T_{1u}, or T_{2u} vibration.

A7.40

Band (cm^{-1})	Polarization	Assignment
16350	z	$^3A_2 \rightarrow {}^3A_1({}^3T_{2g})$
16500	(x, y)	$^3A_2 \rightarrow {}^3E({}^3T_{2g})$
23000	(x, y)	$^3A_2 \rightarrow {}^3E({}^3T_{2g})$

A7.41 (i) Ground state configuration: $(1a_{1g})^2(1a_{2u})^2(1e_g)^4(1e_{1u})^4(1e_{2g})^4(2a_{1g})^2$. Ground state: $^1A_{1g}$.

(ii)

Band (cm^{-1})	ε	Assignment	Comment
18900	7	$2a_{1g} \rightarrow 2e_{1g}$	$^1A_{1g} \rightarrow {}^3E_{1g}$; spin-forbidden
20900	<1	$1e_{2g} \rightarrow 2e_{1g}$	$^1A_{1g} \rightarrow {}^3E_{1g}$ or $^3E_{2g}{}^\dagger$; spin-forbidden
21800	36	$2a_{1g} \rightarrow 2e_{1g}$	$^1A_{1g} \rightarrow {}^1E_{1g}$; spin-allowed
22400	<1	$1e_{2g} \rightarrow 2e_{1g}$	$^1A_{1g} \rightarrow {}^3E_{1g}$ or $^3E_{2g}$; spin-forbidden
24000	72	$1e_{2g} \rightarrow 2e_{1g}$	$^1A_{1g} \rightarrow {}^1E_{1g}$ or $^1E_{2g}{}^\dagger$; spin-allowed
30800	49	$1e_{2g} \rightarrow 2e_{1g}$	$^1A_{1g} \rightarrow {}^1E_{1g}{}^\dagger$ or $^1E_{2g}$; spin-allowed

†Upon quantitative treatment, these assignments are preferred.

Molecular Orbital Theory | 8

PROBLEMS

8.1   The particle-in-a-box wavefunctions and their corresponding energies are:

$$\psi_n(x) = \sqrt{\frac{2}{a}}\sin\left(\frac{n\pi x}{a}\right), \quad 0 \leq x \leq a \ (a \text{ is the length of the box});$$

$$E_n = \frac{n^2 h^2}{8ma^2}; \ n = 1, 2, 3, \ldots.$$

(i)   The conjugated polyenes butadiene (A), vitamin A (B), and carotene (C) are, respectively, colorless, orange-yellow, and ruby red. Explain the colors of these compounds qualitatively in terms of the free-electron model.

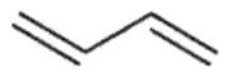

(A) 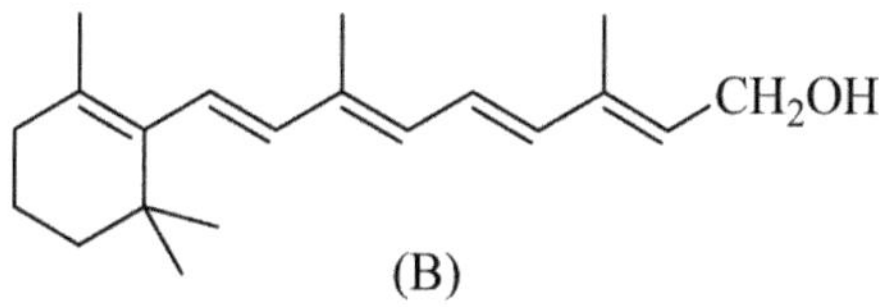   (B)

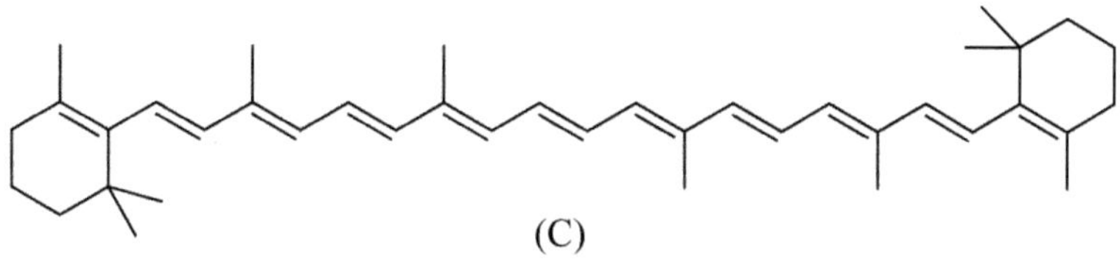

(C)

(ii)   Vitamin A absorbs most strongly at 3320 Å. Estimate the length of the "box".

8.2   Consider the following free-electron model for conjugated polyenes:

(a)   The π electrons of a polyene of N carbon atoms are to be treated as a system of free electrons on a wire of length $(N+1)\ell$, where $\ell = 1.40$ Å, the average length of single and double carbon–carbon bonds, and the positions of the atoms are to be taken as $x = \ell, 2\ell, 3\ell, \ldots, N\ell$.

Problems in Structural Inorganic Chemistry. Second edition. Wai-Kee Li, Hung Kay Lee, Dennis Kee Pui Ng, Yu-San Cheung, Kendrew Kin Wah Mak, and Thomas Chung Wai Mak. © Oxford University Press 2019. Published in 2019 by Oxford University Press. DOI: 10.1093/oso/9780198823902.001.0001

(b) At most two electrons (of opposite spins) can be accommodated in the same π orbital.

For the polyene decapentaene, $C_{10}H_{12}$, in its ground state,

(i) write down expressions for all orbitals $\psi_n(x)$ and their associated energies E_n;

(ii) predict the wavelength of the first absorption band; and

(iii) calculate the charge densities at the center of each carbon–carbon bond, and comment on the results. The overall charge density distribution at any point is given by $\rho(x) = \Sigma 2\psi_n^2(x)$ (summed over doubly occupied orbitals).

8.3 In an extremely crude model of anthracene, $C_{14}H_{10}$, the π electrons are considered to be confined to a rectangular box of dimensions $4\ \text{Å} \times 7\ \text{Å}$. Using the appropriate particle-in-a-box energy expression, calculate the wavelength (in Å) of the transition from the ground state to the first excited state.

8.4 The energies of monocyclic conjugated polyenes can be obtained graphically by means of a Frost–Hückel circle. To achieve this, we first draw a circle with a radius of $2|\beta|$ (recalling that $\beta < 0$), then we place the n-sided regular polygon inside this circle, making sure that one of the vertices touches the bottom of the circle, i.e., the y-coordinate of this vertex is 2β (see the figure on the right). Then the energies of the cyclic polyene can be "read off" from the y-coordinates of the remaining vertices.

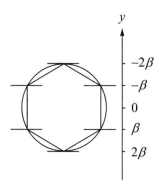

Let us take the simple case of benzene as an example. When we put the hexagon inside our circle, we can readily see that the energies are $\alpha + 2\beta < \alpha + \beta$ (twice) $< \alpha - \beta$ (twice) $< \alpha - 2\beta$.

Based on the Frost–Hückel circle for a general monocyclic polyene,

(i) show that the $4n + 2$ rule for aromaticity is a direct consequence of this circle;

(ii) determine the general formula for the energies of the energy levels of the π molecular orbitals.

8.5 Using the Frost–Hückel circle (see Problem 8.4), deduce the π energy levels of the cyclooctatetraene dianion, $C_8H_8^{2-}$, and hence estimate its delocalization energy. Match the energy levels with sketches of Hückel molecular orbitals.

8.6 In the particle-in-a-ring model for benzene proposed by J. R. Platt, the π electrons are considered to move around the perimeter of a circle of constant potential bounded by infinite walls of the circle.

   (i)   Write down the appropriate Schrödinger equation.

  (ii)   State the boundary condition and solve for the energies and wavefunctions.

 (iii)   Point out any differences between the state energy schemes for the particle-in-a-ring and particle-in-a-one-dimensional-box problems.

REFERENCE: J. R. Platt, Classification of spectra of cata-condensed hydrocarbons. *J. Chem. Phys.* **17**, 484–95 (1949).

8.7 In a very crude bonding model of the hydrogen molecular ion, H_2^+, the electron may be considered as a particle in a one-dimensional box whose width is the same as the internuclear separation r.

   (i)   Express the kinetic energy T of the electron as a function of r. (Hint: Make use of the de Broglie relationship.)

  (ii)   Assuming the average potential energy V of the electron to be that of an electron at rest midway between the nuclei, express the total energy E as a function of r.

 (iii)   Calculate the equilibrium bond distance r_0 in Å and the ground state energy E_0 in eV. The experimental values are 1.06 Å and -16.3 eV, respectively.

8.8 In a simple LCAO–MO treatment of a homonuclear diatomic molecule the two solutions for the energy are $E_{\pm} = \dfrac{H_{11} \pm H_{12}}{1 \pm S}$. When the two atoms are infinitely apart, the total energy of the system will be $2H_{11}$, the energy of the two separated atoms. Using this as the reference state, the potential energy of a two-electron homonuclear diatomic molecule is given by $V(r) = \dfrac{2H_{11} + 2H_{12}}{1 + S} - 2H_{11}$.

   (i)   Express the potential energy $V(r)$ as a function of overlap integral S by making use of Cusachs' approximation for the exchange integral: $H_{12} = \frac{1}{2}S(2 - S)(H_{11} + H_{22})$. Sketch the plot $V(r)$ against S, keeping in mind that H_{11} is negative and is a constant with respect to r and S. Also note that $H_{11} = H_{22}$ for a homonuclear diatomic molecule.

  (ii)   Calculate the value of S at the minimum in the potential. Hence express $V(r_e)$ (r_e is the equilibrium bond distance) in terms of H_{11}.

 (iii)   Koopmans' theorem states that the first ionization energy (IE) of a molecule is the negative value of the energy of the highest filled molecular orbital. Use this to express IE in terms of H_{11}.

(iv) Calculate the energy of the antibonding molecular orbital. Hence obtain the energy of the first spectral transition (Δ) in terms of H_{11}.

(v) For the hydrogen molecule, H_{11} has the value -13.6 eV. Calculate the bond dissociation energy D_e, IE, and Δ. The experimental values are 4.75, 15.42, and 11.4 eV, respectively.

REFERENCE: W. F. Cooper, G. A. Clark and C. R. Hare, A simple, quantitative molecular orbital theory. *J. Chem. Educ.* **48**, 247–51 (1971).

8.9 (i) Construct labeled molecular orbital energy level diagrams for the H_2^+ and H_2 molecules and point out their differences.

(ii) Write down expressions for the normalized molecular orbital (MO) and valence bond (VB) wavefunctions for the ground state of (a) H_2^+ and (b) H_2 in terms of a and b (denoting 1s functions on the two nuclei a and b); and S (the overlap of a and b). Are the MO and VB wavefunctions identical for each of H_2^+ and H_2?

(iii) The VB wavefunction for the ground state of H_2 in (ii), ψ_{VB}, has only covalent terms and no ionic terms. To improve ψ_{VB}, we can mix in an optimal amount of ionic character:

$$\psi_1 = \psi_{VB} + \lambda \psi_i, \qquad \text{where } \psi_i = (2 + 2S^2)^{-1/2}[a(1)a(2) + b(1)b(2)].$$

On the other hand, the MO wavefunction for the ground state of H_2 in (ii), ψ_{MO}, can be improved by mixing in an optimal amount of configuration $(\sigma_{1s}^*)^2$:

$$\psi_2 = \psi(\sigma_{1s}^2) + \mu \psi(\sigma_{1s}^{*2}),$$

where $\psi(\sigma_{1s}^{*2}) = (2 - 2S)^{-1}[a(1) - b(1)][a(2) - b(2)]$.

Note that ψ_1 and ψ_2 are not normalized, S is the overlap integral, and the parameters λ and μ are to be determined variationally.

Show that $\lambda = \dfrac{1 + \mu - (1 - \mu)S}{1 - \mu - (1 + \mu)S}$ if ψ_1 and ψ_2 are identical (apart from a numerical factor).

(iv) The single configuration, $(\sigma_{1s})^2$, is a good approximation for a stable H_2 molecule. However, significant error arises if only this single configuration is employed for the dissociation of H_2. Explain what this configuration represents for dissociating H_2 molecule and hence why significant error arises. Hint: interpret the configuration in terms of ψ_{VB} and ψ_i.

(v)   Explain why the dissociated H_2 can be exclusively 2H (without "mixing" with the ionic form, $H^+ + H^-$) whereas we cannot speak of a pure "$H \cdot \cdot H$" form or a pure "$H^+ H^-$" form for real H_2 molecules.

8.10   (i)   Consider the $^3\Sigma_u^+$ state of H_2, which is the triplet state arising from the configuration $(\sigma_{1s})^1 (\sigma_{1s}^*)^1$. Let a and b denote 1s orbitals on atoms a and b, respectively. Also let $g (= a + b)$ and $u (= a - b)$ denote the wavefunction for σ_{1s} and that for σ_{1s}^*, respectively. Write down the total, but unnormalized, valence bond (VB) and molecular orbital (MO) wavefunctions for this state. Are these two wavefunctions different?

(ii)   Using the aforementioned notation, write down the determinantal MO and VB wavefunctions for the $(\sigma_{1s})^2 (\sigma_{1s}^*)^2$ configuration of He_2. Furthermore, show that these two wavefunctions are identical, aside from a numerical factor, by making use of the following general theorem on determinants: if each column (or row) of a determinant is a linear combination of the columns (or rows) of another determinant, the two determinants are identical, aside from a numerical factor.

(iii)   Why do the MO and VB treatments for the ground state $(^1\Sigma_g^+)$ for hydrogen molecule lead to different results, yet they do not for the problems for the $^3\Sigma_g^+$ state for H_2 and the ground state of He_2?

REFERENCE: W.-K. Li, MO and VB wave functions for He_2. *J. Chem. Educ.* **67**, 131 (1990).

8.11   The positronium atom (Pos) was introduced in Problem 3.2. In this problem, we are concerned with the positronium molecule, $(Pos)_2$. Indirect experimental evidence [D. B. Cassidy and A. P. Mills, Jr., The production of molecular positronium. *Nature* **449**, 195–7 (2007)] for $(Pos)_2$ was reported in 2007, 61 years after the prediction of its existence from theory.

It is noted that this molecule consists of four particles of the same mass, two electrons (e^-) and two positrons (e^+). A simple variational wavefunction for $(Pos)_2$ is:

$$\psi = e^{-ar_{13} - br_{14} - ar_{24} - br_{23}} + e^{-br_{13} - ar_{14} - br_{24} - ar_{23}}.$$

In this expression, a and b are parameters to be optimized. In addition, subscripts "1" and "2" are for the two positrons, while subscripts "3" and "4" are for the two electrons. Hence, r_{13} is simply the distance between e^+ #1 and e^- #3, etc. The inclusion of the second term in the wavefunction ensures the indistinguishability of the two electrons (as well as of the two positrons).

Determine whether the spins of the two electrons are paired or unpaired. Do the same for the spins of the two positrons.

8.12 Consider isoelectronic diatomic molecules LiF and C_2. How would their molecular orbital energy level diagrams contrast?

8.13 Compare the multi-center bonds in XeF_2 with the ones in B_2H_6.

8.14 (i) For $H_2O(C_{2v}$ symmetry), write down the molecular orbitals derived from the valence orbitals and determine the ground electronic configuration and state.

(ii) With the results obtained in (i), rationalize why the ground state of CH_2 is a triplet, while that of CF_2 is a singlet.

REFERENCE: C.-K. Lee and W.-K. Li, INDO study of some substituted methylenes. *Chem. Phys. Lett.* **46**, 523–6 (1977).

8.15 Predict the geometry of sulfur trioxide, SO_3, and describe the bonding in terms of both the valence bond (VB) and molecular orbital (MO) theories.

8.16 The atomic orbitals involved in the bridging BHB bonds in diborane (symmetry D_{2h}) are labeled as in the figure on the right. Give a molecular orbital description of the bonding.

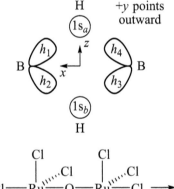

8.17 The compound $K_4[Ru_2Cl_{10}O] \cdot H_2O$ contains a binuclear anion having the structure shown in the accompanying figure. Discuss the bonding in this anion using a simple molecular orbital treatment of the linear Ru–O–Ru moiety by considering the overlap of metal and oxygen orbitals of compatible symmetry. Your answer should include a qualitative energy level diagram and a prediction of the kind of magnetism expected of the anion. For simplicity assume d^2sp^3 hybridization for the ruthenium atoms in the σ framework.

8.18 Sketch the molecular orbital interactions for acetylene coordinated to a transition metal atom. Comment on their relative importance.

8.19 The two-electron bond in gas-phase HF may be described in either of the following ways:

$$\psi_{MO}(1,2) = [a\phi_H(1) + b\phi_F(1)][a\phi_H(2) + b\phi_F(2)],$$
$$\psi_{VB}(1,2) = c[\phi_H(1)\phi_F(2)] + d[\phi_H(2)\phi_F(1)],$$

where ϕ_H is the 1s orbital of hydrogen, ϕ_F is the $2p_z$ orbital of fluorine, and a, b, c, d are coefficients to be optimized.

   (i)   Expanding $\psi_{MO}(1, 2)$, there will be four terms in the wavefunction. Which of the four terms would be the most important? Why?

  (ii)   Which of the two wavefunctions, $\psi_{MO}(1, 2)$ or $\psi_{VB}(1, 2)$, provides a better description of the H–F bond? Justify your choice clearly.

8.20   (i)   Construct the molecular orbital energy level diagram for the OH radical. In doing so, you only need to consider the 1s orbital of H and the 2p orbitals of O. A convenient coordinate system is given on the right.

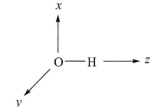

      (ii)   Write down the ground electronic configuration for OH. What type of molecular orbital does the unpaired electron of OH occupy: bonding, nonbonding, or antibonding? If it is a nonbonding orbital, on which atom is this orbital localized?

    (iii)   The first ionization energy of OH is 13.2 eV, while that of HF is 16.1 eV. The difference between these two values (2.9 eV) is very similar to the difference (2.8 eV) between the first ionization energies of O (15.8 eV) and F (18.6 eV). Why?

8.21 Tetrasulfur tetranitride, S_4N_4, has a cradle structure belonging to point group D_{2d}. An understanding of its structure is possible on the basis of transannular interaction between opposite pairs of atoms in a hypothetical planar S_4N_4 system with D_{4h} symmetry.

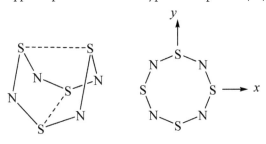

(i) In the planar S_4N_4 system, each atom contributes one valence electron to a σ bond pair and two electrons to an exocyclic lone pair. The π molecular orbitals constructed from the nitrogen $2p_z$ and sulfur $3p_z$ orbitals are illustrated below. (The open and black circles indicate that the positive and negative lobes of the p orbitals point towards and away from the reader, respectively.) Arrange these in a qualitative energy level diagram with appropriate symmetry labels and indicate their occupancies.

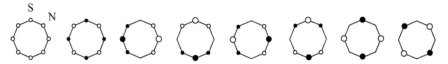

(ii) With reference to the molecular orbital scheme in (i), comment on the instability of planar S_4N_4 and explain how a symmetric distortion yielding a cradle structure can stabilize the system.

(iii) The $p_\sigma - p_\sigma$ overlap integral Σ (for N, P, S, and As) as a function of interatomic distance R (Å) is shown in the following diagram.

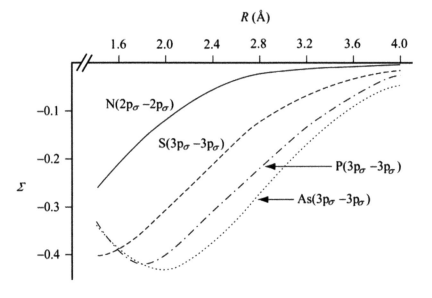

Make use of this information to suggest structures for the related molecules $\alpha - P_4S_4$ and As_4S_4 (realgar). [Hint: consider the relative values of Σ for different pairs of atoms at $R \approx 2.4$ Å.]

REFERENCE: R. Gleiter, Structure and bonding in cyclic sulfur–nitrogen compounds — molecular orbital considerations. *Angew. Chem. Int. Ed.* **20**, 444–52 (1981).

8.22 Walter Yeranos has applied a semi-empirical (extended Hückel) molecular model to XeF_4. It is found that the ground electronic configuration for this compound is

$$\cdots [a_{2g}(-17.5)]^2[3a_{1g}(-12.1)]^2[2a_{2u}(-10.6)]^2[3e_u(-5.2)]^0[4a_{1g}(-2.6)]^0$$
$$[5a_{1g}(-2.0)]^0[3b_{1g}(-1.9)]^0[3e_g(-1.8)]^0[3b_{2g}(-1.5)]^0.$$

The orbital energies, in eV, are given in parentheses in the above expression.

Based on the foregoing results, make a plausible assignment for the following transitions observed in this compound's electronic transition: 5.43, 6.8, 8.3, 9.4, and 11.3 eV. For each proposed assignment, indicate whether the transition is symmetry-allowed or not. Since the results are based on a crude approximation, the calculated and observed values may not match perfectly.

REFERENCE: W. Yeranos, Semi-empirical molecular orbital energy levels of XeF_4. *Mol. Phys.* **11**, 85–92 (1966).

8.23 The extended Hückel molecular orbital energy levels for the trigonal bipyramidal complex $[CuCl_5]^{3-}$ (D_{3h} symmetry) have been calculated by Hatfield and co-workers. The ground configuration for this complex is (orbital energies, in units of $10^3\,cm^{-1}$, are given in parentheses):

$$\cdots [2e'(-120.4)]^4[2a_2''(-115.3)]^2[1e''(-112.2)]^4[3e'(-104.2)]^4[4e'(-103.5)]^4$$
$$[3a_2''(-103.4)]^2[1a_2'(-103.2)]^2[2e''(-103.0)]^4[5e'(-101.7)]^4[3e''(-100.1)]^4$$
$$[5a_1'(-94.0)]^1[4a_2''(14.2)]^0[6a_1'(23.5)]^0[6e'(70.0)]^0.$$

Calculate the wavenumber of the six lowest-energy symmetry-allowed electric dipole transitions.

REFERENCE: W. E. Hatfield, H. D. Bedon, and S. M. Horner, Molecular orbital theory for the pentachlorocuprate(II) ion. *Inorg. Chem.* **4**, 1181–4 (1965).

8.24 Sodium nitroprusside is sometimes used in emergencies to bring down high blood pressure. The nitroprusside anion, $[Fe(CN)_5NO]^{2-}$, is an octahedral complex with C_{4v} symmetry. From the molecular orbital energy level diagram, the ground electronic configuration is (orbital energies, in $10^3\,cm^{-1}$, are given in parentheses):

$$\cdots [6e(-90.7)]^4[2b_1(-86.1)]^2[7e(-65.1)]^0[3b_1(-52.5)]^0$$
$$[5a_1(-51.1)]^0[8e(-36.0)]^0 \cdots$$

The electronic spectrum of $[Fe(CN)_5NO]^{2-}$ exhibits the following bands (in 10^3 cm^{-1}):

20.1 (x, y-polarized), 25.4(z-polarized), 30.3(not polarized),

37.8(x, y-polarized), 42.0(x, y-polarized), 50.0(x, y-polarized; $\varepsilon = 24000$).

(i) Make an assignment for the observed bands. Comment on your assignments. Further information: for the "parent" complex, $[Fe(CN)_6]^{4-}$, the following bands are observed: $31.0 \left[{}^1A_{1g} \left(t_{2g}^6 \right) \rightarrow {}^1T_{1g} \left(t_{2g}^5 e_g^1 \right) \right]$, $37.0 \left[{}^1A_{1g} \left(t_{2g}^6 \right) \rightarrow {}^1T_{2g} \left(t_{2g}^5 e_g^1 \right) \right]$, and $45.0 \left[t_{2g} \rightarrow \pi^*CN \right]$.

(ii) Comment on the ground electronic configuration of $[Fe(CN)_5NO]^{3-}$ and the molecular geometry of this anion.

REFERENCE: P. T. Manoharan and H. B. Gary, Electronic structure of nitroprusside ion. *J. Am. Chem. Soc.* **87**, 3340–8 (1965).

8.25 In a very early extended Hückel molecular orbital treatment for MnO_4^- (T_d symmetry), the ground electronic configuration was determined to be:

$$[1a_1(-276.5)]^2[1t_2(-260.5)]^6[1e(-136.6)]^4[2t_2(-126.3)]^6[2a_1(-112.1)]^2$$
$$[3t_2(-105.7)]^6[t_1(-96.4)]^6[2e(-73.0)]^0[4t_2(-49.3)]^0[5t_2(-18.8)]^0[3a_1(128.8)]^0.$$

In the above expression, the energies of the orbitals, in 10^3 cm^{-1}, are included in parentheses. Based on the data given, make a plausible assignment for the following transitions observed in this ion's electronic spectrum: 14.5 (weak), 18.3 (strong), 28.0 (weak), 32.2 (strong), and 44.0 (strong) (values in 10^3 cm^{-1}). For each proposed transition, indicate the states as well as the configuration involved.

Note: In the extended Hückel approximation, the transition energy is merely the energy difference between the orbitals involved. In other words, in this approximation, the states arising from the same configuration have the same energy. Since the approximation is a crude one, it is not expected that the calculated and observed transition energies match perfectly.

REFERENCE: A. Viste and H. B. Gray, The electronic structure of permanganate ion. *Inorg. Chem.* **3**, 1113–23 (1964).

8.26 Biphenylene has the structure shown on the right. It has D_{2h} symmetry, and a set of Cartesian coordinates is also given in the figure. The 12 π molecular orbitals for this molecule are shown below:

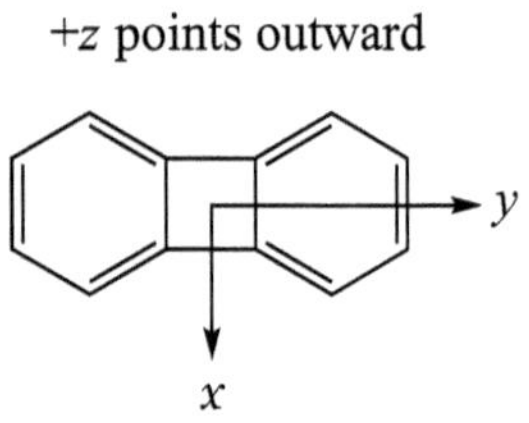

+z points outward

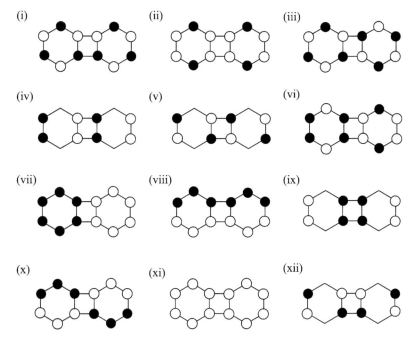

Here the shaded and unshaded circles denote positive and negative atomic orbitals, respectively. Also, and most importantly, only the signs of the upper lobes of the π-bonding p orbitals are shown. Make use of the D_{2h} character table to determine the irreducible representation of each of the 12 molecular orbitals. Show your reasoning clearly.

8.27 Cyclobutadiene iron tricarbonyl has been synthesized by Pettit and co-workers:

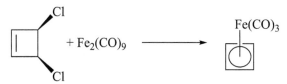

In molecular orbital terminology, give a detailed account of the cyclobutadiene–metal bonding. (The labeling of the carbon atoms of cyclobutadiene as well as the coordinate system chosen for cyclobutadiene and Fe are shown on the right.)

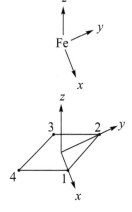

REFERENCE: G. F. Emerson, L. Watts, and R. Pettit, Cyclobutadiene- and benzocyclobutadiene-iron tricarbonyl complexes. *J. Am. Chem. Soc.* **87**, 131–3 (1965).

8.28 A hypothetical molecule AH_9 has the structure of a tricapped trigonal prism (D_{3h} symmetry). The labeling of the hydrogen atoms and the coordinate system chosen for this molecule are shown on the right.

It is noted that the $+x$-axis bisects the a-d distance (i.e., atom h is on the $-x$-axis) and the $+y$-axis is pointing out of the face of $abed$, resulting in a right-handed coordinate system.

Determine the symmetry classifications and the expressions of the nine linear combinations of ligand orbitals.

8.29 A hypothetical molecule AH_5 has a square pyramidal structure as shown below. Also shown in the figure are the coordinates adopted and the labeling of the hydrogens.

(i) Taking advantage of the C_{4v} symmetry of the system, determine the irreducible representations of the five linear combinations of the (five) hydrogen 1s orbitals.

(ii) Match by symmetry the five hydrogen orbital linear combinations with the ns and np orbitals of the central atom A.

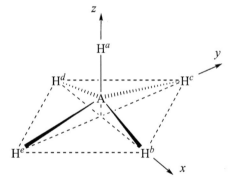

(iii) Construct a schematic energy level diagram for the molecular orbitals formed. In constructing this diagram, you may assume the following energy ordering for the participating atomic orbitals: $E(ns \text{ of A}) < E(1s \text{ of H}) < E(np \text{ of A})$. In addition, it is known that central atom A has the configuration $ns^2 np^3$; determine the ground configuration and the ground electronic state of AH_5.

8.30 Consider the squarate dianion $C_4O_4^{2-}$ having a D_{4h} planar structure and a fully delocalized π system as shown on the right.

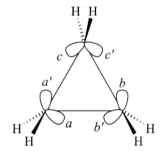

(i) Work out the represention $\Gamma_{C\pi}$ based on the four p_z orbitals of the C atoms.

(ii) Work out the representation $\Gamma_{O\pi}$ based on the four p_z orbitals of the O atoms.

(iii) Sketch the SALCs (symmetry-adapted linear combinations) of $\Gamma_{C\pi}$ and $\Gamma_{O\pi}$ and then combine them to obtain the molecular orbitals of $C_4O_4^{2-}$. Label their symmetries, arrange them in the order of increasing energy, and put in the electrons.

(iv) If a transition metal atom M approaches the $C_4O_4^{2-}$ ion along the z-axis, which atomic orbitals of M can interact with the π molecular orbitals of $C_4O_4^{2-}$?

8.31 The "atomic" orbitals participating in the bonding of the carbon ring of cyclopropane are shown on the right.

(i) Taking advantage of the D_{3h} symmetry of the system, construct the symmetry-adapted wavefunctions of the six molecular orbitals for the ring and arrange them in ascending order of energy. Hint: Construct three localized bonding orbitals (such as $a+b'$) and three localized antibonding orbitals (such as $a-b'$) from the atomic orbitals. Then construct two sets of delocalized symmetry-adapted wavefunctions: one set from the localized bonding orbitals and the other set from the localized antibonding orbitals.

(ii) Assuming Hückel approximation (Coulomb integral equal to α; $H_{ab'} = H_{bc'} = H_{ca'} = \beta$), determine the Hückel energies for the molecular orbitals obtained in (i).

8.32 A hypothetical molecule AH_4 has a "disphenoidal" structure displayed on the right. Also shown in the figure are the coordinates adopted and the labeling of the hydrogen atoms.

(i)   Taking advantage of the C_{2v} symmetry of the system, construct the four linear combinations of the hydrogen 1s orbitals and match them by symmetry with the ns and np orbitals of the central atom A.

(ii) Given that the ground electronic configuration for atom A is ... $(ns)^2(np)^4$, and that the energy ordering of the participating atomic orbital is $E(ns$ of A$) < E(1s$ of H$) < E(np$ of A$)$, construct a schematic energy level diagram for the molecular orbitals formed from the results of (i). Based on your energy level diagram, determine the ground configuration and the ground state of AH_4. In addition, determine the lowest-energy spin- and symmetry-allowed electronic transition of this hypothetical molecule. State clearly the orbitals and states involved as well as the type(s) of polarization of this transition.

8.33 The hydrogen cluster H_8 has a tetragonal structure with D_{4h} symmetry, as displayed in the figure shown below. Here the four upper H atoms are labeled from a to d, and the four lower H atoms are labeled from a' to d', with atom a directly above atom a', and so on. It is straightforward to deduce the symmetries of the linear combinations of the eight H 1s orbitals:

$$\Gamma_H = A_{1g} + B_{1g} + E_g + A_{2u} + B_{2u} + E_u.$$

(i)   Using the projection operator method or otherwise, derive the eight normalized linear combinations of the hydrogen 1s orbitals. Determine the wavefunctions of the eight molecular orbitals and classify them to the symmetry species of the D_{4h} point group. Advisory note: Applying the "formal" projection operator method may not be the quickest way to arrive at your answers.

(ii) Pictorially illustrate the eight molecular orbitals you obtained in (i) and arrange them in order of increasing energy.

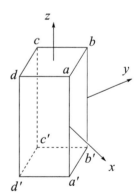

8.34 A convenient coordinate system for diborane $(B_2H_6$, with D_{2h} symmetry$)$ is displayed below. Note that, for the two boron atoms, one coordinate system is right-handed, while the other one is left-handed. Also, hydrogens a, b, c, and d are on the yz-plane, while hydrogens e and f are on the xz-plane.

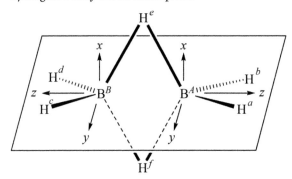

To construct the molecular orbitals of this molecule, we first linearly combine the 2s and 2p orbitals of the two boron atoms in the following manner (where we use subscripts "A" and "B" to distinguish the two boron atoms):

Symmetry	Boron atomic orbitals	Hydrogen atomic orbitals	Molecular orbitals
	$(2)^{-1/2}(2s_A + 2s_B)$		
	$(2)^{-1/2}(z_A + z_B)$		
	$(2)^{-1/2}(2s_A - 2s_B)$		
	$(2)^{-1/2}(z_A - z_B)$		
	$(2)^{-1/2}(x_A + x_B)$		
	$(2)^{-1/2}(x_A - x_B)$		
	$(2)^{-1/2}(y_A + y_B)$		
	$(2)^{-1/2}(y_A - y_B)$		

These eight combinations of boron atomic orbitals are illustrated pictorially below:

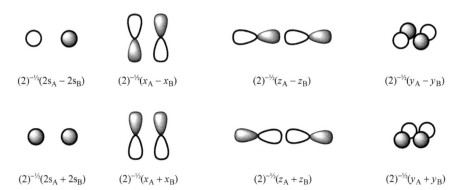

(i) With the aid of the D_{2h} character table, determine the symmetry species of the eight linear combinations given in the above table.

(ii) Using the projection operator method or otherwise, derive the six normalized linear combinations of the hydrogen orbitals and place these combinations in the appropriate slots of the given table. Use the notation $a, b, \ldots, f$ shown in the first figure above to denote the hydrogen 1s orbitals. Finally, write down the molecular orbitals that can be formed in the appropriate column.

8.35 Consider the hypothetical cyclic phosphonitrilic species $[N_3(PCl_2)_2]^{n+}$. After taking care of σ bonding, we may assume that the participating atomic orbitals for the π bonds around the ring are the three nitrogen p_π orbitals and two phosphorus d_π orbitals. These orbitals are shown and labeled on the right, where only the signed lobes above the molecular plane are displayed.

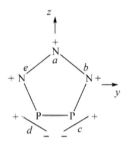

(i) Taking advantage of the C_{2v} symmetry of the system, construct the five π molecular orbitals formed by atomic orbitals $a, b, c, d,$ and e and illustrate them pictorially. For each of the molecular orbitals, be sure that you put in the irreducible representation to which it belongs. Here we take the C_2 axis as the z-axis, and the five-membered ring is on the yz-plane.

(ii) Discuss the nature (bonding, antibonding, nonbonding, etc.) of the five molecular orbitals you have constructed, and arrange them in order of increasing energy. Rationalize your ordering clearly. What is the optimal value of n (charge of the system) in order to achieve the most favorable electronic configuration?

8.36 Consider the hypothetical cyclic phosphonitrilic compound $[N_4(PX_2)_2]$. After taking care of the σ bonding, it is assumed that the participating atomic orbitals for the π bonds around the ring are the four nitrogen p_π orbitals and two phosphorus d_π orbitals. These orbitals are shown and labeled below, where only the signed lobes above the molecular plane are displayed.

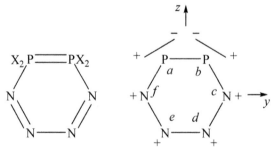

(i) Taking advantage of the C_{2v} symmetry of the system, display the six π molecular orbitals formed by the aforementioned atomic orbitals pictorially and label the orbitals with their symmetry. Also, arrange the molecular orbitals in order of increasing energy. Rationalize your ordering clearly.

(ii) Write down the ground electronic configuration and ground state of this system with six π electrons.

(iii) What is the lowest-energy symmetry-allowed electronic transition for this system? Clearly write out the states as well as the configurations involved.

8.37 Some of the most often studied simple molecules are listed in the following table. Complete this table regarding the bond lengths of these molecules' cations and anions. Your answer may be ">" (for "is longer than"), "<" (for "is shorter than") or "≈" (for "is about the same as"). Show your reasoning clearly.

Note that some of these cations or anions may not exist. Also, in the final two entries of the table, "$B-H_t$" and "$B-H_b$" denote terminal and bridging B–H bonds, respectively.

Molecule	Cation		Anion	
E.g.: O_2	$r(O_2^+) < r(O_2)$		$r(O_2^-) > r(O_2)$	
CH_4	$r(C-H)$ in CH_4^+	$r(C-H)$ in CH_4	$r(C-H)$ in CH_4^-	$r(C-H)$ in CH_4
NH_3	$r(N-H)$ in NH_3^+	$r(N-H)$ in NH_3	$r(N-H)$ in NH_3^-	$r(N-H)$ in NH_3
H_2O	$r(O-H)$ in H_2O^+	$r(O-H)$ in H_2O	$r(O-H)$ in H_2O^-	$r(O-H)$ in H_2O
BF_3	$r(B-F)$ in BF_3^+	$r(B-F)$ in BF_3	$r(B-F)$ in BF_3^-	$r(B-F)$ in BF_3
B_2H_6	$r(B-H_t)$ in $B_2H_6^+$	$r(B-H_t)$ in B_2H_6	$r(B-H_t)$ in $B_2H_6^-$	$r(B-H_t)$ in B_2H_6
B_2H_6	$r(B-H_b)$ in $B_2H_6^+$	$r(B-H_b)$ in B_2H_6	$r(B-H_b)$ in $B_2H_6^-$	$r(B-H_b)$ in B_2H_6

8.38 The ionic compound $[S_2I_4^{2+}][AsF_6^-]_2$ was synthesized about 30 years ago. The structure of the $S_2I_4^{2+}$ cation is shown on the right. This cation has C_2 symmetry and consists of one S_2 unit and two I_2^+ units, joined by weak sulfur–iodine bonds. Also, the value ($\approx 90°$) of the I_a-S-I_b angle suggests that the S–I bonds are formed by two mutually perpendicular π (or π^*) orbitals on the S_2 unit and the π (or π^*) orbitals on each of the I_2^+ units.

(i) Write down the electronic configuration of S_2 (using O_2 as a reference) and its bond order.

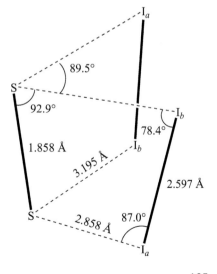

(ii) Write down the electronic configuration of I_2^+ (using F_2^+ as a reference) and its bond order.

(iii) Combine the results of (i) and (ii) to show that the bond formed between the S_2 unit and each I_2^+ unit is a four-center two-electron (4c–2e) bond. In other words, there are two 4c–2e bonds in $S_2I_4^{2+}$.

(iv) If it is assumed that the four electrons in the 4c–2e bonds are equally shared by the three participating units, show that the bond order of the S–S bond in $S_2I_4^{2+}$ is $2\frac{1}{3}$, and that of the I–I bond is $1\frac{1}{3}$.

REFERENCE: W.-K. Li, Fractional bond order. *J. Chem. Educ.* **62**, 605 (1985).

8.39 The oxalate ion, $C_2O_4^{2-}$, with D_{2h} symmetry, has a surprisingly long C–C bond (1.559–1.574 Å in seven compounds with a mean of 1.567 Å, longer than the normal C–C single bond in aliphatic compounds), despite its planar structure with "conjugated" double bonds.

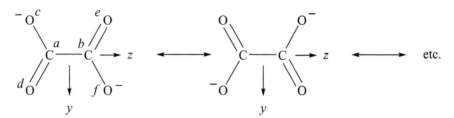

(i) Deduce the irreducible representations generated by the six 2p orbitals (labeled above) which participate in the π bonding of the anion. In addition, derive the four linear combinations of oxygen orbitals and the two combinations of carbon orbitals with the symmetries you have just deduced.

(ii) Illustrate pictorially the six π molecular orbitals of this anion and arrange them in the order of increasing energy.

(iii) Fill in the eight π electrons and attempt to rationalize the exceedingly long C–C bond.

REFERENCE: G. Zhou and W.-K. Li, The abnormally long C–C bond in the oxalate ion. *J. Chem. Educ.* **66**, 572 (1989).

8.40 (i) C. A. Coulson showed that the solutions to the Hückel secular determinant and equations for a linear conjugated polyene skeleton of n carbon atoms can be written in the general forms:

$$x_j = \frac{\alpha - E_j}{\beta} = -2\cos\left(\frac{j\pi}{n+1}\right),$$

$$\psi_j = \sum_k c_{jk}\phi_k, \text{ with } c_{jk} = \sqrt{\frac{2}{n+1}}\sin\left(\frac{jk\pi}{n+1}\right), j = 1, 2, \ldots, n.$$

For the pentadienyl radical, $CH_2{=}CH{-}CH{=}CH{-}CH_2\cdot$,

(a) tabulate the set of Hückel molecular orbitals and associated energies;

(b) calculate the total π energy (E_{TOT}) and delocalization energy (DE) in terms of α and β;

(c) calculate the wavelength in Å of the transition between the highest occupied molecular orbital (HOMO) and lowest unoccupied molecular orbital (LUMO), taking β as –253 kJ mol^{-1};

(d) calculate the π bond order of each C–C bond;

(e) estimate the length of each C–C bond from the accompanying length-order correlation curve for C–C bond (which is redrawn from C. A. Coulson, *Valence*, 2nd edn., Oxford, London, 1961, p. 270). Note that in the figure, "total bond order" includes the σ bond of a C–C bond, e.g., it is equal to 2 for an ideal C–C double bond.

(ii) Consider the following free-electron model for pentadienyl: the five π electrons move over a "box" of length 6ℓ (with $\ell = 1.405$ Å, the average length of single and double carbon–carbon bonds), the carbon atoms being located at $x = \ell, 2\ell, 3\ell, 4\ell,$ and 5ℓ.

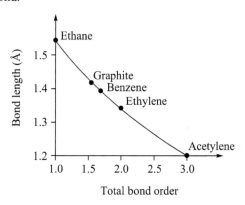

(a) Write down expressions for the HOMO and LUMO and their associated energies.

(b) Calculate the wavelength in Å of the transition between the above two energy levels.

REFERENCE: C. A. Coulson, The electronic structure of some polyenes and aromatic molecules. IV. The nature of the links of certain free radicals. *Proc. R. Soc. Lond.* **A164**, 383–96 (1938).

8.41 This problem deals with the interaction between the 2p orbitals of C and O in forming the CO molecule. To solve the following 2×2 secular determinant:

$$\begin{vmatrix} H_{OO} - E & H_{CO} - ES \\ H_{CO} - ES & H_{CC} - E \end{vmatrix} = 0,$$

where integral S can be S_σ (σ-type overlap) or S_π (π-type overlap), we are given:

$$H_{CC} = -86 \text{ kK} \qquad H_{OO} = -128 \text{ kK}, \qquad \text{where 1 kK} = 10^3 \text{ cm}^{-1},$$
$$H_{CO} = -2S(H_{CC}H_{OO})^{1/2}; \qquad S_\sigma = 0.137, \qquad S_\pi = 0.062.$$

Solve the two (one for σ orbitals and one for π orbitals) determinants to obtain the energies and wavefunctions of the molecular orbitals. The calculation required here is best done on a PC.

8.42 R. Clampitt and L. Gowland have reported the discovery of the hydrogen ion clusters H_n^+ where n is an odd number ranging from 3 to 99 about 40 years ago. They discussed the structure of these ions and made the suggestion that they probably involve the clustering of H_2 molecules around an H_3^+ unit. For example, H_5^+ and H_9^+ are to be regarded as $H_3^+ \cdot H_2$ and $H_3^+ \cdot 3H_2$, respectively. According to detailed theoretical calculations, H_3^+ is in the form of an equilateral triangle with sides of 0.878 Å and a stability of 392 kJ mol^{-1} with respect to H_2 and H^+.

   Carry out a simple LCAO–MO calculation of the total energy of triangular H_3^+ with inclusion of overlap. Express the energy E as a function of the overlap integral S using Cusachs' approximation for the exchange integral:

$$\beta = H_{ij} = S(2 - S)\alpha, \text{ with } \alpha = H_{ii}.$$

Calculate the stability in kJ mol^{-1} of H_3^+ with respect to H_2 and H^+. The dissociation energy of H_2 (D_e) is 4.75 eV and the Coulomb integral α in this case is -13.6 eV.

Hint: Use the substitution $x = \dfrac{\alpha - E}{\beta - ES}$ and find the value of S at minimum E.

REFERENCES: R. Clampitt and L. Gowland, Clustering of cold hydrogen gas on protons. *Nature* **223**, 815–6 (1971); S. W. Harrison, L. J. Massa, and P. Solomon, Binding energy and geometry of the hydrogen clusters H_n^+. *Nature Phys. Sci.* **245**, 31–2 (1972).

8.43 Borane B_5H_9 has a square pyramidal structure for the boron atoms (symmetry C_{4v}). After taking care of B–H bonding, there remain seven boron orbitals and six electrons for the bonding of the boron skeleton. These orbitals are schematically shown below. For convenience, it may be assumed that orbitals b and c are pure $2p_x$ and $2p_y$ orbitals, respectively, while the remaining ones are some kind of s-p hybrids. Note that orbital a points towards the center of the basal plane.

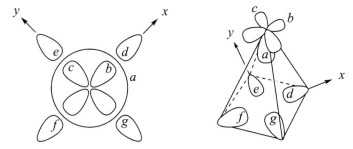

(i)   Determine the normalized linear combination of the basal boron orbitals that would match in symmetry with the apical boron orbitals.

(ii) Determine the Hückel energies of the seven resultant molecular orbitals assuming that

(a)   the Coulomb integrals of all boron orbitals are equal to α;

(b)   the exchange integrals between apical and basal orbitals are equal to β; and

(c)   the exchange integrals between basal orbitals are zero.

(iii) What are the ground configuration and ground state of the system?

(iv) Derive the explicit forms of the filled molecular orbitals.

REFERENCE: W. H. Eberhardt, B. Crawford, and W. N. Lipscomb, The valence structure of the boron hydrides. *J. Chem. Phys.* **22**, 989–1001 (1954).

8.44 (i)   Applying Hückel molecular orbital theory, the secular determinant for the π system of each "island" (P=N–P unit) for phosphazenes has the form:

$$\begin{vmatrix} \alpha_P - E & \beta & 0 \\ \beta & \alpha_N - E & \beta \\ 0 & \beta & \alpha_P - E \end{vmatrix} = 0,$$

where α_P and α_N are the Coulomb integrals for the P 3d and N 2p orbitals, respectively, and β is the resonance integral between these two orbitals. Assuming that $\alpha_N = \alpha_P + k\beta$, where k is a proportionality constant, solve this determinant to obtain the total π electron energy (E_π) of each "island".

(ii)   Write down the secular determinant for the island in the classical picture, i.e., when delocalization is ignored. Again assuming $\alpha_N = \alpha_P + k\beta$, obtain the total (classical) π energy (E_c) of this system.

(iii)   With the results of (i) and (ii), calculate the delocalization energy (DE) of each allylic island. The result of Dewar and co-workers is DE = 0.83β per island. Hence the total DE for $(NPCl_2)_3^*$ is about 2.5β, which is comparable to that found for benzene, 2β. Show clearly how Dewar and his co-workers arrived at this conclusion.

* According to Dewar's theory, phosphonitrilic compound $(NPCl_2)_n$ consists of n "independent" allylic P=N−P islands. In every island, each P atom contributes a d orbital, and the N atom contributes a p orbital for the π bonding of the unit. It is important to note that each P atom takes part in the π bonding of two neighboring islands, using two d orbitals (which are orthogonal to each other) in doing so. In other words, as shown below, the π bonding of $(NPCl_2)_3$ (with three islands) has contributions from a total of nine atomic orbitals. Based on this bonding description, it can be easily seen that the cyclic molecule $(NPCl_2)_n$ does not have to be planar. Furthermore, even if it is planar, the planarity is not as rigid as that found in benzene.

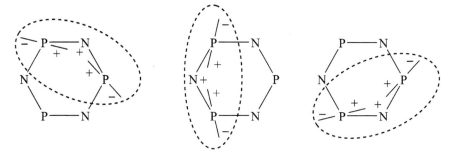

REFERENCE: M. J. S. Dewar, E. A. C. Lucken, and M. A. Whitehead, The structure of the phosphonitrilic halides. *J. Chem. Soc.*, 2423–9 (1960).

8.45   The π atomic orbitals of propenyl radical, $CH_2CHCH_2\cdot$; are labeled on the right. The Hückel secular determinant for this system is

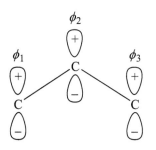

$$\begin{vmatrix} x & 1 & 0 \\ 1 & x & 1 \\ 0 & 1 & x \end{vmatrix} = 0, \text{ where } x = (\alpha - E)/\beta.$$

The solutions of this determinant are $x_1 = -\sqrt{2}$, $x_2 = 0$, and $x_3 = \sqrt{2}$.

Determine the normalized wavefunction of the three π molecular orbitals of this radical. In other words, for the π molecular orbital you have chosen, determine the coefficients c_1, c_2, and c_3 of the wavefunction $\psi = c_1\phi_1 + c_2\phi_2 + c_3\phi_3$. Finally, assuming C_{2v} symmetry for the system, what is the irreducible representation of the orbital whose wavefunction you have just determined?

8.46 There are three resonance structures for the cyclopropenyl cation:

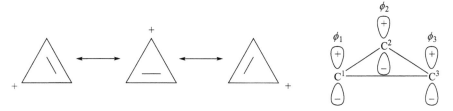

In other words, the two π electrons are delocalized. Since there are three π atomic orbitals, ϕ_1, ϕ_2, and ϕ_3, located at atoms C^1, C^2, and C^3, respectively (as shown above), there are three π molecular orbitals, and their normalized wavefunctions are:

$$\psi_1 = (3)^{-1/2}(\phi_1 + \phi_2 + \phi_3),$$
$$\psi_2 = (6)^{-1/2}(2\phi_1 - \phi_2 - \phi_3),$$
$$\psi_3 = (2)^{-1/2}(\phi_2 - \phi_3).$$

(i) Calculate the Hückel energies of ψ_1, ψ_2, and ψ_3.

(ii) Calculate the total π energy for the two π electrons. Also, determine the delocalization energy of the system.

(iii) Classify ψ_1, ψ_2, and ψ_3 as bonding, antibonding, and antibonding molecular orbitals.

8.47 Consider a hypothetical cyclic molecule X_3 in which π bonds are formed by the d orbitals on the X atoms. The three d orbitals that participate in the π bonding are labeled in the figure on the right, where only the signed lobes above the molecular plane are shown.

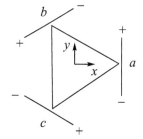

(i) Determine the symmetries and expressions of the three molecular orbitals.

(ii) Assuming Hückel approximation: $H_{aa} = H_{bb} = H_{cc} = \alpha, H_{ab} = H_{bc} = H_{ac} = -\beta > 0$ (since $\beta < 0$ by convention), determine the energies of the three molecular orbitals.

(iii) How many electrons are required in order to obtain the most stable closed-shell configuration for this system? Does the $4n + 2$ rule hold in this case?

8.48 Consider a square planar molecule X_4, whose π bonds are formed by the four d orbitals labeled on the right. Note that only the signed lobes of the orbitals above the molecular plane are shown.

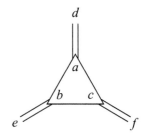

(i) Determine the symmetries and the expressions of the four molecular orbitals.

(ii) Applying the Hückel approximation, determine the energies of the molecular orbitals in terms of α and β. What is the minimum number of electrons required for this system in order to achieve maximum stability?

(iii) Write down the ground configuration and electronic state of the system (with minimum number of electrons) you have determined in (ii). Also, write down the lowest-energy symmetry- and spin-allowed electronic transition of the system. For this transition, you need to specify the states and configurations involved.

(iv) Are there any nonbonding molecular orbitals in the system? If yes, would these orbitals still be nonbonding if we ignored the Hückel approximation? Explain. Finally, what type of overlap (σ, π, or δ) exists between the atomic orbitals in these "nonbonding" molecular orbitals?

8.49 Shown on the right is a planar triene (molecular formula C_6H_6) with three conjugated π bonds. The symmetry point group of this molecule is D_{3h}. Also labeled in the figure are the six π bonding atomic orbitals. It is straightforward to show that the representation generated by atomic orbitals $a, b, \ldots, f$ is:

$$\Gamma_\pi = 2A_2'' + 2E''.$$

(i) Derive the six linear combinations of these atomic orbitals, with the symmetries given by Γ_π.

(ii) Set up the three 2×2 secular determinants with A_2'' and E'' symmetry. Solve the determinants to obtain the Hückel energies of the six π molecular orbitals.

(iii)   Fill in the six electrons to determine the π electron energy (E_π) of the system. In addition, calculate the delocalization energy (DE) of this molecule.

8.50   Consider the planar triene with D_{2h} symmetry shown on the right. In this figure, the six p_z atomic orbitals taking part in the π bonding of the molecule are labeled from a to f. It is straightforward to show that the representation generated by atomic orbitals a to f is:

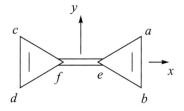

$$\Gamma_\pi = B_{2g} + 2B_{3g} + A_u + 2B_{1u}.$$

(i)   Determine the Hückel energies (in terms of Coulomb integral α and resonance integral β) of the two molecular orbitals with B_{2g} and with A_u symmetry.

(ii)   Set up the two 2×2 secular determinants with B_{3g} and B_{1u} symmetry. Solve these determinants to obtain the Hückel energies of the remaining four molecular orbitals.

(iii)   Fill in the six electrons to determine the π electron energy (E_π) of the system and to calculate the delocalization energy (DE) of this molecule.

8.51   Shown on the right is a bicyclic planar triene (molecular formula C_6H_4) with conjugated π bonds. The symmetry point group of this molecule is D_{2h}. Also labeled in the figure are the six π bonding atomic orbitals. It is straightforward to show that the representation generated by atomic orbitals $a, b, \ldots, f$ is:

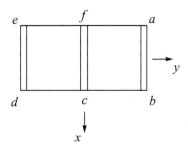

$$\Gamma_\pi = 2B_{2g} + B_{3g} + A_u + 2B_{1u}.$$

(i)   Determine the Hückel energies (in terms of Coulomb integral α and resonance integral β) of the two molecular orbitals with B_{3g} and with A_u symmetry.

(ii)   Set up the two 2×2 secular determinants with B_{2g} and B_{1u} symmetry. Solve these determinants to obtain the Hückel energies (in terms of Coulomb integral α and resonance integral β) of the remaining four molecular orbitals.

(iii)   Fill in the six electrons to determine the π electron energy (E_π) of the system. In addition, calculate the delocalization energy (DE) of this molecule.

8.52 Consider the "hydrogen cluster" H_5^{n+} with a trigonal bipyramidal structure (D_{3h} symmetry), as shown below. Also labeled in the figure are the 1s atomic orbitals participating in the cluster bonding. Here, atoms a, b, and c occupy the equatorial positions, while atoms d and e take up the axial positions.

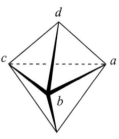

(i) Determine the symmetries and Hückel energies of the five molecular orbitals formed by these atomic orbitals. Assume the Hückel approximation:

$$H_{ii} = \alpha; H_{ij} = \beta, \text{ with } i \neq j \text{ except } H_{de} = 0.$$

(ii) What is/are the likely charge(s) of the cluster in order for it to be most stable electronically?

8.53 Consider the "hydrogen cluster" H_6^{n+} with an octahedral structure (O_h symmetry) shown on the right. Also labeled in the figure are the 1s orbitals participating in the cluster bonding.

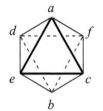

(i) Taking advantage of the O_h symmetry of the cluster and assuming the Hückel approximation, determine the energies and wavefunctions of the six molecular orbitals formed.

(ii) What is the minimum number of electrons required for this system in order to achieve maximum stability? Write down the ground configuration and ground state of the cluster.

(iii) Write down the lowest-energy symmetry- and spin-allowed electronic transition of this cluster. For this transition, you need to specify the states and configurations involved.

8.54 Consider the hypothetical hydrogen cluster H_8, which has the cubic structure shown on the right. Also labeled in the diagram are the hydrogen atomic 1s orbitals taking part in the bonding of the cluster. Note that the unprimed orbitals form a tetrahedral arrangement (ditto for the primed orbitals).

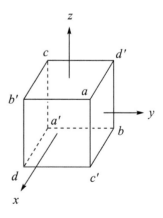

(i) Applying the Hückel approximation, determine the energies and the wavefunctions of the eight molecular orbitals. (Hint: First derive the linear combinations of the unprimed orbitals. Then do the same for the primed orbitals. Finally, further combine these two sets of orbitals.)

(ii) Classify the symmetry of the eight molecular orbitals according to the irreducible representations of the O_h point group.

8.55 Taking advantage of the D_{3h} symmetry of the cyclic trimer $(NPF_2)_3$, determine the Hückel energies (E) and molecular orbitals (MO) of its π electron system. The participating N 2p and P 3d atomic orbitals and their signed lobes above the molecular plane are shown in the figure on the right. Assume the following Hückel approximation:

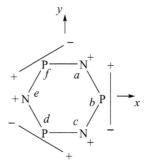

$$H_{aa} = H_{cc} = H_{ee} = \alpha, H_{bb} = H_{dd} = H_{ff} = \alpha + \beta$$

(probably not realistic since N is more electronegative; see 8.44),

$$H_{ab} = -H_{bc} = H_{cd} = -H_{de} = H_{ef} = -H_{af} = \beta.$$

REFERENCES: E. Heilbronner and H. Bock (translated by W. Martin and A. J. Rackstraw), *The HMO Model and Its Application*, vol. 2, Wiley, London, 1976, pp. 132–3. See also, W.-K. Li and T. C. W. Mak, Hückel orbitals and energies of cyclic phosphonitrilic halides. *J. Chem. Educ.* **58**, 362–3 (1981).

8.56 Predict a possible product for the thermal dimerization of butadiene.

8.57 Complete the following reactions.

(i) Ring–closing:

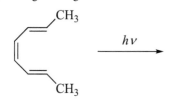

(ii) Ring–opening:

$\overset{\text{CH}_2\text{CH}_3}{\underset{\text{CH}_2\text{CH}_3}{}}$ $\xrightarrow{\Delta}$

(iii)   Ring–closing:

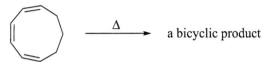

$\xrightarrow{\Delta}$ a bicyclic product

In your answers, specify the mode (conrotatory or disrotatory) of the process. You also need to show clearly the stereochemistry (*cis/trans*) of the products.

8.58 Complete the following ring–opening reactions:

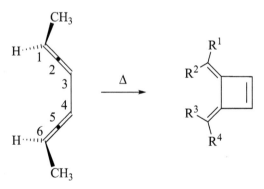

In your answers, specify the mode (conrotatory or disrotatory) of the ring-opening process. Hint: Both products are substituted cyclooctatetraenes.

REFERENCE: R. B. Woodward and R. Hoffmann, *The Conservation of Orbital Symmetry*, Verlag Chemie, Weinheim/Bergstr., 1970, p. 63.

8.59 Identify groups R^1, R^2, R^3, and R^4 in the product of the following reaction. Does this reaction proceed in a conrotatory or disrotatory manner?

8.60 Consider the following reaction:

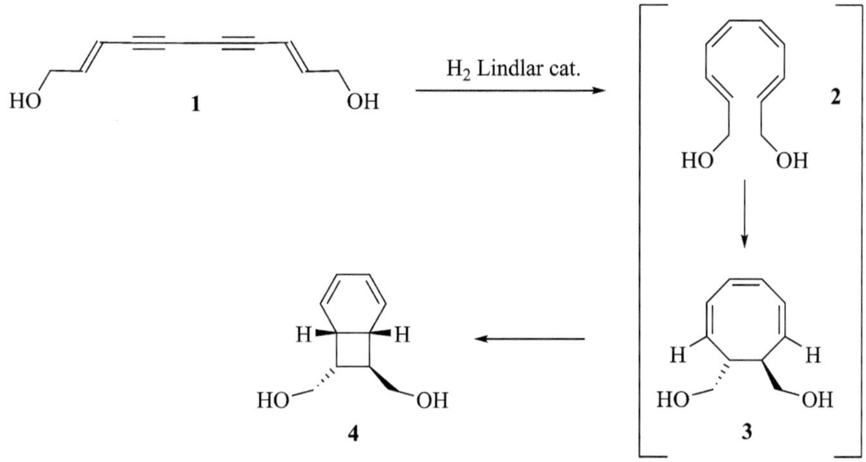

For each of cyclization steps **2** → **3** and **3** → **4**, determine (i) the number of π electrons involved and (ii) whether the step is conrotatory or disrotatory. Finally, is the whole process **1** → **4** carried out thermally or photochemically?

REFERENCE: K. C. Nicolaou, N. A. Petasis, R. E. Zipkin, and J. Uenishi, The endiandric acid cascade. Electrocyclizations in organic synthesis. I. Stepwise, stereocontrolled total synthesis of endiandric acids A and B. *J. Am. Chem. Soc.*, **104**, 5555–7 (1982).

8.61 Consider the following reactions. In both the reactants and products, only the two "most important" hydrogens are drawn. Note that the two reactants are steroisomers: one is optically active and the other one is not. For each of the reactions, determine whether it is a conrotatory or a disrotatory process. In addition, which reaction is allowed thermally? Clearly spell out your reasoning.

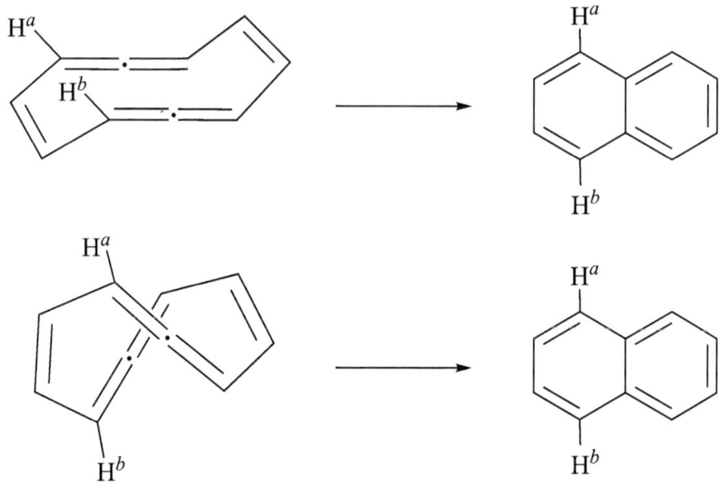

REFERENCE: R. B. Woodward and R. Hoffmann, *The Conservation of Orbital Symmetry*, Verlag Chemie, Weinheim/Bergstr., 1970, pp. 63–4.

8.62 In the following two reactions, denoted as (A) and (B), cyclodecapentaene cyclizes into *trans-* and *cis*-9,10-dihydronaphthalene. Apply Woodward-Hoffmann rules to determine the reaction condition (thermal or photochemical) and reaction pathway (conrotatory or disrotatory) of each process.

trans-9,10-
dihydronaphthalene

(A)   cyclodecapentaene   (B)

cis-9,10-
dihydronaphthalene

REFERENCE: R. B. Woodward and R. Hoffmann, *The Conservation of Orbital Symmetry*, Verlag Chemie, Weinheim/Bergstr., 1970, p. 61.

8.63 For each of the following five reactions, (A) to (E), write down the reaction pathway (conrotatory or disrotatory) of the process. In addition, for bicyclic species I, II, III, and IV, determine the relative orientation (*cis* or *trans*) of the two bridgehead hydrogens bonded to the carbon atoms at which the two rings are fused.

REFERENCE: R. B. Woodward and R. Hoffmann, *The Conservation of Orbital Symmetry*, Verlag Chemie, Weinheim/Bergstr., 1970, pp. 60–1.

8.64 [16]Annulene, $C_{16}H_{16}$, can undergo two successive cyclization steps to form a tricyclic product, as shown below:

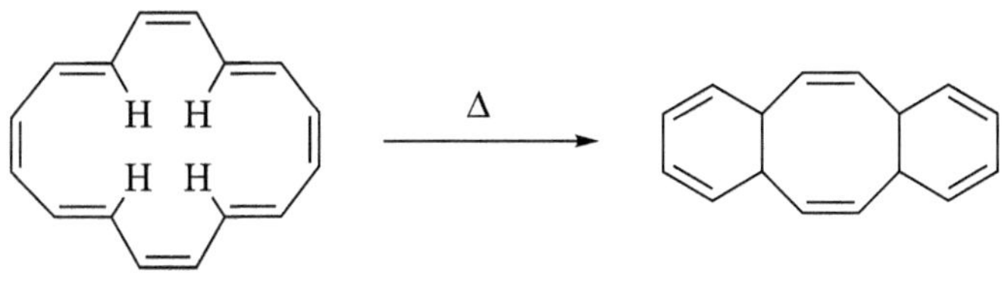

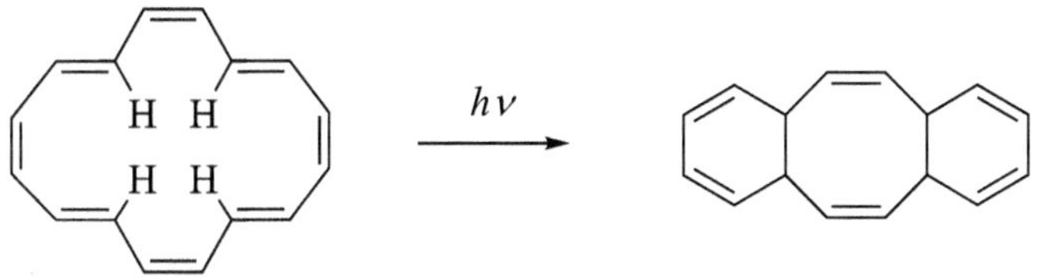

(i)   Clearly, the stereochemistry of the products from the thermal and photochemical reactions are not the same. List all the possible products for each reaction. Particularly, specify the stereochemistry of the four bridgehead hydrogens bonded to the carbon atoms at which the rings are fused.

(ii) For every cyclization step of each reaction, state whether it is a conrotatory or a disrotatory process.

REFERENCES: R. B. Woodward and R. Hoffmann, *The Conservation of Orbital Symmetry*, Verlag Chemie, Weinheim/Bergstr., 1970, p. 62. H.-L. Lee and W.-K. Li, Computational study on the electrocyclic reactions of [16]annulene. *Org. Biomol. Chem.* **1**, 2748–54 (2003).

8.65 Predict the products of the photochemical reactions between butadiene and isobutene:

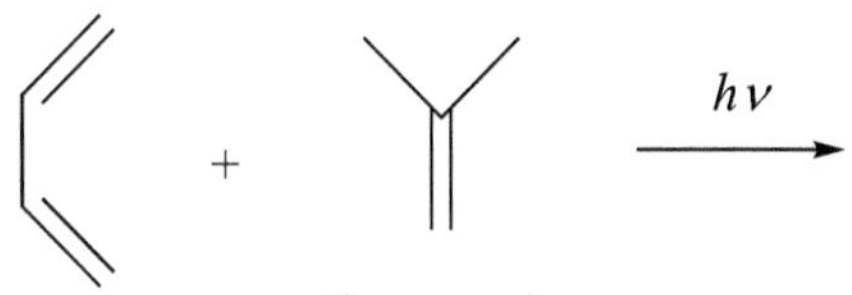

(in excess)

Note that isobutene is in excess here. That means you need to consider (i) the reaction between one molecule of butadiene and one molecule of isobutene and (ii) the reaction between one molecule of each of the products in (i) with another molecule of isobutene. Be sure that you have exhausted all the stereochemical possibilities. Finally, dimerization of either reactant is not to be considered.

8.66 Predict the products of the photochemical reactions in a mixture of *cis*-2-butene, *trans*-2-butene, and isobutene. Note that you need to consider the dimerization reaction of each reactant.

REFERENCE: R. B. Woodward and R. Hoffmann, *The Conservation of Orbital Symmetry*, Verlag Chemie, Weinheim/Bergstr., 1970, p. 74.

8.67 For the propenyl cation, there are two resonance structures:

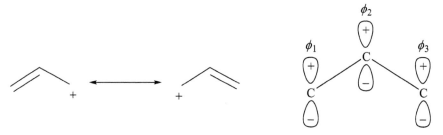

In other words, the two π electrons are delocalized. The three atomic p orbitals participating in the π bonding of this system are labeled above. The three Hückel molecular orbitals for this system are:

$$\psi_1 = \frac{1}{2}\phi_1 + \frac{1}{\sqrt{2}}\phi_2 + \frac{1}{2}\phi_3,$$

$$\psi_2 = \frac{1}{\sqrt{2}}\phi_1 - \frac{1}{\sqrt{2}}\phi_3,$$

$$\psi_3 = \frac{1}{2}\phi_1 - \frac{1}{\sqrt{2}}\phi_2 + \frac{1}{2}\phi_3.$$

Consider the thermal electrocyclic reactions of propenyl cation and propenyl radical:

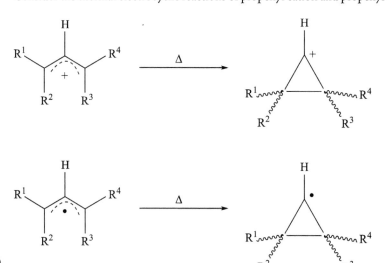

Determine with brief justification, the stereochemistry of groups R^1, R^2, R^3, and R^4 in the two cyclic products.

8.68 The Hückel molecular orbitals of pentadienyl radical $(CH_2{=}CH{-}CH{=}CH{-}CH_2\cdot)$ are:

$$\psi_5 = \frac{1}{\sqrt{12}}\phi_1 - \frac{1}{2}\phi_2 + \frac{1}{\sqrt{3}}\phi_3 - \frac{1}{2}\phi_4 + \frac{1}{\sqrt{12}}\phi_5,$$

$$\psi_4 = \frac{1}{2}(\phi_1 - \phi_2 + \phi_4 - \phi_5),$$

$$\psi_3 = \frac{1}{\sqrt{3}}(\phi_1 - \phi_3 + \phi_5),$$

$$\psi_2 = \frac{1}{2}(\phi_1 + \phi_2 - \phi_4 - \phi_5),$$

$$\psi_1 = \frac{1}{\sqrt{12}}\phi_1 + \frac{1}{2}\phi_2 + \frac{1}{\sqrt{3}}\phi_3 + \frac{1}{2}\phi_4 + \frac{1}{\sqrt{12}}\phi_5.$$

For each of the following processes, predict whether it will proceed in a conrotatory or a disrotatory manner: (i) thermal electrocyclic reaction of the pentadienyl cation, (ii) thermal electrocyclic reaction of the pentadienyl radical, and (iii) photochemical electrocyclic reaction of the pentadienyl anion.

SOLUTIONS

A8.1 (i) Let n be the number of double bonds in the molecule; the number of π electrons is thus $2n$. Let d be the average length of carbon–carbon bonds in the polyene system and assume that the "box" extends half a bond length beyond each terminal carbon atom. Thus $a = 2nd$. Considering that C–C and C=C bonds measure 1.54 and 1.34 Å, respectively, a good estimate for d is 1.40 Å. In the ground state, the first n orbitals are filled. In the first excited state, an electron jumps from the nth orbital to the $(n+1)^{th}$ orbital. According to the free-electron model, the energy of this transition is $E_{n+1} - E_n = \dfrac{h^2(2n+1)}{32md^2n^2}$. The energies of the light absorbed decrease in the order butadiene, vitamin A, and carotene. Since the wavelengths that are not absorbed by a compound account for its color as seen by the human eye, the variation in color is explained in the following way. The absorption of butadiene occurs in the ultraviolet region. If vitamin A absorbs primarily in the high energy (violet/blue) region, the observed color is a consequence of all colors not absorbed: the combination of green, yellow, orange, and red observed as yellow-orange (the complement of the colors most strongly absorbed). Similarly, the observed color of carotene can be described as complementary to the color(s) most strongly absorbed. Carotene absorbs the lower energy green light and its color is a mixture of red and blue.

(ii) For vitamin A ($n = 5$), energy absorbed $= \dfrac{11h^2}{8ma^2}$, or $a^2 = \dfrac{11h\lambda}{8mc}$. Substitution of known data gives $a = 10.5$ Å, in rough agreement with the "expected" value $10d \approx 14.0$ Å.

A8.2 (i) The wavefunctions and the energies of the orbitals are

$$\psi_n = \sqrt{\frac{2}{11\ell}} \sin\left(\frac{n\pi x}{11\ell}\right), 0 \le x \le 11\ell; E_n = \frac{n^2h^2}{8m(11\ell)^2}; n = 1, 2, 3, \dots .$$

(ii) $\lambda = \dfrac{hc}{(E_6 - E_5)} = 7119$ Å.

(iii) The charge densities at the centers of the bonds, counting from one end of the chain, in units of ℓ^{-1}, are 1.219, 0.861, 1.108, 0.905, 1.091, 0.905, 1.108, 0.861, and 1.219. The results clearly show the alternating (single bond–double bond) character of the bonds. Another important conclusion that can be drawn is that the difference between successive bonds becomes less as the center of the molecule is approached. As n increases, the neighboring bond lengths would become more and more alike.

A8.3 For a rectangular box of length a and width b (both measured in Å),

$$E(n_x, n_y) = \left(\frac{h^2}{8m}\right)\left(\frac{n_x^2}{a^2} + \frac{n_y^2}{b^2}\right) = 6.024 \times 10^{-18}\left(\frac{n_x^2}{a^2} + \frac{n_y^2}{b^2}\right) \text{ J.}$$

Level	n_x	n_y	$\frac{n_x^2}{16} + \frac{n_y^2}{49}$	
1	1	1	0.0829	
2	1	2	0.1441	
3	1	3	0.2462	
6	1	4	0.3890	
8	1	5	0.5727	(lowest unoccupied molecular orbital)
10	1	6	0.7972	
4	2	1	0.2704	
5	2	2	0.3316	
7	2	3	0.4337	(highest occupied molecular orbital)
9	2	4	0.5765	

There are 14 π electrons in anthracene. In its ground state, the seven orbitals of lowest energy are filled. In the first excited state, an electron is promoted from the $(2, 3)$ level to the $(1, 5)$ level.

$$\Delta E = 6.024 \times 10^{-18} \times (0.5727 - 0.4337)\,\text{J}.$$
$$\lambda = hc/\Delta E = 2372\,\text{Å}.$$

The absorption spectrum of anthracene shows bands at 2550(vs), 3790(s), and 6700(w) Å.

REFERENCE: S. F. Mason, Molecular electronic absorption spectra. Q. Rev. Chem. Soc. **15**, 287–371 (1961).

A8.4 (i) From the Frost–Hückel circle, we see that the lowest energy level is always non-degenerate, while all the remaining levels are doubly degenerate, except the highest one for a polygon with an even number of vertices. To obtain a stable closed-shell configuration, the lowest $2n + 1$ π molecular orbitals must be full, i.e., $4n + 2$ π electrons are required. This electronic configuration gives rise the $4n + 2$ rule. Finally, it is noted that, strictly speaking, the $4n + 2$ rule applies to monocyclic systems only. In other words, the aromaticity of naphthalene does not arise from this rule.

(ii) The energy of the lowest energy level is given by $E_1 = \alpha + 2\beta$ (non-degenerate). Let m be the number of the atoms in the ring, which is also the number of the vertices in the polygon. If m is even:

$$E_{k+1} = \alpha + 2\beta \cos\left(\frac{2k\pi}{m}\right) \text{ (doubly degenerate)}; k = 1, 2, 3, \ldots, (m/2 - 1);$$

$$E_{(m/2)+1} = \alpha - 2\beta \qquad \text{(non-degenerate)}.$$

If m is odd:

$$E_{k+1} = \alpha + 2\beta \cos\left(\frac{2k\pi}{m}\right) \text{ (doubly degenerate)}; k = 1, 2, 3, \ldots, (m - 1)/2.$$

213

A8.5

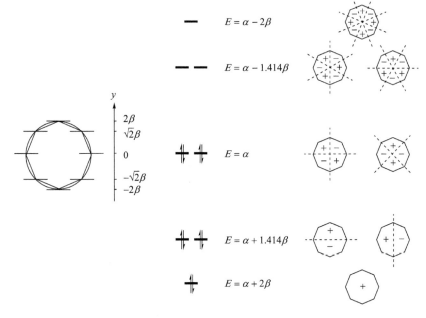

$E = \alpha - 2\beta$

$E = \alpha - 1.414\beta$

$E = \alpha$

$E = \alpha + 1.414\beta$

$E = \alpha + 2\beta$

2β
$\sqrt{2}\beta$
0
$-\sqrt{2}\beta$
-2β

The energy for the 10 delocalized π electrons in $C_8H_8^{2-}$

$$= [4(\alpha) + 4(\alpha + 1.414\beta) + 2(\alpha + 2\beta)] = 10\alpha + 9.656\beta.$$

The classical structure of $C_8H_8^{2-}$ (such as the one shown on the right) contains three "formal" double bonds.

The energy for the 10 π electrons in this structure
$$= 4(\alpha) + 6(\alpha + \beta) = 10\alpha + 6\beta.$$
Delocalization energy $= (10\alpha + 9.656\beta) - (10\alpha + 6\beta) = 3.656\beta.$

A8.6   (i) $\left(\dfrac{1}{r^2}\right)\left(\dfrac{d^2\psi}{d\phi^2}\right) + \dfrac{8\pi^2 mE}{h^2}\psi = 0$, note that r (radius of the ring) is constant,
$\theta = 90°$; V constant, arbitrarily set to zero.

(ii)   Solving the wave equation yields

$$\psi = A\sin n\phi + B\cos n\phi, \text{ with } n = \dfrac{\pi r\sqrt{8mE}}{h}.$$

Since ψ must be single-valued, there is a boundary condition of the form $\psi(\phi) = \psi(\phi + 2\pi)$. Application of this condition shows that n must be an integer. Hence, the energy expression is $E_n = \dfrac{n^2h^2}{8\pi^2 mr^2}, n = 0, 1, 2, 3 \ldots$

(iii) For this system, there are two aspects which do not exist for the particle-in-a-box problem. First, $n = 0$ is an acceptable quantum number. The energy of this state is zero and the corresponding wavefunction $\psi_0 = (2\pi)^{-1/2}$. Secondly, all states with $n > 0$ are doubly degenerate, with wavefunctions $\pi^{-1/2} \sin n\phi$ and $\pi^{-1/2} \cos n\phi$. These two wavefunctions are for the states with the same (kinetic) energy but moving in opposite directions. On the contrary, for the particle-in-a-box problem, translation motions with the same (kinetic) energy but in opposite directions belong to the same state because the direction of motion changes to the opposite one after bouncing back from the wall.

A8.7 (i) $T = \dfrac{p^2}{2m} = \dfrac{h^2}{2m\lambda^2} = \dfrac{h^2}{8mr^2}$. ($\lambda = 2r$ for ground state of particle-in-a-box)

(ii) $V = -\dfrac{2e^2}{r/2} + \dfrac{e^2}{r} = -\dfrac{3e^2}{r}$.

$E = T + V = \dfrac{h^2}{8mr^2} - \dfrac{3e^2}{r}$.

(iii) By differentiating E with respect to r, the energy minimum is found at $r_0 = \dfrac{h^2}{12me^2}$. At r_0, $E_0 = -\dfrac{18me^4}{h^2}$. Substitution of known constants into the above two expressions gives $r_0 = 1.74$ Å and $E_0 = -12.4$ eV.

A8.8 (i) $H_{12} = S(2-S)H_{11}$; $V(r) = 2S\left(\dfrac{1-S}{1+S}\right) \times$ H_{11}. A schematic plot of $V(r)$ against S is shown on the right.

(ii) Differentiating $V(r)$ with respect to S, the minimum of $V(r)$ occurs at $S = \sqrt{2} - 1 = 0.414$; and $V(r_e) = 0.343\,H_{11} = -D_e$.

(iii) At r_e, $H_{12} = S(2 - S)H_{11} = 0.657\,H_{11}$.
$$IE = -E_+ = -\left(\dfrac{H_{11} + H_{12}}{1 + S}\right) = -1.172\,H_{11}.$$

(iv) $E_- = \left(\dfrac{H_{11} - H_{12}}{1 - S}\right) = 0.586\,H_{11}.$
$\Delta = E_- - E_+ = (0.586 - 1.172)\,H_{11} = -0.586\,H_{11}.$

(v) The calculated D_e, IE, and Δ for H_2 are 4.66, 15.94, and 7.97 eV, respectively.

A8.9   (i) H_2^+ H_2

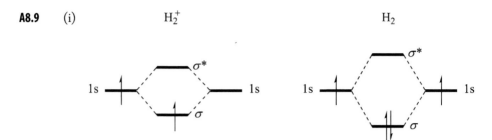

The two energy level diagrams differ in the separation between σ and σ^* and in the number of electrons.

(ii) (a) In the VB picture, the electron is located in the 1s orbital of either atom. The (unnormalized) VB wavefunction is then $\psi_{VB} = (a + b)$.

In the MO picture, the mixing of the two 1s atomic orbitals results in a bonding orbital and an antibonding orbital, the (unnormalized) wavefunctions of which are $(a + b)$ and $(a - b)$, respectively. The (unnormalized) MO wavefunction is then $\psi_{MO} = (a + b)$, which is identical to ψ_{VB}.

The normalized wavefunctions can be determined as: $\psi_{VB} = \psi_{MO} = [2(1 + S)]^{-1/2}(a + b)$.

(b) $\psi_{MO} = (2 + 2S)^{-1}[a(1) + b(1)][a(2) + b(2)] = \psi(\sigma_{1s}^2)$;
$\psi_{VB} = (2 + 2S^2)^{-1/2}[a(1)b(2) + b(1)a(2)]$.

For this system, the two wavefunctions are not the same.

(iii) Expanding ψ_2, we have $\psi_2 = [1 - \mu - (1 + \mu)S](2 - 2S)^{-1}[a(1)b(2) + b(1)a(2)] + [1 + \mu - (1-S)](2-2S)^{-1}[a(1)a(2) + b(1)b(2)]$. Compare this expression with: $\psi_1 = (2 + 2S^2)^{-1/2}[a(1)b(2) + b(1)a(2)] + \lambda(2 + 2S^2)^{-1/2}[a(1)a(2) + b(1)b(2)]$, taking into account that ψ_1 and ψ_2 are identical (apart from a numerical factor) we have:

$$\frac{[1 - \mu - (1 + \mu)S](2 - 2S)^{-1}}{(2 + 2S^2)^{-1/2}} = \frac{[1 + \mu - (1 - \mu)S](2 - 2S)^{-1}}{\lambda(2 + 2S^2)^{-1/2}}$$

$$\text{or } \lambda = \frac{1 + \mu - (1 - \mu)S}{1 - \mu - (1 + \mu)S}.$$

(iv) If only the configuration, $(\sigma_{1s})^2$, is employed, $\mu = 0$ and $\lambda = 1$. The (unnormalized) wavefunction becomes $\psi_{VB} + \psi_i$, which represents equal mixing of two sets of products: 50% of 2H and 50% of $H^+ + H^-$. However, the (lowest-energy) dissociation is expected to result in 2H (two neutral atoms), the energy of which is significantly lower than that of a "mixture" of 50% of 2H and 50% of $H^+ + H^-$.

The correct (unnormalized) wavefunction for the ground state of 2H is, in fact, $\psi(\sigma_{1s}^2) + \psi(\sigma_{1s}^{*2})$.

(v) For dissociated H_2, the atoms are well separated from each other. There is practically zero overlap between the two 1s orbitals. Therefore, the meaning of pure "H·" form is unambiguous: at any instant, each electron stays in the 1s orbital of one of the atoms but not in that of the other atom. In contrast, the two atoms in a stable H_2 molecule are close enough for significant overlap of the two 1s orbitals. The pure "H · ·H" form is ambiguous. Conceptually, this form refers to the condition in which the two electrons stay in different 1s orbitals. In reality, however, any point in the molecular region is actually "covered" by both orbitals. There is no point covered only by one orbital but not the other. Similarly, the pure "H^+H^-" form is also unreal because it implies a situation in which, at any instant, both electrons stay at points belonging to one orbital but not the other.

A8.10 (i) To write down the VB function, it is noted that, since this is a triplet state, the orbital part of the wavefunction must be antisymmetric with respect to the exchange of the coordinates of the two electrons: $a(1)b(2) - b(1)a(2)$.

Hence, the total wavefunction is

$$[a(1)b(2) - b(1)a(2)] \begin{cases} \alpha(1)\alpha(2) \\ \beta(1)\beta(2) \\ [\alpha(1)\beta(2) + \beta(1)\alpha(2)] \end{cases}.$$

Similarly, the orbital part of the MO function has the form $u(1)g(2) - g(1)u(2)$, which, upon expansion and coupling with the spin functions, leads to the same results as those given for the VB function:

$$[u(1)g(2) - g(1)u(2)] \begin{cases} \alpha(1)\alpha(2) \\ \beta(1)\beta(2) \\ [\alpha(1)\beta(2) + \beta(1)\alpha(2)] \end{cases}$$

$$= \{[a(1) - b(1)][a(2) + b(2)] - [a(1) + b(1)][a(2) - b(2)]\} \begin{cases} \alpha(1)\alpha(2) \\ \beta(1)\beta(2) \\ [\alpha(1)\beta(2) + \beta(1)\alpha(2)] \end{cases}$$

$$= 2[a(1)b(2) - b(1)a(2)] \begin{cases} \alpha(1)\alpha(2) \\ \beta(1)\beta(2) \\ [\alpha(1)\beta(2) + \beta(1)\alpha(2)] \end{cases}.$$

(ii)   VB wavefunction for He_2:

$$\begin{vmatrix} a(1)\alpha(1) & a(1)\beta(1) & b(1)\alpha(1) & b(1)\beta(1) \\ a(2)\alpha(2) & a(2)\beta(2) & b(2)\alpha(2) & b(2)\beta(2) \\ a(3)\alpha(3) & a(3)\beta(3) & b(3)\alpha(3) & b(3)\beta(3) \\ a(4)\alpha(4) & a(4)\beta(4) & b(4)\alpha(4) & b(4)\beta(4) \end{vmatrix}$$

MO wavefunction for He_2:

$$\begin{vmatrix} g(1)\alpha(1) & g(1)\beta(1) & u(1)\alpha(1) & u(1)\beta(1) \\ g(2)\alpha(2) & g(2)\beta(2) & u(2)\alpha(2) & u(2)\beta(2) \\ g(3)\alpha(3) & g(3)\beta(3) & u(3)\alpha(3) & u(3)\beta(3) \\ g(4)\alpha(4) & g(4)\beta(4) & u(4)\alpha(4) & u(4)\beta(4) \end{vmatrix}$$

$$= \begin{vmatrix} [a(1)\alpha(1)+b(1)\alpha(1)] & [a(1)\beta(1)+b(1)\beta(1)] & [a(1)\alpha(1)-b(1)\alpha(1)] & [a(1)\beta(1)-b(1)\beta(1)] \\ [a(2)\alpha(2)+b(2)\alpha(2)] & [a(2)\beta(2)+b(2)\beta(2)] & [a(2)\alpha(2)-b(2)\alpha(2)] & [a(2)\beta(2)-b(2)\beta(2)] \\ [a(3)\alpha(3)+b(3)\alpha(3)] & [a(3)\beta(3)+b(3)\beta(3)] & [a(3)\alpha(3)-b(3)\alpha(3)] & [a(3)\beta(3)-b(3)\beta(3)] \\ [a(4)\alpha(4)+b(4)\alpha(4)] & [a(4)\beta(4)+b(4)\beta(4)] & [a(4)\alpha(4)-b(4)\alpha(4)] & [a(4)\beta(4)-b(4)\beta(4)] \end{vmatrix}$$

As illustrated in the last expression, each column of the MO determinant is a linear combination of two columns from the VB determinant. For example, column 1 of the MO determinant is the sum of rows 1 and 3 of the VB determinant. Therefore, these two wavefunctions are identical, aside from a numerical factor.

(iii)   When we compare the MO wavefunction, $[a(1) + b(1)][a(2) + b(2)]$, and the VB counterpart, $a(1)b(2) + b(1)a(2)$, for the $^1\Sigma_g^+$ state for H_2 molecule, it is immediately obvious that the ionic terms $a(1)a(2)$ and $b(1)b(2)$ are included in the former and excluded in the latter. (Also see Problem 8.9.) To put it another way, there is no correction for electron correlation in the MO function; the two electrons are free to move independently of each other, and both are allowed to be on the same hydrogen atom, resulting in an ionic state. However, for the $^3\Sigma_u^+$ state of H_2, the formation of such ionic states is disallowed. It is because the two electrons of the $^3\Sigma_u^+$ ionic states have the same spin and hence having both electrons in the 1s orbital of one of the atom is forbidden by the exclusion principle. The formation of an ion by electron build-up is not possible, unless 2s or other orbitals are considered, which are disregarded in the present approximation.

For He_2, each atom already has two electrons in its 1s orbital. The third electron cannot be put in the 1s orbital and an ion cannot be formed. Therefore, the contributions of the ionic terms are zero and the two wavefunctions are identical for each of the two species.

A8.11 It is evident that the wavefunction is symmetric with respect to the exchange of subscripts "1" and "2" (as well as for the exchange of subscripts "3" and "4"). Hence the spin part of the total wavefunction must be antisymmetric with respect to the exchange of the same two subscripts. In other words, the spins of the two electrons are paired; the same holds for the spins of the two positrons. Mathematically, the spin wavefunctions are $(2)^{-1/2}(\alpha_1\beta_2 - \beta_1\alpha_2)$ for the electrons and $(2)^{-1/2}(\alpha_3\beta_4 - \beta_3\alpha_4)$ for the two positrons.

A8.12 In C_2, since the interacting atomic orbitals have the same energy (excepting the $2s-2p_\sigma$ interaction), the bonding (antibonding) molecular orbitals are significantly lower (higher) in energy than their constituent atomic orbitals. On the other hand, this situation is not found in LiF, whose valence atomic orbital energies have been estimated as (in 10^3 cm^{-1}): −44 for Li 2s; −151 and −374 for F 2p and 2s, respectively. Hence, the bonding orbitals have similar energies to the constituent F atomic orbitals; the antibonding orbitals have similar energies to the constituent Li atomic orbitals. As a result, the bonding orbital electrons reside predominantly on the F atom and a simplified bonding picture may be represented by Li^+F^-.

A8.13 The Xe 5p orbital that lies on the molecular axis and the two F $2p_\sigma$ orbitals form three molecular orbitals (in ascending energy): one bonding, one essentially nonbonding, one antibonding, with the first two filled. So the (single) bond that binds the three atoms together may be called a three-center four-electron bond. On the other hand, for each three-center $(B-H_{brid}-B)$ bond in B_2H_6, only the bonding orbital is filled. Hence, it is a three-center two-electron bond. Hence XeF_2 is known to be electron-rich, while B_2H_6 is called electron-deficient, or electron-poor.

A8.14 (i) The coordinate system chosen has the z-axis as the C_2 axis and the x-axis perpendicular to the molecular plane, as shown on the right.

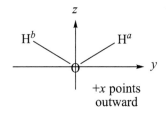

$+x$ points outward

Symmetry	O atomic orbital	H atomic orbital	Molecular orbital
A_1	$2s, 2p_z$	$\frac{1}{\sqrt{2}}(1s_a + 1s_b)$	$1a_1, 2a_1, 3a_1$
B_1	$2p_x$		$1b_1$
B_2	$2p_y$	$\frac{1}{\sqrt{2}}(1s_a - 1s_b)$	$1b_2, 2b_2$

The ground configuration is $(1a_1)^2(1b_2)^2(2a_1)^2(1b_1)^2(2b_2)^0(3a_1)^0$ and the ground state is $^1A_1$.

(ii) For CX_2, in the triplet state, the electronic configuration is $(1a_1)^2(1b_2)^2$ $(2a_1)^1(1b_1)^1$. Since the $1b_1$ orbital is an orbital localized on the C atom, its energy is influenced by the ligands only to a very small extent. On the other hand, the $2a_1$ orbital is localized mainly on the ligands. Therefore, a highly electronegative ligand such as F will greatly stabilize the $2a_1$ orbital, leading to a large energy separation between the $2a_1$ ard $1b_1$ orbitals and thus favoring a singlet ground state. Conversely, for CH_2, the energy gap between these two MOs is quite small, leading to a spin triplet ground state.

A8.15 According to the VSEPR theory, SO_3 should be trigonal planar with D_{3h} symmetry.

VB description:

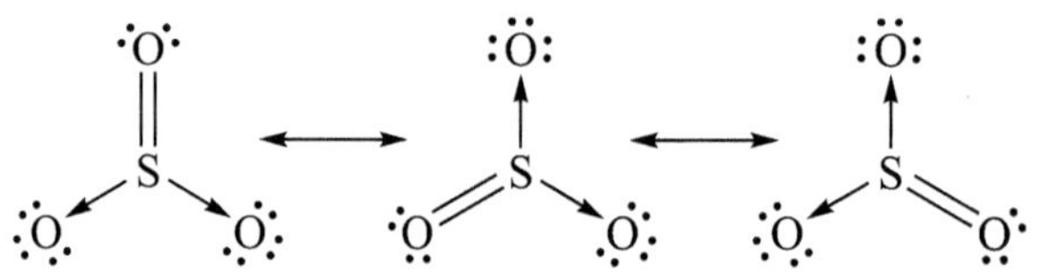

The sulfur atom is sp^2 hybridized; the bond order for the S–O linkage is $1\frac{1}{3}$. In summary, there are three σ bonds, one π bond, and eight lone pairs (including the O 2s electrons) on the oxygen atoms.

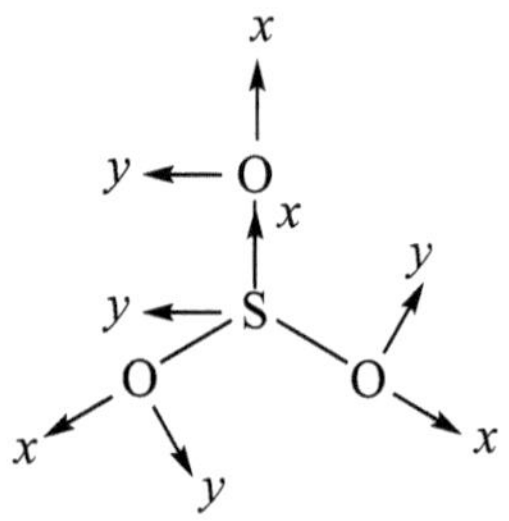

MO description:

The coordinate system chosen is shown on the right. Note that all z-axes are pointing towards the reader.

Symmetry	S atomic orbital	O atomic orbital	Molecular orbital
A_1'	3s	$(3)^{-1/2}(x_a + x_b + x_c)$	$1a_1', 2a_1'$
E'	$(3p_x, 3p_y)$	$\begin{cases} (6)^{-1/2}(2x_a - x_b - x_c) \\ (2)^{-1/2}(x_b - x_c) \end{cases}$ $\begin{cases} (6)^{-1/2}(2y_a - y_b - y_c) \\ (2)^{-1/2}(y_b - y_c) \end{cases}$	$1e', 2e', 3e'$
A_2''	$3p_z$	$(3)^{-1/2}(z_a + z_b + z_c)$	$1a_2'', 2a_2''$
A_2'		$(3)^{-1/2}(y_a + y_b + y_c)$	$1a_2'$
E''		$\begin{cases} (6)^{-1/2}(2z_a - z_b - z_c) \\ (2)^{-1/2}(z_b - z_c) \end{cases}$	$1e''$

The ground configuration and state are $(1a_1')^2(1e')^4(1a_2'')^2(1a_2')^2(1e'')^4(2e')^4(2a_2'')^0$ $(2a_1')^0(3e')^0$ and $^1A_1'$, respectively. In summary, there are three σ bonds $[(1a_1')^2(1e')^4]$, one π bond $[(1a_2'')^2]$, and five nonbonding electron pairs (not including the O 2s electrons).

A8.16

Symmetry	B atomic orbital	H atomic orbital	Bridge molecular orbital
A_g	$\frac{1}{2}(h_1 + h_2 + h_3 + h_4)$	$\frac{1}{\sqrt{2}}(1s_a + 1s_b)$	$1a_g, 2a_g$
B_{1u}	$\frac{1}{2}(h_1 - h_2 - h_3 + h_4)$	$\frac{1}{\sqrt{2}}(1s_a - 1s_b)$	$1b_{1u}, 2b_{1u}$
B_{2g}	$\frac{1}{2}(h_1 - h_2 + h_3 - h_4)$		$1b_{2g}$
B_{3u}	$\frac{1}{2}(h_1 + h_2 - h_3 - h_4)$		$1b_{3u}$

In ascending energy order, the molecular orbitals are $1a_g$, $1b_{1u}$, $1b_{3u}$, $1b_{2g}$, $2a_g$, and $2b_{1u}$, with the first two filled with electrons. Compare these results with those described in A8.13.

A8.17 The electronic configuration of Ru is $4d^6 5s^2$. In building up the σ framework, the Ru and O atoms make use of their d^2sp^3 and sp hybrid orbitals, respectively. The remaining atomic orbitals to be considered are d_{xy}, d_{xz}, and d_{yz} for each Ru and p_x and p_y for O. These eight atomic orbitals may be linearly combined to form eight molecular orbitals which will accommodate 12 electrons. Consider bonding in the xz-plane. Interaction of d_{xz} orbitals of Ru and p_x of O will yield

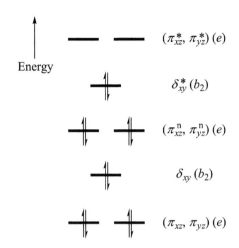

three molecular orbitals denoted π_{xz}, π_{xz}^n, and π_{xz}^*. An equivalent set of molecular orbitals π_{yz}, π_{yz}^n, and π_{yz}^* arises from bonding in the yz-plane. The d_{xy} metal orbitals interact weakly to give two δ molecular orbitals.

Since all electrons are paired, the anion is expected to be diamagnetic. The shapes of the molecular orbitals are not shown here.

A8.18 The side-on coordination of acetylene to a transition metal may be described qualitatively by the following orbital interactions:

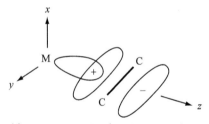

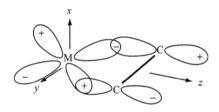

(a) Vacant metal σ (combination of s, p_z, and d_{z^2}) with filled ligand $\pi_{\parallel}$.

(b) Filled metal $\pi_{\parallel}$ (combination of d_{yz} and p_y) with vacant ligand $\pi_{\parallel}^*$.

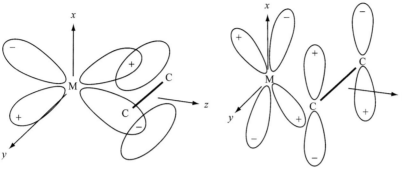

(c) Vacant or filled metal π (combination of d_{xz} and p_x) with filled ligand $\pi_\perp$.

(d) Filled or vacant metal δ (d_{xy}) with vacant ligand $\pi_\perp^*$.

Subscripts "$\parallel$" and "$\perp$" indicate that the orbitals concerned are parallel and perpendicular, respectively, to the M-C_2 plane. The extent of overlap decreases in the order of (a) > (b) > (c) > (d). Interaction (a) is bonding, (b) is normally bonding since most transition metals have d electrons which may occupy $d\pi_\parallel$ orbitals. In early transition metals with vacant $d\pi_\perp$ orbitals (c) is bonding, but in later transition metal complexes, especially those with d^{10} configuration, the interaction should be repulsive and antibonding. Interaction (d) has a negligible effect because of poor overlap.

A8.19 (i) $\psi_{MO}(1,2) = a^2 \phi_H(1)\phi_H(2) + b^2 \phi_F(1)\phi_F(2) + ab\phi_H(1)\phi_F(2) + ab\phi_H(2)\phi_F(1)$, in which the four terms on the right-hand side represent structures H^-F^+, H^+F^-, H–F, and H–F, respectively. As HF is fairly ionic, the second term, representing H^+F^-, should be the most important one. Based on similar reasoning, the first term (representing H^-F^+) should be least important.

(ii) ψ_{VB}, consisting of only the last two terms of ψ_{MO}, does not have a term describing H^+F^-. Hence ψ_{MO} is a better wavefunction.

A8.20 (i) The energy level diagram for OH is similar to that of HF. The only difference is that OH has one less electron. In other words, the ground configuration is $(2s)^2(\sigma_z)^2(\pi_x^n = \pi_y^n)^3$.

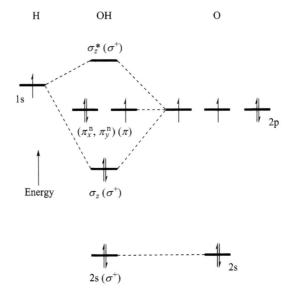

$$H \qquad\qquad OH \qquad\qquad\qquad O$$

$\sigma_z^* (\sigma^+)$

$1s$

$(\pi_x^n, \pi_y^n)(\pi)$

$2p$

Energy

$\sigma_z (\sigma^+)$

$2s$

$2s (\sigma^+)$

(ii) The unpaired electron is located in a nonbonding orbital which is localized on O.

(iii) When OH is ionized, the electron is taken from an oxygen "atomic orbital". Similarly, when HF is ionized, the electron is taken away from a fluorine "atomic orbital". Hence the difference between the ionization energies (IEs) of HF and OH is similar to the difference between the IEs of atomic F and O.

A8.21 (i) The π-MO scheme is shown on the right. The open and black circles indicate that the positive and negative lobes of the p orbitals point towards and away from the reader, respectively. Since N is more electronegative than S, the b_{1u} orbital lies below the b_{2u} orbital. In accordance with Hund's rule, the 12 π electrons in the planar system doubly occupy the a_{2u}, e_g, b_{1u}, and b_{2u} orbitals, while the degenerate higher e_g orbitals are half filled. So S_4N_4 will be paramagnetic if it adopts a planar structure.

a_{2u} ———

e_g

b_{2u}

b_{1u}

e_g

a_{2u}

S

N

224

(ii) The resulting triplet ground state is expected to undergo Jahn–Teller distortion. Symmetric distortion to give the cradle structure leads to strong interaction of two pairs of transannular p orbitals. The top a_{2u} and b_{2u} orbitals are stabilized whereas the higher e_g orbitals become destabilized, resulting in a stable singlet ground state. The symmetry arguments remain valid regardless of whether the two transannular bonds are formed between N or S centers, but molecular orbital calculations and overlap considerations [see part (iii)] indicate that bond formation between S atoms is much more efficient.

(iii) From the figure, it can be seen that, at $R \approx 2.4$ Å, the relevant $p_\sigma - p_\sigma$ overlap integrals for P and As are larger than those for S and N. It is therefore expected that in $\alpha - P_4S_4$ and As_4S_4, the S atoms stay co-planar whereas the transannular bonds are formed between P and As atoms, respectively. The observed structures are illustrated below:

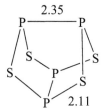 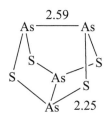

A8.22

Observed transition (eV)	Assignment: calculated value (eV)	Allowed or not
5.43	$^1A_{1g} \rightarrow {}^1E_g\ (2a_{2u} \rightarrow 3e_u),\ 5.4$	No
6.8	$^1A_{1g} \rightarrow {}^1E_u\ (3a_{1g} \rightarrow 3e_u),\ 6.9$	Yes
8.3	$^1A_{1g} \rightarrow {}^1A_{2u}\ (2a_{2u} \rightarrow 4a_{1g}),\ 8.0$	Yes
9.4	$^1A_{1g} \rightarrow {}^1A_{2u}\ (2a_{2u} \rightarrow 5a_{1g}),\ 8.6$	Yes
	$^1A_{1g} \rightarrow {}^1E_u\ (2a_{2u} \rightarrow 3e_g),\ 8.8$	Yes
11.3	$^1A_{1g} \rightarrow {}^1E_u\ (a_{2g} \rightarrow 3e_u),\ 12.3$	Yes

A8.23 In units of 10^3 cm^{-1}, the transitions are:

$^2A_1' \rightarrow {}^2E'\ (5e' \rightarrow 5a_1'),\ 7.7;$ $\qquad$ $^2A_1' \rightarrow {}^2A_2''\ (3a_2'' \rightarrow 5a_1'),\ 9.4;$

$^2A_1' \rightarrow {}^2E'\ (4e' \rightarrow 5a_1'),\ 9.5;$ $\qquad$ $^2A_1' \rightarrow {}^2E'\ (3e' \rightarrow 5a_1'),\ 10.2;$

$^2A_1' \rightarrow {}^2A_2''\ (2a_2'' \rightarrow 5a_1'),\ 21.3;$ $\qquad$ and $\qquad$ $^2A_1' \rightarrow {}^2E'\ (2e' \rightarrow 5a_1'),\ 26.4.$

A8.24 (i)

Observed bands (10^3 cm^{-1})	Assignment	
20.1 ($\perp$)	$^1A_1 \to {}^1E \, (2b_1 \to 7e)$	
25.4 ($\parallel$)	$^1A_1 \to {}^1A_1 \, (6e \to 7e)$	
30.3	$^1A_1 \to {}^1A_2 \, (2b_1 \to 3b_1)$	(forbidden, hence not polarized)
37.8($\perp$)*	$^1A_1 \to {}^1E \, (6e \to 5a_1)$	
42.0($\perp$)*	$^1A_1 \to {}^1E \, (6e \to 3b_1)$	
50.0($\perp$)	$^1A_1 \to {}^1E \, (2b_1 \to 8e)$	(an M $\to$ L charge transfer band)

* If the transition energy is calculated by subtracting the orbital energy of the lower state from that of the upper state, the assignments for these two bands will be reversed. However, this simple method gives only an approximation for the transition energy. To be more accurate, the value thus calculated should be corrected for interelectronic repulsion energy.

(ii) If the given energy level diagram is used for $[\text{Fe(CN)}_5\text{NO}]^{3-}$, the ground configuration is ... $(6e)^4(2b_2)^2(7e)^1$, which leads to a degenerate ground state. Thus, the molecule will undergo a Jahn-Teller distortion to remove the degeneracy. The expected distortion is bending of the "originally linear" Fe–N–O linkage.

A8.25

Band (10^3 cm^{-1})	Assignment	Calculated value (10^3 cm^{-1})	Comment
14.5	$^1A_1 \to {}^1T_1 \, (t_1 \to 2e)$	23.4	Forbidden
18.3	$^1A_1 \to {}^1T_2 \, (t_1 \to 2e)$	23.4	Allowed
28.0	$^1A_1 \to {}^1T_1 \, (3t_2 \to 2e)$	32.7	Forbidden
32.2	$^1A_1 \to {}^1T_2 \, (3t_2 \to 2e)$	32.7	Allowed
44.0	$^1A_1 \to {}^1T_2 \, (3t_2 \to 4t_2)$	47.1	Allowed; transitions from $^1A_1$ to the other states arising from the same excited configuration are forbidden

A8.26

D_{2h}	E	$C_2(z)$	$C_2(y)$	$C_2(x)$	i	$\sigma(xy)$	$\sigma(xz)$	$\sigma(yz)$	Symmetry
(i), (viii), (xii)	1	-1	1	-1	1	-1	1	-1	B_{2g}
(ii), (ix), (xi)	1	1	-1	-1	-1	-1	1	1	B_{1u}
(iii), (v), (x)	1	1	1	1	-1	-1	-1	-1	A_u
(iv), (vi), (vii)	1	-1	-1	1	1	-1	-1	1	B_{3g}

A8.27 Representation (Γ_π) based on the four p_z atomic orbitals of the cyclobutadiene system in point group C_{4v}: $\Gamma_\pi = A_1 + B_1 + E$. Let p_z of atom i be denoted by ϕ_i.

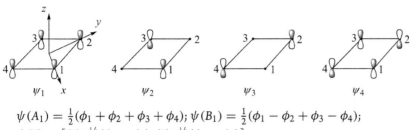

$$\psi(A_1) = \tfrac{1}{2}(\phi_1 + \phi_2 + \phi_3 + \phi_4); \ \psi(B_1) = \tfrac{1}{2}(\phi_1 - \phi_2 + \phi_3 - \phi_4);$$
$$\psi(E) = [(2)^{-1/2}(\phi_1 - \phi_3), (2)^{-1/2}(\phi_2 - \phi_4)].$$

Symmetry	Fe atomic orbital	Cyclobutadiene group orbital	Bond type
A_1	s, p_z, d_{z^2}	$\tfrac{1}{2}(\phi_1 + \phi_2 + \phi_3 + \phi_4)$	σ
B_1	$d_{x^2-y^2}$	$\tfrac{1}{2}(\phi_1 - \phi_2 + \phi_3 - \phi_4)$	δ
B_2	d_{xy}	—	n
E	(p_x, p_y)	$[(2)^{-1/2}(\phi_1 - \phi_3), (2)^{-1/2}(\phi_2 - \phi_4)]$	π
	(d_{xz}, d_{yz})		

A8.28 A_1': $(6)^{-1/2}(a + b + c + d + e + f), (3)^{-1/2}(g + h + i)$.

A_2'': $(6)^{-1/2}(a + b + c - d - e - f)$.

E': $\begin{cases} (12)^{-1/2}(2a - b - c + 2d - e - f) \\ \tfrac{1}{2}(b - c + e - f) \end{cases}$, $\begin{cases} (6)^{-1/2}(2h - g - i) \\ (2)^{-1/2}(g - i) \end{cases}$.

E'': $\begin{cases} (12)^{-1/2}(2a - b - c - 2d + e + f) \\ \tfrac{1}{2}(b - c - e + f) \end{cases}$.

A8.29 (i) & (ii)

Symmetry	A. O. on A	A. O. on H	M. O.
A_1	2s	a	$1a_1, 2a_1, 3a_1, 4a_1$
	$2p_z$	$\tfrac{1}{2}(b + c + d + e)$	
B_1	—	$\tfrac{1}{2}(b - c + d - e)$	$1b_1$
E	$\begin{cases} 2p_x \\ 2p_y \end{cases}$	$\begin{cases} (2)^{-1/2}(b - d) \\ (2)^{-1/2}(c - e) \end{cases}$	$1e, 2e$

(iii)

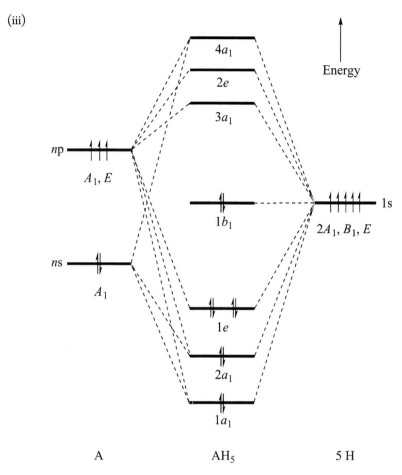

A   AH$_5$ 5 H

The ground configuration is $(1a_1)^2(2a_1)^2(1e)^4(1b_1)^2$ and the ground state is $^1A_1$.

A8.30 (i)

D_{4h}	E	$2C_4$	C_2	$2C_2'$	$2C_2''$	i	$2S_4$	σ_h	$2\sigma_v$	$2\sigma_d$
$\Gamma_{C\pi}$	4	0	0	0	-2	0	0	-4	0	2

$\Gamma_{C\pi} = E_g + A_{2u} + B_{1u}$.

(ii) $\Gamma_{O\pi} = \Gamma_{C\pi} = E_g + A_{2u} + B_{1u}$.

(iii) SALCs of $\Gamma_{C\pi}$:

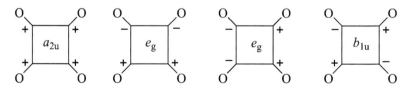

SALCs of $\Gamma_{O\pi}$:

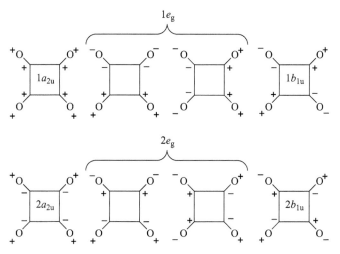

Combined SALCs:

Energy order: $1a_{2u} < 1e_g < 2a_{2u} \approx 1b_{1u} < 2e_g < 2b_{1u}$.

There are 10 π electrons and the following π MOs are filled: $1a_{2u}$, $1e_g$, $2a_{2u}$, and $1b_{1u}$.

(iv)

π MOs of $C_4O_4^{2-}$*	Valency AOs of M
$1a_{2u}$, $2a_{2u}$	s, p_z, d_{z^2}
$1e_g$ $2e_g$ $1e_g$ $2e_g$	$\begin{Bmatrix} p_x \\ p_y \end{Bmatrix}$, $\begin{Bmatrix} d_{xz} \\ d_{yz} \end{Bmatrix}$
$1b_{1u}$, $2b_{1u}$	d_{xy}

* Note that the symmetry classification for the $C_4O_4^{2-}$ ring is no longer applicable when the ring is capped by metal ion M.

A8.31 (i) The three localized bonding orbitals are:

$$\sigma_1 = (2)^{-1/2}(a + b');$$
$$\sigma_2 = (2)^{-1/2}(b + c');$$
$$\sigma_3 = (2)^{-1/2}(c + a').$$

The three localized antibonding orbitals are:

$$\sigma_1{}^* = (2)^{-1/2}(a - b');$$
$$\sigma_2{}^* = (2)^{-1/2}(b - c');$$
$$\sigma_3{}^* = (2)^{-1/2}(c - a').$$

The symmetry-adapted wavefunctions are:

$$\phi_1 = (3)^{-1/2}(\sigma_1 + \sigma_2 + \sigma_3) = (6)^{-1/2}(a + b + c + a' + b' + c')(\text{symmetry: } A_1');$$

$$\left.\begin{aligned}
\phi_2 &= (6)^{-1/2}(2\sigma_1 - \sigma_2 - \sigma_3) \\
&= (12)^{-1/2}(2a - b - c - a' + 2b' - c') \\
\phi_3 &= (2)^{-1/2}(\sigma_2 - \sigma_3) \\
&= (2)^{-1}(b - c - a' + c');
\end{aligned}\right\} \quad (\text{symmetry: } E');$$

$$\phi_4 = (3)^{-1/2}(\sigma_1^* + \sigma_2^* + \sigma_3^*) = (6)^{-1/2}(a + b + c - a' - b' - c')(\text{symmetry: } A_2');$$

$$\left.\begin{aligned}
\phi_5 &= (6)^{-1/2}(2\sigma_1^* - \sigma_2^* - \sigma_3^*) \\
&= (12)^{-1/2}(2a - b - c + a' - 2b' + c') \\
\phi_6 &= (2)^{-1/2}(\sigma_2^* - \sigma_3^*) \\
&= (6)^{1/2}(b - c + a' - c');
\end{aligned}\right\} \quad (\text{symmetry: } E');$$

(ii) $E(1a_1') = H_{11} = \int \phi_1 \hat{H} \phi_1 d\tau = \alpha + \beta.$
$E(1a_2') = H_{44} = \int \phi_4 \hat{H} \phi_4 d\tau = \alpha - \beta.$
For $1e'$ and $2e'$, ϕ_2 and ϕ_6 are symmetric with respect to $\sigma_v(yz)$, whereas ϕ_3 and ϕ_5 are antisymmetric with respect to $\sigma_v(yz)$.

$$H_{22} = \int \phi_2 \hat{H} \phi_2 d\tau = \alpha + \beta, H_{66} = \int \phi_6 \hat{H} \phi_6 d\tau = \alpha - \beta, H_{26} = \int \phi_2 \hat{H} \phi_6 d\tau = 0.$$

Hence, ϕ_2 and ϕ_6 do not mix with each other under Hückel approximation. Similarly,

$$H_{33} = \int \phi_3 \hat{H} \phi_3 d\tau = \alpha + \beta, H_{55} = \int \phi_5 \hat{H} \phi_5 d\tau = \alpha - \beta, H_{35} = \int \phi_3 \hat{H} \phi_5 d\tau = 0.$$

Therefore, $E(1a'_1) = E(1e') < E(1a'_2) = E(2e')$. The ground configuration is $(1a'_1)^2(1e')^2$.

[If the less significant exchange integrals (H_{ab}, H_{ac}, and $H_{ac'}$, which are ignored under the Hückel approximation specified in the problem) are taken into account, $1a'_1$ is lower than $1e'$ in energy. But it is not obvious whether $2a'_2$ is lower than $2e'$ in energy.]

A8.32 (i)

Symmetry	A. O. on A	A. O. on H	M. O.
A_1	2s	$(2)^{-1/2}(a+b)$	$1a_1, 2a_1, 3a_1, 4a_1$
	$2p_z$	$(2)^{-1/2}(c+d)$	
B_1	$2p_x$	$(2)^{-1/2}(a-b)$	$1b_1, 2b_1$
B_2	$2p_y$	$(2)^{-1/2}(c-d)$	$1b_2, 2b_2$

(ii)

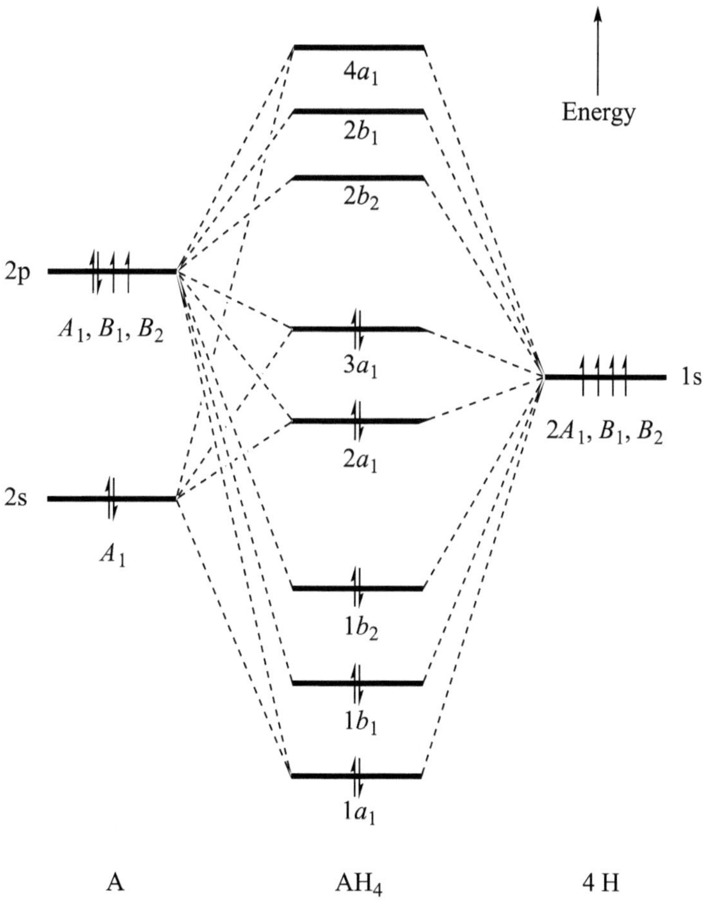

231

The ground configuration is $(1a_1)^2(1b_1)^2(1b_2)^2(2a_1)^2(3a_1)^2$ and the ground state is $^1A_1$.

The lowest-energy transition is: $\ldots (3a_1)^2; {}^1A_1 \rightarrow \ldots (3a_1)^1(2b_2)^1; {}^1B_2$.

Since $A_1 \times \Gamma_x \times B_2 = A_1 \times B_1 \times B_2 = A_2 \neq A_1$,

$$A_1 \times \Gamma_y \times B_2 = A_1 \times B_2 \times B_2 = A_1,$$

$$A_1 \times \Gamma_z \times B_2 = A_1 \times A_1 \times B_2 = B_2 \neq A_1,$$

the transition is both spin- and symmetry-allowed and it is y-polarized.

A8.33 (i)

D_{4h}	E	$2C_4$	C_2	$2C_2'$	$2C_2''$	i	$2S_4$	σ_h	$2\sigma_v$	$2\sigma_d$
Γ_H	8	0	0	0	0	0	0	0	4	0

$\Gamma_H = A_{1g} + B_{1g} + E_g + A_{2u} + B_{2u} + E_u$.

$A_{1g}: (8)^{-1/2}(a + b + c + d + a' + b' + c' + d')$.
$A_{2u}: (8)^{-1/2}(a + b + c + d - a' - b' - c' - d')$.
$B_{1g}: (8)^{-1/2}(a - b + c - d + a' - b' + c' - d')$.
$B_{2u}: (8)^{-1/2}(a - b + c - d - a' + b' - c' + d')$.

$$E_g: \begin{cases} \dfrac{1}{2}(a - c - a' + c') \\ \dfrac{1}{2}(b - d - b' + d') \end{cases} \qquad E_u: \begin{cases} \dfrac{1}{2}(a - c + a' - c') \\ \dfrac{1}{2}(b - d + b' - d') \end{cases}$$

(The linear combinations may be obtained in the following manner. First, derive the linear combinations of the unprimed orbitals. Then do the same for the primed orbitals. Finally, combine each of the linear combinations of the unprimed orbitals with the linear combination of the unprimed orbitals with appropriate symmetry with respect to the σ_v and σ_d planes.)

(ii) Molecular orbitals arranged in order of increasing energy:

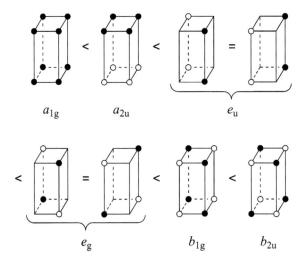

A8.34 (i) & (ii)

Symmetry	Boron atomic orbitals	Hydrogen atomic orbitals	Molecular orbitals
A_g	$(2)^{-1/2}(2s_A + 2s_B)$	$(2)^{-1/2}(e + f)$	$1a_g, 2a_g, 3a_g, 4a_g$
	$(2)^{-1/2}(z_A + z_B)$	$\frac{1}{2}(a + b + c + d)$	
B_{1u}	$(2)^{-1/2}(2s_A - 2s_B)$	$\frac{1}{2}(a + b - c - d)$	$1b_{1u}, 2b_{1u}, 3b_{1u}$
	$(2)^{-1/2}(z_A - z_B)$		
B_{3u}	$(2)^{-1/2}(x_A + x_B)$	$(2)^{-1/2}(e - f)$	$1b_{3u}, 2b_{3u}$
B_{2g}	$(2)^{-1/2}(x_A - x_B)$	—	$1b_{2g}$
B_{2u}	$(2)^{-1/2}(y_A + y_B)$	$\frac{1}{2}(a - b + c - d)$	$1b_{2u}, 2b_{2u}$
B_{3g}	$(2)^{-1/2}(y_A - y_B)$	$\frac{1}{2}(a - b - c + d)$	$1b_{3g}, 2b_{3g}$

A8.35 (i) & (ii)

C_{2v}	E	$C_2(z)$	$\sigma_v(xz)$	$\sigma_v(yz)$	
Γ_a	1	−1	1	−1	$= B_1.$
Γ_{be}	2	0	0	−2	$= B_1 + A_2.$
Γ_{cd}	2	0	0	−2	$= B_1 + A_2.$

B_1 functions: a, $(2)^{-1/2}(b + e)$, $(2)^{-1/2}(c + d)$.
A_2 functions: $(2)^{-1/2}(b - e)$, $(2)^{-1/2}(c - d)$.

$E \longrightarrow$

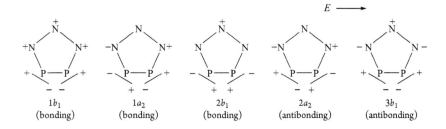

To achieve the most stable configuration and to maximize the bonding interaction, the molecular charge should be −1. Then we have the configuration $(1b_1)^2(1a_2)^2(2b_1)^2$, which satisfies the $4n + 2$ rule.

233

A8.36 (i) B_1 basis functions: $(2)^{-1/2}(a+b)$; $(2)^{-1/2}(c+f)$; $(2)^{-1/2}(d+e)$. They are linearly combined to form the following three molecular orbitals:

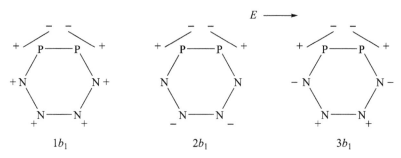

$1b_1$ $2b_1$ $3b_1$

A_2 basis functions: $(2)^{-1/2}(a-b)$; $(2)^{-1/2}(f-c)$; $(2)^{-1/2}(e-d)$. They are linearly combined to form the following three molecular orbitals:

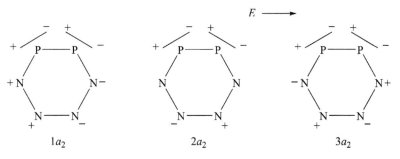

$1a_2$ $2a_2$ $3a_2$

By considering the nodal characters of the molecular orbitals, we have the overall ordering of $1b_1 < 1a_2 < 2b_1 < 2a_2 \approx 3b_1 < 3a_2$.

(ii) With six π electrons, the ground configuration and ground state of this system are $(1b_1)^2(1a_2)^2(2b_1)^2$ and $^1A_1$, respectively.

(iii) The lowest-energy transition is: $\ldots (2b_1)^2; {}^1A_1 \rightarrow \ldots (2b_1)^1(2a_2)^1; {}^1B_2$. It is a spin- and symmetry-allowed transition and the transition is y-polarized. [See A8.32(ii).]

A8.37

Molecule	Cation	Anion
O_2	$r(O_2^+) < r(O_2)$	$r(O_2^-) > r(O_2)$
CH_4	$r(\text{C–H})$ in $CH_4^+ > r(\text{C–H})$ in CH_4	$r(\text{C–H})$ in $CH_4^- > r(\text{C–H})$ in CH_4
NH_3	$r(\text{N–H})$ in $NH_3^+ \approx r(\text{N–H})$ in NH_3	$r(\text{N–H})$ in $NH_3^- > r(\text{N–H})$ in NH_3
H_2O	$r(\text{O–H})$ in $H_2O^+ \approx r(\text{O–H})$ in H_2O	$r(\text{O–H})$ in $H_2O^- > r(\text{O–H})$ in H_2O
BF_3	$r(\text{B–F})$ in $BF_3^+ \approx r(\text{B–F})$ in BF_3	$r(\text{B–F})$ in $BF_3^- > r(\text{B–F})$ in BF_3
B_2H_6	$r(\text{B–H}_t)$ in $B_2H_6^+ \approx r(\text{B–H}_t)$ in B_2H_6	$r(\text{B–H}_t)$ in $B_2H_6^- \approx r(\text{B–H}_t)$ in B_2H_6
B_2H_6	$r(\text{B–H}_b)$ in $B_2H_6^+ > r(\text{B–H}_b)$ in B_2H_6	$r(\text{B–H}_b)$ in $B_2H_6^- \approx r(\text{B–H}_b)$ in B_2H_6

General guidelines:

(a) Taking an electron away from a bonding orbital will lengthen the bond.

(b) Adding an electron to a bonding orbital will shorten the bond.

(c) Taking an electron away from an antibonding orbital will shorten the bond.

(d) Adding an electron to an antibonding orbital will lengthen the bond.

(e) Either taking an electron away or adding an electron to a nonbonding orbital will not have much effect on the bond length.

A8.38 (i) to (iii)

$$I_2^+: \ldots(\sigma_z)^2(\pi_x{=}\pi_y)^4(\pi_x^*)^2(\pi_y^*)^1$$

four-center
two-electron bond $\uparrow\downarrow$

$$S_2: \ldots(\sigma_z)^2(\pi_x{=}\pi_y)^4(\pi_x^*)^1(\pi_y^*)^1$$

four-center
two-electron bond $\downarrow$

$$I_2^+: \ldots(\sigma_z)^2(\pi_x{=}\pi_y)^4(\pi_x^*)^1(\pi_y^*)^2$$

(iv) If the two four-center two-electron bonds are assumed to be equally shared by the three units, each unit would carry $\frac{4}{3}$ antibonding electrons, or $-\frac{2}{3}$ bond order. So,

$$\text{bond order of S–S} = 3 - \tfrac{2}{3} = 2\tfrac{1}{3},$$
$$\text{bond order of I–I} = 2 - \tfrac{2}{3} = 1\tfrac{1}{3}.$$

Refer to cited reference for additional details.

A8.39 (i) $\Gamma_C = B_{2g} + B_{3u};$ $B_{2g}: (2)^{-1/2}(a - b)$ and $B_{3u}: (2)^{-1/2}(a + b).$

$\Gamma_O = B_{1g} + B_{2g}$ $B_{1g}: \frac{1}{2}(c - d + e - f),$ $B_{2g}: \frac{1}{2}(c + d - e - f),$

    $+ A_u + B_{3u};$ $A_u: \frac{1}{2}(c - d - e + f),$ and $B_{3u}: \frac{1}{2}(c + d + e + f).$

(ii) The six π molecular orbitals are illustrated pictorially below in the order of increasing energy. The "+" and "−" signs denote the signs of the upper lobes of the π-bonding p orbitals. The energies of a_u and b_{1g} should be similar.

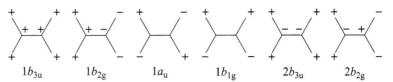

$1b_{3u}$ $1b_{2g}$ $1a_u$ $1b_{1g}$ $2b_{3u}$ $2b_{2g}$

(iii) There are 34 valence electrons in the $C_2O_4^{2-}$ ion. Ten electrons are used in the formation of the σ bonds, 16 electrons are used in the in-plane lone pairs, leaving eight electrons for the out-of-plane π orbitals. Thus the first four illustrated molecular orbitals are filled, and the C orbitals take part in only the lowest two molecular orbitals. While there is bonding interaction between the C orbitals in $1b_{3u}$, there is antibonding interaction between the C orbitals in $1b_{2g}$. In other words, there is no net π bonding between the carbon atoms and the C–C bond is only a σ bond. If we consider that the antibonding effect is larger than the bonding effect, the C–C bond length should be longer than a typical C–C single bond.

A8.40 (i) (a)

Hückel MO	Coefficients c_{jk}					Energy
	ϕ_1	ϕ_2	ϕ_3	ϕ_4	ϕ_5	
ψ_1	$\sqrt{3}/6$	$\frac{1}{2}$	$\sqrt{3}/3$	$\frac{1}{2}$	$\sqrt{3}/6$	$E_1 = \alpha + \sqrt{3}\beta$
ψ_2	$\frac{1}{2}$	$\frac{1}{2}$	0	$-\frac{1}{2}$	$-\frac{1}{2}$	$E_2 = \alpha + \beta$
ψ_3	$\sqrt{3}/3$	0	$-\sqrt{3}/3$	0	$\sqrt{3}/3$	$E_3 = \alpha$
ψ_4	$\frac{1}{2}$	$-\frac{1}{2}$	0	$\frac{1}{2}$	$-\frac{1}{2}$	$E_4 = \alpha - \beta$
ψ_5	$\sqrt{3}/6$	$-\frac{1}{2}$	$\sqrt{3}/3$	$-\frac{1}{2}$	$\sqrt{3}/6$	$E_5 = \alpha - \sqrt{3}\beta$

(b) $E_{TOT} = 5\alpha + 5.464\beta$; DE $= 1.464\beta$.

(c) $\lambda = 4727$ Å.

(d) π bond orders:

$C_1-C_2 = 2[(\sqrt{3}/6)(\frac{1}{2})(1) + (\frac{1}{2})(\frac{1}{2})(1) + (\sqrt{3}/3)(0)(\frac{1}{2})] = 0.789$;

$C_2-C_3 = 2[(\frac{1}{2})(\sqrt{3}/3)(1) + (\frac{1}{2})(0)(1) + (0)(-\sqrt{3}/3)(\frac{1}{2})] = 0.577$.

(e) Total bond orders (including σ bond of C–C bonds):

$C_1-C_2 = 0.789 + 1 = 1.789$; $C_2-C_3 = 1.577$.

Estimated bond lengths from the correlation curve:

$C_1-C_2 = 1.38$ Å and $C_2-C_3 = 1.41$ Å.

(ii) (a) HOMO: $\psi_3(x) = \sqrt{\dfrac{1}{3\ell}} \sin\left(\dfrac{\pi x}{2\ell}\right), E_3 = \dfrac{h^2}{32m\ell^2}$;

LUMO: $\psi_4(x) = \sqrt{\dfrac{1}{3\ell}} \sin\left(\dfrac{2\pi x}{3\ell}\right), E_4 = \dfrac{h^2}{18m\ell^2}$.

(b) $\lambda = 3347$ Å.

A8.41 For the σ orbitals: $H_{CO} = -2S(H_{CC}H_{OO})^{1/2} = -28.748$ kK.

Expanding the secular determinant gives: $0.981\, E^2 + 206.123\, E + 10181.563 = 0$.

Upon solving the equation, $E = -130.638$ kK or -79.428 kK.

The secular equations are: $(H_{OO} - E)a_O + (H_{CO} - ES)a_C = 0$

and $(H_{CO} - ES)a_O + (H_{CC} - E)a_C = 0$,

where a_O and a_C are the coefficients in the wavefunction for the molecular orbital:

$\psi = a_O(2p_O) + a_C(2p_C)$.

Putting $E = -130.638$ kK into either secular equation, we obtain: $a_C/a_O = 0.243$ and $\psi = a_O[(2p_O) + 0.243\,(2p_C)]$. Normalization of ψ gives $a_O = 0.943$ and hence $a_C = 0.229$.

The above process can be repeated for the other orbitals. The results are summarized below.

Orbital	σ_z	σ_z^*	π_x, π_y	π_x^*, π_y^*
Energy (kK)	-130.638	-79.428	-128.596	-84.611
a_O	0.943	0.362	0.986	0.178
a_C	0.229	-0.983	0.117	-0.995

A8.42 Let ϕ_1, ϕ_2, and ϕ_3 denote the atomic orbitals of the hydrogens used for forming molecular orbitals. By symmetry, the three molecular orbitals are:

$$\psi_1 = N_1(\phi_1 + \phi_2 + \phi_3);$$
$$\psi_2 = N_2(2\phi_1 - \phi_2 - \phi_3);$$
$$\psi_3 = N_3(\phi_2 - \phi_3).$$

Wavefunction ψ_1 has the lowest energy and is filled with the two delocalized electrons. The normalization constant is $N_1 = (3 + 6S)^{-1/2}$. Using Cusachs' approximation for β and minimizing $E_1[= (1 + 2S)^{-1}(1 + 4S - 2S^2)\alpha]$ with respect to S, the overlap S for the minimum energy is $\frac{1}{2}(-1 + \sqrt{3})$, or 0.366. At this minimum, $\beta = 0.598\alpha$ and $E = 2.536\alpha$. Stability of H_3^+ with respect to H_2 and H^+ is $E - (2\alpha - D_e) = -245$ kJ mol^{-1}.

A8.43 (i)

Symmetry	Apical B orbital	Basal B orbitals
A_1	a	$\frac{1}{2}(d+e+f+g)$
E	(b,c)	$\left[\frac{1}{\sqrt{2}}(d-f), \frac{1}{\sqrt{2}}(e-g)\right]$
B_1		$\frac{1}{2}(d-e+f-g)$

(ii) Secular determinants:

$$A_1: \quad \begin{vmatrix} \alpha-E & 2\beta \\ 2\beta & \alpha-E \end{vmatrix} = 0 = \begin{vmatrix} x & 2 \\ 2 & x \end{vmatrix}, \text{ with } x = \frac{\alpha-E}{\beta}.$$

$$E(1a_1) = \alpha + 2\beta; E(2a_1) = \alpha - 2\beta.$$

$$B_1: \quad |\alpha - E| = 0; \ E(1b_1) = \alpha.$$

$$E: \quad \begin{vmatrix} \alpha-E & \sqrt{2}\beta \\ \sqrt{2}\beta & \alpha-E \end{vmatrix} = 0 = \begin{vmatrix} x & \sqrt{2} \\ \sqrt{2} & x \end{vmatrix}, \text{ with } x = \frac{\alpha-E}{\beta}.$$

$$E(1e) = \alpha + \sqrt{2}\beta; E(2e) = \alpha - \sqrt{2}\beta.$$

(iii) In ascending order of energy, the order of the molecular orbitals are: $1a_1$, $1e$, $1b_1$, $2e$, $2a_1$. Hence, the ground configuration is $(1a_1)^2(1e)^4$ and the ground state is $^1A_1$.

(iv) Wavefunctions for the filled orbitals are:

$$1a_1 = \frac{1}{\sqrt{2}}a + \frac{1}{\sqrt{2}}\left[\frac{1}{2}(d+e+f+g)\right];$$

$$1e = \begin{cases} \frac{1}{\sqrt{2}}b + \frac{1}{\sqrt{2}}\left[\frac{1}{\sqrt{2}}(d-f)\right] = \frac{1}{\sqrt{2}}b + \frac{1}{2}(d-f); \\ \frac{1}{\sqrt{2}}c + \frac{1}{\sqrt{2}}\left[\frac{1}{\sqrt{2}}(e-g)\right] = \frac{1}{\sqrt{2}}c + \frac{1}{2}(e-g). \end{cases}$$

A8.44 (i) The determinant expands to give:

$$(E - \alpha_P)[(E^2 - (\alpha_P + \alpha_N)E - (\alpha_P\alpha_N - 2\beta^2)] = 0.$$

Solving for E and substituting $\alpha_N = \alpha_P + k\beta$, we have:

$$E = \alpha_P \text{(nonbonding)},$$

$$\frac{1}{2}[2\alpha_P + k\beta \pm (k^2 + 8)^{1/2}\beta](+ : \text{bonding}; - : \text{antibonding}).$$

There are two π electrons delocalized in the π system.

$$E_\pi = 2\alpha_P + k\beta + (k^2 + 8)^{1/2}\beta.$$

(ii) Since the π bond is localized between N and one of the P atoms, the secular determinant is:

$$\begin{vmatrix} \alpha_P - E & \beta \\ \beta & \alpha_N - E \end{vmatrix} = 0,$$

$E = \frac{1}{2}[2\alpha_P + k\beta \pm (k^2 + 4)^{1/2}\beta]$ (+: bonding; −: antibonding).

$E_c = 2\alpha_P + k\beta + (k^2 + 4)^{1/2}\beta.$

(iii) DE per island $= E_\pi - E_c = [(k^2 + 8)^{1/2} - (k^2 + 4)^{1/2}]\beta$

$\approx [(8)^{1/2} - (4)^{1/2}]\beta$ (assuming $k^2 << 4$)

$= 0.83\beta.$

DE for the whole $(NPCl_2)_3$ molecule $= 3 \times 0.83\beta = 2.5\beta.$

A8.45 The secular equations are: $xc_1 + c_2 = 0$

$c_1 + xc_2 + c_3 = 0$

$c_2 + xc_3 = 0$

x	E	Wavefunction	Symmetry
$-\sqrt{2}$	$\alpha + \sqrt{2}\beta$	$\psi(\pi) = \frac{1}{2}\phi_1 + \frac{1}{\sqrt{2}}\phi_2 + \frac{1}{2}\phi_3$	B_1
0	α	$\psi(n) = \frac{1}{\sqrt{2}}(\phi_1 - \phi_3)$	A_2
$\sqrt{2}$	$\alpha - \sqrt{2}\beta$	$\psi(\pi^*) = \frac{1}{2}\phi_1 - \frac{1}{\sqrt{2}}\phi_2 + \frac{1}{2}\phi_3$	B_1

A8.46 (i) $E_1 = \frac{1}{3}\int(\phi_1 + \phi_2 + \phi_3)\hat{H}(\phi_1 + \phi_2 + \phi_3)d\tau$

$= \frac{1}{3}[\int\phi_1\hat{H}\phi_1 d\tau + \int\phi_1\hat{H}\phi_2 d\tau + \int\phi_1\hat{H}\phi_3 d\tau + \int\phi_2\hat{H}\phi_1 d\tau + \int\phi_2\hat{H}\phi_2 d\tau$

$+ \int\phi_2\hat{H}\phi_3 d\tau + \int\phi_3\hat{H}\phi_1 d\tau + \int\phi_3\hat{H}\phi_2 d\tau + \int\phi_3\hat{H}\phi_3 d\tau]$

$= \frac{1}{3}(3\alpha + 6\beta) = \alpha + 2\beta.$

Similarly, $E_2 = \frac{1}{6}\int(2\phi_1 - \phi_2 - \phi_3)\hat{H}(2\phi_1 - \phi_2 - \phi_3)d\tau = \alpha - \beta.$

$E_3 = \frac{1}{2}\int(\phi_2 - \phi_3)\hat{H}(\phi_2 - \phi_3)d\tau = \alpha - \beta.$

(ii) $E_\pi = 2\alpha + 4\beta$; DE $= 2\alpha + 4\beta - (2\alpha + 2\beta) = 2\beta.$

(iii) ψ_1, ψ_2, and ψ_3 are bonding, nonbonding, and antibonding molecular orbitals, respectively.

A8.47 (i)

D_{3h}	E	$2C_3$	$3C_2$	σ_h	$2S_3$	$3\sigma_v$	
Γ_d	3	0	1	-3	0	-1	$= A_1'' + E''$
$R(a)$	a	b, c	a, b, c	$-a$	$-b, -c$	$-a, -b, -c$	
$R(b)$	b	a, c	a, b, c	$-b$	$-a, -c$	$-a, -b, -c$	
$R(c)$	c	a, b	a, b, c	$-c$	$-a, -b$	$-a, -b, -c$	

$P^{A_1''} a: (3)^{-1/2}(a + b + c)$.
$P^{E''} a: 2a - b - c$ (1), chosen!
$P^{E''} b: 2b - a - c$ (2),
$P^{E''} c: 2c - a - b$ (3),
     since (2) + (3) = (1), so we take (2) − (3): $b - c$.

After normalization, the E'' combinations are: $(6)^{-1/2}(2a - b - c)$ and $(2)^{-1/2}(b - c)$.

(ii) The Hückel energies are:

$$E(a_1'') = \alpha - 2\beta \text{ (antibonding)},$$
$$E(e'') = \alpha + \beta \text{ (bonding)}.$$

(iii) The most stable configuration is $(e'')^4$, which does not follow the $4n + 2$ rule. Indeed, for such Möbius systems, there is a $4n$ rule for the stable species.

A8.48 (i) $a_{1u} = \frac{1}{2}(a + b + c + d)$.

$e_g = \left[\frac{1}{\sqrt{2}}(a - c), \frac{1}{\sqrt{2}}(b - d)\right]$.

$b_{1u} = \frac{1}{2}(a - b + c - d)$.

(ii) The Hückel energies are: $E(a_{1u}) = \alpha - 2\beta$ (antibonding),

$$E(e_g) = \alpha \text{ (nonbonding)},$$
$$E(b_{1u}) = \alpha + 2\beta \text{ (bonding)}.$$

Maximum stability is achieved with two electrons in the system.

(iii) The ground configuration is $(b_{1u})^2$ and the ground state is $^1A_{1g}$. The lowest-energy transition is $[(b_{1u})^2]^1A_{1g} \rightarrow [(b_{1u})^1(e_g)^1]^1E_u$.

(iv) Yes, the two e_g orbitals are nonbonding. If we ignore the Hückel approximation, the atomic orbitals of each component of the e_g molecular orbitals has long-range (positive) δ-overlap and hence the e_g orbitals are slightly bonding in nature.

A8.49 (i) A_2'' basis functions: $(3)^{-1/2}(a + b + c), (3)^{-1/2}(d + e + f)$.

E'' basis functions : $\begin{cases} (6)^{-1/2}(2a - b - c) \\ (2)^{-1/2}(b - c) \end{cases}$, $\begin{cases} (6)^{-1/2}(2d - e - f) \\ (2)^{-1/2}(e - f) \end{cases}$.

(ii) A_2'' determinant: $H_{11} = \frac{1}{3} \int (a + b + c)\hat{H}(a + b + c)d\tau = \alpha + 2\beta$.

$$H_{22} = \frac{1}{3} \int (d + e + f)\hat{H}(d + e + f)d\tau = \alpha.$$

$$H_{12} = \frac{1}{3} \int (a + b + c)\hat{H}(d + e + f)d\tau = \beta.$$

$$\begin{vmatrix} \alpha + 2\beta - E & \beta \\ \beta & \alpha - E \end{vmatrix} = \begin{vmatrix} x + 2 & 1 \\ 1 & x \end{vmatrix} = x^2 + 2x - 1 = 0.$$

$x = 0.414, -2.414$.

$E(1a_2'') = \alpha + 2.414\beta; E(2a_2'') = \alpha - 0.414\beta$.

E'' determinant $[\text{for } (2)^{-1/2}(b - c) \text{ and } (2)^{-1/2}(e - f)]$:

$H_{11} = \frac{1}{2} \int (b - c)\hat{H}(b - c)d\tau = \alpha - \beta$.

$H_{22} = \frac{1}{2} \int (e - f)\hat{H}(e - f)d\tau = \alpha$.

$H_{12} = \frac{1}{2} \int (b - c)\hat{H}(e - f)d\tau = \beta$.

$$\begin{vmatrix} \alpha - \beta - E & \beta \\ \beta & \alpha - E \end{vmatrix} = \begin{vmatrix} x - 1 & 1 \\ 1 & x \end{vmatrix} = x^2 - x - 1 = 0.$$

$x = 1.618, -0.618$.

$E(1e'') = \alpha + 0.618\beta; E(2e'') = \alpha - 1.618\beta$.

(iii) In ascending energy order, the molecular orbitals are $1a_2''$, $1e''$, $2a_2''$, $2e''$, with the ground electronic configuration being $(1a_2'')^2(1e'')^4$.

$E_\pi = 2(\alpha + 2.414\beta) + 4(\alpha + 0.618\beta) = 6\alpha + 7.300\beta$.

$DE = (6\alpha + 7.300\beta) - 6(\alpha + \beta) = 1.300\beta$.

A8.50 (i) $B_{2g}: \frac{1}{2}(a - b + c - d); E(1b_{2g}) = \alpha - \beta$.

$A_u: \frac{1}{2}(a - b - c + d); E(1a_u) = \alpha - \beta$.

(ii) B_{3g}: $\psi_1 = \frac{1}{2}(a + b - c - d)$; $\psi_2 = \frac{1}{\sqrt{2}}(e - f)$.

$$\begin{vmatrix} \alpha + \beta - E & \sqrt{2}\beta \\ \sqrt{2}\beta & \alpha - \beta - E \end{vmatrix} = \begin{vmatrix} x+1 & \sqrt{2} \\ \sqrt{2} & x-1 \end{vmatrix};$$

$E(1b_{3g}) = \alpha + 1.732\beta.$

$E(2b_{3g}) = \alpha - 1.732\beta.$

B_{1u}: $\psi_3 = \frac{1}{2}(a + b + c + d)$; $\psi_4 = \frac{1}{\sqrt{2}}(e + f)$.

$$\begin{vmatrix} \alpha + \beta - E & \sqrt{2}\beta \\ \sqrt{2}\beta & \alpha + \beta - E \end{vmatrix} = \begin{vmatrix} x+1 & \sqrt{2} \\ \sqrt{2} & x+1 \end{vmatrix};$$

$E(1b_{1u}) = \alpha + 2.414\beta.$

$E(2b_{1u}) = \alpha - 0.414\beta.$

(iii) In ascending energy order, the molecular orbitals are $1b_{1u}$, $1b_{3g}$, $2b_{1u}$, $1b_{2g} = 1a_u$, $2b_{3g}$, with the first three filled with six π electrons.

$$E_\pi = 2(\alpha + 2.414\beta) + 2(\alpha + 1.732\beta) + 2(\alpha - 0.414\beta) = 6\alpha + 7.464\beta.$$

$$DE = (6\alpha + 7.464\beta) - 6(\alpha + \beta) = 1.464\beta.$$

A8.51 (i) $1b_{3g} = \frac{1}{2}(a + b - d - e)$; $E(1b_{3g}) = \alpha + \beta.$

$1a_u = \frac{1}{2}(a - b + d - e)$; $E(1a_u) = \alpha - \beta.$

(ii) B_{2g}: $\psi_1 = \frac{1}{2}(a - b - d + e)$; $\psi_2 = \frac{1}{\sqrt{2}}(f - c).$

$H_{11} = \alpha - \beta$; $H_{22} = \alpha - \beta$; $H_{12} = \sqrt{2}\beta.$

$$\begin{vmatrix} \alpha - \beta - E & \sqrt{2}\beta \\ \sqrt{2}\beta & \alpha - \beta - E \end{vmatrix} = \begin{vmatrix} x-1 & \sqrt{2} \\ \sqrt{2} & x-1 \end{vmatrix} = 0;$$

$E(1b_{2g}) = \alpha + 0.414\beta.$

$E(2b_{2g}) = \alpha - 2.414\beta.$

B_{1u}: $\psi_3 = \frac{1}{2}(a + b + d + e)$; $\psi_4 = \frac{1}{\sqrt{2}}(f + c).$

$H_{33} = \alpha + \beta$; $H_{44} = \alpha + \beta$; $H_{34} = \sqrt{2}\beta.$

$$\begin{vmatrix} \alpha + \beta - E & \sqrt{2}\beta \\ \sqrt{2}\beta & \alpha + \beta - E \end{vmatrix} = \begin{vmatrix} x+1 & \sqrt{2} \\ \sqrt{2} & x+1 \end{vmatrix} = 0;$$

$E(1b_{1u}) = \alpha + 2.414\beta.$

$E(2b_{1u}) = \alpha - 0.414\beta.$

(iii) In ascending energy order, the molecular orbitals are $1b_{1u}$, $1b_{3g}$, $1b_{2g}$, $2b_{1u}$, $1a_u$, $2b_{2g}$, with the first three filled with six π electrons.

$$E_\pi = 2(\alpha + 2.414\beta) + 2\alpha + 2(\alpha + 0.414\beta) = 6\alpha + 7.656\beta.$$

$$DE = (6\alpha + 7.656\beta) - 6(\alpha + \beta) = 1.656\beta.$$

A8.52 (i) A_1' basis functions: $\psi_1 = \frac{1}{\sqrt{3}}(a + b + c)$; $\psi_2 = \frac{1}{\sqrt{2}}(d + e)$.

E'' basis functions: $\begin{cases} (6)^{-1/2}(2a - b - c) \\ (2)^{-1/2}(b - c) \end{cases}$.

A_2'' basis function: $\frac{1}{\sqrt{2}}(d - e)$.

For A_1', $H_{11} = \alpha + 2\beta$, $H_{22} = \alpha$, and $H_{12} = \sqrt{6}\beta$.

$$\begin{vmatrix} \alpha + 2\beta - E & \sqrt{6}\beta \\ \sqrt{6}\beta & \alpha - E \end{vmatrix} = \begin{vmatrix} x + 2 & \sqrt{6} \\ \sqrt{6} & x \end{vmatrix} = 0, \text{ with } x = \frac{\alpha - E}{\beta}.$$

Solving this determinant yields $x = -1 \pm \sqrt{7}$ and hence $E = \alpha + (1 \pm \sqrt{7})\beta$.

$E(1a_1') = \alpha + 3.646\beta$; $E(2a_1') = \alpha - 1.646\beta$.

$E(1e'') = \alpha - \beta$.

$E(1a_2'') = \alpha$.

(ii) In ascending energy order, the molecular orbitals are $1a_1'$, $1a_2''$, $1e''$, $2a_1'$. The most stable configurations for H_5^{n+} should be

(a) $(1a_1')^2$ for H_5^{3+},

(b) $(1a_1')^2(1a_2'')^1$ for H_5^{2+}, and

(c) $(1a_1')^2(1a_2'')^2$ for H_5^+.

A8.53 (i) From group theory, $\Gamma_H = A_{1g} + T_{1u} + E_g$.

Symmetry	Orbital	Wavefunction	Hückel energy
A_{1g}	$1a_{1g}$	$\frac{1}{\sqrt{6}}(a + b + c + d + e + f)$	$\alpha + 4\beta$
T_{1u}	$1t_{1u}$	$\begin{cases} \frac{1}{\sqrt{2}}(a - b) \\ \frac{1}{\sqrt{2}}(c - d) \\ \frac{1}{\sqrt{2}}(e - f) \end{cases}$	α
E_g	$1e_g$	$\begin{cases} \frac{1}{\sqrt{12}}(2a + 2b - c - d - e - f) \\ \frac{1}{2}(c + d - e - f) \end{cases}$	$\alpha - 2\beta$

(ii) A stable configuration with the least number of electrons is $(1a_{1g})^2$ and the ground state is $^1A_{1g}$.

(iii) The lowest symmetry-allowed and spin-allowed transition is

$$(1a_{1g})^2; {}^1A_{1g} \rightarrow (1a_{1g})^1(1t_{1u})^1; {}^1T_{1u}.$$

A8.54

Unprimed orbitals	Primed orbitals	Molecular orbitals[‡]		Hückel energy	Symmetry in O_h
$\frac{1}{2}(a+b+c+d)$	$\frac{1}{2}(a'+b'+c'+d')^†$	$(8)^{-1/2}(a+b+c+d+a'+b'+c'+d')$		$\alpha + 3\beta$	A_{1g}
		$(8)^{-1/2}(a+b+c+d-a'-b'-c'-d')$		$\alpha - 3\beta$	A_{2u}
$\frac{1}{2}(a-b-c+d)^*$	$\frac{1}{2}(a'-b'-c'+d')^†$	$(8)^{-1/2}(a-b-c+d+a'-b'-c'+d')$			
		$(8)^{-1/2}(a-b+c-d+a'-b'+c'-d')$		$\alpha - \beta$	T_{2g}
		$(8)^{-1/2}(a+b-c-d+a'+b'-c'-d')$			
$\frac{1}{2}(a-b+c-d)^*$	$\frac{1}{2}(a'-b'+c'-d')^†$				
		$(8)^{-1/2}(a-b-c+d-a'+b'+c'-d')$			
$\frac{1}{2}(a+b-c-d)^*$	$\frac{1}{2}(a'+b'-c'-d')^†$	$(8)^{-1/2}(a-b+c-d-a'+b'-c'+d')$		$\alpha + \beta$	T_{1u}
		$(8)^{-1/2}(a+b-c-d-a'-b'+c'+d')$			

* These combinations are constructed by taking two "negatives" among b, c, and d.

† These combinations are constructed by replacing a, b, c, and d with a', b', c', and d', respectively. By symmetry operation i, a, b, c, and d are turned to a', b', c', and d', respectively.

‡ The t_{2g} orbitals are the sums of the primed and unprimed combinations. The t_{1u} orbitals are the differences of the primed and unprimed combinations.

A8.55 There are two sets of equivalent atoms: $\{a, c, e\}$ for nitrogen and $\{b, d, f\}$ for phosphorus. For each set, symmetry-adapted wavefunctions may be obtained in a straightforward manner:

Molecular orbital	Symmetry
$\psi_1 = \frac{1}{\sqrt{3}}(a+c+e)$	A_2''
$\psi_2 = \frac{1}{\sqrt{2}}(a-c), \psi_3 = \frac{1}{\sqrt{6}}(2e-a-c)$	E''
$\psi_4 = \frac{1}{\sqrt{3}}(b+d+f)$	A_1''
$\psi_5 = \frac{1}{\sqrt{2}}(d-f), \psi_6 = \frac{1}{\sqrt{6}}(2b-d-f)$	E''

To mix the two sets of e'' orbitals, ψ_2 is mixed with ψ_6 [both of which being antisymmetric with respect to $\sigma_v(xz)$] and ψ_3 is mixed with ψ_5 [both of which being symmetric with respect to $\sigma_v(xz)$].

The Hückel secular determinant for ψ_2 and ψ_6 is:

$$\begin{vmatrix} \alpha - E & \sqrt{3}\beta \\ \sqrt{3}\beta & (\alpha + \beta) - E \end{vmatrix} = 0,$$

and $E = \alpha + \frac{1}{2}(1 \pm \sqrt{13})\beta$. The same result is obtained for mixing ψ_3 and ψ_5.

The six resultant orbital energies and wavefunctions are summarized in the following table:

MO	$1e''$		$1a_1''$	$1a_2''$	$2e''$	
Energy	$\alpha + \frac{1}{2}(1 + \sqrt{13})\beta$		$\alpha + \beta$	α	$\alpha + \frac{1}{2}(1 - \sqrt{13})\beta$	
a	$-p(6)^{-1/2}$	$p(2)^{-1/2}$	0	$(3)^{-1/2}$	$-q(6)^{-1/2}$	$q(2)^{-1/2}$
b	0	$2q(6)^{-1/2}$	$(3)^{-1/2}$	0	0	$-2p(6)^{-1/2}$
c	$-p(6)^{-1/2}$	$-p(2)^{-1/2}$	0	$(3)^{-1/2}$	$-q(6)^{-1/2}$	$-q(2)^{-1/2}$
d	$-q(2)^{-1/2}$	$-q(6)^{-1/2}$	$(3)^{-1/2}$	0	$p(2)^{-1/2}$	$p(6)^{-1/2}$
e	$2p(6)^{-1/2}$	0	0	$(3)^{-1/2}$	$2q(6)^{-1/2}$	0
f	$q(2)^{-1/2}$	$-q(6)^{-1/2}$	$(3)^{-1/2}$	0	$-p(2)^{-1/2}$	$p(6)^{-1/2}$

where $p^2 = \frac{1}{2}(1 - 13^{-1/2})$ and $q^2 = \frac{1}{2}(1 + 13^{-1/2})$.

A8.56

Δ

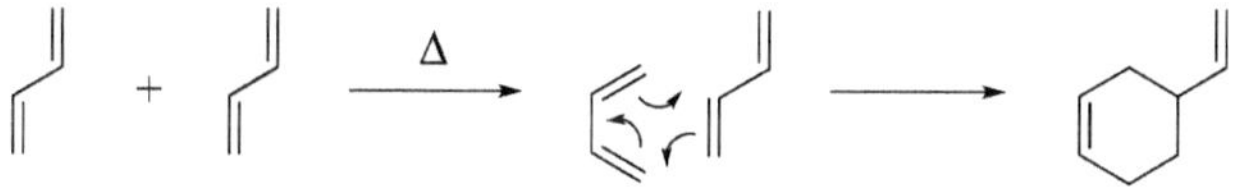

4-vinylcyclohexane

A8.57 (i)

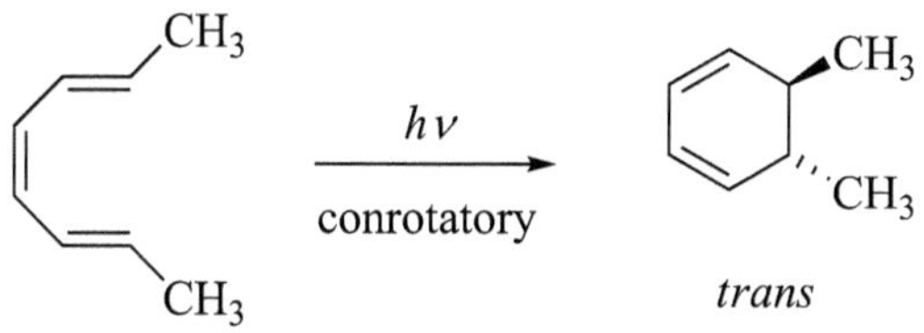

CH3
CH3
hν
conrotatory
CH3
CH3
trans

(ii)

CH2CH3
CH2CH3
Δ
disrotatory
CH2CH3
CH2CH3

trans-cis-cis

(iii)

H
H
Δ
disrotatory
cis

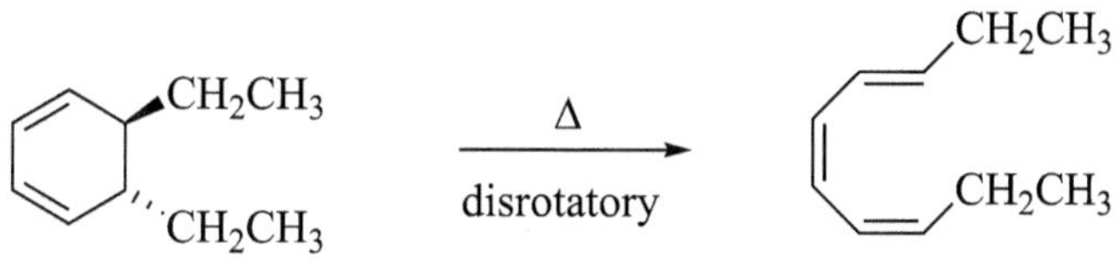

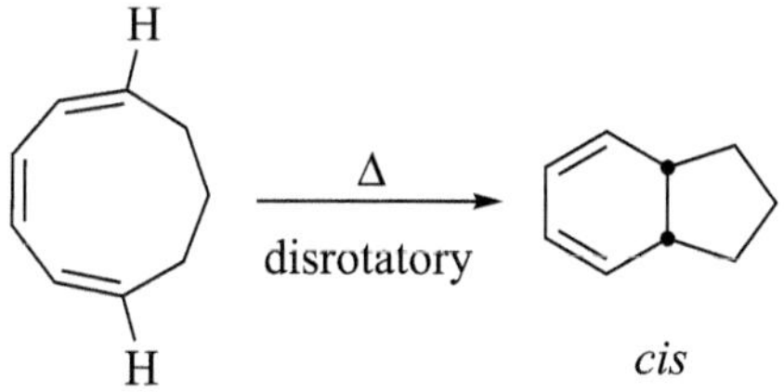

A8.58

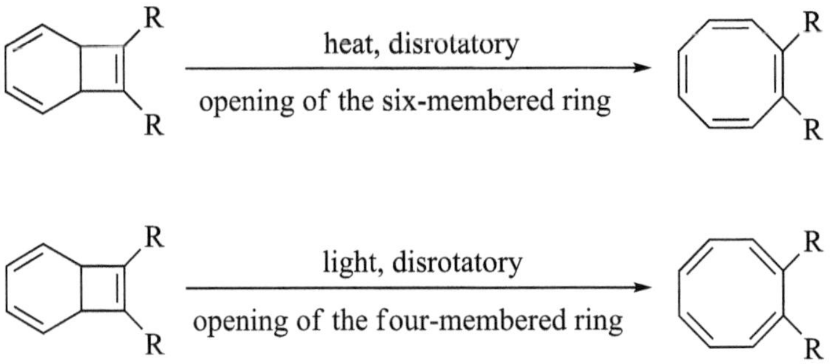

heat, disrotatory

opening of the six-membered ring

light, disrotatory

opening of the four-membered ring

Note: Conrotatory opening is unlikely.

A8.59

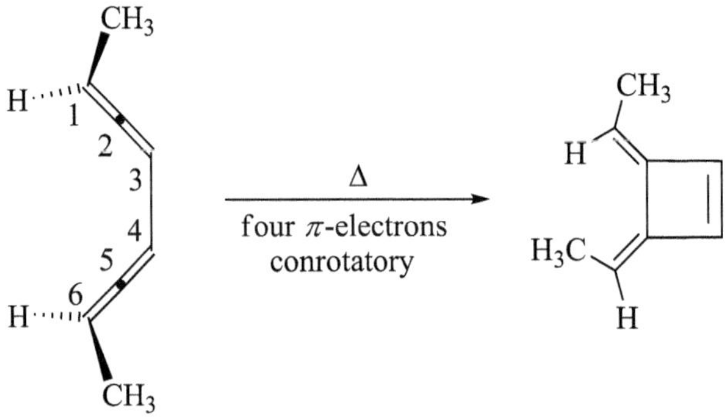

Δ

four π-electrons
conrotatory

A8.60 (i)   **2 → 3**: eight π electrons; **3 → 4**: six π electrons.

(ii)   **2 → 3**: conrotatory; **3 → 4**: disrotatory.

Finally, the whole process **1 → 4** is carried out thermally.

A8.61 The first reaction follows a disrotatory pathway, which is thermally allowed. The second reaction is a conrotatory process, which is thermally forbidden (or photochemically allowed).

A8.62

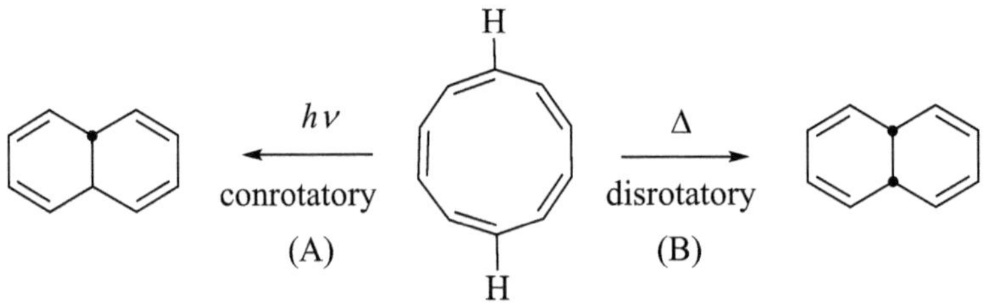

$h\nu$

conrotatory

(A)

Δ

disrotatory

(B)

A8.63

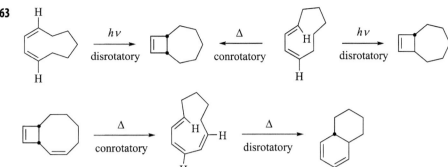

A8.64 The thermal and photochemical electrocyclic reactions consist of two successive disrotatory processes and two successive conrotatory processes, respectively.

A8.65

Altogether, there are five products, (A) to (E).

A8.66

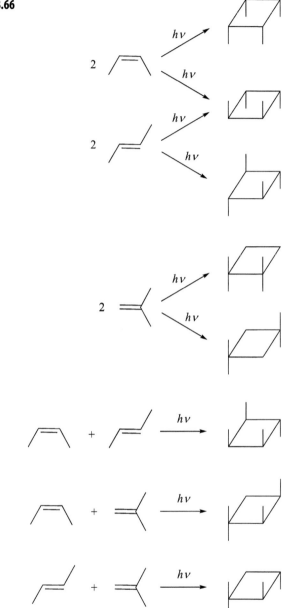

A8.67 The HOMOs for the propenyl cation and propenyl radical are ψ_1 and ψ_2, respectively.

A8.68   (i)   Cation: thermal electrocyclic reaction, HOMO = ψ_2, conrotatory.
  (ii)   Radical: thermal electrocyclic reaction, HOMO = ψ_3, disrotatory.
  (iii)   Anion: photochemical electrocyclic reaction, HOMO = ψ_4, conrotatory.

9 | Vibrational Spectroscopy

PROBLEMS

9.1 The normal modes of vibration of C_2H_4 are shown below. By inspection of the D_{2h} character table, deduce the symmetry species to which the normal modes belong. The coordinate system adopted is shown on the right.

9.2 Phosphorus oxychloride, $OPCl_3$, is a colorless, moisture-sensitive liquid. This molecule has C_{3v} symmetry and its infrared and Raman spectra exhibit the following bands (in cm^{-1}):

Infrared	Raman
1292	1290 (pol)
580	581
487	486 (pol)
340	337
267	267 (pol)
Not observed	193

250   *Problems in Structural Inorganic Chemistry*. Second edition. Wai-Kee Li, Hung Kay Lee, Dennis Kee Pui Ng, Yu-San Cheung, Kendrew Kin Wah Mak, and Thomas Chung Wai Mak. © Oxford University Press 2019. Published in 2019 by Oxford University Press. DOI: 10.1093/oso/9780198823902.001.0001

(i) Determine the symmetry and activity of the nine vibrational modes of this molecule.

(ii) Pictorially illustrate the stretching modes of this molecule and assign them to the above observed bands. Do not attempt to illustrate and assign the bending modes.

9.3 The noble gas compound XeO_3F_2 has a trigonal bipyramidal structure with D_{3h} symmetry.

(i) Determine the representation Γ_{vib} of this molecule and reduce it to the irreducible representations of the D_{3h} group.

(ii) Determine the spectral activities of the vibrational modes deduced.

(iii) Pictorially illustrate the stretching modes of this molecule.

(iv) The vibrational frequencies of this compound are shown in the following table (in cm^{-1}). Make an assignment of these bands and justify your answers.

Raman (cm^{-1})	Infrared (cm^{-1})
892	896
807 (pol)	—
—	632
567 (pol)	—
—	375
361	—
316	321
190	≈190*

*Expected, but unobserved due to instrumental limitation.

9.4 Consider the square pyramidal molecule xenon oxide tetrafluoride, $XeOF_4$, with C_{4v} symmetry. A convenient coordinate system for this molecule is shown on the right. In our convention, each symmetry plane σ_v passes through two F atoms, while the σ_d planes do not.

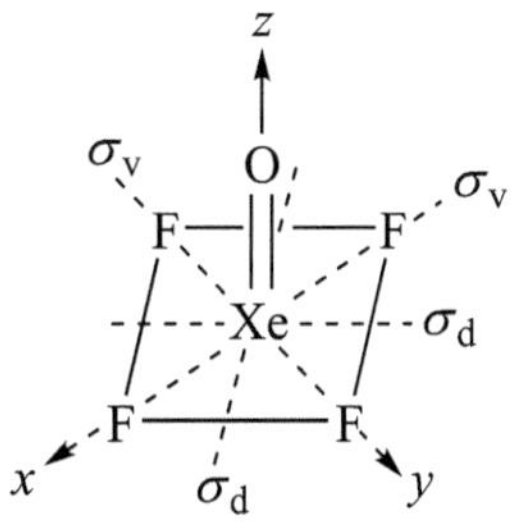

(i) Taking advantage of the symmetry of the system, determine the symmetry and activity of the vibrational modes of this molecule.

(ii) Determine the symmetry and activity of the stretching modes of this molecule. Also, illustrate these stretching motions pictorially.

(iii) The infrared and Raman spectra of $XeOF_4$ exhibit the bands (in cm^{-1}) tabulated below. Make assignments for the stretching vibrational bands and justify your assignments clearly. Do you expect significant coupling (or interaction) between the Xe=O and Xe—F stretching modes? Explain your answer succinctly.

Infrared	928(s)	609(vs)	578(vs)	—	362(ms)	288 (s)	—	N.O.
Raman	919(s)(pol)	N.O.	566(vs)(pol)	530(s)	364(mw)	286(vw)	231(w)	161(vw)

Note: "N.O." denotes "not observed" (but expected!); "s" for "strong"; "w" for "weak"; "m" for "medium"; "v" for "very". In the Raman spectrum, one more band is expected, but not observed.

9.5 The Raman spectrum of the BiI_4^- anion exhibits the following absorption frequencies (cm^{-1}): 141s (polarized), 129s, 107m (polarized), 80w, and 53br. Use these data to deduce the probable molecular geometry (among the possible T_d, D_{4h}, and C_{2v} structures) and explain your reasoning. For the probable structures you have deduced, sketch the normal modes that give rise to polarized Raman lines.

9.6 The ReF_7 molecule has a pentagonal bipyramidal structure with D_{5h} symmetry.

(i) Work out Γ_{vib} and the symmetry of the normal modes and comment on the infrared and Raman spectral activities of these normal modes.

(ii) Work out Γ_{ax} (the representation based on stretching vibration of the two axial Re–F bonds) and Γ_{eq} (the representation based on stretching vibration of the five equatorial Re–F bonds).

(iii) Which normal modes are mainly involved with deformation of the ReF_7 molecule?

Given: $\cos 72° = \frac{1}{4}(\sqrt{5} - 1)$; $\cos 144° = -\frac{1}{4}(\sqrt{5} + 1)$.

9.7 Consider a hypothetical metal carbonyl complex $M(CO)_8$ with a square antiprismatic structure (D_{4d} symmetry). A bird's-eye view of this complex is displayed below. Carbonyl groups a to h are labeled in the figure.

(i) Taking advantage of the symmetry of this system, determine the symmetry and activity of the eight CO stretching normal modes of this molecule.

(ii) Describe the eight CO stretching normal modes with positive sign indicating stretching and negative sign indicating contraction. Thus the expression $a - b + c - d + e - f + g - h$ denotes a vibrational motion in which CO bonds a, c, e, and g are stretching and bonds b, d, f, and h are contracting. Warning: This expression may not be one of the answers.

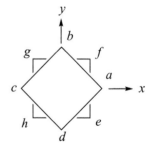

9.8 The first argon compound, HArF, was prepared in 2000. This linear molecule has the following two (calculated) bond lengths: $H-Ar = 1.329\,\text{Å}$ and $Ar-F = 1.969\,\text{Å}$.

(i) Write out the Lewis structure for this linear triatomic molecule.

(ii) It can be shown that, aside from the lone pairs residing in the "atomic orbitals" of Ar and F, we are left with three orbitals (1s on H, a 3p orbital on Ar, and a 2p orbital on F), as shown below, to form the molecular orbitals. Construct three molecular orbitals from these three atomic orbitals and show the three molecular orbitals pictorially. Characterize each molecular orbital as bonding, antibonding, or nonbonding, and arrange the three molecular orbitals in ascending energy order.

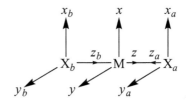

(iii) Among the three molecular orbitals you have formed, the lower two should be filled with electrons. Are the linkages $H-Ar$ and $Ar-F$ in HArF stronger or weaker than "normal two-electrons bonds"? Explain your answer clearly.

(iv) Two "stretching" bands, at 687.0 and 1969.5 cm^{-1}, are observed in the infrared spectrum of HArF. Which band is for the H–Ar bond (stretching) vibration and which is for the Ar–F bond (stretching) vibration?

REFERENCE: L. Khriachtchev, M. Pettersson, N. Runeberg, J. Lundell, and M. Räsänen, A stable argon compound. *Nature* **406**, 874–6 (2000).

9.9 A coordinate system for a d^9 linear complex MX$_2$ ($D_{\infty h}$ symmetry) is shown on the right.

(i) Deduce the crystal field splitting pattern for this compound's metal d orbitals. Briefly justify your results.

(ii) Construct the suitable linear combinations of ligand orbitals which match in symmetry with the metal orbitals. In addition, construct a schematic molecular orbital energy level diagram for MX$_2$. The orbitals participating in bonding are 3d(-86), 4s(-62) and 4p(-32) for M and 3s(-204) and 3p(-111) for X, where the energies of the atomic orbitals, in $10^3\ \text{cm}^{-1}$, are given in parentheses.

(iii) Write down the lowest-energy d-d and L $\rightarrow$ M charge transfer transitions of MX$_2$, clearly indicating the orbitals and states involved.

(iv) Deduce the vibrational mode(s) which is (are) responsible for making the aforementioned d-d transition vibronically allowed. Illustrate the deduced mode(s) pictorially.

9.10 The aromatic species $(CH)_3^+$, $(CH)_4^{2+}$, $(CH)_5^-$, and $(CH)_6$ have only one infrared-active C–H stretching band. Write down the symmetry of the band for each of the species. [Note: There is no need to determine the symmetries of all the C–H stretching modes.]

9.11 The structure of ethane with D_{3d} symmetry is shown on the right. In this figure, the seven single bonds are labeled a to g.

(i) Determine the symmetry species of all the stretching motions of this molecule.

(ii) Describe the stretching motions with positive sign indicating stretching and negative sign indicating contraction. Thus the expression $a - b + c - d + e - f + g$ denotes a vibrational motion in which bonds a, c, e, and g are stretching and bonds b, d, and f are contracting. Warning: This expression may not be one of the answers!

(iii) The stretching modes of this molecule are exhibited in its Raman and infrared spectra:

$$\text{Raman (cm}^{-1}\text{)}: \quad 2969 \text{ (depol)}, 2954 \text{ (pol)}, 995 \text{ (pol)};$$
$$\text{Infrared (cm}^{-1}\text{)}: \quad 2985 \text{ (m)}, 2896 \text{ (s)}.$$

Assign these spectral lines to the vibrational motions you described in (ii).

(iv) Are the stretching motions of the C–H and C–C bonds strongly coupled? Justify your answer.

9.12 Consider the linear molecule acetylene, C_2H_2, which has $D_{\infty h}$ symmetry.

(i) Sketch the normal modes of this molecule and label their symmetries with the aid of the $D_{\infty h}$ character table.

(ii) Attempt an assignment of the observed data (in cm^{-1}) listed in the following table.

IR	Raman
	611.8
729.1	
	1973.8 (pol)
3287	
	3373.7 (pol)

9.13 A convenient coordinate system for formaldehyde is shown on the right.

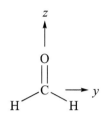

(i) Determine whether the $\pi \to \pi^*$ and $n \to \pi^*$ transitions are electric-dipole-allowed or not. If the transitions are allowed, are the transition moments lying along the C=O bond?

(ii) Determine the symmetry of this molecule's vibrational modes. Which of these modes is (are) responsible for making the electronic transition $^1A_1 \to {}^1A_2$ vibronically allowed?

9.14 Hexachlorobenzene is isostructural with benzene with D_{6h} symmetry.

(i) Deduce the representation of Γ_{C-Cl} based on the six exocyclic C–Cl bonds and their infrared and Raman activities.

(ii) Deduce the representation Γ_{C-C} based on the six C–C bonds in the aromatic ring and their infrared and Raman activities.

(iii) Write an expression for Γ_{def}, the symmetry species of those normal modes that mainly exhibit deformation (or angular distortion) of the molecule. Furthermore, break down Γ_{def} into $\Gamma_{in\text{-}plane\text{-}def}$ (the symmetry species of in-plane deformation modes) and $\Gamma_{out\text{-}of\text{-}plane\text{-}def}$ (the symmetry species of out-of-plane deformation modes).

(iv) Experimental data (cm^{-1}) on the vibration spectrum of hexachlorobenzene are tabulated below.

	IR (vapor)	Raman (liquid)	Symmetry and description
$\tilde{\nu}_3$		1522m	E_{2g}:C–C
$\tilde{\nu}_4$	1350vs		E_{1u}:C–C
$\tilde{\nu}_1$		1225m (pol)	A_{1g}:C–C breathing mode
$\tilde{\nu}_5$	696vs		E_{1u}:C–Cl
$\tilde{\nu}_2$		372s (pol)	A_{1g}:C–Cl breathing mode
$\tilde{\nu}_6$		340w	E_{2g}:C–Cl
$\tilde{\nu}_7$		322m	
$\tilde{\nu}_8$		213w	

Noting that the C–C bond is stronger than the C–Cl bond in hexachlorobenzene, attempt an assignment of the bond stretching mode [as deduced in parts (i) and (ii)] to the observed fundamentals. Ignore those frequencies that are concerned mainly with molecular deformation.

(v) Determine the symmetry type of the following normal modes of hexachloro benzene.

(a)   (b)   (c)   (d)

9.15 The carbonyl compound $Mn(CO)_5$ is rather unstable and its structure has been specu- lated as either square pyramidal (C_{4v} symmetry) or trigonal bipyramidal (D_{3h}).

(i) For each of the two possible structures, deduce the symmetry and activity of all the CO stretching modes.

(ii) The infrared spectrum of $Mn(CO)_5$ exhibits three CO stretching bands: 2060 (0.141), 1938 (1.000), 1911(0.378) cm^{-1}, with the relative intensities of the bands given in the parentheses. Based on this information, which of the two possible structures appears to be correct?

(iii) Assign each observed band to a vibrational mode. Also, illustrate these modes pictorially.

(iv) Why is the intensity of the band at 2060 cm^{-1} particularly low?

REFERENCE: H. Huber, E. P. Kundig, G. A. Ozin, and A. J. Poe, Reactions of monatomic and diatomic manganese with carbon monoxide. Matrix infrared spectroscopic evidence for pentacarbonylman- ganese $Mn(CO)_5$ and the binuclear carbonyls $Mn_2(CO)_n$ (where $n = 1$ or 2). *J. Am. Chem. Soc.* **97**, 308–14 (1975).

9.16 The carbonyl compound $Fe(CO)_5$ has a trigonal bipyramidal structure with D_{3h} symmetry.

(i) Deduce all the possible structures when two of the carbonyl groups are substituted by two L ligands, assuming that the substituted species still has a trigonal bipyrami- dal structure. Also, for each possible structure deduce the symmetry and activity of all the CO stretching modes. Illustrate all the CO stretching modes pictorially.

(ii) The two compounds, $(\phi_3P)_2Fe(CO)_3$ and $(MeNC)_2Fe(CO)_3$, are expected to have the same symmetry. The following CO stretching bands are found in their IR spectra:

$$(\phi_3P)_2Fe(CO)_3: \qquad 1887\ cm^{-1},$$
$$(MeNC)_2Fe(CO)_3: \qquad 2009\ cm^{-1}\ (0.5),\ 1927\ cm^{-1}\ (10),$$

with the relative intensities being given in parentheses. In view of these data, which of the structures deduced in (i) is most likely? Make assignments for the bands observed. Also, comment on why only one band is observed in $(\phi_3 P)_2 Fe(CO)_3$ but two bands are observed in $(MeNC)_2 Fe(CO)_3$, though the two compounds are expected to have the same symmetry.

REFERENCE: W.-K. Li, CO stretching in metal carbonyls. *J. Chem. Educ.* **57**, 722 (1980).

9.17 Di-iron nonacarbonyl, $Fe_2(CO)_9$ (point group D_{3h} with six terminal CO and three bridging CO as shown on the right), exhibits $\tilde{\nu}_{CO}$ infrared absorption at 2080, 2034, and 1828 cm^{-1}.

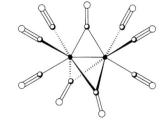

(i) Deduce Γ_{vib} for the whole molecule and the infrared and Raman activities of the normal modes.

(ii) Deduce Γ_{Fe-Fe} and Γ_{Fe-C} based on the Fe–Fe and Fe–C stretching vibrations.

(iii) Deduce Γ_{termCO} (for terminal CO) and Γ_{bridCO} (for bridging CO) based on the C–O stretching vibrations.

(iv) Which normal modes are involved mainly with deformation of the molecules?

(v) Assign the observed $\tilde{\nu}_{CO}$ infrared bands to appropriate normal modes and sketch them.

REFERENCE: F. A. Cotton and J. M. Troup, Accurate determination of a classic structure in the metal carbonyl field: nonacarbonyldi-iron. *J. Chem. Soc., Dalton Trans.* 800–2 (1974).

9.18 For the octahedral complex *trans*-M(CO)$_4$L$_2$ (with D_{4h} symmetry), it can be deduced that:

$$\Gamma_{CO}[trans-M(CO)_4L_2] = A_{1g}(R) + B_{1g}(R) + E_u(IR).$$

In other words, we expect to see only one C–O stretching band in the infrared spectrum of this complex. Displayed on the right is the infrared spectrum of *trans*–Mo(CO)$_4$[P(OC$_6$H$_5$)$_3$]$_2$. It is clear that the prominent band near 1950 cm^{-1} should be taken as the E_u mode. On the other hand, we also see two much weaker bands near 2020 and 1980 cm^{-1}. Explain succinctly why we see these two spectral lines at all.

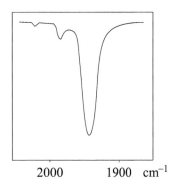

2000   1900   cm^{-1}

REFERENCE: D. J. Darensbourg and M. Y. Darensbourg, Infrared determination of stereochemistry in metal complexes: an application of group theory. *J. Chem. Educ.* **47**, 33–5 (1970).

9.19 A bird's-eye view of the S_8 molecule (point group D_{4d}) is displayed below, with the eight S–S bonds labeled. The infrared and Raman frequencies (cm^{-1}) observed for this molecule are listed in the table below.

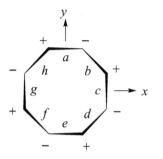

"+" sign denotes atoms that are above the paper plane and "–" sign indicates atom below.

Infrared	Raman
	475s* (pol)
471m*	
	437w
	248w
243s	
	218s (pol)
191s	
	152s
	86s

* Note that 471m and 475s are not coincidences.

(i) Deduce the symmetry types of the vibrational modes, Γ_{vib}, and their infrared and Raman activities.

(ii) Describe the eight normal modes mainly involving bond stretching, Γ_{S-S} with positive sign indicating stretching and negative sign indicating contraction. Thus the expression, $a - b + c - d + e - f + g - h$, denotes a vibrational motion in S–S bonds a, c, e, and g are stretching and bonds b, d, f, and h are contracting. Warning: This expression may not be one of the answers.

(iii) Attempt an assignment of the observed frequencies to the normal modes associated with Γ_{S-S}.

9.20 The infrared spectra of three related metal nitrosyl carbonyl complexes are shown on the right. Assign these spectra to the complexes $Fe(CO)_2(NO)_2$, $[Mn(CO)_2(NO)_2]^-$, and $[Mn(CO)_3(NO)]^-$. Justify your answer clearly.

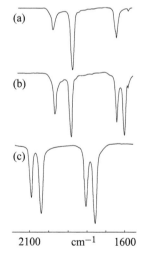

9.21 Two important idealized geometries for seven-coordination are (a) the monocapped trigonal prism (C_{2v} symmetry) and (b) the pentagonal bipyramid (D_{5h} symmetry) as shown below. The seven-coordinate complex $Mo(CN)_7^{4-}$ has been examined both

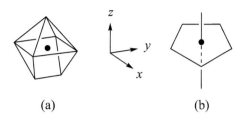

(a)   (b)

as solid $K_4Mo(CN)_7 \cdot 2H_2O$ and in aqueous solution. In the C–N stretching region, the infrared spectrum has bands at 2119, 2115, 2090, 2080, 2074, and 2059 cm^{-1} for the solid and at 2080 and 2040 cm^{-1} for solutions.

(i) For each geometry, deduce the symmetry types of the C–N stretching modes and their infrared and Raman activities. Also write down the number of infrared bands for each geometry.

(ii) Based on the observed data, determine the geometry of $[Mo(CN)_7]^{4-}$ in solid and in aqueous solution.

REFERENCE: G. R. Rossman, F. D. Tsay, and H. B. Gray, Spectroscopic and magnetic properties of heptacyanomolybdate(III). Evidence for pentagonal-bipyramidal and monocapped trigonal-prismatic structures. *Inorg. Chem.* **12**, 824–9 (1973).

9.22 Addition of molybdenum hexacarbonyl $Mo(CO)_6$ to $MeSi(CH_2SMe)_3$ (a tripodal thioether ligand as shown on the right) led to formation of a molybdenum tricarbonyl complex of formula $[\{MeSi(CH_2SMe)_3\}$ $Mo(CO)_3]$. The IR spectrum (KBr pellet) of the $[\{MeSi(CH_2SMe)_3\}Mo(CO)_3]$ complex shows two very strong carbonyl vibrational bands at 1918 cm^{-1} and 1785 cm^{-1}.

(i) Predict the coordination environment around the Mo center in $[\{MeSi(CH_2SMe)_3\}Mo(CO)_3]$ based on this spectroscopic information.

(ii) In addition to the two carbonyl vibrational bands mentioned above, a weak shoulder band at $\sim$1820 cm^{-1} was also observed and assignable to a carbonyl vibration of the $[\{MeSi(CH_2SMe)_3\}Mo(CO)_3]$ complex. Assume this complex is the only product present in the product mixture, suggest one plausible reason for the occurrence of this weak carbonyl vibration. [Hint: The thioether arms of the $MeSi(CH_2SMe)_3$ ligand are not "rigid" moieties.]

REFERENCE: H. W. Yim, L. M. Tran, E. D. Dobbin, D. Rabinovich, L. M. Liable-Sands, C. D. Incarvito, K.-C. Lam, and A. L. Rheingold, Tris[(alkylthio)methyl]silanes: Synthesis and structures of chromium, molybdenum, and tungsten complexes with a tripodal thioether ligand. *Inorg. Chem.* **38**, 2211–5 (1999).

SOLUTIONS

A9.1

D_{2h}	E	$C_2(z)$	$C_2(y)$	$C_2(x)$	i	$\sigma(xy)$	$\sigma(xz)$	$\sigma(yz)$	Symmetry
(i), (iii), (x)	1	1	1	1	1	1	1	1	A_g
(iv)	1	1	1	1	−1	−1	−1	−1	A_u
(vi), (vii)	1	1	−1	−1	1	1	−1	−1	B_{1g}
(v)	1	1	−1	−1	−1	−1	1	1	B_{1u}
(viii)	1	−1	1	−1	1	−1	1	−1	B_{2g}
(ii), (xi)	1	−1	1	−1	−1	1	−1	1	B_{2u}
(ix), (xii)	1	−1	−1	1	−1	1	1	−1	B_{3u}

A9.2 (i)

C_{3v}	E	$2C_3$	$3\sigma_v$	
$\Gamma(N_0)$	5	2	3	
$f(R)$	3	0	1	
Γ_{3N}	15	0	3	$= 4A_1 + A_2 + 5E.$

$$\Gamma_{vib} = \Gamma_{3N} - \Gamma_{trans} - \Gamma_{rot} = 3A_1(IR/R) + 3E(IR/R).$$

(ii) $\Gamma_{stretch} = 2A_1 + E.$

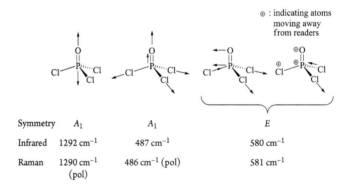

⊕ : indicating atoms moving away from readers

Symmetry	A_1	A_1	E
Infrared	1292 cm^{-1}	487 cm^{-1}	580 cm^{-1}
Raman	1290 cm^{-1} (pol)	486 cm^{-1} (pol)	581 cm^{-1}

A9.3 (i)&(ii)

D_{3h}	E	$2C_3$	$3C_2$	σ_h	$2S_3$	$3\sigma_v$
$\Gamma(N_0)$	6	3	2	4	1	4
$f(R)$	3	0	−1	1	−2	1
Γ_{3N}	18	0	−2	4	−2	4

$$\Gamma_{3N} = 2A_1' + A_2' + 4E' + 3A_2'' + 2E''.$$

Hence, $\Gamma_{vib} = \Gamma_{3N} - \Gamma_{trans} - \Gamma_{rot} = 2A_1'(R) + 3E'(IR/R) + 2A_2''(IR) + E''(R).$

(iii)

D_{3h}	E	$2C_3$	$3C_2$	σ_h	$2S_3$	$3\sigma_v$	
Γ_{stretch}	5	2	1	3	0	3	$= 2A_1' + A_2'' + E'$.

Stretching modes:

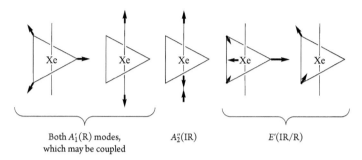

Both A_1'(R) modes, which may be coupled A_2''(IR) E'(IR/R)

(iv)

Raman (cm^{-1})	Infrared (cm^{-1})	Assignment	
892	896	E'	Stretching
807 (pol)	—	A_1'	Stretching
—	632	A_2''	Stretching
567 (pol)	—	A_1'	Stretching
—	375	A_2''	Bending
361	—	E''	Bending
316	321	E'	Bending
190	≈ 190	E'	Bending

A9.4   (i)

C_{4v}	E	$2C_4$	C_2	$2\sigma_v$	$2\sigma_d$	
$\Gamma(N_0)$	6	2	2	4	2	
$f(\mathbf{R})$	3	1	-1	1	1	
Γ_{3N}	18	2	-2	4	2	$= 4A_1 + A_2 + 2B_1 + B_2 + 5E$.

$\Gamma_{\text{vib}} = \Gamma_{3N} - \Gamma_{\text{trans}} - \Gamma_{\text{rot}} = 3A_1(\text{IR/R}) + 2B_1(\text{R}) + B_2(\text{R}) + 3E(\text{IR/R})$.

(ii)

C_{4v}	E	$2C_4$	C_2	$2\sigma_v$	$2\sigma_d$	
Γ_{stretch}	5	1	1	3	1	$= 2A_1(\text{IR/R}) + B_1(\text{R}) + E(\text{IR/R})$.

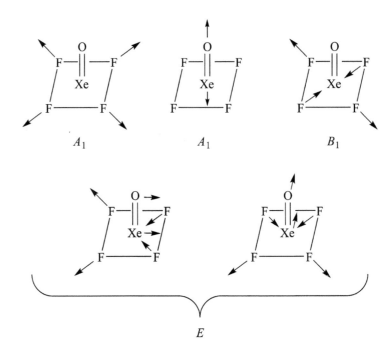

E

(iii)

Infrared (cm^{-1})	Raman (cm^{-1})	Assignment
928	919 (pol)	A_1(Xe=O)
609	N.O.	E
578	566 (pol)	A_1(Xe—F)
530	—	B_1

There should be little coupling between the Xe=O and Xe—F stretching modes, as Xe=O is a double bond, and Xe—F is a single bond. Their stretching frequencies are more than 300 cm^{-1} apart and hence there is little coupling between the two stretching modes.

A9.5

Point group	Γ_{vib}	No. of bands	No. of Raman active bands	No. of polarized Raman bands
T_d	A_1(R) + E(R) + $2T_2$(R, IR)	4	4	1
D_{4h}	A_{1g}(R) + B_{1g}(R) + B_{2g}(R) + A_{2u}(IR) + B_{2u} + $2E_u$(IR)	7	3	1
C_{2v}	$4A_1$(R, IR) + A_2(R) + $2B_1$(R, IR) + $2B_2$(R, IR)	9	9	4

The observed Raman spectrum shows five lines including two polarized ones. Since polarized lines originate from totally symmetric normal modes, and the number of observed fundamentals exceeds those deduced for point group T_d and D_{4h}, the BiI_4^- ion must have C_{2v} symmetry. Note that this geometry is consistent with the prediction of VSEPR theory. The normal modes that give rise to polarized Raman lines are sketched below.

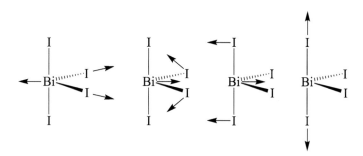

A9.6 (i)

D_{5h}	E	$2C_5$	$2C_5^2$	$5C_2$	σ_h	$2S_5$	$2S_5^3$	$5\sigma_v$
$\Gamma(N_0)$	8	3	3	2	6	1	1	4
$f(R)$	3	$1-\eta^-$	$1-\eta^+$	-1	1	$-1-\eta^-$	$-1-\eta^+$	1
Γ_{3N}	24	$3-3\eta^-$	$3-3\eta^+$	-2	6	$-1-\eta^-$	$-1-\eta^+$	4

N.B.: $\eta^\pm = (1 \pm \sqrt{5})/2$.

$\Gamma_{3N} = 2A_1' + A_2' + 4E_1' + 2E_2' + 3A_2'' + 2E_1'' + E_2''$.

$\Gamma_{vib} = \Gamma_{3N} - \Gamma_{trans} - \Gamma_{rot} = 2A_1'(R, pol) + 3E_1'(IR) + 2E_2'(R) + 2A_2''(IR) + E_1''(R) + E_2''$.

To sum up: There are five IR bands and also five Raman bands (two of which polarized).

(ii)

D_{5h}	E	$2C_5$	$2C_5^2$	$5C_2$	σ_h	$2S_5$	$2S_5^3$	$5\sigma_v$	
Γ_{ax}	2	2	2	0	0	0	0	2	$= A_1' + A_2''$.
Γ_{eq}	5	0	0	1	5	0	0	1	$= A_1' + E_1' + E_2'$.

(iii) The normal modes that are mainly involved with deformation and have little stretching motions are:

$$\Gamma_{def} = \Gamma_{vib} - \Gamma_{ax} - \Gamma_{eq} = 2E_1' + E_2' + A_2'' + E_1'' + E_2''.$$

A9.7 (i)

D_{4d}	E	$2S_8$	$2C_4$	$2S_8^3$	C_2	$4C_2'$	$4\sigma_d$
Γ_{CO}	8	0	0	0	0	0	2

$\Gamma_{CO} = A_1(R) + B_2(IR) + E_1(IR) + E_2(R) + E_3(R).$

A_1	1	1	1	1	1	1	1
B_2	1	-1	1	-1	1	-1	1
E_1	2	$\sqrt{2}$	0	$-\sqrt{2}$	-2	0	0
E_2	2	0	-2	0	2	0	0
E_3	2	$-\sqrt{2}$	0	$\sqrt{2}$	-2	0	0
$R(a)$	a	e,f	b,d	g,h	c	e,f,g,h	a,b,c,d
$R(b)$	b	f,g	a,c	e,h	d	e,f,g,h	a,b,c,d
$R(e)$	e	a,d	f,h	b,c	g	a,b,c,d	e,f,g,h

(ii)

A_1: $\quad P^{A_1}(a) = 2(a+b+c+d+e+f+g+h) \Rightarrow a+b+c+d+e+f+g+h.$

B_2: $\quad P^{B_2}(a) = 2(a+b+c+d-e-f-g-h) \Rightarrow a+b+c+d-e-f-g-h.$

E_1: $\quad P^{E_1}(a) = 2(a-c) + \sqrt{2}(e+f-g-h) \Rightarrow \sqrt{2}(a-c) + (e+f-g-h).$

$\qquad P^{E_1}(b) = 2(b-d) - \sqrt{2}(e-f-g+h) \Rightarrow \sqrt{2}(b-d) - (e-f-g+h).$

E_2: $\quad P^{E_2}(a) = 2(a-b+c-d) \Rightarrow a-b+c-d.$

$\qquad P^{E_2}(e) = 2(e-f+g-h) \Rightarrow e-f+g-h.$

E_3: $\quad P^{E_3}(a) = 2(a-c) - \sqrt{2}(e+f-g-h) \Rightarrow \sqrt{2}(a-c) - (e+f-g-h).$

$\qquad P^{E_3}(b) = 2(b-d) + \sqrt{2}(e-f-g+h) \Rightarrow \sqrt{2}(b-d) + (e-f-g+h).$

A9.8 (i) H—Är—F̈:

(ii)

H	Ar	F		
⊕	⊖×⊕	⊖×⊕	Antibonding	—
⊕		⊕×⊖	Nonbonding	⥮
⊕	⊕×⊖	⊖×⊕	Bonding	⥮

(iii) Since the three atoms are bound by only two bonding electrons, the bonds linking the three atoms are weaker than normal two-electron bonds. Indeed, this is called a three-center four-electron bond.

(iv) $\tilde{\nu}(H–Ar) = 1968.5\ cm^{-1}$; $\tilde{\nu}(F–Ar) = 687.0\ cm^{-1}$. Since an H atom is much lighter than F or Ar, stretching modes involving H atoms usually have (much) higher wavenumbers.

A9.9 (i) By considering the electrostatic interaction between the metal ion and the two ligands, the following splitting pattern for the metal 3d orbitals can be readily obtained: $\delta_g(x^2-y^2, xy) < \pi_g(xz, yz) < \sigma_g^+(z^2)$.

(ii)

M orbital		X orbital	Molecular orbital
Σ_g^+	s, z^2	$(2)^{-1/2}(s_a + s_b), (2)^{-1/2}(z_a + z_b)$	$1\sigma_g^+, 2\sigma_g^+, 3\sigma_g^+, 4\sigma_g^+$
Σ_u^+	z	$(2)^{-1/2}(s_a - s_b), (2)^{-1/2}(z_a - z_b)$	$1\sigma_u^+, 2\sigma_u^+, 3\sigma_u^+$
Π_u	(x, y)	$\left[(2)^{-1/2}(x_a + x_b), (2)^{-1/2}(y_a + y_b)\right]$	$1\pi_u, 2\pi_u$
Π_g	(xz, yz)	$\left[(2)^{-1/2}(x_a - x_b), (2)^{-1/2}(y_a - y_b)\right]$	$1\pi_g, 2\pi_g$
Δ_g	$(x^2 - y^2, xy)$		$1\delta_g$

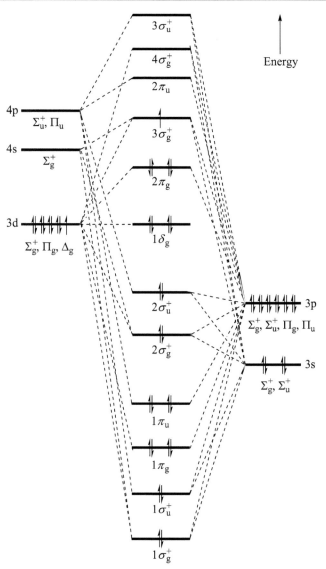

(iii)   d–d transition: $^2\Sigma_g^+[\ldots(2\pi_g)^4(3\sigma_g^+)^1] \to\ ^2\Pi_g[\ldots(2\pi_g)^3(3\sigma_g^+)^2]$.
Charge transfer:

$$^2\Sigma_g^+\left[\ldots(2\sigma_u^+)^2(1\delta_g)^4(2\pi_g)^4(3\sigma_g^+)^1\right] \to\ ^2\Sigma_u^+\left[\ldots(2\sigma_u^+)^1(1\delta_g)^4(2\pi_g)^4(3\sigma_g^+)^2\right].$$

(iv)   For z-polarization: $\Sigma_g^+\times\Sigma_u^+\times\Pi_g = \Pi_u$, i.e., Π_u mode is responsible for making the transition vibronically allowed.

For x, y-polarization: $\Sigma_g^+\times\Pi_u\times\Pi_g = \Sigma_u^+\times\Sigma_u^-\times\Delta_u$, since there is no vibration of either Σ_u^- or Δ_u symmetry, Σ_u^+ is responsible for making the transition vibronically allowed.

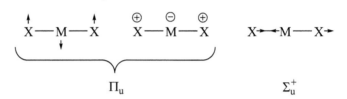

A9.10   The infrared-active vibrational modes of a molecule must have the symmetry of x, y, or z. The point groups of the specified species are D_{nh}, $n = 3, 4, 5$, and 6. The vibrational modes with the same symmetry as z are out-of-plane modes for molecular deformation. Hence the symmetry of the band must be the same as that for the degenerate set for x and y, which is E', E_u, E_1', and E_{1u} for $C_3H_3^+$, $C_4H_4^{2+}$, $C_5H_5^-$, and C_6H_6, respectively.

A9.11   (i)

D_{3d}	E	$2C_3$	$3C_2$	i	$2S_6$	$3\sigma_d$	
Γ_{C-C}	1	1	1	1	1	1	$= A_{1g}$.
Γ_{C-H}	6	0	0	0	0	2	$= A_{1g} + A_{2u} + E_g + E_u$.

$\Gamma_{C-C} = A_{1g}(R)$.
$\Gamma_{C-H} = A_{1g}(R) + A_{2u}(IR) + E_g(R) + E_u(IR)$.

(ii) & (iii)

Γ_{C-C}	A_{1g}	g	995 cm^{-1} (pol)
Γ_{C-H}	A_{1g}	$a + b + c + d + e + f$	2954 cm^{-1} (pol)
	A_{2u}	$a + b + c - d - e - f$	2896 cm^{-1} (m)
	E_g	$\begin{cases} 2a - b - c + 2d - e - f \\ b - c - e + f \end{cases}$	2969 cm^{-1} (depol)
	E_u	$\begin{cases} 2a - b - c - 2d + e + f \\ b - c + e - f \end{cases}$	2985 cm^{-1} (s)

(iv)   The C–C and C–H stretching modes with A_{1g} symmetry are not strongly coupled because their energies are significantly different.

A9.12 (i) IR-active modes: Σ_u^+, Π_u, two bands.

Raman-active modes: Σ_g^+ (polarized), Π_g, three lines.

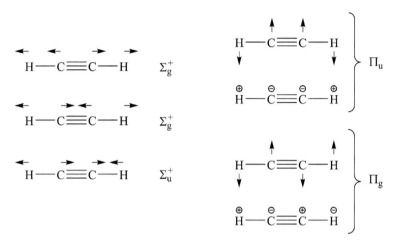

(ii)

IR	Raman	Assignment
	611.8	Π_g
729.1		Π_u
	1973.8 (pol)	Σ_g^+ $\left(\text{H—C}\equiv\text{C—H} \right)$
3287		Σ_u^+
	3373.7 (pol)	Σ_g^+ $\left(\text{H—C}\equiv\text{C—H} \right)$

A9.13 (i)

C_{2v}	E	C_2	$\sigma_v(xz)$	$\sigma_v'(yz)$	
$\Gamma_\pi = \Gamma_{\pi*}$	1	-1	1	-1	$= B_1$.
Γ_n	2	0	0	2	$= A_1 + B_2$.

$\pi \to \pi^*$: allowed, z-polarized, transition moment lying along the C=O
bond;

$n(A_1) \to \pi^*$: allowed, x-polarized, transition moment lying perpendicular to the
molecular plane;

$n(B_2) \to \pi^*$: forbidden.

(ii)

C_{2v}	E	C_2	$\sigma_v(xz)$	$\sigma_v'(yz)$
$\Gamma(N_0)$	4	2	2	4
$f(\boldsymbol{R})$	3	-1	1	1
Γ_{3N}	12	-2	2	4

$= 4A_1 + A_2 + 3B_1 + 4B_2.$

$\Gamma_{vib} = \Gamma_{3N} - \Gamma_{trans} - \Gamma_{rot} = 3A_1 + B_1 + 2B_2.$

$A_1 \times A_2 \times A_1 = A_2 \text{(forbidden)};$

$A_1 \times A_2 \times B_1 = B_2 \text{(}y\text{-polarized)}; \text{ and}$

$A_1 \times A_2 \times B_2 = B_1 \text{(}x\text{-polarized)}.$

So either B_1 or B_2 mode can make the transition $^1A_1 \rightarrow {}^1B_2$ vibronically allowed.

A9.14 (i) & (ii)

D_{6h}	E	$2C_6$	$2C_3$	C_2	$3C_2'$	$3C_2''$	i	$2S_3$	$2S_6$	σ_h	$3\sigma_d$	$3\sigma_v$
Γ_{C-Cl}	6	0	0	0	2	0	0	0	0	6	0	2
Γ_{C-C}	6	0	0	0	0	2	0	0	0	6	2	0

$\Gamma_{C-Cl} = A_{1g}(R) + B_{1u} + E_{1u}(IR) + E_{2g}(R).$

$\Gamma_{C-C} = A_{1g}(R) + B_{2u} + E_{1u}(IR) + E_{2g}(R).$

(iii)

D_{6h}	E	$2C_6$	$2C_3$	C_2	$3C_2'$	$3C_2''$	i	$2S_3$	$2S_6$	σ_h	$3\sigma_d$	$3\sigma_v$
N_0	12	0	0	0	4	0	0	0	0	12	0	4
$f(\boldsymbol{R})$	3	2	0	-1	-1	-1	-3	-2	0	1	1	-1
Γ_{3N}	36	0	0	0	-4	0	0	0	0	12	0	-4

$\Gamma_{3N} = 2A_{1g} + 2A_{2g} + 2B_{2g} + 2E_{1g} + 4E_{2g} + 2A_{2u} + 2B_{1u} + 2B_{2u}$
$\quad\quad + 4E_{1u} + 2E_{2u}.$

$\Gamma_{vib} = \Gamma_{3N} - \Gamma_{trans} - \Gamma_{rot}$
$\quad\quad = 2A_{1g} + A_{2g} + 2B_{2g} + E_{1g} + 4E_{2g} + A_{2u} + 2B_{1u} + 2B_{2u}$
$\quad\quad + 3E_{1u} + 2E_{2u}.$

$\Gamma_{def} = \Gamma_{vib} - \Gamma_{C-Cl} - \Gamma_{C-C} = A_{2g} + 2B_{2g} + E_{1g} + 2E_{2g} + A_{2u} + B_{1u}$
$\quad\quad + B_{2u} + E_{1u} + 2E_{2u}.$

Since in-plane and out-of-plane motions are symmetric and antisymmetric with respect to the symmetry element, σ_h, respectively. Hence,

$\Gamma_{\text{in-plane-def}} = A_{2g} + E_{1g} + 2E_{2g} + B_{1u} + B_{2u} + E_{1u};$

$\Gamma_{\text{out-of-plane-def}} = 2B_{2g} + A_{2u} + 2E_{2u}.$

(iv)

	IR (vapor)	Raman (liquid)	Symmetry and description
$\tilde{\nu}_3$		1522m	E_{2g}: C–C
$\tilde{\nu}_4$	1350vs		E_{1u}: C–C
$\tilde{\nu}_1$		1225m (pol)	A_{1g}: C–C breathing mode
$\tilde{\nu}_5$	696vs		E_{1u}: C–Cl
$\tilde{\nu}_2$		372s (pol)	A_{1g}: C–Cl breathing mode
$\tilde{\nu}_6$		340w	E_{2g}: C–Cl
$\tilde{\nu}_7$		322m	
$\tilde{\nu}_8$		213w	

(v) (a) A_{2u}, (b) B_{1u}, (c) A_{2g}, (d) B_{2g}.

A9.15 (i)

D_{3h}	E	$2C_3$	$3C_2$	σ_h	$2S_3$	$3\sigma_v$	
Γ_{CO}	5	2	1	3	0	3	$= 2A_1(R) + E'(IR/R) + A_2''(IR)$.

C_{4v}	E	$2C_4$	C_2	$2\sigma_v$	$2\sigma_d$	
Γ_{CO}	5	1	1	3	1	$= 2A_1(IR) + B_1(R) + E(IR/R)$.

(ii) Since three infrared bands are observed, the square pyramidal structure appears to be more likely.

(iii)

$\tilde{\nu}(cm^{-1})$	Assignment	Symmetry	Comment
2060	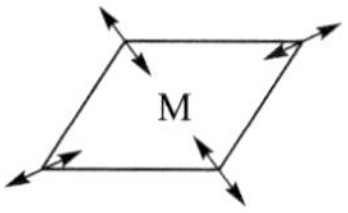	A_1	A breathing mode, with high energy
1938	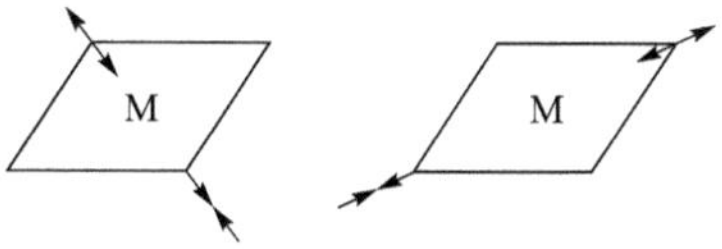	E	Largest change in dipole moment, with high intensity
1911	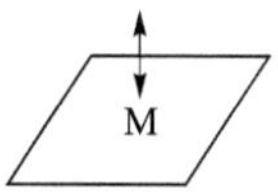	A_1	"By elimination"

(iv) The two A_1 modes appear not to couple strongly. As a result, there is a pseudo inversion center for the A_1 mode at 2060 cm^{-1}. This pseudo inversion center is responsible for this band's low intensity.

A9.16 (i) There are three possible structures: both Ls are in axial positions (D_{3h}), both Ls in equatorial positions (C_{2v}), or one L in axial position and one L in equatorial position (C_s).

D_{3h}	E	$2C_3$	$3C_2$	σ_h	$2S_3$	$3\sigma_v$	
Γ_{CO}	3	0	1	3	0	1	$= A_1'(R) + E'(IR/R)$.

C_{2v}	E	C_2	$\sigma_v(xz)$	$\sigma_v'(yz)$	
Γ_{CO}	3	0	1	3	$= 2A_1(IR/R) + B_1(IR/R)$.

C_s	E	$\sigma_h(xz)$	
Γ_{CO}	3	1	$= 2A'(IR/R) + A''(IR/R)$.

The CO stretching modes are illustrated in the figure below.

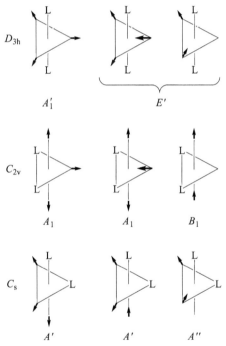

(ii) Since only one intense CO stretch band is observed in each of the spectra, the D_{3h} structure is most likely. In $(MeNC)_2Fe(CO)_3$, the totally symmetric A_1' mode may gain a little IR intensity since the non-linearity of the Me–N–C group slightly perturbs the D_{3h} symmetry of the MeNC–Fe(CO)$_3$–CNMe grouping. Another reason has been suggested by Cotton and Parish [F. A. Cotton and R. V. Parish, The stereochemistry of five-coordinate compounds. Part I. Infrared spectra of some iron(0) compounds. *J. Chem. Soc.,* 1440–6 (1960)].

A9.17 (i)

D_{3h}	E	$2C_3$	$3C_2$	σ_h	$2S_3$	$3\sigma_v$
N_0	20	2	2	6	0	8
$f(R)$	3	0	-1	1	-2	1
Γ_{3N}	60	0	-2	6	0	8

$\Gamma_{3N} = 7A'_1 + 4A'_2 + 11E' + 2A''_1 + 7A''_2 + 9E''$.

$\Gamma_{vib} = \Gamma_{3N} - \Gamma_{trans} - \Gamma_{rot}$
$= 7A'_1(R) + 3A'_2 + 10E'(IR/R) + 2A''_1 + 6A''_2(IR) + 8E''(R)$.

(ii) & (iii)

D_{3h}	E	$2C_3$	$3C_2$	σ_h	$2S_3$	$3\sigma_v$	
Γ_{Fe-Fe}	1	1	1	1	1	1	$= A'_1(R)$.
Γ_{Fe-C}	12	0	0	0	0	4	$= 2A'_1(R) + 2E'(IR/R)$ $\quad + 2A''_2(IR) + 2E''(R)$.
Γ_{termCO}	6	0	0	0	0	2	$= A'_1(R) + E'(IR/R)$ $\quad + A''_2(IR) + E''(R)$.
Γ_{bridCO}	3	0	1	3	0	1	$= A'_1(R) + E'(IR/R)$.

(iv) $\Gamma_{def} = \Gamma_{vib} - \Gamma_{Fe-Fe} - \Gamma_{Fe-C} - \Gamma_{termCO} - \Gamma_{bridCO}$
$= 2A'_1(R) + 3A'_2 + 6E'(IR/R) + 2A''_1 + 3A''_2(IR) + 5E''(R)$.

$\Gamma_{def} = \Gamma_{vib} - \Gamma_{Fe-Fe} - \Gamma_{Fe-C} - \Gamma_{termCO} - \Gamma_{bridCO}$.

(v) $\tilde{\nu}_{CO}$ for terminal CO stretching: E' and A''_2: 2080 and 2034 cm^{-1}.

$\tilde{\nu}_{CO}$ for bridging CO stretching: E': 1828 cm^{-1}.

A9.18 The two forbidden bands show up weakly in the spectrum because when the axial ligands are $P(OC_6H_5)_3$ (which destroys the C_4 axis of the *trans*–$M(CO)_4L_2$ system), the symmetry of the complex is not strictly D_{4h}, as shown on the right.

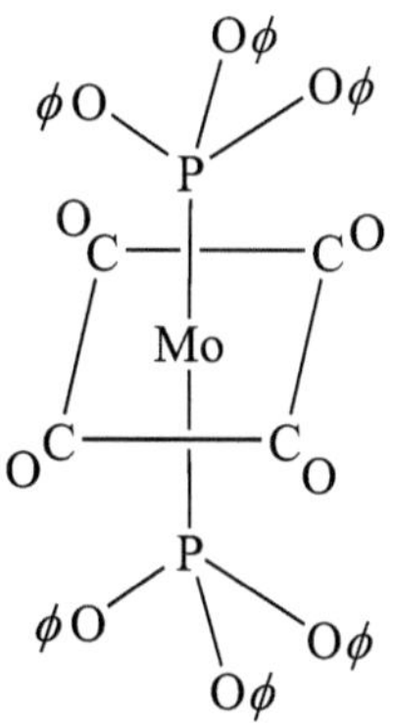

A9.19 (i)

D_{4d}	E	$2S_8$	$2C_4$	$2S_8^3$	C_2	$4C'_2$	$4\sigma_d$
N_0	8	0	0	0	0	0	2
$f(R)$	3	$\sqrt{2}-1$	1	$-\sqrt{2}-1$	-1	-1	1
Γ_{3N}	24	0	0	0	0	0	2

$\Gamma_{3N} = 2A_1 + A_2 + B_1 + 2B_2 + 3E_1 + 3E_2 + 3E_3$.

$\Gamma_{vib} = \Gamma_{3N} - \Gamma_{trans} - \Gamma_{rot} = 2A_1(R) + B_1 + B_2(IR) + 2E_1(IR)$
$+ 3E_2(R) + 2E_3(R)$.

271

(ii)

D_{4d}	E	$2S_8$	$2C_4$	$2S_8^3$	C_2	$4C_2'$	$4\sigma_d$
Γ_{S-S}	8	0	0	0	0	2	0

$\Gamma_{S-S} = A_1(R) + B_1 + E_1(IR) + E_2(R) + E_3(R).$

A_1	1	1	1	1	1	1	1
B_1	1	-1	1	-1	1	1	-1
E_1	2	$\sqrt{2}$	0	$-\sqrt{2}$	-2	0	0
E_2	2	0	-2	0	2	0	0
E_3	2	$-\sqrt{2}$	0	$\sqrt{2}$	-2	0	0
$R(a)$	a	b,h	c,g	d,f	e	a,c,e,g	b,d,f,h
$R(b)$	b	a,c	d,h	e,g	f	b,d,f,h	a,c,e,g
$R(c)$	c	b,d	a,e	f,h	g	a,c,e,g	b,d,f,h

A_1: $\quad P^{A_1}(a) = 2(a+b+c+d+e+f+g+h) \Rightarrow a+b+c+d+e+f+g+h.$

B_1: $\quad P^{B_2}(a) = 2(a-b+c-d+e-f+g-h) \Rightarrow a-b+c-d+e-f+g-h.$

E_1: $\quad \begin{cases} P^{E_1}(a) = 2(a-e) + \sqrt{2}(b-d-f+h) \Rightarrow \sqrt{2}(a-e) + (b-d-f+h). \\ P^{E_1}(c) = 2(c-g) - \sqrt{2}(b+d-f-h) \Rightarrow \sqrt{2}(c-g) - (b+d-f-h). \end{cases}$

E_2: $\quad \begin{cases} P^{E_2}(a) = 2(a-c+e-g) \Rightarrow a-c+e-g. \\ P^{E_2}(b) = 2(b-d+f-h) \Rightarrow b-d+f-h. \end{cases}$

E_3: $\quad \begin{cases} P^{E_3}(a) = 2(a-e) - \sqrt{2}(b-d-f+h) \Rightarrow \sqrt{2}(a-e) - (b-d-f+h). \\ P^{E_3}(c) = 2(c-g) + \sqrt{2}(b+d-f-h) \Rightarrow \sqrt{2}(c-g) + (b+d-f-h). \end{cases}$

(iii)

Infrared	Raman	Assignment and comment
	475s (pol)	A_1: The only totally symmetric Raman active bond stretching mode.
471m		E_1: The only IR active bond stretching mode.
	437w	E_2 or E_3: The remaining IR active bond stretching mode.

A9.20

Spectrum	Complex	Symmetry	Comment
(a)	$[Mn(CO)_3(NO)]^-$	C_{3v}	One N–O and two C–O bands (one of which is a degenerate E mode).
(b)	$[Mn(CO)_2(NO)_2]^-$	C_{2v}	Four bands are expected and observed. With a negative charge, back bonding is more efficient, strengthening the M–L bonds and weakening the C–O and N–O bonds.
(c)	$Fe(CO)_2(NO)_2$	C_{2v}	Again there should be four bands. But now the C–O and N–O bonds are weaker.

A9.21 (i) C_{2v}: $\Gamma_{CN} = 3A_1(IR/R) + A_2(R) + B_1(IR/R) + 2B_2(IR/R)$, with six IR bands.
D_{5h}: $\Gamma_{CN} = 2A_1'(R) + E_1'(IR) + E_2'(R) + A_2''(IR)$, with two IR bands.

(ii) Based on the spectral data given, it appears that $[Mo(CN)_7]^{4-}$ has a mono-capped trigonal prismatic structure in solid. It becomes a pentagonal bipyramid in solution.

A9.22 (i)

C_{3v}	E	$2C_3$	$3\sigma_v$
Γ_{CO}	3	0	1

$\Gamma_{CO} = A_1(IR/R) + E(IR/R)$

Therefore, the $[\{MeSi(CH_2SMe)_3\}Mo(CO)_3]$ complex conforms closely to C_{3v} molecular symmetry with the $MeSi(CH_2SMe)_3$ ligand coordinating to the Mo atom in a facial manner.

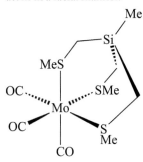

(ii) The weak carbonyl vibrational band is attributed to a slight distortion of the complex from idealized C_{3v} molecular symmetry.

10 Crystal Structure

PROBLEMS

10.1   In a close-packed layer each sphere (atom) is in contact with six others such that the neighboring six spheres just touch one another. Why are six spheres, not five, or seven or some other number, required for close packing around each sphere?

10.2   List the contents (number and type of atoms, ions, and/or molecules) of the conventional unit cell in each of the following crystalline solids:

(i)	Cubic close-packed metal	(ii)	Hexagonal close-packed metal
(iii)	CsCl	(iv)	NaCl (rock salt)
(v)	CaF_2 (fluorite)	(vi)	Cubic ZnS (zinc blende)
(vii)	Hexagonal ZnS (wurtzite)	(viii)	TiO_2 (rutile)
(ix)	Diamond	(x)	Graphite
(xi)	I_2	(xii)	Hexagonal ice
(xiii)	Dry ice (solid CO_2)	(xiv)	Benzene

10.3   Crystalline CaS (density 2.58 g cm^{-3}) has been shown by the powder method to have the NaCl type of structure.

(i)   Which of the following are allowed reflections?

100, 110, 111, 200, 210, 211, 220, and 222.

(ii)   Calculate the length of the unit cell edge.

(iii)   Calculate the smallest observable Bragg angle for Cu K_α radiation ($\lambda = 1.5418\,\text{Å}$).

(iv)   Derive values for the radii of the Ca^{2+} and S^{2-} ions. Assume that ionic radius is inversely proportional to effective nuclear charge estimated from Slater's rules (see Problem 3.6).

 Problems in Structural Inorganic Chemistry. Second edition. Wai-Kee Li, Hung Kay Lee, Dennis Kee Pui Ng, Yu-San Cheung, Kendrew Kin Wah Mak, and Thomas Chung Wai Mak. © Oxford University Press 2019. Published in 2019 by Oxford University Press. DOI: 10.1093/oso/9780198823902.001.0001

10.4 The cubic unit cell of an alloy is illustrated in the figure on the right.

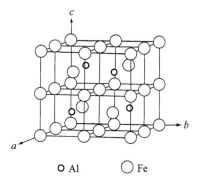

- (i) Deduce the stoichiometry (i.e., the formula) of this alloy.
- (ii) What is the lattice type $(P, C, I, \text{ or } F)$?
- (iii) For $a = 5.780$ Å, calculate the density in g cm^{-3} and the shortest Fe–Al interatomic distance.
- (iv) Describe the coordination (i.e., the number and type of nearest neighbors and their geometrical arrangement) around each of the following types of crystallographically distinct atoms:

  (a) Fe(1) at $(0, 0, 0)$, (b) Fe(2) at $(\frac{1}{2}, 0, 0)$, (c) Fe(3) at $(\frac{1}{4}, \frac{1}{4}, \frac{1}{4})$, (d) Al at $(\frac{1}{4}, \frac{1}{4}, \frac{3}{4})$.

- (v) Deduce the diffraction indices of a reflection observed at $\theta = 36.6°$ using Cu K$_\alpha$ radiation ($\lambda = 1.5418$ Å).

10.5 Aluminum is cubic with $a = 4.0415$ Å. The following reflections were observed with Cu K$_\alpha$ radiation ($\lambda = 1.5418$ Å): 111, 200, 220, 311, 222, 400, 331, 420, 422, 333, and 511.

- (i) What is the Bravais lattice type?
- (ii) Calculate the size of the reciprocal unit cell referred to base vectors $\{\lambda a_i^*\}$.
- (iii) Calculate the interplanar spacings corresponding to the first two reflections.
- (iv) What is the highest order $h00$ reflection observable with Mo K$_\alpha$ radiation ($\lambda = 0.7107$ Å)?

10.6 The mineral perovskite has a cubic unit cell of edge 3.84 Å with Ca, Ti, and O atoms located at its corners, body-center, and face-centers, respectively.

- (i) Give the stoichiometry of perovskite.
- (ii) Calculate the density of the crystal.
- (iii) Describe the coordination of the three types of ions.
- (iv) Arrange, with brief justification, the following X-ray reflections in order of decreasing intensity: (a) 100, 200, and 400; (b) 110 and 220.

10.7 A binary compound of calcium (represented by larger black circles) and carbon (open circles) has the tetragonal unit cell shown on the right with $a = b = 3.87$ Å and $c = 6.37$ Å. The fractional coordinates of the carbon atom directly above the calcium atom at the origin are $x = 0, y = 0, z = 0.406$.

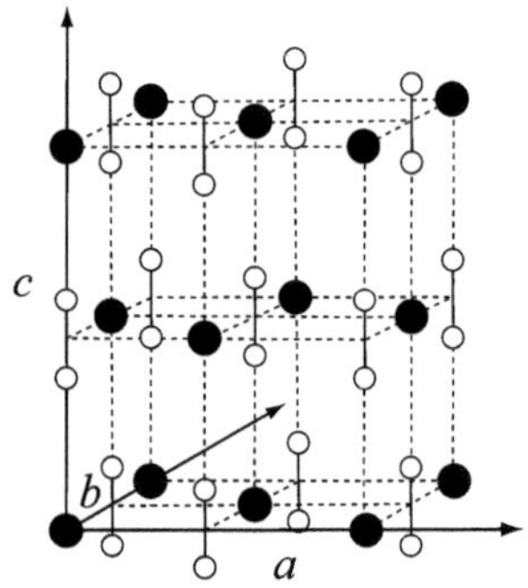

(i) Deduce the stoichiometric formula for this compound, the number of formula units in the unit cell, and the charge on the C_2 group.

(ii) Calculate the C–C bond distance within a C_2 group, the two shortest Ca–C distances, and the nonbonded (van der Waals) C $\cdots$ C separation.

(iii) The present structure is closely related to that of a simple ionic compound. Name that crystalline compound and discuss the relationship between the two structures.

10.8 One crystalline form of disodium tetraborate, $Na_2B_4O_7$, is monoclinic with unit cell dimensions $a = 11.858, b = 10.674, c = 12.197$ Å, $\beta = 106°41'$. The measured density is 1.713 g cm^{-3}. Are the crystals hydrated? If so, what is the hydration number?

10.9 (i) *Trans*-dichlorobis(2,4-dimethylthiazole)copper(II) crystallizes in space group $C2/c$ with $a = 12.320(13), b = 8.760(7), c = 14.592(11)$ Å, $\beta = 105.81(6)°$, $V = 1515.2$ Å^3, and $Z = 4$. Estimate the maximum number of independent reflections in a data set collected with Mo K$_\alpha$ radiation ($\lambda = 0.71069$ Å) out to $\theta \leq 30°$.

(ii) Derive a general expression for the angle $\theta(\vec{a}, \vec{a}^*)$ between $\vec{a}$ and $\vec{a}^*$ in terms of direct unit cell parameters. Verify that it reduces to $\theta(\vec{a}, \vec{a}^*) = \beta - 90°$ in the monoclinic case (b-axis unique, β obtuse).

(iii) In a tetragonal crystal the angle between the normals to the faces (130) and (112) is 44.0°. Find the axial ratio c/a.

REFERENCE: D. P. Gavel and D. J. Hodgson, *Trans*-dichlorobis(2,4-dimethylthiazole)copper(II). *Acta Cryst.* **B35**, 2704–7 (1979).

10.10 The crystal structure of dichlorobis(4-vinylpyridine)zinc(II) was erroneously described in space group $P\bar{1}$ with two molecules in a unit cell of dimensions $a = 7.501(4)$, $b = 7.522(5), c = 14.482(6)$ Å, $\alpha = 90.41(4), \beta = 90.53(4), \gamma = 105.29(5)°$. All reflections $hh\ell$ with ℓ odd were unobserved except six which were reported as very weak. Calculate the dimensions of a C-centered unit cell based on the vectors $[110], [\bar{1}10]$, and

[001] and show that, with allowance for experimental error, the data are consistent with space group $C2/c$.

REFERENCE: R. E. Marsh and V. Schomaker, Some incorrect space groups in Inorganic Chemistry, Volume 16. *Inorg. Chem.* **18**, 2331–6 (1979).

10.11 X-ray and neutron diffraction have led to the following data for the crystal structure of ordinary hexagonal ice.

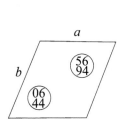

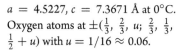

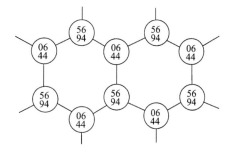

$a = 4.5227, c = 7.3671$ Å at 0°C. Oxygen atoms at $\pm(\frac{1}{3}, \frac{2}{3}, u; \frac{2}{3}, \frac{1}{3}, \frac{1}{2} + u)$ with $u = 1/16 \approx 0.06$.

Structure of ice projected along c. Oxygen atoms at $z = 56/100$ and $z = 94/100$ have the same pair of x and y fractional coordinates.

In this three-dimensional network, each oxygen atom is hydrogen-bonded to four other oxygen atoms in a tetrahedral environment. Note the open channels running through the centers of the puckered six-membered rings which account for the low density of ice. Each oxygen atom has therefore four tetrahedrally placed hydrogen neighbors: two closer together at 1.0 Å and two further away at 1.8 Å. The orientations of the water molecules are, however, entirely random, so that an ice crystal can exist in a large number of configurations.

(i) Calculate the density of hexagonal ice and the lengths of the two kinds of chemically identical but crystallographically distinct O–H $\cdots$ O hydrogen bonds.

(ii) The enthalpy of vaporization of water is 2.258 kJ g^{-1} at 100°C. Estimate the energy (in kJ mol^{-1}) required to break O–H $\cdots$ O hydrogen bonds.

(iii) In one mole of ice, there are $2N_0$ hydrogen atoms. Since, as indicated above, each has two choices of position along an O–O axis, there would be a total of 2^{2N_0} "possible" configurations. However, only a fraction of these are "allowed". Determine the number of "allowed" configurations and, hence, calculate the residual entropy for hexagonal ice.

10.12 A unit cell of a clathrate inclusion compound is shown in the figure on the right. [Note that only half of each benzene ring belongs to the unit cell.]

(i) Deduce the stoichiometry of the compound.

(ii) Briefly describe the structure.

(iii) Would you expect to obtain similar compounds by replacing benzene with (a) pyridine and (b) naphthalene? Why?

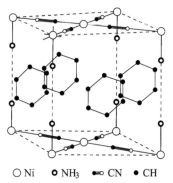

○ Ni   ○ NH₃   ●○ CN   ● CH

REFERENCE: J. H. Rayner and H. M. Powell, Structure of molecular compounds. Part X. Crystal structure of the compound of benzene with an ammonia-nickel cyanide complex. *J. Chem. Soc.* 319–28 (1952).

10.13 Crystals of the complex $[Ni(H_2O)_2(piaH)_2]Cl_2$ are blue needles and have a magnetic moment of 3.25 BM. Pyridine-2-carboxamide, abbreviated piaH, may be expected to act as a bidentate ligand in two ways as shown.

Deduce as much information as you can from the following crystal data: monoclinic, $a = 13.567$, $b = 10.067$, $c = 6.355$ Å, $\beta = 113.7°$; FW $= 409.8$; measured density $= 1.72$ g cm^{-3}; systematic absences: $h0\ell$ with h odd, $0k0$ with k odd.

REFERENCE: A. Masuko, T. Nomura, and Y. Saito, The crystal structure of *trans*-diaquo-bis(pyridine-2-carboxamide)-nickel(II) chloride, [Ni(H₂O)₂(piaH)₂]Cl₂. *Bull. Chem. Soc. Japan* **41**, 511–5 (1967).

10.14 A 1:2 complex between Ni^{2+} and N,N-dimethyl-β-mercaptoethylamine, $Me_2NCH_2CH_2SH$, has been synthesized. It is desired to establish the configuration (*cis* or *trans*) of the complex and the coordination (square planar or tetrahedral) of Ni^{2+}. Explain how the following information enables one to elucidate the stereochemistry. Also state the point group to which it belongs and sketch the bis(N,N-dimethyl-β-mercaptoethylamine)nickel(II) molecule.

(a) The compound is diamagnetic.

(b) It crystallizes in space group $P2_1/n$ (the same as $P2_1/c$ except for a different choice of the a- and c-axes) with two molecules per unit cell.

REFERENCE: R. L. Girling and E. L. Amma, The crystal and molecular structure of bis(N, N-dimethyl-β-mercaptoethylamine)nickel(II). *Inorg. Chem.* **6**, 2009–12 (1967).

10.15 Crystals of *p*-chloronitrobenzene are monoclinic with $a = 3.84, b = 6.80, c = 13.35$ Å, and $\beta = 97°31'$. The systematic absences are $h0\ell$ with ℓ odd, $0k0$ with k odd. The density measured by flotation in aqueous potassium iodide is 1.52 g cm^{-3}. What information can be deduced from these data?

REFERENCE: T. C. W. Mak and J. Trotter, The crystal structure of *p*-chloronitrobenzene. *Acta Cryst.* **15**, 1078–80 (1962).

10.16 Nitronium perchlorate forms monoclinic crystals belonging to symmetry class $2/m$. The unit cell dimensions are $a = 9.16, b = 7.08, c = 7.30$ Å, $\beta = 112.1°$. The density of the crystals is 2.20 g cm^{-3}. The systematic extinctions are $hk\ell$ absent for $(h + k)$ odd, and $h0\ell$ absent for h odd or ℓ odd. Assuming that the compound is composed of NO_2^+ and ClO_4^- ions, what conclusions can be drawn about their positions in the unit cell?

REFERENCE: M. R. Truter, D. W. J. Cruickshank, and G. A. Jeffrey, The crystal structure of nitronium perchlorate. *Acta Cryst.* **13**, 855–62 (1960).

10.17 Identify the space group for each of the crystal structures shown in projection in (i)–(vi). The atomic coordinates with reference to the axis of projection are given as fractions and/or in one-hundredths.

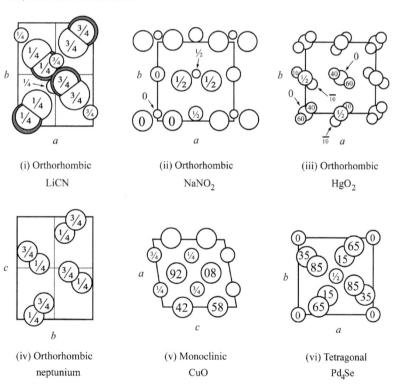

(i) Orthorhombic
LiCN

(ii) Orthorhombic
NaNO$_2$

(iii) Orthorhombic
HgO$_2$

(iv) Orthorhombic
neptunium

(v) Monoclinic
CuO

(vi) Tetragonal
Pd$_4$Se

279

10.18 List all possible space groups which are consistent with the observed Laue symmetry and systematic absences (extinctions):

    (i)   Monoclinic, $hk\ell$ with $(h + k)$ odd, $h0\ell$ with ℓ odd.

    (ii)   Orthorhombic, $hk\ell$ with $(h + k)$ odd, $h0\ell$ with ℓ odd.

    (iii) Tetragonal, $4/m$, $hk\ell$ with $(h + k + \ell)$ odd.

    (iv) Tetragonal, $4/mmm$, $hk\ell$ with $(h + k + \ell)$ odd.

    (v)   Rhombohedral, $\bar{3}m$, $hk\ell$ [referred to triply primitive hexagonal unit cell] with $(-h + k + \ell) \neq 3n$.

    (vi) Hexagonal, $6/mmm$, $h\bar{k}\ell$ with ℓ odd.

    (vii) Cubic, $m3m$, $hk\ell$ with $(h + k)$ odd or $(k + \ell)$ odd. [Reflection is observable only if indices are all odd or all even.]

10.19 The diagram on the right shows the symmetry elements in a certain orthorhombic space group (origin located at $\bar{1}$).

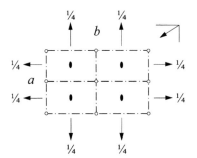

    (i)   Deduce the space group symbol.

    (ii)   Deduce the fractional coordinates of the general equivalent positions in this space group.

    (iii) What are the systematic absences $(hk\ell)$ in this space group?

    (iv) Is this space group uniquely determined from the systematic absences?

10.20 The diagram on the right shows the symmetry elements in a certain orthorhombic space group (origin located at $\bar{1}$).

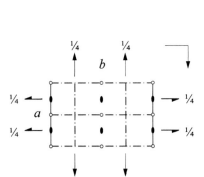

    (i)   Deduce the space group symbol.

    (ii)   Deduce the fractional coordinates of the general equivalent positions in this space group.

    (iii) What are the systematic absences $(hk\ell)$ in this space group?

    (iv) Is this space group uniquely determined from the systematic absences?

10.21 (i) A diagram showing the symmetry elements in a certain orthorhombic space group (origin located at $\bar{1}$) is shown below. Deduce the space group symbol.

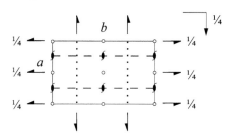

(ii) Write the coordinates of the general equivalent positions labeled 1 to 8 in the following diagram for this space group:

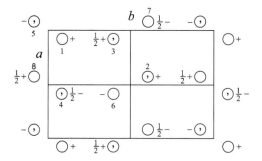

10.22 Shown in the diagram on the right are the symmetry elements in space group *Pmmn* (origin at $\bar{1}$). Deduce the coordinates of the general equivalent positions.

What space groups are consistent with the following systematic absences?

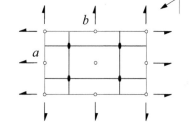

(i) Monoclinic, $0k0$ with k odd.

(ii) Orthorhombic, $hk\ell$ with $(h+k)$ odd, $0k\ell$ with ℓ odd, $h0\ell$ with ℓ odd, and $hk0$ with h odd.

(iii) Tetragonal, Laue symmetry $4/m$, $hk0$ with $(h+k)$ odd, and 00ℓ with ℓ odd.

10.23 A hydrate of copper sulfate, $CuSO_4 \cdot nH_2O$ crystallizes in the triclinic space group $P\bar{1}$ with $a = 6.113$, $b = 10.712$, $c = 5.958$ Å, $\alpha = 82.30$, $\beta = 107.29$, and $\gamma = 102.57°$. The measured density of the crystal is 2.284 g cm^{-3}. Given: FW = 159.61 for $CuSO_4$.

(i) Calculate the volume of the unit cell.

(ii) Deduce values of Z and n.

(iii) Calculate the density of the unit cell for comparison with the measured value.

10.24 The illustration shows the crystal structure of silver azide, AgN_3, projected along the b-axis. The tetramolecular unit cell is orthorhombic. Deduce the space group by inspection and indicate the symmetry elements which characterize the space group in another diagram. What positions do the silver and nitrogen atoms occupy in the space group?

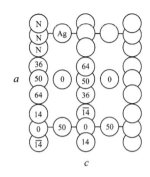

10.25 The iron carbide cementite is an important component of steel. The crystals are orthorhombic with $a = 4.524, b = 5.089, c = 6.743$ Å. The crystal structure projected along b is shown on the right, in which iron atoms are represented by larger circles and the origin is at the lower right corner.

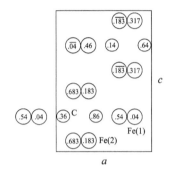

(i) Deduce the stoichiometry of cementite.

(ii) Sketch a symmetry diagram using standard graphic symbols and name the space group.

(iii) List the general equivalent positions in the space group.

(iv) Describe the coordination around the carbon atom and calculate all crystallographically distinct Fe–C distances using the following positional parameters:

	x	y	z
C	0.03	0.36	$\frac{1}{4}$
Fe(1)	−0.167	0.040	$\frac{1}{4}$
Fe(2)	0.333	0.183	0.065

10.26 The orthorhombic structure of trimethylamine-N-oxide hydro-chloride projected along its b-axis is shown in the figure on the right ($a = 14.27$, $b = 5.40, c = 7.61$ Å):

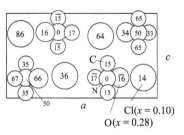

(i) Sketch a symmetry diagram using standard graphic symbols and hence determine the space group.

(ii) Deduce the general and special equivalent positions in the space group and present your results under the headings: positions, point symmetry, and coordinates.

(iii) Calculate the $O \cdots Cl$ distance and comment on any significant difference between this and 3.2 Å, the sum of the van der Waals radii of O and Cl.

REFERENCES: C. Rérat, Structure cristalline du chlorhydrate de triméthylaminoxyde. *Acta Cryst.* **13**, 63–71 (1960); A. Caron and J. Donohue, Refinement of the crystal structure of trimethylamine oxide hydrochloride $(CH_3)_3NO \cdot HCl$. *Acta Cryst.* **15**, 1052–3 (1962).

10.27 The crystal and molecular structure of xenon difluoride, XeF_2, has been determined by a three-dimensional neutron diffraction study. The crystals are tetragonal with $a = 4.315$, $c = 6.990$ Å. The space group is $I4/mmm$ and there are two molecules in the unit cell. Special positions of four-fold and two-fold multiplicity are:

Position	Point symmetry	Coordinates $(0, 0, 0; \frac{1}{2}, \frac{1}{2}, \frac{1}{2})+$
4(e)	$4mm$	$0, 0, z; 0, 0, \bar{z}$
4(d)	$\bar{4}m2$	$0, \frac{1}{2}, \frac{1}{4}; \frac{1}{2}, 0, \frac{1}{4}$
4(c)	mmm	$0, \frac{1}{2}, 0; \frac{1}{2}, 0, 0$
2(b)	$4/mmm$	$0, 0, \frac{1}{2}$
2(a)	$4/mmm$	$0, 0, 0$

(i) Assuming that the length of the Xe–F bond is 2.00 Å (1.93 to 1.94 Å in XeF_4), assign positions for the atoms and deduce values for the positional parameters, if any. Explain your deductions concisely.

(ii) Draw a legible illustration (perspective or in projection) of the crystal structure. Calculate the shortest nonbonded $F \cdots F$ and $Xe \cdots F$ contacts.

REFERENCES: H. H. Hyman (ed.), *Noble-Gas Compounds*, University of Chicago Press, Chicago, 1963, pp. 221–5; S. Siegel and E. Gebert, Crystallographic studies of XeF_2 and XeF_4. *J. Am. Chem. Soc.* **85**, 240 (1963).

10.28 Tin(IV) tetrafluoride, SnF_4, crystallizes in space group $I4/mmm$ with $a = 4.04$, $c = 7.93$ Å, and $Z = 2$. The atoms occupy the following special positions:

Atom	Position	Fractional coordinates $(0, 0, 0; \frac{1}{2}, \frac{1}{2}, \frac{1}{2})+$
Sn	2(a)	$0, 0, 0$
F'	4(c)	$0, \frac{1}{2}, 0; \frac{1}{2}, 0, 0$
F''	4(e)	$0, 0, z; 0, 0, \bar{z}$ with $z = 0.237$

(i) Give the conditions for systematic absences, if any.

(ii) Sketch the structure of crystalline SnF_4 showing the contents of a unit cell.

(iii) Deduce the point symmetry of special positions 2(a), 4(c), and 4(e).

(iv) Calculate the Sn–F interatomic distances and describe the coordination around Sn.

(v) Describe the crystal structure of SnF_4.

REFERENCE: R. Hoppe and W. Dähne, Die Kristallstruktur von SnF_4 und PbF_4. *Naturwiss.* **49**, 254–5 (1962).

10.29 The halogens chlorine, bromine, and iodine crystallize in the same space group *Bmab*. [This is identical to *Cmca* (No. 64) in *The International Tables for X-ray Crystallography*, vol. I, except for an interchange of the *b*- and *c*-axes.] The orthorhombic unit cell contains eight atoms in special position 8(f).

Position	Point symmetry	Coordinates $(0, 0, 0; \frac{1}{2}, 0, \frac{1}{2})+$
8(f)	*m*	$0, y, z; 0, \bar{y}, \bar{z}; \frac{1}{2}, \frac{1}{2}-y, z; \frac{1}{2}, \frac{1}{2}+y, \bar{z}$
4(a)	$2/m$	$0, 0, 0; 0, \frac{1}{2}, \frac{1}{2}$

Atom	Positional parameters		Unit cell dimensions (Å)			
	y	*z*	*a*	*b*	*c*	
Cl	0.100	0.130	6.24	8.26	4.48	$(-160°C)$
Br	0.110	0.135	6.67	8.72	4.48	$(-150°C)$
I	0.1156	0.1493	7.27007	9.79344	4.79004	$(26°C)$

The figure on the right shows the orthorhombic structure of solid iodine projected along its *a*-axis. Shaded and unshaded atoms lie at $x = \frac{1}{2}$ and $x = 0$, respectively. The I_2 molecules are easily discernible. The required point symmetry of the I_2 molecule in the crystal is $2/m$, although its full molecular symmetry is ∞/mm ($D_{\infty h}$).

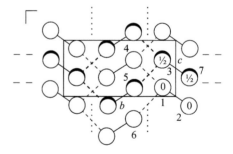

(i) Referring to the figure, tabulate the coordinates of atoms 1–7 and calculate all interatomic distances from atom 1 to its nearest neighbors.

(ii) Describe the crystal structure of solid I_2.

10.30 Palladous selenide, PdSe, is tetragonal with $a = 6.72$, $c = 6.90$ Å. The space group is $P4_2/m$, and there are eight formula units per unit cell.

Atom	Position	Point symmetry	Coordinates	Parameters
Pd(1)	2(e)	$\bar{4}$	$\pm(0, 0, \frac{1}{4})$	
Pd(2)	2(c)	$2/m$	$0, \frac{1}{2}, 0; \frac{1}{2}, 0, \frac{1}{2}$	
Pd(3)	4(j)	m	$\pm(x, y, 0; y, \bar{x}, \frac{1}{2})$	$x = 0.455, y = 0.235$
Se	8(k)	1	$\pm(x, y, z; x, y, \bar{z}; \bar{y}, x, z + \frac{1}{2};$ $y, \bar{x}, z + \frac{1}{2})$	$x = 0.20, y = 0.325,$ $z = 0.230$

  (i) Make a scaled drawing of the crystal structure of PdSe projected down its c-axis, i.e., on (001). An atom may be represented as a circle of suitable radius enclosing its fractional coordinate above or below the plane of the paper.
 (ii) Calculate all crystallographically independent Pd–Se bond distances, and Se–Pd–Se and Pd–Se–Pd bond angles. Describe the environment of the Pd and Se atoms.

REFERENCE: K. Schubert, H. Breiner, W. Burkhardt, E. Günzel, R. Haufler, H. L. Lukas, H. Vetter, J. Wegst, and M. Wilkens, Einige strukturelle Ergebnisse an metallischen Phasen II. *Naturwiss.* **44**, 229–30 (1957).

In Problems 10.31–10.35, sufficient data are provided to enable the reader to make scaled drawings of crystal structures. The axial directions are specified in each case so that a good projection of the structure may be obtained.

  An atom may be represented as a circle of suitable radius enclosing its fractional coordinate (often expressed in one-hundredths for simplicity) above or below the plane of the paper. Values for various kinds of atomic radii are given in A. F. Wells, *Structural Inorganic Chemistry*, 4th edn., Clarendon, Oxford, 1975, pp. 236, 259 and 1022.

  The completed drawing should reveal the presence of atomic aggregates, namely complex ions and molecules. Bond distances and valence angles, as well as the shorter van der Waals contacts, may then be calculated. In structures which can be described as a packing of simple ions, the coordination around each ionic species should be noted, and its nearest-neighbor separations determined.

10.31 The rhombohedral form of sulfur has a rhombohedral unit cell containing six atoms:

$$a_R = 6.46 \text{ Å}, \alpha = 115°8'.$$

The corresponding hexagonal unit cell containing 18 atoms has the edges:

$$a_H = 10.818 \text{ Å}, C_H = 4.280 \text{ Å}.$$

Space group $R\bar{3}$ (hexagonal setting).

Atom	Position	Point symmetry	Coordinates $(0, 0, 0; \frac{1}{3}, \frac{2}{3}, \frac{2}{3}; \frac{2}{3}, \frac{1}{3}, \frac{1}{3}) +$
S	18(f)	1	$x, y, z; \bar{y}, x - y, z; y - x, \bar{x}, z$
			$\bar{x}, \bar{y}, \bar{z}; y, x, \bar{z}; x - y, x, \bar{z}$
	3(a)	$\bar{3}$	$0, 0, 0$

Parameters: $x = 0.1454, y = 0.1882, z = 0.1055$.

Drawing: Project down c, with a pointing to the right and b upwards.

REFERENCE: J. Donohue, A. Caron and E. Goldish, The crystal and molecular structure of S_6 (sulfur-6). *J. Am. Chem. Soc.* **83**, 3748–51 (1961).

10.32 The form of chlorine trifluoride, ClF_3, which is stable below $-82.5°C$, is orthorhombic with a tetramolecular unit cell:

Space group *Pnma*, $a = 8.825, b = 6.09, c = 4.52$ Å.

Atom	Position	x	y	z
Cl	4(c)	0.1582	$\frac{1}{4}$	0.3790
F(1)	4(c)	0.0422	$\frac{1}{4}$	0.1010
F(2)	8(d)	0.1517	0.5315	0.3634

Drawing: Project down c, with origin at the lower right corner, a pointing to the left and b upwards.

REFERENCE: R. D. Burbank and F. N. Bensey, The structures of the interhalogen compounds. I. Chlorine trifluoride at $-120°C$. *J. Chem. Phys.* **21**, 602–8 (1953).

10.33 Crystals of phosphorus pentachloride, PCl_5, are tetragonal with a tetramolecular cell: Space group $P4/n$, $a = 9.22, c = 7.44$ Å.

Position	Point symmetry	Coordinates
8(g)	1	$x, y, z; \frac{1}{2} - x, \frac{1}{2} - y, z; \bar{y}, \frac{1}{2} + x, z; \frac{1}{2} + y, \bar{x}, z$
		$\bar{x}, \bar{y}, \bar{z}; \frac{1}{2} + x, \frac{1}{2} + y, \bar{z}; y, \frac{1}{2} - x, \bar{z}; \frac{1}{2} - y, x, \bar{z}$
2(c)	4	$\frac{1}{4}, \frac{3}{4}, z; \frac{3}{4}, \frac{1}{4}, \bar{z}$
2(a)	$\bar{4}$	$\frac{1}{4}, \frac{1}{4}, 0; \frac{3}{4}, \frac{3}{4}, 0$

Atom	Position	x	y	z
P(1)	2(a)	$\frac{1}{4}$	$\frac{1}{4}$	0
P(2)	2(c)	$\frac{1}{4}$	$\frac{3}{4}$	0.62
Cl(1)	2(c)	$\frac{1}{4}$	$\frac{3}{4}$	0.905
Cl(2)	2(c)	$\frac{1}{4}$	$\frac{3}{4}$	0.34
Cl(3)	8(g)	0.31	0.084	0.15
Cl(4)	8(g)	0.334	0.954	0.62

Drawing: Project down b, with a pointing to the right and c upwards.

REFERENCE: D. Clark, H. M. Powell, and A. F. Wells, The crystal structure of phosphorus pentachloride. *J. Chem. Soc.* 642–5 (1942).

10.34 Phosphorus pentabromide, PBr_5, is orthorhombic with four molecules in the unit cell: Space group *Pbcm*, $a = 5.63$, $b = 16.94$, $c = 8.31$ Å.

Position	Point symmetry	Coordinates
8(e)	1	$x, y, z; \bar{x}, \bar{y}, \frac{1}{2} + z; x, \frac{1}{2} - y, \bar{z}; \bar{x}, \frac{1}{2} + y, \frac{1}{2} - z$
		$\bar{x}, \bar{y}, \bar{z}; x, y, \frac{1}{2} - z; \bar{x}, \frac{1}{2} + y, z; x, \frac{1}{2} - y, \frac{1}{2} + z$
4(d)	m	$x, y, \frac{1}{4}; \bar{x}, \bar{y}, \frac{3}{4}; \bar{x}, \frac{1}{2} + y, \frac{1}{4}; x, \frac{1}{2} - y, \frac{3}{4}$

Atom	Position	x	y	z
P	4(d)	0.035	0.125	$\frac{1}{4}$
Br(1)	4(d)	−0.211	0.038	$\frac{1}{4}$
Br(2)	4(d)	−0.154	0.237	$\frac{1}{4}$
Br(3)	8(e)	0.258	0.123	0.040
Br(4)	4(d)	0.605	0.400	$\frac{1}{4}$

Drawing: Project down a, with c pointing to the right and b upwards.

REFERENCE: M. van Driel and C. H. MacGillavry, The crystal structure of phosphorus pentabromide. *Rec. Trav. Chim. Pays-Bas* **62**, 167–71 (1943).

10.35 Nitrogen pentoxide (nitric acid anhydride) N_2O_5, is hexagonal with a bimolecular unit cell:

Space group $P6_3/mmc$, $a = 5.410$, $c = 6.570$ Å at $-60°$C.

Position	Point symmetry	Coordinates
6(h)	mm	$x, 2x, \frac{1}{4}; 2\bar{x}, \bar{x}, \frac{1}{4}; x, \bar{x}, \frac{1}{4}; \bar{x}, 2\bar{x}, \frac{3}{4}; 2x, x, \frac{3}{4}; \bar{x}, x, \frac{3}{4}$
4(f)	$3m$	$\frac{1}{3}, \frac{2}{3}, z; \frac{2}{3}, \frac{1}{3}, \bar{z}; \frac{2}{3}, \frac{1}{3}, \frac{1}{2} + z; \frac{1}{3}, \frac{2}{3}, \frac{1}{2} - z$
2(d)	$\bar{6}m2$	$\frac{1}{3}, \frac{2}{3}, \frac{3}{4}; \frac{2}{3}, \frac{1}{3}, \frac{1}{4}$
2(c)	$\bar{6}m2$	$\frac{1}{3}, \frac{2}{3}, \frac{1}{4}; \frac{2}{3}, \frac{1}{3}, \frac{3}{4}$
2(b)	$\bar{6}m2$	$0, 0, \frac{1}{4}; 0, 0, \frac{3}{4}$

Atom	Position	Parameters
N(1)	2(b)	
N(2)	2(d)	
O(1)	6(h)	$x = 0.1327$
O(2)	4(f)	$z = -0.0743$

Drawing: Project down c, with a pointing to the right and b upwards at a slant.

REFERENCE: E. Grison, K. Eriks, and J. L. de Vries, Structure cristalline de l'anhydride azotique, N_2O_5. *Acta Cryst.* **3**, 290–4 (1950).

10.36 Rutile (TiO_2) is tetragonal with unit-cell dimensions $a = 4.58$, $c = 2.95$ Å. The titanium atoms are in positions $0, 0, 0; \frac{1}{2}, \frac{1}{2}, \frac{1}{2}$, and the oxygen atoms are in positions $u, u, 0$; $\bar{u}, \bar{u}, 0; \frac{1}{2} + u, \frac{1}{2} - u, \frac{1}{2}; \frac{1}{2} - u, \frac{1}{2} + u, \frac{1}{2}$ with $u = 0.31$.

(i) What is the arrangement of oxygen atoms about a titanium atom, and vice versa?

(ii) Sketch the crystal structure and determine the space group.

(iii) Calculate the contributions of the Ti and O atoms to the structure factors. What systematic absences occur?

10.37 X-ray analyses of [3]- and [4]-rotane (polycyclopropylidenes) reveal that bond length shortening occurs in the inner rings, which may be attributed to a π bond order enhancement resulting from the disposition of the adjacent (paddle-like) peripheral cyclopropane rings.

[3]-rotane

[4]-rotane

Crystal data

[3]-rotane: C_9H_{12}, hexagonal, $a = 6.826(4)$, $c = 9.738(5)$ Å, $Z = 2$. Space group
   $P6_3/m$.

[4]-rotane: $C_{12}H_{16}$, monoclinic, $a = 7.613(4)$, $b = 7.126(4)$, $c = 9.738(3)$ Å,
   $\beta = 112°1'(1)$, $Z = 2$. Space group $P2_1/n$.

For each rotane, write down

(i)   the systematic absences;
(ii) the Laue symmetry of the crystal;
(iii) the required molecular symmetry in the crystal;
(iv) the idealized molecular symmetry; and
(v)   the number of crystallographically distinct C–C bonds.

In space group $P6_3/m$, Wyckoff positions of multiplicity 2 are:

2(d)	$\bar{6}$	$\frac{2}{3}, \frac{1}{3}, \frac{1}{4}; \frac{1}{3}, \frac{2}{3}, \frac{3}{4}$	
2(c)	$\bar{6}$	$\frac{1}{3}, \frac{2}{3}, \frac{1}{4}; \frac{2}{3}, \frac{1}{3}, \frac{3}{4}$	
2(b)	$\bar{3}$	$0, 0, 0; 0, 0, \frac{1}{2}$	
2(a)	$\bar{6}$	$0, 0, \frac{1}{4}; 0, 0, \frac{3}{4}$	

REFERENCE: C. Pascard, T. Prangé, A. de Meijere, W. Weber, J.-P. Barnier, and J.-M. Conia, "Paddle-wheel" hydrocarbons. Intracyclic C–C bond length shortening in rotanes. X-Ray crystal structures of [3]- and [4]-rotane. *J. Chem. Soc., Chem. Commun.*, 425–6 (1979).

10.38 Deep blue crystals of "cuprammonium sulfate", $[Cu(NH_3)_4]SO_4 \cdot H_2O$, are orthorhombic with a unit cell of dimensions $a = 7.08$, $b = 12.14$, $c = 10.68$ Å, and $Z = 4$. The space group has the following sets of general and special positions:

Position	Point symmetry	Coordinates
8(d)	1	$\pm(x, y, z; \frac{1}{2} - x, \frac{1}{2} - y, \frac{1}{2} + z; \frac{1}{2} - x, y, z; x, \frac{1}{2} - y, \frac{1}{2} + z)$
4(c)	m	$\pm(\frac{1}{4}, y, z; \frac{1}{4}, \frac{1}{2} - y, \frac{1}{2} + z)$
4(b)	$\bar{1}$	$0, \frac{1}{2}, 0; \frac{1}{2}, \frac{1}{2}, 0; 0, 0, \frac{1}{2}; \frac{1}{2}, 0, \frac{1}{2}$
4(a)	$\bar{1}$	$0, 0, 0; \frac{1}{2}, 0, 0; 0, \frac{1}{2}, \frac{1}{2}; \frac{1}{2}, \frac{1}{2}, \frac{1}{2}$

The atoms in $[Cu(NH_3)_4]^{2+}$ (except hydrogens) are assigned positions and parameters in the following table:

Atom	Position	Parameters
Cu	4(c)	y_{Cu}, z_{Cu}
N(1)	8(d)	$x_{N(1)}, y_{N(1)}, z_{N(1)}$
N(2)	8(d)	$x_{N(2)}, y_{N(2)}, z_{N(2)}$
...	...	...

(i) Identify the space group and give the conditions for systematic absences of X-ray reflections.

(ii) What is the idealized point group for $[Cu(NH_3)_4]^{2+}$ (neglecting hydrogen atoms)? Which of its symmetry element(s) is (are) utilized in the crystal?

(iii) Assign, with justification, positions to the remaining atoms (i.e., complete the above table) and list the positional parameters defining the crystal structure.

(iv) Give the number of chemically equivalent but crystallographically distinct bond distances and angles in the $[Cu(NH_3)_4]^{2+}$ and SO_4^{2-} ions.

REFERENCE: F. Mazzi, The crystal structure of cupric tetrammine sulfate monohydrate, $Cu(NH_3)_4SO_4 \cdot H_2O$. *Acta Cryst.* **8**, 137–41 (1955).

10.39 Beryl, $Be_3Al_2(SiO_3)_6$, is hexagonal with $a = 9.206$, $c = 9.205$ Å, $Z = 2$. The space group is $P6/mcc$ and the equipoint sets are:

Position	Point symmetry	Coordinates
24(m)	ℓ	x, y, z; etc.
12(l)	m	$x, y, 0$; etc.
12(k)	2	$x, 2x, \frac{1}{4}$; etc.
12(j)	2	$x, 0, \frac{1}{4}$; etc.
12(i)	2	$\frac{1}{2}, 0, z$; etc.
8(h)	3	$\frac{1}{3}, \frac{2}{3}, z$; etc.
6(g)	$2/m$	$\frac{1}{2}, 0, 0$; etc.
6(f)	222	$\frac{1}{2}, 0, \frac{1}{4}$; etc.
4(e)	6	$0, 0, z$; etc.
4(d)	$\bar{6}$	$\frac{1}{3}, \frac{2}{3}, 0$; etc.
4(c)	32	$\frac{1}{3}, \frac{2}{3}, \frac{1}{4}$; etc.
2(b)	$6/m$	$0, 0, 0$; $0, 0, \frac{1}{2}$
2(a)	62	$0, 0, \frac{1}{4}$; $0, 0, \frac{3}{4}$

(i) The $[(SiO_3)_6]^{12-}$ group consists of six corner-sharing SiO_4 tetrahedra arranged in a ring with overall symmetry $6/m$. Sketch its structure.

(ii) Discuss the possible modes of coordination of the Be and Al atoms.

(iii) Assign positions to all atoms and list the positional parameters which uniquely define the crystal structure. Explain your deductions concisely.

REFERENCE: W. L. Bragg and J. West, The structure of beryl, $Be_3Al_2Si_6O_{18}$. *Proc. R. Soc. Lond.* **A111**, 691–714 (1926).

10.40 Hexamethylenetetramine, $(CH_2)_6N_4$, is the first organic molecule to be subjected to crystal structure analysis. The crystals are cubic, with $a = 7.02$ Å, $\rho = 1.33$ g cm^{-3}. The Laue symmetry is $m3m$ and the systematic absences are $hk\ell$ with $(h + k + \ell)$ odd.

(i) How many molecules are there in the unit cell?

(ii) Which space groups are consistent with the observed systematic absences and Laue symmetry?

(iii) Establish the true space group unequivocally by considering the molecular symmetry of hexamethylenetetramine.

(iv) Assign positions to all atoms, including hydrogens, in the unit cell. What positional parameters are required to describe the crystal structure?

(v) Deduce fractional coordinates for carbon and nitrogen by making the following assumptions:

Separation between carbon atoms in neighboring molecules in axial direction = 3.68 Å; length of the C–N bond = 1.48 Å.

REFERENCE: L. N. Becka and D. W. J. Cruickshank, The crystal structure of hexamethylenetetramine. I. X-ray studies at 298, 100 and 34°K. *Proc. R. Soc. Lond.* **A273**, 435–54 (1963).

10.41 Basic beryllium acetate, $Be_4O(CH_3COO)_6$, has the properties of a molecular solid rather than an ionic salt. The crystal structure is cubic, with $a = 15.74$ Å and $\rho_x = 1.39$ g cm^{-3}. The diffraction (Laue) symmetry is $m3$ and only the following reflections are observed:

(a) $hk\ell$ when h, k, ℓ are all even or all odd;

(b) $0k\ell$ when $(k + \ell) = 4n$.

(i) Determine the space group.

(ii) Assign positions to all atoms except hydrogens and list the positional parameters which define the crystal and molecular structure. Explain your deductions clearly.

(iii) What is the idealized molecular point group? Which molecular symmetry elements are not utilized in the space group?

(iv) Sketch and describe the molecular structure.

REFERENCE: A. Tulinsky and C. R. Worthington, An accurate analysis of basic beryllium acetate by three-dimensional Fourier methods. *Acta Cryst.* **10**, 748–9 (1957).

10.42 When equivalent amounts of NH_4Cl and $HgCl_2$ are heated together in a sealed tube, crystals of NH_4HgCl_3 are formed. The space group is $P4/mmm$ with $a = 4.19$, $c = 7.94$ Å, and $Z = 1$. With the assumptions listed below, deduce the coordinates of all the non-hydrogen atoms in the unit cell. Describe the coordination around Hg^{2+}, and calculate the Hg^{2+} to Cl^- distances. Present your reasoning succinctly.

(a) Hg^{2+} lies at the origin. NH_4^+ (considered as a spherical ion) is as far away as possible from it.

(b) Atoms cannot approach each other closer than 2.3 Å and the $Cl^- \cdots Cl^-$ contact must be greater than 3.0 Å.

(c) Each NH_4^+ is surrounded by eight Cl^- at distances of about 3.35 Å (i.e., the interionic distance found in NH_4Cl).

In space group P4/mmm, Wyckoff positions of multiplicities 1 and 2 are:

Position	Point symmetry	Coordinates
2(h)	4mm	$\frac{1}{2}, \frac{1}{2}, z; \frac{1}{2}, \frac{1}{2}, \bar{z}$
2(g)	4mm	$0, 0, z; 0, 0, \bar{z}$
2(f)	mmm	$0, \frac{1}{2}, 0; \frac{1}{2}, 0, 0$
2(e)	mmm	$0, \frac{1}{2}, \frac{1}{2}; \frac{1}{2}, 0, \frac{1}{2}$
1(d)	4/mmm	$\frac{1}{2}, \frac{1}{2}, \frac{1}{2}$
1(c)	4/mmm	$\frac{1}{2}, \frac{1}{2}, 0$
1(b)	4/mmm	$0, 0, \frac{1}{2}$
1(a)	4/mmm	$0, 0, 0$

REFERENCE: E. J. Harmsen, Ammonium mercury(II) trichloride monohydrate. *Z. Krist.* **100A**, 208–11 (1938).

10.43 Diammonium tellurium(VI) dioxide tetrahydroxide, $(NH_4)_2[TeO_2(OH)_4]$, is monoclinic with $a = 8.019(2)$, $b = 6.568(1)$, $c = 6.352(2)$ Å, $\beta = 103.89(2)°$, and $Z = 2$. The observed systematic absences are: $hk\ell$ with $(h + k)$ odd.

(i) According to VSEPR theory, the anion has the structure as shown on the right. Which bond length is expected to be shorter, Te–O^- or Te–OH? Justify your answer.

(ii) Assume that (a) the space group is C2/m; (b) the H atoms of the OH groups do not lie in the equatorial plane of the anion; and (c) the formally negative terminal O atom acts as an acceptor in strong $O \cdots H$–O and $O \cdots H$–N hydrogen bonds. Assign positions to all non-hydrogen atoms in the asymmetric unit and explain your deductions clearly.

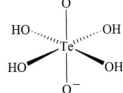

The general and special equivalent positions in space group C2/m are:

Position	Point symmetry	Coordinates$(0, 0, 0; \frac{1}{2}, \frac{1}{2}, 0) +$
8(j)	1	$x, y, z; x, \bar{y}, z; \bar{x}, y, \bar{z}; \bar{x}, \bar{y}, \bar{z}$
4(i)	m	$x, 0, z; \bar{x}, 0, \bar{z}$
4(h)	2	$0, y, \frac{1}{2}; 0, \bar{y}, \frac{1}{2}$
4(g)	2	$0, y, 0; 0, \bar{y}, 0$
4(f)	$\bar{1}$	$\frac{1}{4}, \frac{1}{4}, \frac{1}{2}; \frac{1}{4}, \frac{3}{4}, \frac{1}{2}$
4(e)	$\bar{1}$	$\frac{1}{4}, \frac{1}{4}, 0; \frac{1}{4}, \frac{3}{4}, 0$
2(d)	$2/m$	$0, \frac{1}{2}, \frac{1}{2}$
2(c)	$2/m$	$0, 0, \frac{1}{2}$
2(b)	$2/m$	$0, \frac{1}{2}, 0$
2(a)	$2/m$	$0, 0, 0$

REFERENCE: G. B. Johansson, O. Lindqvist, and J. Moret, Diammonium tellurium(VI) dioxide tetrahydroxide. *Acta Cryst.* **B35**, 1684–6 (1979).

10.44 A 1:1 molecular adduct of 2,5-dihydroxy-1,4-benzoquinone (DHBQ) and 4, 4′-bipyridine (BPY), $C_6H_4O_4 \cdot C_{10}H_8N_2$ (FW = 296.28), crystallizes in the monoclinic system with $a = 20.868, b = 7.0151, c = 9.1087$ Å, and $\beta = 92.843°$. The systematic absences are $hk\ell$ with $(h+k)$ odd; and $(h0\ell)$ with ℓ odd. The measured density of the crystalline sample is $1.478 \, \mathrm{g \, cm^{-3}}$. The structural formulas of DHBQ and BPY are shown on the right.

(i) Calculate the geometric parameters of the reciprocal unit cell $(a^*, b^*, c^*, \alpha^*, \beta^*, \gamma^*,$ and $V^*)$.

(ii) Estimate the number of independent reflections that can be observed using Mo K_α radiation $(\lambda = 0.71073 \, \text{Å})$.

(iii) Calculate the value of Z (the number of formula units per unit cell) and draw conclusions about the site symmetries of the two kinds of molecular ions in the unit cell.

(iv) What space group(s) is (are) consistent with the systematic absences?

(v) If the space group is centrosymmetric, make deductions about the site symmetries of the two kinds of molecules in the unit cell.

(vi) What kind of interaction would you expect to exist between DHBQ and BPY? Draw a sketch to illustrate your answer.

REFERENCE: J. A. Cowan, J. A. K. Howard, and M. A. Leech, The 1:1 adduct of 2,5-dihydroxy-1,4-benzoquinone with 4,4′-bipyridine. *Acta Cryst.* **C57**, 302–3 (2001).

10.45 The unit cells of two high-temperature super-conductors of the ternary oxide system, tetragonal phase I (space group $P4/mmm$) and orthorhombic phase II (space group $Pmmm$) are shown in the figures on the right. In each case, write the stoichiometric formula and deduce the value of Z (the number of formula units per unit cell).

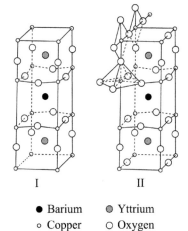

I   II

REFERENCE: figures from J. W. Lynn (ed.), *High Temperature Superconductivity*, Springer-Verlag, New York, 1990, p. 110.

● Barium   ◉ Yttrium
○ Copper   ○ Oxygen

10.46 N, N'-Diphenylbenzamidinium nitrate, $[Ph(H)NC(Ph)N(H)Ph]^+ \cdot NO_3^-$, FW $= 335.36$, crystallizes in the monoclinic system with $a = 15.804$, $b = 12.889$, $c = 11.450$ Å, $\beta = 132.20°$. The measured density of the crystalline sample is 1.30 g cm^{-3}. The systematic absences are: $hk\ell$ with $(h + k)$ odd and $h0\ell$ with ℓ odd. The structural formula of the organic cation is shown above; note that the two nitrogen atoms are equivalent by resonance.

   (i)   Calculate the value of Z (the number of formula units per unit cell).

  (ii)   What space group(s) is (are) consistent with the systematic absences?

 (iii)   If the space group is centrosymmetric, what is the composition of the asymmetric unit?

 (iv)   Make deductions about the site symmetries of the two kinds of molecular ions in the unit cell.

  (v)   If the hydrogen atoms are ignored, how many atomic positional parameters are required to define the crystal and molecular structure?

 (vi)   Assuming that hydrogen bonding exists between the cation and anion, sketch a diagram showing the intermolecular interaction between them.

REFERENCE: J. Barker, W. Errington, and M. G. H. Wallbridge, *N, N'*-Diphenylbenzamidinium nitrate. *Acta Cryst.* **C55**, 1583–5 (1999).

10.47 Cyanogen chloride, $N\equiv C-Cl$, crystallizes in the orthorhombic space group *Pmmn* with $a = 5.684$, $b = 3.977$, $c = 5.740$ Å, and $Z = 2$. The molecule occupies the following special position in the space group (origin at $\bar{1}$):

Wyckoff position	Point symmetry	Coordinates of equivalent positions
2(a)	*mm*	$\frac{1}{4}, \frac{1}{4}, z; \frac{3}{4}, \frac{3}{4}, -z$

The z parameters of the atoms as determined by X-ray analysis are: $C(1)$, 0.4239; $N(1)$, 0.6258; $Cl(1)$, 0.1499.

(i) Calculate the $C(1)$–$N(1)$ and $C(1)$–$Cl(1)$ bond distances.

(ii) Sketch a packing drawing of the crystal structure projected along the *a*-axis, showing that the molecules are stacked in linear chains.

(iii) Calculate the separation between the closest atoms of adjacent molecules in the same chain.

(iv) Calculate the shortest nonbonded distance between carbon and nitrogen atoms belonging to adjacent chains running in opposite directions.

SOLUTIONS

A10.1 The centers of three identical spheres in contact with one another form an equilateral triangle. In a close-packed layer, the center of each sphere is the common vertex of $360°/60° =$ six equilateral triangles.

A10.2
(i)	Four metal atoms	(ii)	Two metal atoms
(iii)	One Cs^+ and one Cl^-	(iv)	Four Na^+ and four Cl^-
(v)	Four Ca^{2+} and eight F^-	(vi)	Four Zn^{2+} and four S^{2-}
(vii)	Two Zn^{2+} and two S^{2-}	(viii)	Two Ti^{4+} and four O^{2-}
(ix)	Eight carbon atoms	(x)	Four carbon atoms
(xi)	Four I_2	(xii)	Four H_2O
(xiii)	Four CO_2	(xiv)	Two C_6H_6

A10.3
(i) For an *F*-type lattice, the diffraction indices must be all even or all odd. Hence, the allowed reflections are 111, 200, 220, and 222.

(ii) $a = 5.705$ Å.

(iii) For the 111 reflection, $\theta = 13.54°$ from the formula $d_{hk\ell} = \dfrac{a}{\sqrt{h^2 + k^2 + \ell^2}}$ for the cubic system.

(iv) Ca^{2+}: $Z_{eff} = 8.40$; S^{2-}: $Z_{eff} = 4.40$.

$r_{Ca^{2+}}/r_{S^{2-}} = 4.40/8.40$; $r_{Ca^{2+}} + r_{S^{2-}} = 5.705/2$.

Solution of the two equations yields $r_{Ca^{2+}} = 0.98$ Å and $r_{S^{2-}} = 1.87$ Å.

A10.4 (i) Fe_3Al.

(ii) F. The lattice is not body-centered since the Fe atoms at the corner and center of the unit cell are in crystallographically distinct though stereo chemically equivalent environments.

(iii) $$\rho = \frac{4[3(55.85) + 26.98]}{6.022 \times 10^{23}(5.780 \times 10^{-8})^3} = 6.69 \text{ g cm}^{-3}.$$

Shortest Fe–Al distance $= \frac{\sqrt{3}a}{4} = 2.503$ Å.

(iv) All atoms are in cubic environments with eight nearest neighbors.

Fe(1) and Fe(2) each has four Fe neighbors and four Al neighbors.

Fe(3) and Al each has eight Fe neighbors.

(v) Use of $2d_{hk\ell} \sin \theta = \lambda$ and $d_{hk\ell} = \dfrac{a}{\sqrt{h^2 + k^2 + \ell^2}}$ leads to $\sqrt{h^2 + k^2 + \ell^2} = 2(5.780/1.5418) \sin 36.6° = 4.47$, or $h^2 + k^2 + \ell^2 \approx 20$. The observed reflection therefore has indices 420.

A10.5 (i) The Bravais lattice type is F since the diffraction indices are all even or all odd.

(ii) $V = 0.05552$ reciprocal lattice units (dimensionless).

(iii) $d_{111} = \dfrac{a}{\sqrt{h^2 + k^2 + \ell^2}} = 2.3334$ Å; $d_{200} = 2.0208$ Å.

(iv) From the Bragg equation, $2 \sin \theta = \lambda/d_{h00} = h\lambda/a$, hence $h\lambda/a \leq 2$. The largest value of h satisfying this condition is 11 for Mo K_α radiation, but as the lattice is F-centered, the highest order $h00$ reflection observable is $(10\,0\,0)$.

A10.6 (i) $CaTiO_3$.

(ii) $\rho = 3.99 \text{ g cm}^{-3}$.

(iii) Ca^{2+} surrounded dodecahedrally by 12 O^{2-};

Ti^{4+} surrounded octahedrally by six O^{2-};

O^{2-} surrounded octahedrally by two Ti^{4+} and four Ca^{2+}.

(iv) (a) 200 > 400 > 100 in intensity;

(b) 220 > 110.

Scattering of X-rays by atoms lying in 100 and 110 planes is out of phase with that by atoms located in between. Reflection 400 is weaker than 200 because intensities generally fall off with increasing Bragg angle.

A10.7 (i) Formula CaC_2; $Z = 4$; charge on C_2 group is -2.

(ii)   C–C bond distance $= 1.20\,\text{Å}$; Ca–C distances $= 2.02\,\text{Å}$ (shortest) and $2.58\,\text{Å}$ (next shortest); $\text{C}\cdots\text{C}$ nonbonded distances $= 2.74\,\text{Å}$ and $2.77\,\text{Å}$.

(iii) The crystal structure of CaC_2 is closely related to that of NaCl. In CaC_2, the symmetry is reduced from cubic to tetragonal and $c > a = b$ because of the parallel alignment of C_2^{2-} anions in the direction of the c-axis.

A10.8   Since $\rho = Z(\text{FW})/N_0 V$, substitution of known values yields $Z(\text{FW}) = 1526$. The compound is hydrated, for if it were anhydrous, Z would have the unacceptable value of 7.6. Let n be the number of moles of water of crystallization, $(201.2 + 18.02n)Z = 1526$.

For $Z = 2$, $n = 31.2$; $Z = 4$, $n = 10.0$; $Z = 6$, $n = 3.0$. The most likely hydration number is 10, though the possibility that $n = 3$ cannot be eliminated on the basis of the given data. The actual molecular formula is, of course, $Na_2B_4O_7 \cdot 10H_2O$.

A10.9   (i)   Volume of reduced limiting sphere $= \left(\tfrac{4}{3}\right)\pi(2\sin\theta_{max})^3 = 33.51\sin^3\theta_{max}$.
Volume of reciprocal unit cell $= \lambda^3/V$.
Approximate number of observable independent reflections
$= (33.51)(\sin^3\theta_{max})(V)/\lambda^3(2)(4) = 2210$.
The numbers 2 and 4 in the denominator allow for lattice centering and Laue multiplicity, respectively.

(ii) $\cos\theta(\vec{a}, \vec{a}^*) = \vec{a}\cdot\vec{a}^*/aa^*$, $\vec{a}^* = \vec{b}\times\vec{c}/V$, $a^* = bc\sin\alpha/V$.
Hence $\cos\theta = (\vec{a}\cdot\vec{b}\times\vec{c}/V)(V/abc\sin\alpha)$.
$V = abc(1 - \cos^2\alpha - \cos^2\beta - \cos^2\gamma + 2\cos\alpha\cos\beta\cos\gamma)^{1/2}$.
Hence $\theta = \cos^{-1}[(1 - \cos^2\alpha - \cos^2\beta - \cos^2\gamma + 2\cos\alpha\cos\beta\cos\gamma)^{1/2}/\sin\alpha]$.
In the monoclinic case (b-axis unique, β obtuse), $\theta = \cos^{-1}[(1 - \cos^2\beta)^{1/2}]$, hence $\theta = \beta - 90°$.

(iii) For the tetragonal system, $a = b$, $a^* = 1/a$, $c^* = 1/c$.
$\vec{\rho}_{hk\ell} = h\lambda\vec{a} + k\lambda\vec{b}^* + \ell\lambda\vec{c}^*$ is normal to $(hk\ell)$.
$\cos 44° = \vec{\rho}_{130}\cdot\vec{\rho}_{112}/|\vec{\rho}_{130}||\vec{\rho}_{112}| = 4/10^{1/2}[2 + 4(c^*/a^*)^2]^{1/2}$.
$c/a = (c^*/a^*)^{-1} = 1.914$.

A10.10

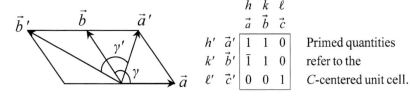

		h	k	ℓ
		$\vec{a}$	$\vec{b}$	$\vec{c}$
h'	$\vec{a}'$	1	1	0
k'	$\vec{b}'$	$\bar{1}$	1	0
ℓ'	$\vec{c}'$	0	0	1

Primed quantities refer to the C-centered unit cell.

Dimensions of the C-centered unit cell are: $a' = 9.115$, $b' = 11.942$, $c'(= c) = 14.482$ Å, $\alpha' = 89.93$, $\beta' = 90.77$, $\gamma' = 89.83°$.

Sample calculation:

$a' = $ length of $[110]$ vector referred to the P unit cell $= (a^2 + b^2 + 2ab \cos \gamma)^{1/2}$

$= 9.115$ Å.

$$\gamma' = \cos^{-1} \frac{|\vec{a}'|^2 + |\vec{b}'|^2 - |\vec{a}' - \vec{b}'|^2}{2 |\vec{a}'| |\vec{b}'|} = 89.83°.$$

$$\left.\begin{array}{l} h' = h + k \\ k' = -h + k \\ \ell' = \ell \end{array}\right\} \quad h' + k' = 2k, \text{ hence } (h' + k') \text{ must be even.}$$

$$\left.\begin{array}{l} h = \frac{1}{2}(h' - k') \\ k = \frac{1}{2}(h' + k') \end{array}\right\} \quad \begin{array}{l} k - h = k', \text{ i.e., if } h = k, k' = 0. \text{ Hence, } hh\ell \text{ reflections with } \ell \text{ odd} \\ \text{in the } P \text{ lattice correspond to } h'0\ell' \text{ with } \ell' \text{ odd in the } C \text{ lattice.} \end{array}$$

Referred to the unit cell based on $[110]$, $[\bar{1}10]$, and $[001]$, the systematic absences are $h'k'\ell'$ with $(h' + k')$ odd, and $h'0\ell'$ with ℓ' odd if the six very weak ($hh\ell$ with ℓ odd) reflections are assumed to be actually absent. If α' and γ' are taken to be exactly 90°, the observed data would be entirely consistent with space group $C2/c$.

A10.11   (i) $Z = 4$, $\rho = 0.917$ g cm^{-3}.

Bond between atoms at $(\frac{1}{3}, \frac{2}{3}, 0.06)$ and $(\frac{1}{3}, \frac{2}{3}, 0.44) = 2.80$ Å.

Bond between atoms at $(\frac{1}{3}, \frac{2}{3}, 0.44)$ and $(\frac{2}{3}, \frac{1}{3}, 0.56) = 2.76$ Å.

(ii)   Each oxygen forms two donor bonds and two acceptor bonds. Since each bond involves two oxygen atoms, the total number of hydrogen bonds broken is equal to twice the number of water molecules. Hence, the energy required to break the hydrogen bonds is $(2.258)(18.01)/2 = 20.33$ kJ mol^{-1}.

(iii)   Around one oxygen atom, there are $4^2 = 16$ "possible" configurations:

     (a)   four hydrogen bonds, one configuration;

     (b)   three hydrogen bonds and one covalent bond, four configurations;

     (c)   two hydrogen bonds and two covalent bonds, six configurations;

     (d)   one hydrogen bonds and three covalent bonds, four configurations;

     (e)   four covalent bonds, one configuration.

It is obvious that only the configurations from (c) are "allowed". Hence, the total "allowed" configurations for one mode of ice is $W = 2^{2N_0} \times (6/16)^{N_0} = (3/2)^{N_0}$. The residual entropy is $S = k \ln W = N_0 k \ln(3/2) = $

3.37 J K^{-1} mol^{-1}, in agreement with the experimental value of 3.4 J K^{-1} mol^{-1}. For further details, see L. Pauling, *Nature of the Chemical Bond*, 3rd edn., Cornell University Press, Ithaca, 1960, pp. 466–8.

A10.12 (i)   Ni(CN)$_2$ · NH$_3$ · C$_6$H$_6$.

(ii) The nickel atoms are linked by cyanide groups to form an extended two-dimensional planar network. The nickel atoms are of two kinds, half being coordinated by four cyanide carbon atoms and half by four nitrogen atoms in a square planar arrangement. Each of the latter nickel atoms is additionally bonded to two *trans* ammonia molecules. The networks are superposed so that the ammonia groups point directly towards one another. The benzene rings are located in cavities arising from stacking of the layers.

(iii) Benzene may be replaced by pyridine which has practically the same size, but not by naphthalene which is too large to fit into such a cavity.

A10.13 The space group is $P2_1/a$. Since $Z = 2$, the complex ion is required to have a center of symmetry. The possible structures are:

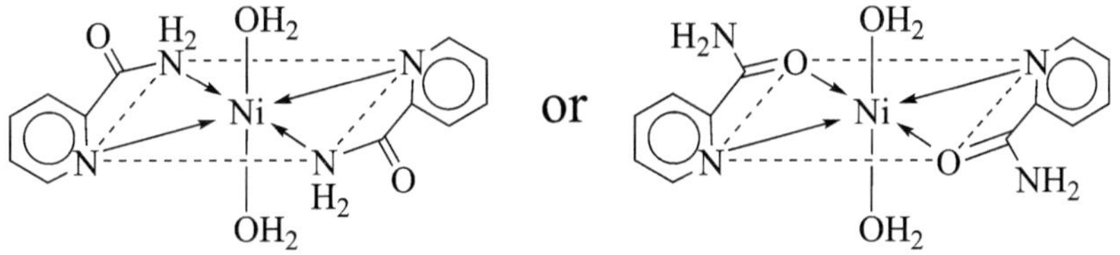

The measured magnetic moment of 3.25 BM is consistent with the presence of two unpaired electrons in a high-spin octahedral complex.

A10.14 (i)   Ni^{2+} has eight 3d electrons, and only the strong field case for a square planar structure is consistent with the observed diamagnetic behavior.

(ii) The molecule must possess a center of symmetry. The expected molecular symmetry is $\bar{1}$, since $2/m$ would require exactly planar chelate rings. The molecule is sketched on the right.

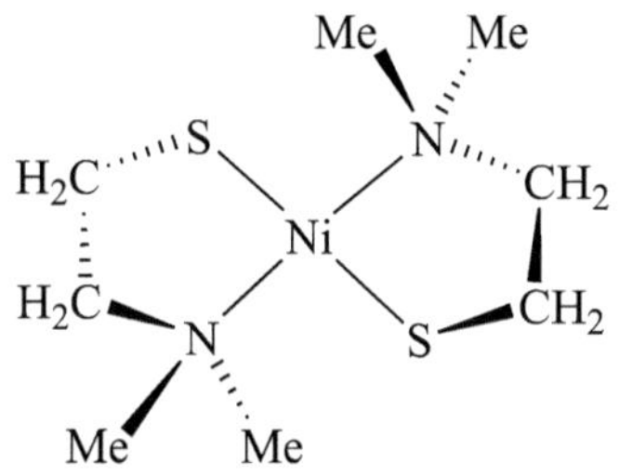

A10.15 Space group $P2_1/c$; $Z = 2$. Other than lattice translations, the symmetry elements in this space group are 2_1, c glide, and $\bar{1}$. Since screw axes and glide planes are

not permissible symmetry elements for finite molecules, it would seem that the
p-chloronitrobenzene molecule possesses a center of symmetry.

In reality, the molecules in the crystalline state are randomly oriented with complete
equivalence of the chlorine and nitro groups, which do not differ much in size. In
other words, the molecule makes no distinction between its head and its tail. Since the
measured atomic positions are averages in time and space over an enormous number of
unit cells, apparent centrosymmetry results. This is an example of statistical disorder of
the static type.

A10.16 The systematic absences indicate C-centering and a c glide. Since the crystal class is
given as $2/m$, the space group is $C2/c$. The given data indicate $Z = 4$; hence both the
NO_2^+ and ClO_4^- groups are located in special positions of four-fold multiplicity:

Position	Point symmetry	Coordinates $(0, 0, 0; \frac{1}{2}, \frac{1}{2}, 0) +$
4(e)	2	$0, y, \frac{1}{4}; 0, \bar{y}, \frac{3}{4}$
4(d)	$\bar{1}$	$\frac{1}{4}, \frac{1}{4}, \frac{1}{2}; \frac{3}{4}, \frac{1}{4}, 0$
4(c)	$\bar{1}$	$\frac{1}{4}, \frac{1}{4}, 0; \frac{3}{4}, \frac{1}{4}, \frac{1}{2}$
4(b)	$\bar{1}$	$0, \frac{1}{2}, 0; 0, \frac{1}{2}, \frac{1}{2}$
4(a)	$\bar{1}$	$0, 0, 0; 0, 0, \frac{1}{2}$

Since ClO_4^- is tetrahedral, it must occupy position 4(e). The NO_2^+ group, which is
expected to be linear, is compatible with either site symmetry $\bar{1}$ or 2. Although packing
consideration does not really eliminate any of the five special positions, it shows that if
NO_2^+ is also located in 4(e), then its O atoms must not lie on the two-fold axis, and the
molecular cation is not required to be exactly linear.

X-ray analysis showed that NO_2^+ does occupy 4(e), being slightly distorted ($\angle O–N–$
$O = 175°$) by the electrostatic field of the surrounding ions.

A10.17

	Crystalline compound	Space group	Remarks*
(i)	LiCN	$Pbnm$	$= Pnma$ (permutation cab)
(ii)	$NaNO_2$	$Im2m$	$= Imm2$ (permutation bca)
(iii)	HgO_2	$Pbca$	
(iv)	Neptunium	$Pmcn$	$= Pnma$ (permutation bca)
(v)	CuO	$C2/c$	
(vi)	Pd_4Se	$P\bar{4}2_1c$	

Steps in identifying the space group of a structure shown in projection:

(a) Look for lattice centering.

(b) *Monoclinic system*: Look for symmetry elements with respect to the unique axis.

   Orthorhombic system: Look for symmetry elements with respect to *a*, *b*, and *c*, respectively. Mirror and glide planes take precedence over rotation and screw axes.

   Tetragonal system: Look for symmetry elements with respect to *c*, *a*, and $[110]$, respectively.

* To arrive at the standard space group symbols, consult "Index of Three-dimensional Space-group Symbols for Various Settings" in *International Tables for X-ray Crystallography*, vol. I, 3rd edn., pp. 543–53.

A10.18 (i) *Cc* or *C2/c*.

(ii) *Cmcm*, *Cmc2₁*, or *C2cm* (= *Ama2* by the permutation $\bar{c}\,ba$, i.e., by relabeling the old *a*, *b*, and *c* axes as −*c*, *b*, and *a* axes, respectively).

(iii) *I4*, *I$\bar{4}$*, or *I4/m*.

(iv) *I422*, *I4mm*, *I$\bar{4}$m2*, *I$\bar{4}$2m*, or *I4/mmm*.

(v) *R32*, *R3m*, or *R$\bar{3}$m*.

(vi) *P$\bar{6}$c2*, *P6₃m*, or *P6₃/mcm*.

(vii) *F432*, *F$\bar{4}$3m*, or *Fm3m*.

A10.19 (i) *Pnnn* (No. 48, origin at $\bar{1}$).

(ii)

Equivalent position	Atomic coordinates
1	x, y, z
2	$-x, \frac{1}{2}+y, \frac{1}{2}+z$
3	$\frac{1}{2}+x, -y, \frac{1}{2}+z$
4	$\frac{1}{2}+x, \frac{1}{2}+y, -z$
5	$-x, -y, -z$
6	$x, \frac{1}{2}-y, \frac{1}{2}-z$
7	$\frac{1}{2}-x, y, \frac{1}{2}-z$
8	$\frac{1}{2}-x, \frac{1}{2}-y, z$

(iii) $0k\ell$ with $(k+\ell)$ odd;
   $h0\ell$ with $(h+\ell)$ odd;
   $hk0$ with $(h+k)$ odd.

(iv) Yes.

A10.20 (i) *Pnna* (No. 52).

(ii)

Equivalent position	Atomic coordinates
1	x, y, z
2	$\frac{1}{2} - x, -y, z$
3	$x, \frac{1}{2} - y, \frac{1}{2} - z$
4	$\frac{1}{2} - x, \frac{1}{2} + y, \frac{1}{2} - z$
5	$-x, -y, -z$
6	$\frac{1}{2} + x, y, -z$
7	$-x, \frac{1}{2} + y, \frac{1}{2} + z$
8	$\frac{1}{2} + x, \frac{1}{2} - y, \frac{1}{2} + z$

(iii) $0k\ell$ with $(k + \ell)$ odd;

$h0\ell$ with $(h + \ell)$ odd;

$hk0$ with h odd.

(iv) Yes.

A10.21 (i) *Pbca* (No. 61).

(ii)

Equivalent position	Atomic coordinates
1	x, y, z
2	$\frac{1}{2} - x, \frac{1}{2} + y, z$
3	$x, \frac{1}{2} - y, \frac{1}{2} + z$
4	$\frac{1}{2} + x, y, \frac{1}{2} - z$
5	$-x, -y, -z$
6	$\frac{1}{2} + x, \frac{1}{2} - y, -z$
7	$-x, \frac{1}{2} + y, \frac{1}{2} - z$
8	$\frac{1}{2} - x, -y, \frac{1}{2} + z$

A10.22 $x, y, z; \frac{1}{2} - x, y, z; x, \frac{1}{2} - y, z; \frac{1}{2} - x, \frac{1}{2} - y, z;$

$\bar{x}, \bar{y}, \bar{z}; \frac{1}{2} + x, \bar{y}, \bar{z}; \bar{x}, \frac{1}{2} + y, \bar{z}; \frac{1}{2} + x, \frac{1}{2} + y, \bar{z}.$

(i) $P2_1$ or $P2_1/m$.

(ii) *Ccca*.

(iii) $P4_2/n$.

A10.23 (i) $V = abc(1 - \cos^2 \alpha - \cos^2 \beta - \cos^2 \gamma + 2 \cos \alpha \cos \beta \cos \gamma)^{1/2} = 362.6 \, \text{Å}^3$.

(ii) $\rho = Z(\text{FW})/N_0 V; \ 2.284 = \dfrac{Z(159.61 + 18.02n)}{(6.022 \times 10^{23}) \times (362.6 \times 10^{-24})}.$

For $Z = 2, n = 4.98$; the exact value of n is 5.

(iii) Calculated density $= \dfrac{2 \times (159.61 + 18.02 \times 5)}{(6.022 \times 10^{23}) \times (362.6 \times 10^{-24})}$ g cm^{-3}

$\qquad\qquad\qquad\qquad = 2.287$ g cm^{-3}.

A10.24 The space group is *Ibam* (No. 72). The original reference gave *Ibma* with a different choice of axes.

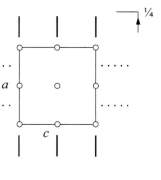

Atom	Position	Point symmetry	Coordinates $(0, 0, 0; \tfrac{1}{2}, \tfrac{1}{2}, \tfrac{1}{2}) +$
Ag	4(b)	222	$\tfrac{1}{2}, 0, \tfrac{1}{4}; \tfrac{1}{2}, 0, \tfrac{3}{4}$
N(1)	4(c)	$2/m$	$0, 0, 0; 0, 0, \tfrac{1}{2}$
N(2)	8(j)	m	$x, y, 0; \bar{x}, \bar{y}, 0;$ $\bar{x}, y, \tfrac{1}{2}; x, \bar{y}, \tfrac{1}{2}$

A10.25 (i) Fe_3C.

(ii) Space group is *Pbnm* (No. 62).

(iii) $x, y, z; \tfrac{1}{2} - x, \tfrac{1}{2} + y, z; x, y, \tfrac{1}{2} - z;$
$\tfrac{1}{2} - x, \tfrac{1}{2} + y, \tfrac{1}{2} - z;$
$\bar{x}, \bar{y}, \bar{z}; \tfrac{1}{2} + x, \tfrac{1}{2} - y, \bar{z}; \bar{x}, \bar{y}, \tfrac{1}{2} + z;$
$\tfrac{1}{2} + x, \tfrac{1}{2} - y, \tfrac{1}{2} + z.$

(iv) The carbon atom is coordinated by six neighboring iron atoms in the form of a trigonal prism. The shortest independent iron–carbon distances are:

Fe(1)–C $= 1.86$ and 1.88 Å, Fe(2)–C $= 2.06$ and 2.16 Å.

A10.26 (i) The space group is *Pnam* (= No. 62, permutation $a\bar{c}b$).

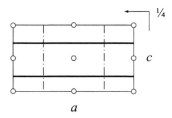

(ii)

Position	Point symmetry	Fractional coordinates
8(d)	1	

$$x, y, z; \tfrac{1}{2} - x, \tfrac{1}{2} + y, \tfrac{1}{2} + z; \tfrac{1}{2} + x, \tfrac{1}{2} - y, z; x, y, \tfrac{1}{2} - z;$$

$$\bar{x}, \bar{y}, \bar{z}; \tfrac{1}{2} + x, \tfrac{1}{2} - y, \tfrac{1}{2} - z; \tfrac{1}{2} - x, \tfrac{1}{2} + y, \bar{z};$$

$$\bar{x}, \bar{y}, \tfrac{1}{2} + z$$

Position	Point symmetry	Fractional coordinates
4(c)	m	$x, y, \tfrac{1}{4}; \bar{x}, \bar{y}, \tfrac{3}{4}; \tfrac{1}{2} - x, \tfrac{1}{2} + y, \tfrac{3}{4}; \tfrac{1}{2} + x, \tfrac{1}{2} - y, \tfrac{1}{4}$
4(b)	$\bar{1}$	$\tfrac{1}{2}, 0, 0; 0, \tfrac{1}{2}, \tfrac{1}{2}; 0, \tfrac{1}{2}, 0; \tfrac{1}{2}, 0, \tfrac{1}{2}$
4(a)	$\bar{1}$	$0, 0, 0; 0, 0, \tfrac{1}{2}; \tfrac{1}{2}, \tfrac{1}{2}, 0; \tfrac{1}{2}, \tfrac{1}{2}, \tfrac{1}{2}$

(iii)   The interatomic distance between the O atom at $(0.28, -0.16, \tfrac{1}{4})$ and Cl atom at $(0.10, 0.14, \tfrac{1}{4})$ is $[(0.18 \times 14.27)^2 + (0.30 \times 5.40)^2]^{1/2} = 3.04\,\text{Å}$.

This is significantly shorter than 3.2 Å (sum of the van der Waals radii of O and Cl) because of hydrogen bonding.

A10.27 (i) Since $Z = 2$, the XeF_2 molecule must occupy a site of symmetry $4/mmm$. The molecule is thus linear, in agreement with the prediction of VSEPR theory. The Xe atom may be placed in position 2(a), and the F atom in 4(e), with $z = 2.00/6.99 = 0.286$. [The refined value of z is 0.2837 ± 3.]

(ii) The shortest $F \cdots F$ and $Xe \cdots F$ non-bonded distances are indicated in the figure on the right.

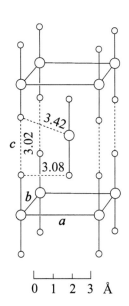

A10.28 (i) Systematic absences: $(h + k + \ell)$ odd.

(ii)

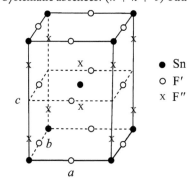

● Sn
○ F′
× F″

(iii)

Special position	Point symmetry
2(a)	$4/mmm$
4(c)	mmm
4(e)	$4mm$

(iv) Sn–F$'$ = $a/2$ = 2.02 Å; Sn–F$''$ = $0.237c$ = 1.88 Å.

The Sn atom is in a tetragonally distorted octahedral environment, with a planar array of four long Sn–F$'$ bonds and two shorter *trans* Sn–F$''$ bonds.

(v) The crystal structure consists of layers of corner-sharing SnF$_6$ octahedra stacked normal to the c-axis.

A10.29 (i)

Atom no.	Fractional coordinates		
1	0	y	z
2	0	$\bar{y}$	$\bar{z}$
3	$\frac{1}{2}$	y	$\frac{1}{2}+z$
4	0	$\frac{1}{2}-y$	$\frac{1}{2}+z$
5	$\frac{1}{2}$	$\frac{1}{2}-y$	z
6	0	$\frac{1}{2}-y$	$-\frac{1}{2}+z$
7	$\frac{1}{2}$	$\bar{y}$	$\frac{1}{2}-z$

Let r_{ij} be the distance from atom i to atom j.

I–I bond distance $= r_{12} = [(2yb)^2 + (2zc)^2]^{1/2}$
$$= [(2 \times 0.1156 \times 9.793)^2 + (2 \times 0.1493 \times 4.790)^2]^{1/2}$$
$$= 2.68 \text{ Å.}$$

Intermolecular distances:

$r_{13} = 4.35$, $r_{14} = 3.46$, $r_{15} = 4.48$, $r_{16} = r_{14}$, $r_{17} = 4.37$ Å.

Note that r_{14} is much shorter than r_{13}, r_{15}, and r_{17}.

(ii) The crystal structure of solid iodine is composed of staggered layers of discrete I$_2$ molecules normal to the a-axis, with each I$_2$ molecule lying in between hollows of neighboring layers. The large interlayer separation is indicative of van der Waals bonding and is reflected in the ease of crystal cleavage parallel to the layers. Within a layer each iodine atom has two nearest neighbors at 3.56 Å. This indicates considerable orbital overlap to form intermolecular, many-center σ bonds spread through the layer and populated with delocalized electrons. Such a bonding picture is consistent with the color, metallic luster, and electrical conductivity of solid iodine.

A10.30 (i)

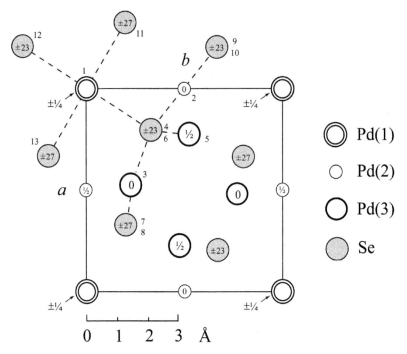

(ii)

No.	Atom	x	y	z	Derived from
1	Pd(1)	0	0	$\frac{1}{4}$	
2	Pd(2)	0	$\frac{1}{2}$	0	
3	Pd(3)	0.455	0.235	0	$x, y, 0$
4	Se	0.200	0.325	0.230	x, y, z
5	Pd(3)	0.235	0.545	$\frac{1}{2}$	$y, 1-x, \frac{1}{2}$
6	Se	0.200	0.325	−0.230	$x, y, \bar{z}$
7	Se	0.675	0.200	0.270	$1-y, x, \frac{1}{2}-z$
8	Se	0.675	0.200	−0.270	$1-y, x, -\frac{1}{2}+z$
9	Se	−0.200	0.675	0.230	$\bar{x}, 1-y, z$
10	Se	−0.200	0.675	−0.230	$\bar{x}, 1-y, \bar{z}$
11	Se	−0.325	0.200	0.270	$\bar{y}, x, \frac{1}{2}-z$
12	Se	−0.200	−0.325	0.230	$\bar{x}, \bar{y}, z$
13	Se	0.325	−0.200	0.270	$y, \bar{x}, \frac{1}{2}-z$

Central Atom		Coordinate by		Coordination	[N2, N1, N3] triplet specifying required bond lengths and angles
No.	Type	No.	Type		
1	Pd(1)	4, 11, 12, 13	Se	Distorted square planar	4–1–11
2	Pd(2)	4, 6, 9, 10	Se	Distorted square planar	4–2–6
3	Pd(3)	4, 6, 7, 8	Se	Distorted square planar	4–3–6, 7–3–8, 4–3–7
4	Se	1	Pd(1)	Distorted tetrahedral	1–4–2, 1–4–3
		2	Pd(2)		1–4–5, 2–4–3
		3, 5	Pd(3)		2–4–5, 3–4–5

N2	N1	N3	N1–N2	N1–N3	N2–N1–N3
4	1	11	2.568 Å	2.568 Å	90.2°
4	2	6	2.389	2.389	83.3
4	3	6	2.413	2.413	82.3
7	3	8	2.390	2.390	102.4
4	3	7	2.413	2.390	87.2
1	4	2	2.568	2.389	99.2
1	4	3	2.568	2.413	101.2
1	4	5	2.568	2.390	122.4
2	4	3	2.389	2.413	94.9
2	4	5	2.389	2.390	105.6
3	4	5	2.413	2.390	126.7

A10.31 The crystal structure is composed of a packing of S_6 molecules in a chair conformation, the torsion angle being 74.5°. The full molecular symmetry is $\bar{3}m$, though only $\bar{3}$ is utilized in the crystal. The bond length and bond angle are 2.057 Å and 102.2°, respectively, which may be compared with the corresponding values 2.059 Å and 107.8° for the S_8 molecule in orthorhombic sulfur. The shortest nonbonded interatomic distance is 3.501 Å.

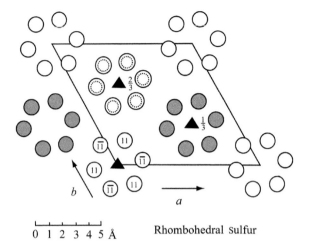

0 1 2 3 4 5 Å   Rhombohedral sulfur

A10.32 The ClF_3 molecule has a bent T-shape with dimensions: Cl–F(1) = 1.621, Cl–F(2) = 1.716 Å, and F(1)–Cl–F(2) = 87.0°.

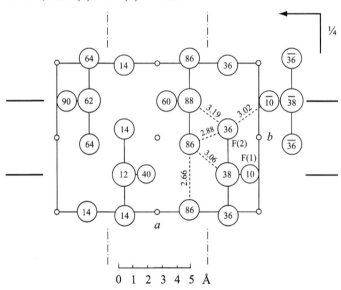

0 1 2 3 4 5 Å

A10.33 Crystalline PCl_5 actually consists of PCl_4^+ and PCl_6^- units packed together in much the same way as CsCl. In the PCl_4^+ tetrahedron, P(1)–Cl(3) = 1.97 Å, bond angles = 111° and 119°. In the PCl_6^- octahedron, P(2)–Cl(1) = 2.08, P(2)–Cl (2) = 2.08, P(2)–Cl(4) = 2.04 Å.

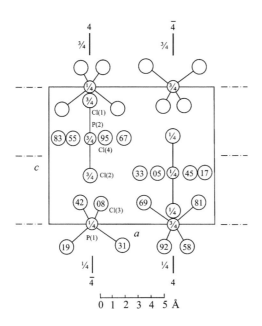

A10.34 In contrast to crystalline PCl$_5$ (see previous problem), PBr$_5$ crystallizes as a packing of PBr$_4^+$ groups and Br$^-$ ions. The molecular dimensions are:

P − Br(1) = 2.02 Å, Br(1) − P − Br(2) = 107.6°,

P − Br(2) = 2.18 Å, Br(1) − P − Br(3) = 112.9°,

P − Br(3) = 2.15 Å, Br(2) − P − Br(3) = 107.4°, Br(3) − P − Br(3′) = 108.5°.

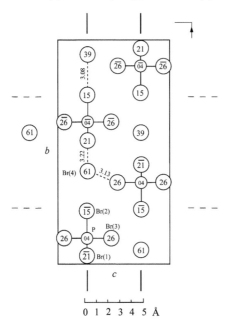

A10.35 The structure is built up of linear NO_2^+ groups arranged end-to-end parallel to c, and trigonal planar NO_3^- groups stacked in a staggered fashion normal to the c-axis. The central N atoms of the two molecular ions are arranged in the same way as the C atoms in graphite.

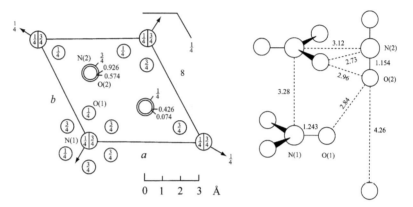

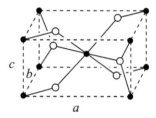

A10.36 (i) The coordination of Ti by O atoms is octahedral, and the arrangement of Ti atoms around O is trigonal planar.

(ii) The space group is $P4_2/mnm$ (No. 136).

(iii) The structure factor F may be expressed as $F = \sum_{j=1}^{p} f_j S_j$ where S, the symmetry factor, is defined as the wave scattered by a set of unit scattering particles occupying the general equivalent positions of the space group.

$$S = A + iB$$
$$= \sum_n \cos[2\pi(hx_n + ky_n + \ell z_n)] +$$
$$i\sum_n \sin[2\pi(hx_n + ky_n + \ell z_n)].$$

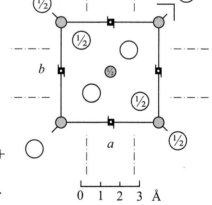

For special position 4(f) (i.e., O),

$$A = \cos[2\pi(hu + ku)] + \cos[2\pi(-hu - ku)] + \cos\left\{2\pi\left[h\left(\frac{1}{2} + u\right)\right.\right.$$
$$\left.\left. + k\left(\frac{1}{2} - u\right) + \ell\left(\frac{1}{2}\right)\right]\right\} + \cos\left\{2\pi\left[h\left(\frac{1}{2} - u\right) + k\left(\frac{1}{2} + u\right)\right.\right.$$
$$\left.\left. + \ell\left(\frac{1}{2}\right)\right]\right\}$$
$$= 2\cos[2\pi(hu + ku)] + 2\cos[2\pi(hu - ku) + \pi(h + k + \ell)].$$

$B = 0$, since the contributions of centrosymmetrically related atoms cancel one another.

For $(h + k + \ell)$ even, $A = 4\cos(2\pi hu)\cos(2\pi ku)$.

For $(h + k + \ell)$ odd, $A = -4\sin(2\pi hu)\sin(2\pi ku)$.

Special position 2(a) (i.e., Ti) may be derived from position 4(f) by setting $u = 0$. The structure factor expressions are:

For $(h + k + \ell)$ even, $F = 2f_{Ti} + 4f_O \cos(2\pi hu)\cos(2\pi ku)$.

For $(h + k + \ell)$ odd, $F = -4f_O \sin(2\pi hu)\sin(2\pi ku)$.

Systematic absences: $0k\ell$ absent for $(k + \ell)$ odd, and $h0\ell$ absent for $(h + \ell)$ odd.

A10.37

		[3]-rotane	[4]-rotane
(i)	Systematic absences:	00ℓ with ℓ odd	$h0\ell$ with $(h + \ell)$ odd; $0k0$ with k odd
(ii)	Laue symmetry:	$6/m$	$2/m$
(iii)	Required molecular symmetry:	6	$\bar{1}$ (cyclobutane ring is planar)
(iv)	Idealized molecular symmetry:	$\bar{6}m2$	$4/mmm$
(v)	No. of crystallographically distinct C–C bonds:	3	8

A10.38 (i) Space group is *Pmcn*. Conditions for systematic absences: $h0\ell$ with ℓ odd; $hk0$ with $(h + k)$ odd.

(ii) The idealized point group for $[Cu(NH_3)_4]^{2+}$ is $4/mmm$ (D_{4h}). Only one of the mirror planes is utilized in the crystal.

(iii)

Atom	Position	Parameters
Cu	4(c)	y_{Cu}, z_{Cu}
N(1)	8(d)	$x_{N(1)}, y_{N(1)}, z_{N(1)}$
N(2)	8(d)	$x_{N(2)}, y_{N(2)}, z_{N(2)}$
S	4(c)	y_S, z_S
O(1)	4(c)	$y_{O(1)}, z_{O(1)}$
O(2)	4(c)	$y_{O(2)}, z_{O(2)}$
O(3)	8(d)	$x_{O(3)}, y_{O(3)}, z_{O(3)}$
H_2O	4(c)	y_{H_2O}, z_{H_2O}

The sulfate group must be placed in position 4(c) since it can possess a plane of symmetry but not a center of symmetry. The central sulfur atom and two of the oxygen atoms, namely O(1) and O(2), therefore lie on the mirror plane. H_2O (or more precisely, the oxygen atom of the water molecule) is expected to occupy position 4(c) so that it may act as a donor in hydrogen bonding to neighboring sulfate and as a ligand for neighboring copper cations.

(iv) Number of crystallographically distinct bond lengths and angles:

$[Cu(NH_3)_4]^{2+}$ with Cu on mirror plane: two independent Cu–N bonds and three N–Cu–N angles.

SO_4^{2-} with symmetry m: three independent S–O bonds and four O–S–O angles.

A10.39 (i) Only one SiO_4 tetrahedron is shown for the sake of simplicity. The $[(SiO_3)_6]^{12-}$ group has idealized symmetry $6/mmm$, but only $6/m$ is utilized in the crystal. The oxygen atoms shared by neighboring SiO_4 tetrahedra lie on the mirror plane containing the silicon atoms. The two exocyclic oxygen atoms belonging to each SiO_4 tetrahedron are related by reflection symmetry.

(ii) Be is expected to be tetrahedrally coordinated; it is therefore assigned to position 6(f). Similarly, Al is expected to be octahedrally coordinated occupying position 4(c).

(iii)

Atom	Position	Point symmetry	Parameters
Al	4(c)	32	none
Be	6(f)	222	none
Si	12(l)	m	x_{Si}, y_{Si}
O(1)	12(l)	m	$x_{O(1)}, y_{O(1)}$
O(2)	24(m)	1	$x_{O(2)}, y_{O(2)}, z_{O(2)}$

A10.40 (i) $Z = 2$.

(ii) Space group $I432$, Laue symmetry 432; $I\bar{4}3m, \bar{4}3m$; $Im3m, m3m$: All three point groups belong to the same Laue class $m3m$.

(iii) The $(CH_2)_6N_4$ molecule has $\bar{4}3m$ symmetry, which is compatible only with $I\bar{4}3m$.

(iv) The special positions in space group $I\bar{4}3m$ are:

Position	Point symmetry	Coordinates$(0, 0, 0; \frac{1}{2}, \frac{1}{2}, \frac{1}{2})+$
24(g)	m	x, x, z; etc.
24(f)	2	$x, \frac{1}{2}, 0$; etc.
12(e)	mm	$x, 0, 0$; etc.
12(d)	$\bar{4}$	$\frac{1}{4}, \frac{1}{2}, 0$; etc.
8(c)	3m	x, x, x; etc.
6(b)	$\bar{4}2m$	$0, \frac{1}{2}, \frac{1}{2}$; etc.
2(a)	$\bar{4}3m$	$0, 0, 0$

The structure consists of a body-centered arrangement of $(CH_2)_6N_4$ molecules.

Assignment	Parameters
N in 8(c)	x_N
C in 12(e)	x_C
H in 24(g)	x_H, z_H

(v)   Carbon atoms with fractional coordinates $(x_C, 0, 0)$ and $(1 - x_C, 0, 0)$ are separated by 3.68 Å.

$$1 - 2x_C = 3.68/7.02, \text{ hence } x_C = 0.238.$$

Consider now the bond from C at $(x_C, 0, 0)$ to N at (x_N, x_N, x_N).

$$[(0.238 - x_N)^2 + x_N^2 + x_N^2)]^{1/2} = 1.48/7.02.$$

Solution of this equation yields $x_N = 0.127$ or 0.032. Consideration of the molecular geometry of $(CH_2)_6N_4$ shows that the first answer is the correct one.

A10.41   (i)   Laue symmetry $m3$; $hk\ell$ present for h, k, ℓ all even or all odd indicates F lattice; $0k\ell$ present for $(k + \ell) = 4n$ indicates the presence of a d glide plane. Hence, the space group is uniquely established as $Fd3$ (No. 203).

(ii)   The general and special equivalent positions in space group $Fd3$ (origin at 23) are:

Position	Point symmetry	Coordinates$(0, 0, 0; 0, \frac{1}{2}, \frac{1}{2}; \frac{1}{2}, 0, \frac{1}{2}; \frac{1}{2}, \frac{1}{2}, 0) +$
96(g)	1	x, y, z; etc.
48(f)	2	$x, 0, 0$; etc.
32(e)	3	x, x, x; etc.
16(d)	$\bar{3}$	$\frac{5}{8}, \frac{5}{8}, \frac{5}{8}$; etc.
16(c)	$\bar{3}$	$\frac{1}{8}, \frac{1}{8}, \frac{1}{8}$; etc.
8(b)	23	$\frac{1}{2}, \frac{1}{2}, \frac{1}{2}; \frac{3}{4}, \frac{3}{4}, \frac{3}{4}$
8(a)	23	$0, 0, 0; \frac{1}{4}, \frac{1}{4}, \frac{1}{4}$

FW of $Be_4O(CH_3COO)_6 = 406$.

$$Z = \rho N_0 V/FW = (1.39)(6.022 \times 10^{23})(15.74 \times 10^{-8})^3/406 \approx 8.$$

Since $Z = 8$, the asymmetric unit consists of one molecule located in special positions 8(a) or 8(b). The former may be taken for the sake of simplicity. The molecular symmetry utilized in the space group is therefore 23.

Assignment of atomic positions:
O(1) in 8(a);
Be in 32(e), not in 16(c)+16(d) because its coordination geometry is tetrahedral rather than octahedral, and the molecule possesses symmetry element 3 but not $\bar{3}$;

C(1)[acetate carbon] in 48(f)
C(2)[methyl carbon] in 48(f)   } both lying on two-fold axis;
O(2)[acetate oxygen] in 96(g).

The crystal and molecular structure is therefore determined by six parameters: x of Be; x of C(1); x of C(2); x, y, z of O(2).

(iii) The idealized molecular point group is $\bar{4}3m(T_d)$. The molecular symmetry in the crystal is $23(T)$. Symmetry elements not utilized are $\bar{4}$ and m.

(iv) In the $Be_4O(CH_3COO)_6$ molecule, a central O atom is surrounded tetrahedrally by four Be atoms. Each pair of Be atoms is bridged symmetrically by an acetate group such that the coordination around each Be is tetrahedral.

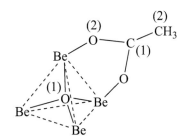

A10.42 Assignment of atomic positions:

Hg^{2+} in 1(a) : $0, 0, 0$;

NH_4^+ in 1(d): $\frac{1}{2}, \frac{1}{2}, \frac{1}{2}$ such that the cations are well separated from each other;

$Cl^-(1)$ in 1(c): $\frac{1}{2}, \frac{1}{2}, 0$ [position 1(b) would be too close (at $a/\sqrt{2} = 2.96\,\text{Å}$) to NH_4^+];

$Cl^-(2)$ in 2(g): $0, 0, z; 0, 0, \bar{z}$.

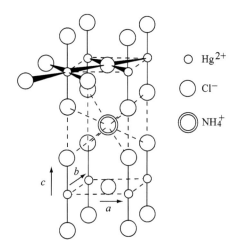

The NH_4^+ ion is thus surrounded by a cubic array of eight $Cl^-(2)$ ions. Positions 2(e) and 2(f) are ruled out because they are too close to NH_4^+ and $Cl^-(1)$, respectively.

Position 2(h) is also unacceptable since there is not enough room for a linear packing of NH_4^+ and $Cl^-(2)$ ions along the c-axis.

The z coordinate of $Cl^-(2)$ may be estimated by setting the $NH_4^+ \cdots Cl^-(2)$ distance to 3.35 Å:

$$\left(\frac{1}{2}a\right)^2 + \left(\frac{1}{2}a\right)^2 + \left[\left(\frac{1}{2} - z\right)c\right]^2 = (3.35)^2, \text{ which yields } z = 0.303.$$

With the above assignment, the coordination around Hg^{2+} is distorted octahedral (symmetry $4/mmm$), with two axial $Cl^-(2)$ ions at $0.303c = 2.41\,\text{Å}$ and four equatorial $Cl^-(1)$ ions at $a/\sqrt{2} = 2.96\,\text{Å}$. The $Cl^-(1) \cdots Cl^-(2)$ and $Cl^-(2) \cdots Cl^-(2)$ contacts are 3.82 and 4.81 Å, respectively.

A10.43 (i) Te–O⁻ is expected to be shorter than Te–OH. The empty 4d orbitals of Te may accept electrons from the formally negative terminal O atoms so that the Te–O⁻ bond actually has an appreciable amount of double bond character.

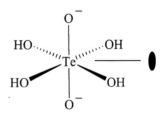

(ii) Since $Z = 2$, the $[TeO_2(OH)_4]^{2-}$ anion must occupy a special position of point symmetry $2/m$. Position 2(a) may be chosen for convenience. In accordance with assumption (b), a two-fold symmetry axis of the space group must pass through Te and bisect the HO–Te–OH bond angle. For NH_4^+, 4(e) and 4(f) (point symmetry $\bar{1}$) are clearly impossible, while 4(g) is ruled out by packing considerations. Of the remaining possibilities, both packing and hydrogen-bond interactions strongly favor 4(i) over 4(h). The assignments are therefore:

Atom	Position	Point symmetry
Te	2(a)	$2/m$
O	4(i)	m
OH	8(j)	1
N	4(i)	m

A10.44 (i) $a^* = 1/a \sin \beta = 0.04798 \, \text{Å}^{-1}$.

$b^* = 1/b = 0.14253 \, \text{Å}^{-1}$.

$c^* = 1/c \sin \beta = 0.10992 \, \text{Å}^{-1}$.

$\cos \alpha^* = (\cos \beta \cos \gamma - \cos \alpha)/\sin \beta \sin \gamma = 0$; hence $\alpha^* = 90°$.

$\cos \beta^* = -\cos \beta$; hence $\beta^* = 180° - \beta = 87.167°$.

$\gamma^* = 90°$.

$V = abc \sin \beta = 1331.8 \, \text{Å}^3; \ V^* = a^* b^* c^* \sin \beta^* = 1/V = 7.509 \times 10^{-4} \, \text{Å}^{-3}$.

(ii) No. of observable reflections $\approx (32\pi V)/3\lambda^3 \times \frac{1}{4} \approx 31077$.

(iii) $\rho = Z(\text{FW})/N_0 V$.

$Z = (1.478 \times 6.022 \times 10^{23} \times 1331.8 \times 10^{-24})/296.28 = 4.0015 \approx 4$.

(iv) The possible space groups are Cc and $C2/c$.

(v) In space group $C2/c$, DHBQ may occupy a site of symmetry 2 or $\bar{1}$, (though $\bar{1}$ is more likely); BPY is expected to have site symmetry 2, as $\bar{1}$ would imply a less stable planar conformation.

(vi) DHBQ and BPY are expected to be connected by a O–H $\cdots$ N hydrogen bond. Actually a weak C–H $\cdots$ O hydrogen bond is also formed to link the two types of molecules into an infinite zigzag ribbon.

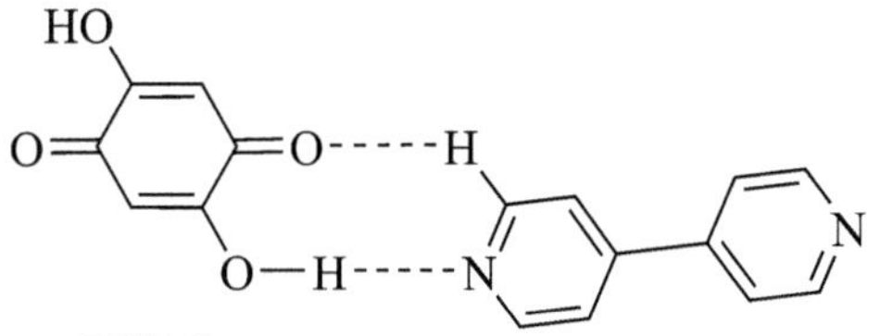

DHBQ   BPY

A10.45 Compound I: $BaY_2Cu_3O_6$, $Z = 1$.
Compound II: $BaY_2Cu_3O_7$, $Z = 1$.

A10.46 (i) $\rho = Z(FW)/N_0V$.

$V = abc \sin \beta = 15.805 \times 12.889 \times 11.450 \times \sin 132.20° = 1727.8\,\text{Å}^3$.

$Z = (1.30 \times 6.022 \times 10^{23} \times 1727.8 \times 10^{-24})/335.36 = 4.03 \approx 4$.

(ii) The possible space groups are Cc and $C2/c$.

(iii) If the centrosymmetric space group $C2/c$ is assumed, the asymmetric unit consists of one half of the structural formula.

(iv) Both cation and anion must occupy Wyckoff (special) positions of symmetry 2 (i.e., special position of the type $0, y, \frac{1}{4}$).

(v) For the cation, a two-fold axis passes through the central C atom and two *para*-related C atoms in the central phenyl ring. The number of positional parameters (omitting the H atoms) required to define the molecular and crystal structure is 35:

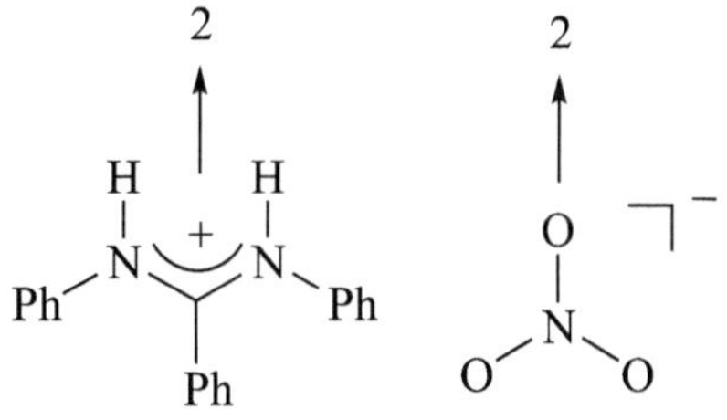

Cation: $(9 \times 3) + (3 \times 1) = 30$.
Anion: $(1 \times 3) + (2 \times 1) = 5$.

(vi) Hydrogen bonds of the type $N\text{–}H \cdots O$ can be formed to generate an ion pair, as shown below:

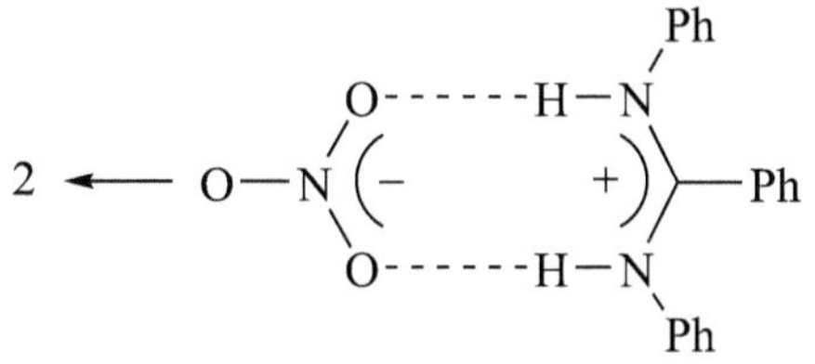

A10.47   (i) $C(1)–N(1) = (0.6258 − 0.4239) \times 5.740 = 1.159\,Å.$

$C(1)–Cl(1) = (0.4239 − 0.1499) \times 5.740 = 1.573\,Å.$

(ii)

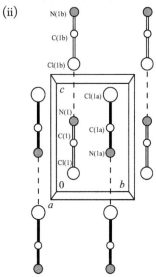

See above figure (molecule represented by open bond at $x = \frac{1}{4}$; molecule represented by filled bonds at $x = \frac{3}{4}$).

	x	y	z
Cl(1b)	$\frac{1}{4}$	$\frac{1}{4}$	1.1499
C(1a)	$\frac{3}{4}$	$\frac{3}{4}$	0.5761

(iii)   Separation between molecules in the same chain is

$$N(1)\cdots Cl(1b) = (1.1499 − 0.6258) \times 5.740 = 3.008\,Å.$$

(iv)   Shortest nonbonded distance between atoms belonging to chains running in opposite directions:

$$C(1)\cdots C(1a) = \sqrt{\left(\frac{1}{2}a\right)^2 + \left(\frac{1}{2}b\right)^2 + [(0.5761 − 0.4239)\,c]^2} = 3.577\,Å.$$

$$N(1)\cdots C(1a) = \sqrt{\left(\frac{1}{2}a\right)^2 + \left(\frac{1}{2}b\right)^2 + [(0.6258 − 0.5761)\,c]^2} = 3.480\,Å.$$

Transition Metal Chemistry | 11

PROBLEMS

11.1 (i) Comment on the following:

  (a) In a transition metal carbonyl complex, the coordinating atom linking the metal and CO ligand is always C despite the fact that O is more electronegative and there are lone pairs on both C and O in CO.

  (b) While CO and N_2 are isoelectronic, dinitrogen complexes are of poorer stability in general.

  (c) In metal carbonyls, the central metal is usually in a low oxidation state.

  (d) Titanium compounds containing CO and NO as ligands are poorly established.

  (ii) Bearing in mind that carbon monoxide forms numerous complexes with transition metals of low oxidation states, do you think that carbonyls of the lanthanides are likely to be stable? Give reasons.

11.2 In a carbonyl complex containing the (linear) OC–M–CO group, how would the CO stretching frequency change when

  (i) one CO is replaced by triethylamine;

  (ii) one CO is replaced by NO;

  (iii) the metal center carries an additional positive charge;

  (iv) the metal center carries an additional negative charge?

11.3 The perchlorate ion shows very little tendency to form complexes with metal ions. Comment.

11.4 $Pd(PF_3)_2Cl_2$ is much more stable than $Pd(NH_3)_2Cl_2$, whereas BF_3NH_3 is much more stable than BF_3PF_3. Comment.

Problems in Structural Inorganic Chemistry. Second edition. Wai-Kee Li, Hung Kay Lee, Dennis Kee Pui Ng, Yu-San Cheung, Kendrew Kin Wah Mak, and Thomas Chung Wai Mak.
© Oxford University Press 2019. Published in 2019 by Oxford University Press.
DOI: 10.1093/oso/9780198823902.001.0001

11.5 Although $Ni(CO)_4$ exists, there is no corresponding palladium or platinum carbonyl complex. On the other hand, carbonyl halides of Pd^{2+} and Pt^{2+} commonly occur but there is no known carbonyl halide of Ni^{2+}. Comment.

11.6 For the Group 11 coinage metals in the Periodic Table, the Cu^{2+} ion is the most stable oxidation state of copper, forming predominantly tetragonal complexes. However, gold ion is known almost exclusively as Au^+ and Au^{3+}, existing predominantly in linear and square planar complexes, respectively. Comment.

11.7 Explain the following facts:

(i) The fluorides of the lanthanides are insoluble in water;

(ii) CuS is insoluble in water.

11.8 Explain why hydrated Pd^{2+} and Pt^{2+} ions are very little known as compared with Ni^{2+}, yet the halide complexes of the former are more common than those of the latter.

11.9 Complexes with chelate ligands are usually more stable than those bearing monodentate ligands, provided that electronic effects of the ligands are similar. Comment.

11.10 Stepwise stability constants, K_n, expressed as common logarithms for the system $Hg^{2+}-NH_3$ at 25 °C, are: $\log K_1 = 8.8$, $\log K_2 = 8.7$, $\log K_3 = 1$, and $\log K_4 = 0.9$.

(i) Why do successive values of K_n decrease?

(ii) Why is the ratio of K_2/K_3 so large?

11.11 Given the following stepwise equilibrium constants (at 25 °C) in aqueous solution,

	$\log K_1$	$\log K_2$	$\log K_3$	$\log K_4$	$\log K_5$
$Cu^{2+} + 5 NH_3$	4.15	3.50	2.89	2.13	−0.52
$Cu^{2+} + 3$ en	10.72	9.31	−1.0		
$Cu^{2+} + 2$ dien	16.0	5.0			
$Cu^{2+} + 2$ ptn	11.1	9.0			

where en, dien, and ptn represent $H_2NCH_2CH_2NH$, $H_2NCH_2CH_2NHCH_2CH_2NH_2$, and $H_2NCH_2CH(NH_2)CH_2NH_2$, respectively.

(i) Among the complexes $[Cu(NH_3)_n(H_2O)_{6-n}]^{2+}$, which one is the most stable?

(ii) Among all the complexes given above, which one is the most stable?

(iii) Why is $[Cu(en)_3]^{2+}$ particularly unstable in comparison with $[Cu(H_2O)_4$ $(en)_2]^{2+}$ and $[Cu(H_2O)_2(en)_2]^{2+}$?

(iv) Which of the two geometrical isomers of $[Cu(H_2O)_2(en)_2]^{2+}$ would dominate?

11.12 (i) $NC-CH_2-NC$ is an ambidentate ligand. Draw a Lewis structure for this compound. Would you expect the cyano (–CN) and isocyano (–NC) groups of this ligand to have linear or bent M–C–N and M–N–C bonds?

(ii) Treatment of pentacarbonyl(η^2-*cis*-cyclooctene)chromium(0) with an excess of isocyanoacetonitrile gives the monosubstituted complex $[Cr(CO)_5(NCCH_2NC)]$, in which the chromium center is coordinated by the isocyano group exclusively. The molecular structure of this complex is given below along with some important bond lengths. On the basis of these data, compare the σ-donor and π-acceptor character of carbon monoxide and isocyanide.

$$Cr-(CO)_{cis} = 1.899 - 1.911 \text{ Å}$$

(iii) Would you expect $[Cr(CO)_5(NCCH_2NC)]$ to have a higher or lower CO stretching frequency in the infrared spectrum than that of $[Cr(CO)_6]$?

REFERENCE: J. Buschmann, D. Lentz, P. Luger, G. Perpetuo, D. Scharn, and S. Willemsen, Synthesis, structural investigation, and ligand properties of isocyanoacetonitrile. *Angew. Chem. Int. Ed. Engl.* **34**, 914–5 (1995).

11.13 (i) Describe briefly the bonding of metal carbonyls.

(ii) The homoleptic carbonyl dication $[Fe(CO)_6]^{2+}$, the first of its type formed by a 3d metal, was prepared by Bley *et al.* Compare its CO stretching frequency with those of its analogues $[V(CO)_6]^-$, $[Cr(CO)_6]$, and $[Mn(CO)_6]^+$.

(iii) For the isoelectronic series $[V(CO)_6]^-$, $[Cr(CO)_6]$, and $[Mn(CO)_6]^+$, would you expect the energy of the metal to ligand charge-transfer band to increase or decrease with increasing charge on the complex? Why?

REFERENCES: B. Bley, H. Willner, and F. Aubke, Synthesis and spectroscopic characterization of hexakis(carbonyl)iron(II) undecafluorodiantimonate(V), [Fe(CO)₆][Sb₂F₁₁]₂. *Inorg. Chem.* **36**, 158–60 (1997). K. Pierloot, J. Verhulst, P. Verbeke, and L. G. Vanquickenborne, Electronic spectra of the d⁶ binary carbonyl complexes Mn(CO)₆ ⁺, Cr(CO)₆, and V(CO)₆ ⁻: an *ab initio* analysis. *Inorg. Chem.* **28**, 3059–63 (1989).

11.14 Compound **A** was synthesized as a ligand for the $[M(CO)_3]^+$ core ($M = {}^{99m}Tc$ or Re). It reacted with $[Re(H_2O)_3(CO)_3]Br$ in refluxing methanol to give a monomeric octahedral complex **B**. Given that **A** is a tridentate ligand which can bind to the $[Re(CO)_3]^+$ core in a *N,N,N-* or *N,N,O-fac* manner, draw a possible structure of **B** for each of these binding modes.

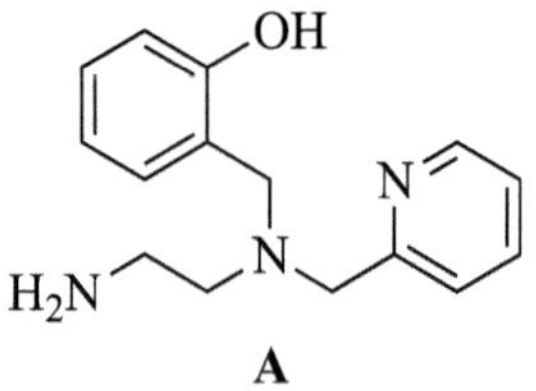

REFERENCE: N. C. Lim, C. B. Ewart, M. L. Bowen, C. L. Ferreira, C. A. Barta, M. J. Adam, and C. Orvig, Pyridine–*tert*-nitrogen–phenol ligands: N,N,O-type tripodal chelates for the [M(CO)₃]⁺ core (M = Re, Tc). *Inorg. Chem.* **47**, 1337–45 (2008).

11.15 The platinum(0) complex $[(Cy_3P)_2Pt(CO)]$ (Cy = cyclohexyl) has a trigonal planar structure, while the analogue $[(Cy_3P)_2Pt(CO)_2]$ adopts a tetrahedral structure. Describe the bonding nature of the two ligands PCy_3 and CO, and predict with justification which complex has a lower CO stretching frequency.

REFERENCE: S. Bertsch, H. Braunschweig, M. Forster, K. Gruss, and K. Radacki, Carbonyl complexes of platinum(0): synthesis and structure of [(Cy₃P)₂Pt(CO)] and [(Cy₃P)₂Pt(CO)₂]. *Inorg. Chem.* **50**, 1816–9 (2011).

11.16 The bipyridine derivative Bipy* (shown below) reacts with a series of late first-row transition metal(II) chlorides to give the corresponding complexes $[Bipy^*MCl_2]$ as below:

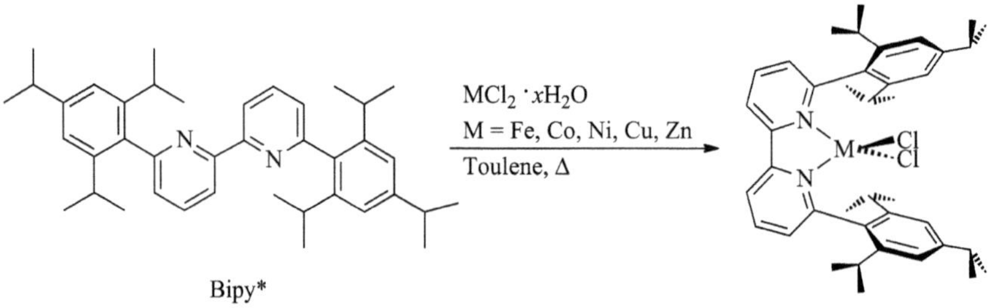

Bipy*

(i) The electronic absorption spectra of the iron and copper complexes in dichloromethane show a ligand-based $\pi \rightarrow \pi^*$ transition near 310 nm ($\varepsilon = 11{,}000\text{-}17{,}000$ L mol^{-1} cm^{-1}). In addition, the former shows another band at 505 nm ($\varepsilon = 260$ L mol^{-1} cm^{-1}), while the latter shows another two bands at 450 nm ($\varepsilon = 860$ L mol^{-1} cm^{-1}) and 890 nm ($\varepsilon = 100$ L mol^{-1} cm^{-1}). Assign all these electronic transitions assuming that both complexes have a perfect tetrahedral structure.

(ii) Compared with the other four complexes, the copper analogue crystallizes in a different manner and is severely distorted. Explain.

(iii) For the nickel complex, the molar magnetic susceptibility (χ_M) measured at 300 K was found to be 3.4×10^{-3} $(BM)^2 K^{-1}$. Determine the magnetic moment and the number of unpaired electron of this complex.

REFERENCE: E. E. Benson, A. L. Rheingold, and C. P. Kubiak, Synthesis and characterization of 6,6'-(2,4,6-triisopropylphenyl)-2,2'-bipyridine (tripbipy) and its complexes of the late first row transition metals. *Inorg. Chem.* **49**, 1458–64 (2010).

11.17 (i) Suppose a compound $C_5H_5AgNH_3$ were isolated. How would the cyclopentadienyl group be bound to the metal?

(ii) The complex $[Cu(en)_2]^{2+}$, where en denotes ethylenediamine, is more stable than $[Cu(NH_3)_4]^{2+}$ but $[Ag(en)]^+$ is less stable than $[Ag(NH_3)_2]^+$. Comment.

11.18 The periodate (IO_6^{5-}) ion forms a stable complex with either Ag^{3+} or Cu^{3+} ions. Sketch the most likely structure of the complex. Would magnetic susceptibility measurements clarify the structure?

11.19 Pyridine-2-carboxamide, abbreviated to piaH, can act as a bidentate ligand in two ways:

Type A   Type B

(i) If coordination occurs in the Type B fashion, sketch the possible isomers of the $[Ni(H_2O)_2(piaH)_2]^{2+}$ ion and indicate their optical activity.

(ii) What can be said about the number and type of possible isomers in the case of N,N-coordination (Type A)?

(iii) The experimental magnetic moment for $[Ni(H_2O)_2(piaH)_2]Cl_2$ has a value of 3.25 BM. Comment on the agreement between theory and experiment.

11.20 Ultraviolet irradiation of a solution of $F_2C=CCl_2$ in light petroleum with $Fe(CO)_5$ gives a complex $(F_2CCCl_2)Fe(CO)_4$. Sketch plausible structures for the complex. For each proposed structure,

(i) state the (idealized) point group to which it belongs;

(ii) state the probable hybridization of the Fe atom;

(iii)   discuss briefly the nature of the bonding between the halogenated olefin and the Fe atom.

REFERENCE: R. Fields, M. M. Germain, R. N. Haszeldine, and P. W. Wiggins, Metal carbonyl chemistry. Part X. Mono(fluoro-olefin) complexes from pentacarbonyliron. *J. Chem. Soc. A.* 1969–74 (1970).

11.21   Vaska discovered the following reaction:

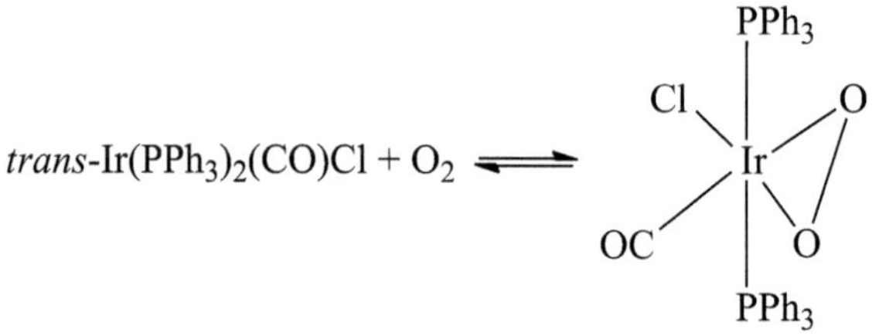

$$\textit{trans-}Ir(PPh_3)_2(CO)Cl + O_2 \rightleftharpoons$$

The reaction is reversible and the product is a dioxygen complex. X-ray diffraction analysis shows that iridium and the dioxygen ligand form an isosceles triangle with the O–O bond distance varying with the other ligands present. Explain why

(i)   with chlorine as co-ligand, the O–O bond distance is 1.30 Å, whereas it is changed to 1.51 Å when chlorine is replaced by iodine;

(ii)   the reaction is reversible as shown above but irreversible when chlorine is replaced by iodine.

REFERENCES: L. Vaska, Oxygen-carrying properties of a simple synthetic system. *Science* **140**, 809–10 (1963); R. W. Horn, E. Weissberger, and J. P. Collman, Oxygen-18 study of the reaction between iridium- and platinum-oxygen complexes and sulfur dioxide to form coordinated sulfate. *Inorg. Chem.* **9**, 2367–71 (1970).

11.22   The preparation of the $[(H_3N)_5M(ONO)]^{2+}$ isomer $(M = Co$ or Rh) from $[(H_3N)_5M(OH_2)]^{3+}$ and HNO_2 is successful even though $[(H_3N)_5M(NO_2)]^{2+}$ is the more stable linkage isomer. Explain.

REFERENCES: R. G. Pearson, P. M. Henry, J. G. Bergmann, and F. Basolo, Mechanism of substitution reactions of complex ions. VI. Formation of nitrito- and nitrocobalt(III) complexes. *J. Am. Chem. Soc.* **76**, 5920–3 (1954); R. K. Murman and H. Taube, *ibid.* **78**, 4886–90 (1956).

11.23   Assuming that the *trans* effect is the only guiding principle and that there is a hypothetical *trans* influence ordering: A < B < C < D, devise procedures for synthesizing each of the three isomers of the square planar complex M(ABCD). Starting chemicals available are MX_4, where X = A, B, C, or D.

11.24 Field *et al.* reported a series of low-valent iron and ruthenium dinitrogen complexes with a bulky tetradentate ligand, namely $P(CH_2CH_2PCy_2)_3$ (Cy = cyclohexyl) labeled as L*. The compounds $M(N_2)L^*$ (M = Fe, Ru) were prepared by treating $[MClL^*]^+$ (M = Fe, Ru) with potassium graphite under an atmosphere of nitrogen. The $^{31}P\{^1H\}$ NMR spectra of these compounds revealed two distinct phosphorus environments.

  (i)   Describe how this reaction proceeded.

  (ii)   Predict and draw out the structure of $M(N_2)L^*$ (M = Fe, Ru).

  (iii) Draw a Lewis structure of a nitrogen molecule. Using this information, predict the bond angle of M–N–N.

  (iv)   In the synthesis of $Fe(N_2)L^*$, the iron dihydride $Fe(H)_2L^*$ was formed as a side product, which has a distorted octahedral geometry. Would you expect the two hydrides to be in a mutually *cis* or *trans* arrangement? Why?

  (v)   Treatment of $M(N_2)L^*$ (M = Fe, Ru) with one equiv. of a weak organic acid resulted in protonation of the metal center giving $[M(N_2)HL^*]^+$ (M = Fe, Ru). Compare their $\nu(N{\equiv}N)$ stretching frequency with that of their neutral counterparts.

REFERENCE: R. Gilbert-Wilson, L. D. Field, S. B. Colbran, and M. M. Bhadbhade, Low oxidation state iron(0), iron(I), and ruthenium(0) dinitrogen complexes with a very bulky neutral phosphine ligand. *Inorg. Chem.* **52**, 3043–53 (2013).

11.25 (i)   Draw a Lewis structure of diisocyanomethane, $H_2C(NC)_2$, and predict whether the isocyano groups form linear or bent M–C–N bonds with metals.

  (ii)   When $[Mn(\eta^5\text{-}C_5H_5)(CO)_3]$ was dissolved in tetrahydrofuran and photolyzed, CO was evolved and compound **A** was formed. Treatment of **A** with diisocyanomethane in CH_2Cl_2 at $-40\,^\circ C$ led to the formation of compound **B**, which has the following spectral data.

  -    ^1H NMR (CD_2Cl_2): δ 4.71 (5 H), 5.01 (2 H)

  -    $^{13}C\{^1H\}$ NMR (CD_2Cl_2): δ 50.1, 83.4, 162.0, 210.5, 228.1

  -    IR: 2147, 2086, 2010, 1903 cm^{-1}

  Predict the structures of **A** and **B**. Also explain how they are formed in the above reactions.

REFERENCE: J. Buschmann, T. Bartolmäs, D. Lentz, P. Luger, I. Neubert, and M. Röttger, Synthesis, structure, and coordination chemistry of diisocyanomethane. *Angew. Chem. Int. Ed. Engl.* **36**, 2372–4 (1997).

11.26 (i) The reaction of $[Zr(\eta^7\text{-}C_7H_7)Cl(TMEDA)]$ (TMEDA = tetramethylethylene-1,2-diamine) with $Li(C_5H_4SiMe_3)$ gives compound **A** as a purple solid. Based on its spectral data given below, suggest the structure of **A**.

- ^1H NMR (C_6D_6): δ 0.08 (s, 9 H), 5.19 (s, 7 H), 5.40 (m, 2 H), 5.48 (m, 2 H)
- ^{13}C NMR (C_6D_6): δ 0.01 (s), 80.3 (d), 104.3 (d), 106.8 (d), 114.2 (s)

(ii) After recrystallization of **A** from a mixture of tetrahydrofuran and hexane at $-60\,^\circ C$, compound **B** is formed. Predict the structure of **B** and determine the total number of valence electrons in this complex.

REFERENCE: A. Glöckner, M. Tamm, A. M. Arif, and R. D. Ernst, A new versatile approach to substituted cyclopentadienyl-cycloheptatrienyl complexes of zirconium (trozircenes). *Organometallics* **28**, 7041–6 (2009).

11.27 Propose structural formulas for compounds **A** to **D** with reference to the following spectral data:

$$[Fe_2(\eta^5\text{-}C_5H_5)_2(CO)_4] \xrightarrow{\text{Na/Hg}} A \xrightarrow{\text{Br}_2} B \xrightarrow{\text{LiAlH}_4} C$$

$$\xrightarrow{C_6H_5Na} A + \text{Hydrocarbon } D$$

- $\nu_{CO} = 1961, 1942, 1790$ cm^{-1} for $[Fe_2(\eta^5\text{-}C_5H_5)_2(CO)_4]$.
- **A** has strong IR bands at 1880 and 1830 cm^{-1}.
- The ^1H NMR spectrum of **C** shows two singlets at δ -12 and 5 ppm with a relative intensity of 1:5.

11.28 Treatment of $[Cu(cyclooctyne)_2Br]$ with $Ag[SbF_6]$ and CO in dichloromethane gives a colorless complex **A**. Its infrared spectrum displays a strong band at 2171 cm^{-1}, which is due to the CO stretch, and a weak band at 2070 cm^{-1}, which is attributed to the C$\equiv$C stretch of the alkyne moieties.

(i) Propose the structure of **A**.
(ii) Assign the oxidation number and count the valence electrons of the copper center in **A**.

(iii)   Given that the CO stretch of free CO is 2143 cm^{-1} and the C≡C stretch of free cyclooctyne is 2216 cm^{-1}, comment on the interactions of these ligands with the copper center.

(iv)   When **A** is further reacted with a substituted aniline (e.g. 4-^{t}BuC$_6$H$_4$NH$_2$), what would you expect to be the product? Why?

REFERENCE: A. Das, C. Dash, M. Yousufuddin, and H. V. R. Dias, Coordination and ligand substitution chemistry of bis(cyclooctyne)copper(I). *Organometallics* **33**, 1644–50 (2014).

11.29   Paz-Sandoval *et al.* studied the substitution reaction of LMn(CO)$_3$ ($L =$ 2,4-dimethyl-η^5-pentadienyl) with PMe$_3$ as shown below:

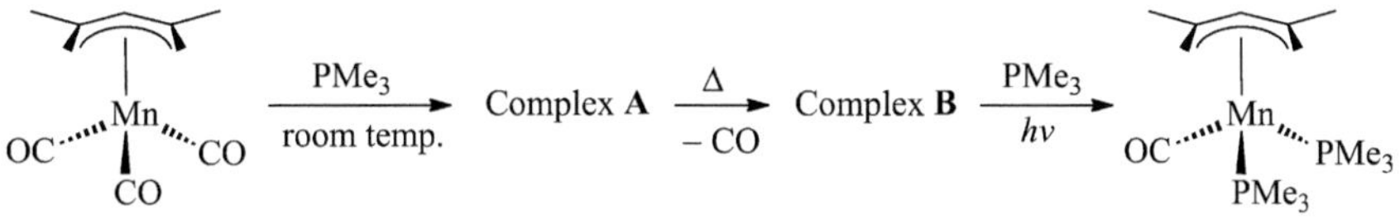

(i)   Given the following ^{13}C{^{1}H} NMR data, propose the structures of complexes **A** and **B**. Do they follow the 18-electron rule? Name the reaction in the first step.
  A: δ 20.9 (PMe$_3$), 24.9, 29.5, 47.3, 66.0, 100.8, 103.1, 148.1, 224.0 (CO).
  B: δ 22.7 (PMe$_3$), 28.7, 56.8, 89.6, 112.2, 224.0 (CO).

(ii)   Would you expect the CO stretching frequency of LMn(CO)(PMe$_3$)$_2$ to be higher or lower than those of LMn(CO)$_3$? Why?

REFERENCE: J. I. de la Cruz Cruz, P. Juárez-Saavedra, B. Paz-Michel, M. A. Leyva-Ramirez, A. Rajapakshe, A. K. Vannucci, D. L. Lichtenberger, and M. A. Paz-Sandoval, Phosphine-substituted (η^5-pentadienyl) manganese carbonyl complexes: geometric structures, electronic structures, and energetic properties of the associative substitution mechanism, including isolation of the slipped η^3-pentadienyl associative intermediate. *Organometallics* **33**, 278–88 (2014).

11.30   2-Methylisothiazol-3(2H)-one (MIO) has a planar five-membered ring and two donor sites, namely a hard oxygen atom and a soft sulfur atom. It reacts with [Ru(NH$_3$)$_5$(OH$_2$)]$^{2+}$ and [Ru(NH$_3$)$_5$(CF$_3$SO$_3$)]$^{2+}$ to give [Ru(NH$_3$)$_5$(MIO-S)]$^{2+}$ (complex **A**) and [Ru(NH$_3$)$_5$(MIO-O)]$^{3+}$ (complex **B**), respectively.

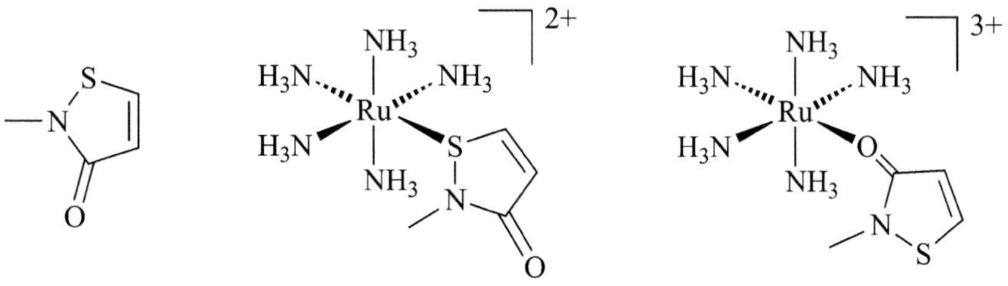

MIO   A   B

(i) MIO can bind to metals with two different atoms. What is the name of this type of ligand?

(ii) Provide an explanation for the different binding modes in **A** and **B**

(iii) In **A**, the Ru–S bond and the Ru–N bond *trans* to MIO are 2.242(2) Å and 2.151(7) Å long, respectively. The corresponding distances for $[Ru(NH_3)_5 (DMSO\text{-}S)]^{2+}$ (DMSO = dimethylsulfoxide) are 2.188(3) Å and 2.209(8) Å. Compare the *trans* influence of MIO-S and DMSO-S, and explain the difference.

REFERENCE: M. Kato, K. Unoura, T. Takayanagi, Y. Ikeda, T. Fujihara, and A. Nagasawa, Preferential behavior on donating atoms of an ambidentate ligand 2-methylisothiazol-3(2*H*)-one in its metal complexes. *Inorg. Chem.* **52**, 13375–83 (2013).

11.31 Treatment of $Fe[N(SiMe_3)_2]_2$ (**A**) with the strong reducing agent KC_8 in the presence of 18-crown-6 in diethyl ether gives complex **B**, which has a linear N–Fe–N arrangement. This compound can also be prepared from the trigonal complex **C** prepared by treating **A** with tricyclohexylphosphine (PCy_3) according to the following scheme:

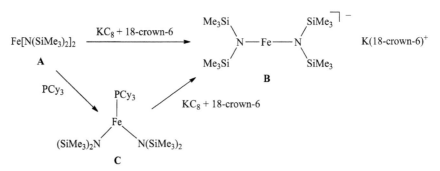

(i) What is the oxidation state of the iron center in **B** and **C**?

(ii) Classify the reactions from **A** to **B** and **C**.

(iii) What are the factors favoring complexes with a low coordination number? Does **B** fulfill these criteria?

(iv) Predict the spin-only magnetic moment for high-spin **B**.

REFERENCE: C. G. Werncke, P. C. Bunting, C. Duhayon, J. R. Long, S. Bontemps, and S. Sabo-Etienne, Two-coordinate iron(I) complex $Fe\{N(SiMe_3)_2\}_2^-$: synthesis, properties, and redox activity. *Angew. Chem. Int. Ed.* **54**, 245–8 (2015).

11.32 Propose a mechanism for each of the following reactions:

(i)

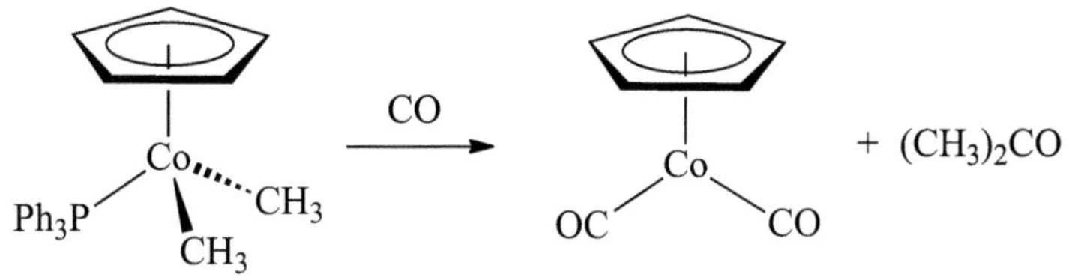

+ (CH₃)₂CO

(ii)

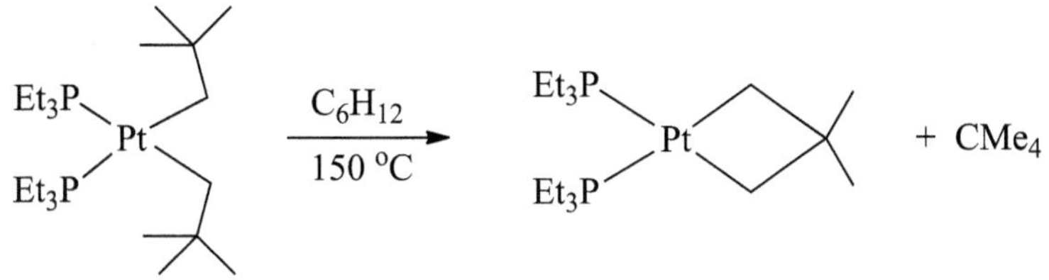

+ CMe₄

(iii)

→ EtCHO + H₂C=CH₂

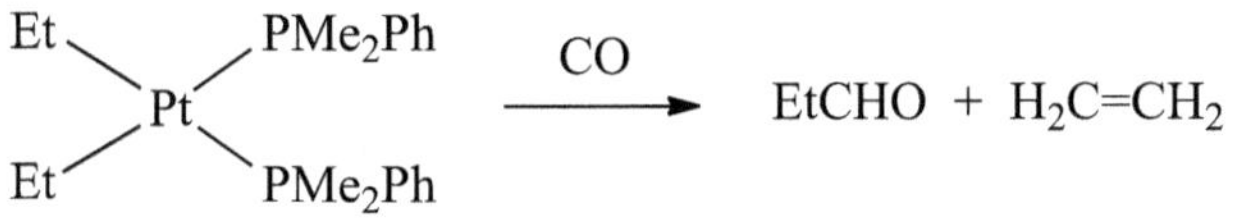

SOLUTIONS

A11.1 (i) (a) Metal carbonyls owe their stability, among other things, to back bonding between metal d_π orbitals and π^* orbitals of CO. Since the dominant part of the π^* orbital of CO resides on C, C is always the coordinating atom in metal carbonyls.

(b) The reasoning given in (a) also applies here. Furthermore, the σ donor orbital in N_2 is bonding in nature, while that of CO is slightly antibonding. Thus N_2 is a poorer σ donor.

(c) When a transition metal is highly charged, the d orbitals are contracted to a large extent and, hence, not efficient in back bonding.

(d) Back bonding, which is an important factor in the stability of metal carbonyls and nitrosyls, is ineffective due to the small number of d electrons in titanium.

(ii) Carbon monoxide is a ligand of relatively poor σ donor character. The stability of the transition metal carbonyls comes from metal d_{π^*} back donation to the vacant π^* orbital of ligand. Hence low oxidation state of the transition metals is a necessity. In the lanthanide series, the 4f orbitals are so well shielded by the outer orbitals that effective back donation is not likely. Hence, the chemistry of these elements is predominantly ionic.

A11.2 The CO stretching frequency will (i) decrease; (ii) increase; (iii) increase; (iv) decrease.

A11.3 The inertness of the perchlorate ion can be attributed to a high degree of double bond character in the Cl–O bonds, which drain the negative charge from the surface of the ion and localize it on the chlorine atom.

A11.4 Phosphorus has vacant d_π orbitals which can accept back donation from the d_π electrons of the Pd (d^8) atom. However, in the boron compound, B has only one vacant orbital for σ bonding and the basicity of NH_3 is much greater than that of PF_3. In addition, the electronegativity factor plays an important role. Owing to the high electronegativity of fluorine, PF_3 cannot donate its electron pair readily.

A11.5 In the zero oxidation state, the 4d (in Pd) and 5d (in Pt) orbitals are too diffuse for effective π-type overlap with the CO ligands, which is responsible for the non-existence of $Pd(CO)_4$ and $Pt(CO)_4$. On the other hand, in the $+2$ oxidation state, the 4d and 5d orbitals are now contracted to such an extent that effective π bonding can occur, whereas the 3d orbitals of Ni^{2+} have become too compact for efficient π overlap.   ·

A11.6 Group 11 elements of $+2$ oxidation state have a d^9 electronic configuration, which is subject to strong Jahn-Teller effect, and the single electron would occupy the $d_{x^2-y^2}$ orbital. For Cu^{2+}, because of its low oxidation state and position in the first transition series, its crystal field splitting is rather small. Orbital $d_{x^2-y^2}$ is only slightly unstable, and it is not easy to remove one more electron from Cu^{2+} due to the high third ionization potential. Therefore the $+2$ oxidation state of copper is stable and common. On the other hand, since gold is a third transition series element, it experiences a splitting of the d orbitals some 80% greater than copper does. Therefore the ninth electron entering the $d_{x^2-y^2}$ orbital will correspondingly be raised higher in energy. It is therefore much easier to ionize this odd electron than in Cu^{2+} so that Au^{2+} disproportionates into Au^+ (d^{10}) and Au^{3+} (d^8) species. Thus, Cu^{2+} exists predominantly in tetragonal complexes because of the Jahn-Teller effect, whereas Au^+ complexes have a linear structure with coordination number two, and Au^{3+} complexes are square planar.

A11.7 (i) The lanthanide cations are triply charged and the fluoride ion, though singly charged, is quite small, so that the entropy changes of solvation for both the cations and the anion, and hence the entropy changes of dissolution of their salts, are very negative. Therefore, though the negative enthalpy changes of dissolution may imply at least sparing solubility by themselves, yet the very negative entropy changes of dissolution result in virtual insolubility.

(ii) Cu^{2+} and S^{2-}, being divalent ions, have negative entropy changes of solvation. The anion-cation polarization between the highly polarizable S^{2-} and polarizing Cu^{2+} will also not favor dissolution of CuS. Similar considerations account for the insolubility of the sulfides of other transition metals.

A11.8 Pd^{2+} and Pt^{2+} ions are typical soft acids and their hydrated ions, i.e., $[M(H_2O)_n]^{2+}$, are thus quite unstable due to the hard base character of the ligand. On the other hand, the halides, Cl^-, Br^-, and I^-, are quite soft in comparison with the oxygen (H_2O) donor, and form quite stable complexes with Pd^{2+} and Pt^{2+}.

A11.9   This extra stability is termed the chelate effect. Consider the following equilibrium:

$$[M(NH_3)_6]^{3+}(aq) + 3\,en(aq) \rightleftharpoons [M(en)_3]^{3+}(aq) + 6\,NH_3(aq)$$

where en $= NH_2CH_2CH_2NH_2$.
In the well known thermodynamic relationship

$$\Delta G = \Delta H - T\Delta S,$$

a more negative ΔG can result from making ΔH more negative or ΔS more positive. The chelate effect is entirely an entropy effect if the ligands have similar electronic effects. Here it takes only three ethylenediamine molecules to displace six ammonia molecules, thus providing a net increase in the number of unbound molecules.

Another way to look at the problem is to visualize a chelate ligand with one end attached to the metal ion. The other end cannot then get very far away, and the probability of it, too, becoming attached to the metal atom is greater than if this ligand site were instead taken up by another independent molecule (which would then have access to a much larger volume of the solution).

A11.10   (i)   There are several reasons for a steady decrease in K_n values as the number of ligands increases: (a) statistical factors; (b) increased steric hindrance as the number of ligands increases if they are bulkier than the H_2O molecules they replace; (c) Coulombic factors, mainly in complexes with charged ligands.

(ii)   The ratio of K_2/K_3 is equal to $[ML_2^{2+}]^2/[ML^{2+}][ML_3^{2+}]$ (where M and L denote Ag and NH_3, respectively). It is well-known that the most stable species in the Hg^{2+}-NH_3 system is ML_2^{2+}. Hence, it is expected that $[ML_2^{2+}] \gg [ML^{2+}]$ and $[ML_3^{2+}]$, i.e., K_2/K_3 is very large.

A11.11 (i) $[Cu(NH_3)_4(H_2O)_2]^{2+}$

(ii) $[Cu(dien)_2]^{2+}$

(iii) Octahedral Cu^{2+} complexes tend to undergo Jahn-Teller effect. However, with three bidentate ligands in $[Cu(en)_3]^{2+}$, the distortion would strain the chelate.

(iv) For the same reason as given in (iii), the *trans* isomer should dominate.

A11.12 (i) The Lewis structure of $NC–CH_2–NC$ is shown below. Since both the carbon and nitrogen ends of the ligands have a linear arrangement of the triple bond and the lone pair, we would expect both the cyano and isocyano groups to form linear bonds with a metal.

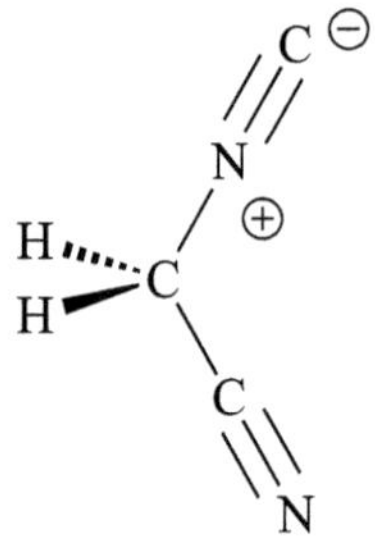

(ii) Compared with CO, isonitrile CNR displays stronger σ-donor and weaker π-acceptor character. Since $Cr–(CO)_{trans}$ bond is shorter than $Cr–CNR$ and $Cr–(CO)_{cis}$ bonds, $(CO)_{trans}$ receives more π-electron density from the metal.

(iii) CO bond order decreases as the extent of back bonding increases. So ν_{CO} of $[Cr(CO)_5(CNCH_2CN)]$ should be lower than that of $[Cr(CO)_6]$.

Experimental data: $[Cr(CO)_5L](KBr)$: 1986, 1974, 1926 cm^{-1}; $[Cr(CO)_6]$: 2000 cm^{-1}.

A11.13 (i) Carbon monoxide is a σ-donor and a π-acceptor. The σ-donating property arises from the filled weak antibonding 5σ orbital of CO, which donates electron density to an empty d orbital of the metal center. The π-accepting property is due to the π^* orbital of CO which accepts electron density from an occupied d orbital of the metal center.

(ii) $[Fe(CO)_6]^{2+} > [Mn(CO)_6]^+ > Cr(CO)_6 > [V(CO)_6]^-$.

Experimental data for ν_{CO}: 2204, 2090, 2000, 1860 cm^{-1}.

All of them are 3d metal complexes with d^6 configuration. Increasing positive charge leads to contraction of the metal d orbitals with attendant decrease in the overlap $M(d_\pi)–CO(\pi^*)$, and hence higher CO stretching frequency.

(iii) All metal ions belong to the first transition series. The higher the charge on the complex, the more contracts will be the 3d orbitals and the lower will be their energy. So the energy of the metal to ligand charge-transfer band will be increased with increasing charge on the complex.

A11.14

A11.15 PCy_3 is a good donor, particularly due to the three electron donating cyclohexyl groups. At the same time, it is also a d_π-d_π ligand which serves as a π-acceptor. CO is a σ donor and a $d_{\pi\text{-}\pi^*}$ π-acceptor. Both ligands bind to the metal centers in a synergistic manner.

Since there is only one CO ligand in $[(Cy_3P)_2Pt(CO)]$ to receive the electron density from the $(Cy_3P)_2Pt$ moiety through back bonding, the extent should be more substantial for this complex compared with $[(Cy_3P)_2Pt(CO)_2]$, which has two CO ligands. Hence the former complex should have a lower CO stretching frequency.

A11.16 (i) $[Bipy^*FeCl_2]$ has a d^6 T_d metal center.

505 nm: a d-d transition assigned to the $^5E \rightarrow {}^5T_2$

$[Bipy^*CuCl_2]$ has a d^9 T_d metal center.

450 nm: a charge-transfer transition from metal to ligand

890 nm: a d-d transition assigned to the $^2T_2 \rightarrow {}^2E$

(ii) $[Bipy^*CuCl_2]$ has a d^9 metal center, which is subject to Jahn-Teller effect. This makes the complex severely distorted.

(iii) The magnetic moment $\mu = 2.84\sqrt{\chi_M T} = 2.84\sqrt{(3.4 \times 10^{-3}) \times 300} = 2.87$ BM. According to the spin-only formula, $\mu_s = \sqrt{n(n+2)}$, we have: $\sqrt{n(n+2)} = 2.87$. Solving: $n = 2.04$. Hence, the complex consists of two unpaired electrons.

A11.17 (i) Since Ag^+ commonly bonds to two ligands to form a "linear" complex, e.g., $[Ag(NH_3)_2]^+$, the structure with the formula $(\eta^1\text{-}C_5H_5)Ag(NH_3)$ is proposed.

(ii) The enhanced stability of the $[Cu(en)_2]^{2+}$ complex over $[Cu(NH_3)_4]^{2+}$ can be explained by the so called "chelate effect" (see A11.9). On the other hand,

A11.18 The periodate ion has an octahedral structure and therefore can be used as a monodentate or bidentate ligand. Considering the bulkiness of the ligand and the chelate effect of
complex formation, the periodate ion would be more stable as a bidentate ligand. The
probable formula may therefore be $[M(IO_6)_2]^{7-}$ or $[M(IO_6)_3]^{12-}$.

   The high oxidation state, $+3$, of the metal together with the suitable electronic
configuration, d^8, tend to favour the formation of a square planar complex. Here large
crystal field splitting and consequently large CFSE is the main factor favoring the
structure. Magnetic susceptibility measurements would clarify the structure: square
planar complexes would give rise to a zero magnetic moment whereas octahedral or
tetrahedral complexes should have paramagnetic properties. The four-coordinate silver
complex has been found to have the following structure:

$$
\begin{bmatrix} & & \end{bmatrix}^{7-}
$$

A11.19 (i)

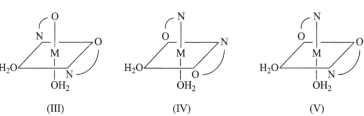

(I)   (II)

(III)   (IV)   (V)

   Isomers (I) and (II) are optically inactive; the others are optically active.

   (ii)   The number and type of possible isomers are exactly the same for both Types A
and B coordination, since the N atoms in the ligand are non-equivalent.

   (iii) Ni^{2+} has the d^8 configuration with two unpaired electrons.

$\mu_s(\text{spin-only}) = \sqrt{n(n+2)} = 2.90 = 2.83 \text{ BM}.$

μ_{s+1} (spin and full orbital contribution) $= \sqrt{4S(S+1) + L(L+1)} = \sqrt{n(n+2) + L(L+1)} = 4.47 \text{ BM}.$

The spin-only magnetic moment is significantly lower than the observed value. The orbital motion of the electrons makes some contribution to the total magnetic moment of Ni^{2+} in the complex.

A11.20

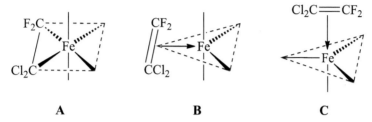

     A   **B**   **C**

(i)   For all three structures, the point group is C_s.

(ii)   For **A**, the hybridization is d^2sp^3 (octahedral model); for **B** and **C**, it is dsp^3 (trigonal bipyramidal model).

(iii) For **A**, normal σ bonds are formed by the overlap of sp^3 hybrid orbitals of carbon with d^2sp^3 hybrid orbitals of Fe. For **B** or **C**, a σ coordinate bond is formed by the overlap of the filled π molecular orbital of the olefin with an empty dsp^3 hybrid orbital of Fe. Back donation from a filled metal d orbital to the empty π^* molecular orbital of the olefin also occurs. There is probably no real difference between **A** and **B** though their formal bonding descriptions differ.

Steric repulsion would rule out a structure in which the olefin ligand occupies an equatorial position of the trigonal bipyramid with its double bond parallel to the axial direction.

A11.21 (i) In the dioxygen complex, iridium receives an electron pair from the π-bond in O_2; back bonding from a filled d orbital on iridium to the empty π^* orbital on the O_2 group probably also occurs. The extent of back bonding depends on the available electronic density at the iridium atom and is expected to be smaller when the more electronegative chlorine ligand is present than when iodine is present in the complex. Hence the O–O bond distance is greater for the iodo-complex.

(ii) As mentioned above, the extent of d_π-π^* back bonding is smaller in the chloro-complex, and this will give rise to a weaker complex and its formation is reversible. The iodo-complex has stronger metal-dioxygen bonding and is formed irreversibly.

A11.22 This is due to the fact that the formation of the nitrito complex occurs without the rupture of the Co–O bond in $[(H_3N)_5Co(OH_2)]^{3+}$. The reaction sequence is:

$$2\ HNO_2 \rightleftharpoons N_2O_3 + H_2O,$$

$$[(H_3N)_5Co-OH_2]^{3+} + H_2O \xrightarrow{\text{fast}} [(H_3N)_5Co-OH]^{2+} + H_3O^+,$$

$$[(H_3N)_5Co-OH]^{2+} + N_2O_3 \xrightarrow{\text{slow}} \left[\begin{array}{c} (H_3N)_5Co-O\cdots\cdot H \\ \vdots \qquad \vdots \\ O{=}N\cdots\cdot O{-}N{=}O \end{array} \right]^{2+}$$

$$\downarrow \text{fast}$$

$$[(H_3N)_5Co-ONO]^{2+} + HNO_2,$$

with a rate law of: rate $= k[\text{complex}][HNO_2]^2$.

Confirmation that the nitrite complex is formed without rupture of the Co–O bond was provided by isotope labelling:

$$[(H_3N)_5Co(O^*H_2)]^{3+} + NO_2^- \rightarrow [(H_3N)_5Co(O^*NO)]_2^+ + H_2O.$$

A11.23

Row 1:
C–C square (C,C / C,C) →A→ A,C square (C,C) →D→ A,D square (C,C) →B→ A,D square (B,C)

Row 2:
C,C square (C,C) →B→ B,C square (C,C) →D→ B,D square (C,C) →A→ B,D square (A,C)

Row 3:
B,B square (B,B) →A→ A,B square (B,B) →D→ A,D square (B,B) →C→ A,D square (C,B)

A11.24 (i) Potassium reduces the divalent metal center to zero valent, then N_2 adds to the vacant bind site.

(ii)   Trigonal bipyramidal structure.

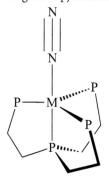

(iii)   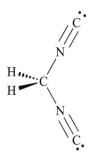   The bonding scheme in N_2 is sp hybridization. Hence, the bond angle of M–N–N is expected to be about 180°. (Experimental values: 177.1° - 179.8°)

(iv)   They should be in a *cis* arrangement as the four P binding sites of L* need to be in adjacent position.

(v)   Back donation from metal d orbital to N_2 antibonding orbital results in decrease of $\nu(N\equiv N)$ stretching frequency. Since the metal centers in $[M(N_2)HL^*]^+$ (M = Fe, Ru) are less electron rich than their neutral counterparts, their $\nu(N\equiv N)$ stretching frequency is lower.

Experimental data:	$\nu(N\equiv N)$ (cm^{-1})
$Fe(N_2)L^*$	1996
$Ru(N_2)L^*$	2083
$[Fe(N_2)HL^*]^+$	2170
$[Ru(N_2)HL^*]^+$	2172

A11.25   (i)   The Lewis structure is shown below.

      Given the fact that both the carbon and nitrogen ends of the isocyano group are sp hybridized and have a linear arrangement of the triple bond and the lone pair, it is expected that it will form linear N–C–M bonds with metals.

(ii)   If CO is liberated from $[Mn(\eta^5\text{-}C_5H_5)(CO)_3]$, it must be replaced by some other ligand if the 18-electron rule is to be maintained. In this case, the new ligand is tetrahydrofuran (THF), a cyclic ether that can act as a σ donor, and compound **A** is $[Mn(\eta^5\text{-}C_5H_5)(CO)_2(THF)]$:

$$[Mn(\eta^5\text{-}C_5H_5)(CO)_3] + THF \rightarrow [Mn(\eta^5\text{-}C_5H_5)(CO)_2(THF)] + CO$$

Each NC group in $H_2C(NC)_2$ has a lone pair and can act as donors to transition metals. In the formation of **B**, the weakly bound THF is replaced by a $H_2C(NC)_2$ ligand.

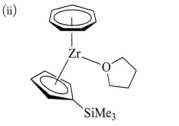

A11.26 (i)

$^1H/^{13}C$

δ 5.19/80.3

δ 0.08/0.01

Zr

SiMe$_3$

δ 5.40 & 5.48 /104.3 & 106.8

δ --/114.2

(ii)

Zr—O

SiMe$_3$

18 valent electrons.

A11.27 **A**: $[Fe(\eta^5\text{-}C_5H_5)(CO)_2]^-$
B: $[Fe(\eta^5\text{-}C_5H_5)(CO)_2Br]$
C: $[Fe(\eta^5\text{-}C_5H_5)(CO)_2H]$
D: benzene

A11.28 (i)

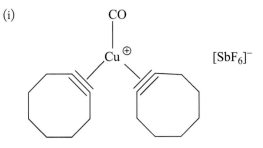

(ii) Cu(I), d^{10}, 16 electrons.

(iii) The CO stretch of **A** (2171 cm^{-1}) is substantially higher than that of the free CO (2143 cm^{-1}). It is a sign of dominant σ-type OC$\rightarrow$Cu bonding interaction with only a low level of Cu$\rightarrow$CO π-back donation. Furthermore, it indicates that the copper site supported by two cyclooctynes in **A** is very electrophilic.

The C$\equiv$C stretch of the alkyne moieties of **A** (2070 cm^{-1}) is lower than the value observed for free cyclooctyne (2216 cm^{-1}). In this complex, (C$\equiv$C)$\rightarrow$Cu σ-donation as well as Cu$\rightarrow$(C$\equiv$C) π-back donation weakens the C$\equiv$C bond, and both could contribute to the lower C$\equiv$C stretch in **A**.

(iv) 'Bu

The Cu–CO bond in **A** is weak (as shown by the high CO stretching frequency and long Cu–CO bond). Hence, displacement of CO, rather than addition to CO, occurs.

A11.29 (i) **A** has seven signals in its $^{13}C\{^1H\}$ NMR spectrum for the 2,4-dimethyl-pentadiethyl ligand, while **B** has only four. The results are consistent with a η^3-binding mode in **A** and η^5-binding mode in **B**. Both complexes obey the 18-electron rule.

A: $3 + 7 + 2 \times 4 = 18$.

B: $5 + 7 + 2 \times 3 = 18$.

The first step is ligand association followed by change in hapticity.

A   **B**

(ii) $LMn(CO)_3$: 2021, 1954, 1934 cm^{-1}.

$LMn(CO)(PMe_3)_2$: 1839 cm^{-1}.

The shift to lower stretching frequency is consistent with increased electron density at the Mn center from replacement of carbonyl ligands (better π-acceptor) with phosphines (better σ-donor).

A11.30 (i) MIO is an ambidentate ligand.

(ii) Because **A** and **B** have the same $Ru(NH_3)_5$ moiety, electronic rather than steric factors should be considered. It is known that the low-spin d^6 Ru(II) and d^5 Ru(III) complexes of amine act as a π-electron donor and a π-electron acceptor, respectively. Thus, the sulfur and oxygen atoms of MIO act as π-electron acceptor and a π-electron donor, respectively. The soft Ru(II) center prefers to bind to the soft sulfur atom of MIO, while the hard Ru(III) center prefers to bind to the hard oxygen atom of MIO.

(iii) The variation in the Ru–S bond length is attributed, in part, to variation in the amounts of π-back bonding of filled Ru d_π orbitals to vacant antibonding orbitals of π-acceptor ligands. On the basis of the Ru–S bond lengths, the π-acceptor ability of sulfur increases in the order: MIO-S < DMSO-S. By comparing the two Ru–N bond lengths, it can be concluded that the *trans* influence of MIO-S is weaker than that of DMSO-S, which is in agreement with the order of the π-acceptor ability based on Ru–S bond lengths.

A11.31 (i) **B**: iron (I), **C**: iron(II).

(ii) **A** to **B**: reduction, **A** to **C**: addition.

(iii) Soft ligands and metals in low oxidation states; large and bulky ligands; counterions of low basicity.

The two ligands in **B** are moderate in size. Hence, the steric rule governing a two-coordinate geometry for iron(II) species is not mandatory in the case of anionic iron(I) compounds.

(iv) For linear complexes, the crystal field splitting pattern would be $3d_{z^2} > (3d_{xz}, 3d_{yz}) > (3d_{xy}, 3d_{x^2-y^2})$. For an iron(I) d^7 high-spin complex, there are three unpaired electrons. Hence, $\mu_s = \sqrt{3(3+2)} = 3.87$ BM.

A11.32 (i)

(ii)

(iii)

12 | Metal Clusters: Bonding and Reactivity

PROBLEMS

12.1 The binuclear chromium carboxylate complex **A** has a paddlewheel structure with a very short Cr–Cr bond distance of 1.9662(5) Å.

(i) Predict the bond order of the Cr–Cr bond in **A**. Write down the electronic configuration of the "Cr$_2$" core. Hint: Carboxylate ligands can be bonded to metals in a bidentate fashion. This coordination mode can be visualized by the following resonance structures.

Problems in Structural Inorganic Chemistry. Second edition. Wai-Kee Li, Hung Kay Lee, Dennis Kee Pui Ng, Yu-San Cheung, Kendrew Kin Wah Mak, and Thomas Chung Wai Mak. © Oxford University Press 2019. Published in 2019 by Oxford University Press. DOI: 10.1093/oso/9780198823902.001.0001

(ii) Addition of CH_3CN to a solution of **A** in dichloromethane results in the isolation of acetonitrile-adduct **B**:

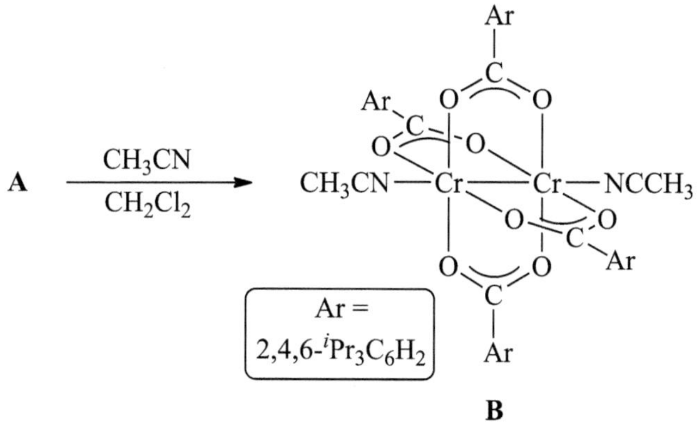

The Cr–Cr bond distance in **B** is found to be 2.3892(2) Å. Account for the difference in the Cr–Cr bond distances in **A** and **B**.

REFERENCE: F. A. Cotton, E. A. Hillard, C. A. Murillo, and H.-C. Zhou, After 155 years, a crystalline chromium carboxylate with a supershort Cr–Cr Bond. *J. Am. Chem. Soc.* **122**, 416–7 (2000).

12.2 The reaction of complex **A** with $^iPrOO^iPr$ is shown as follows:

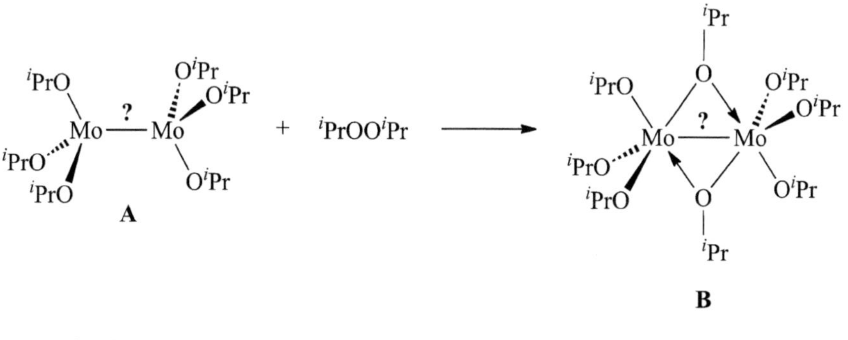

Using the electron counting method, predict the bond order of the Mo–Mo linkage in **A** and **B**, respectively.

REFERENCE: M. H. Chisholm, C. C. Kirkpatrick, and J. C. Huffman, Reactions of metal-metal multiple bonds. 7. Addition of the halogens Cl_2, Br_2, and I_2 and diisopropyl peroxide to hexaisopropoxy-dimolybdenum (M≡M). Dinuclear oxidative-addition reactions accompanied by metal-metal bond-order changes from 3 to 2 to 1. *Inorg. Chem.* **20**, 871–6 (1981).

12.3 Complex **A** was synthesized and structurally characterized:

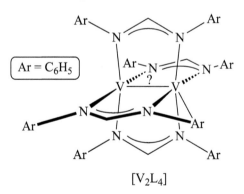

$$Cr_2(O_2CCH_3)_4 \ + \ 2 \quad \text{(o-tBuO-C}_6\text{H}_4\text{)Li} \quad \xrightarrow[\text{25°C, 5h}]{\text{THF}} \quad \textbf{A}$$

(i) Using the electron counting method, predict the bond order of the Cr–Cr linkage in **A**.

(ii) Experimentally, the Cr–Cr bond distance in **A** measures 1.862(1) Å, which is slightly longer than the Cr–Cr bond distance in the related tetrakis(2-methoxy-5-methylphenyl)dichromium complex, $[Cr_2(2\text{-MeOC}_6H_3\text{-}5\text{-Me})_4]$. Suggest a plausible reason to account for the difference in the Cr–Cr bond distances in these two complexes.

REFERENCE: F. A. Cotton and M. Millar, An exceedingly short metal-metal bond in a bis(*o*-alkoxyphenyl)dicarboxylatodichromium compound. *Inorg. Chem.* **17**, 2014–7 (1978).

12.4 The structure of a paddlewheel-type divanadium(II) amidinate complex, $[V_2L_4]$, is shown in the figure below.

Ar = C_6H_5

$[V_2L_4]$

(i) Using the electron counting method, predict the bond order of the V–V bond in the V_2^{4+} core. Also, write down the electronic configuration of the V_2^{4+} core. (Hint: The bidentate coordination mode of the amidinate ligands can be visualized as follows.)

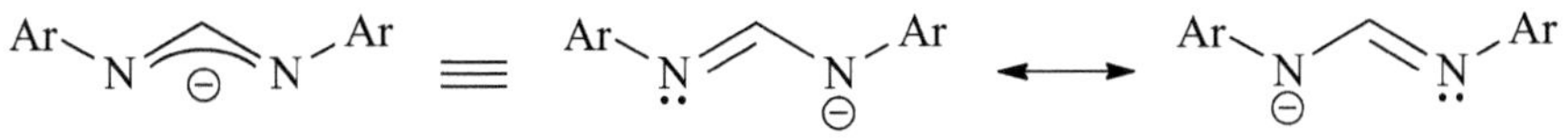

(ii) Treatment of the $[V_2L_4]$ complex with potassium graphite led to isolation of the corresponding anionic complex $[V_2L_4]^-$, which also has a similar paddlewheel structure. What is the bond order of the V–V bond in $[V_2L_4]^-$? Do you expect any variation in the V–V bond distances of these two complexes? Explain your answers.

(iii) If a more basic bidentate ligand (instead of L) is used to prepare a structurally similar complex analogous to $[V_2L_4]$, how do you expect the V–V bond distance changes? Explain your answers.

REFERENCE: F. A. Cotton, E. A. Hillard, C.A. Murillo, and X. Wang, New chemistry of the triply bonded divanadium (V_2^{4+}) unit and reduction to an unprecedented V_2^{3+} core. *Inorg. Chem.* **42**, 6063–70 (2003).

12.5 The coordination behavior of a series of N,N'-di(aryl)formamidinate ligands (aryl $= 2$–XC_6H_4 where $X=$ Me, OMe, Cl, Br) towards $CrCl_2$ was studied by Cotton and co-workers.

X = Me, OMe, Cl, Br

The synthesis of the 2-MeC_6H_4 substituted formamidinate complex is outlined in the following reaction scheme:

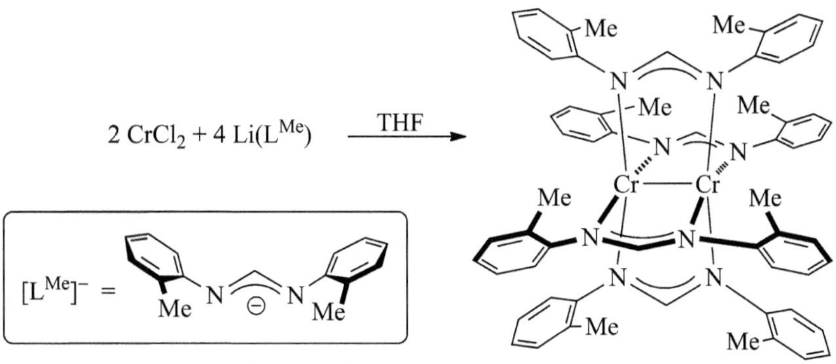

$2\,CrCl_2 + 4\,Li(L^{Me})\ \xrightarrow{\ THF\ }$

$[L^{Me}]^- =$

Cr–Cr = 1.925(1) Å

(i) Using the electron counting method, predict the bond order of the Cr–Cr bond in the complex.

(ii) Do you expect the $[Cr(L^{Me})_2]_2$ complex to be diamagnetic or paramagnetic? Suggest with explanation one single physical method to support your answer.

(iii) Successive replacement of the *ortho* Me substituent in the aryl ring by OMe, Cl, and Br substituents led to a gradual increase in the Cr–Cr bond distance of the corresponding dichromium complex from 1.925(1) Å (X = Me) to 2.140(2) (X = OMe), 2.208(2) (X = Cl), and 2.272(2) Å (X = Br). The solid-state structure of the *ortho* MeO-substituted complex is shown. Two short Cr···OMe interactions of 2.635(2) Å and 2.402(2) Å, respectively, are observed in this structure. Comment on the order of increase in the Cr–Cr bond distances in these complexes.

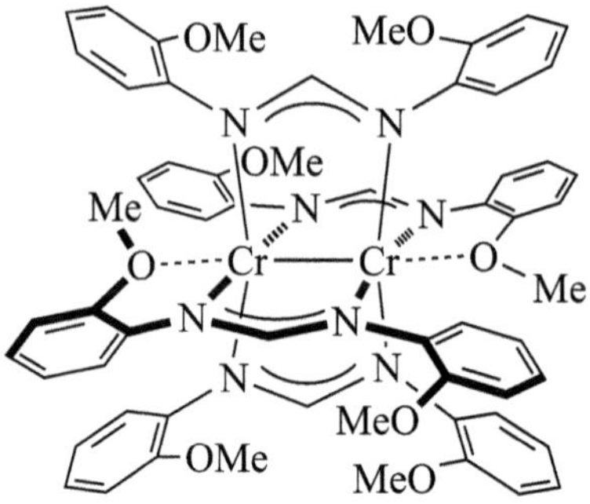

REFERENCE: F. A. Cotton, L. M. Daniels, C. A. Murillo, and P. Schooler, Chromium(II) complexes bearing 2-substituted *NN′*–diarylformamidinate ligands. *J. Chem. Soc., Dalton Trans.* 2007–12 (2000).

12.6 The reactivity of metal-metal bonded cluster compounds has attracted considerable research interest in recent years. The quintuply bonded dichromium(I) complex **A** reacts readily with AlMe₃, yielding a carboalumination product, **B**.

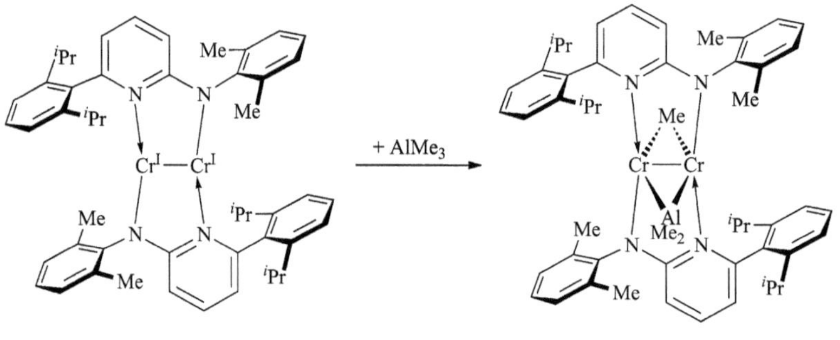

Cr–Cr = 1.750(1) Å

A

+ AlMe₃

Cr–Cr = 1.8365(8) Å

B

(i)   Using the electron counting method, predict the bond order of the Cr–Cr linkage in **B**.

(ii) Addition of PhC≡CPh to **A** afforded complex **C**.

$$\mathbf{A} + \text{PhC} \equiv \text{CPh} \rightarrow \mathbf{C}.$$

Write down the structure of **C**.

REFERENCES: H. K. Lee and W.-K. Li, Bonding and reactivity of metal clusters. *Chem. Educator* **20**, 1–8 (2015); A. Noor, E. S. Tamne, S. Qayyum, T. Bauer, and R. Kempe, Cycloaddition reactions of a chromium–chromium quintuple bond. *Chem. Eur. J.* **17**, 6900–3 (2011).

12.7   The metal complex ion MAl_4^-, having a square pyramidal structure with C_{4v} symmetry, was prepared in 2001. The structure of MAl_4^- may be described as a square Al_4^{2-} base, capped by a M^+ cation. This problem is mainly concerned with the novel Al_4^{2-} anion.

The seven molecular orbitals occupied by the 14 valence electrons of Al_4^{2-} are shown below.

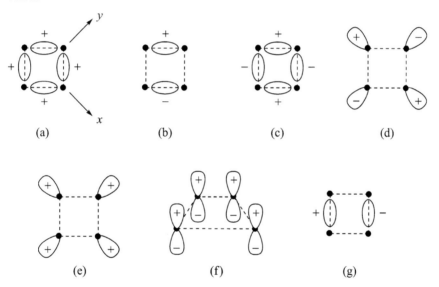

(i)   Determine the symmetry and the nature (bonding or nonbonding) of the molecular orbitals labeled (a) to (g). In addition, identify the (out-of-plane) π bonding molecular orbital(s) of the Al_4^{2-} anion.

(ii)   Arrange these seven molecular orbitals in ascending energy order.

(iii) Sketch the most important resonance structures of Al_4^{2-} and hence determine the bond order of each Al–Al bond. Is this cyclic anion aromatic? Explain. Finally, compare the bonding pictures provided by the molecular orbitals and by the resonance structures.

(iv) If a Cu^+ cation approaches the Al_4^{2-} anion along the z-axis to form complex ion $CuAl_4^-$, list the orbitals of Cu^+ and Al_4^{2-} that participate in the bonding interaction.

REFERENCE: X. Li, A. E. Kuznetsov, H.-F. Zhang, A. I. Boldyrev, and L.-S. Wang, Observation of all-metal aromatic molecules. *Science* **291**, 859–61 (2001).

12.8 The structures of $Mo_6Cl_8^{4+}$ (left) and $Ta_6Cl_{12}^{2+}$ (right) are shown below.

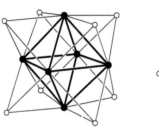 

In both of these metal cluster ions, the six metal atoms form a regular octahedron. In $Mo_6Cl_8^{4+}$, the eight Cl atoms are above the eight faces of the metal octahedron; in $Ta_6Cl_{12}^{2+}$, the 12 Cl atoms are above the 12 edges of the octahedron.

(i) Determine the bond order of the M–M linkage in each of these ions.

(ii) Are the structures of these ions in accord with the VSEPR theory? Explain.

12.9 The di-molybdenum cluster $[Mo_2(SO_4)_4]^{3-}$ has a paddlewheel structure as illustrated below.

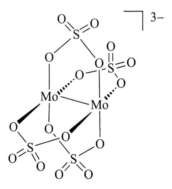

(i) Using the electron counting method, predict the bond order of the Mo–Mo linkage in $[Mo_2(SO_4)_4]^{3-}$.

(ii) Why is the $K_3[Mo_2(SO_4)_4]$ complex EPR active?

REFERENCE: F. A. Cotton, B. A. Frenz, and T. R. Webb, Structure of a compound with a molybdenum-to-molybdenum bond of order three and one-half. *J. Am. Chem. Soc.* **95**, 4431–2 (1973).

12.10 As outlined in the following reaction scheme, reduction of quadruply bonded tantalum(IV) dimer **A** with two molar equivalents of sodium amalgam produced di-tantalum dihydride complex **B**.

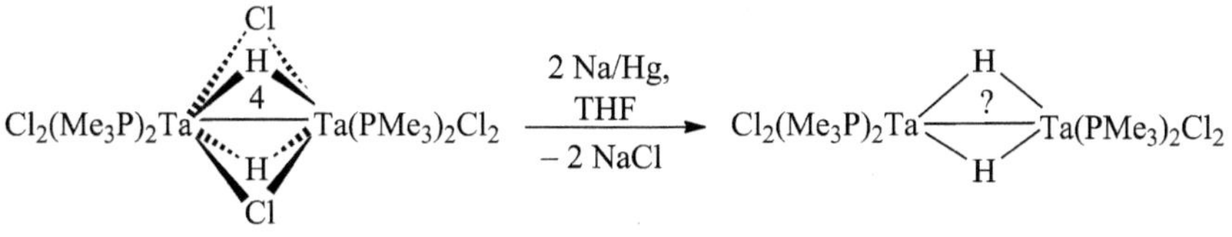

A   B

(i) Using the electron counting method, predict the bond order of the Ta–Ta linkage in **B**.

(ii) The following reactions of **B** were observed. Write down an appropriate reagent for each of these reactions.

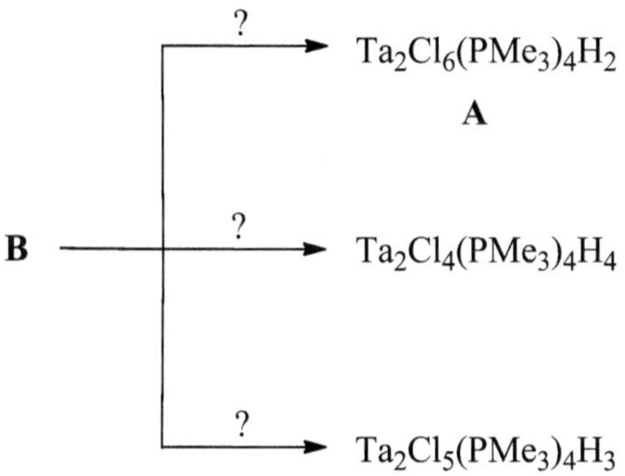

REFERENCE: A. J. Scioly, M. L. Luetkens, Jr., R. B. Wilson, Jr., J. C. Huffman, and A. P. Sattelberger, Synthesis and characterization of binuclear tantalum hydride complexes. *Polyhedron* **6**, 741–57 (1987).

12.11 Treatment of the quintuply bonded di-chromium(I) complex **A** with 3-hexyne afforded di-chromium complex **B**:

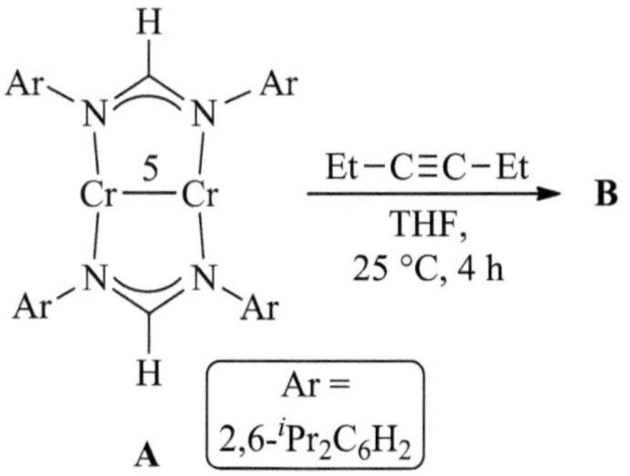

Predict the structure of **B**. Write down the oxidation state of the Cr atoms and the bond order of the Cr–Cr linkage in **B**.

REFERENCE: H.-Z. Chen, S.-C. Liu, C.-H. Yen, J.-S. K. Yu, Y.-J. Shieh, T.-S. Kuo, and Y.-C. Tsai, Reactions of metal–metal quintuple bonds with alkynes: [2+2+2] and [2+2] cycloadditions. *Angew. Chem. Int. Ed.* **51**, 10342–6 (2012).

12.12 Treatment of the quintuply bonded dimolybdenum amidinate complex **A** with two molar equivalents of 1-pentyne in diethyl ether yielded diamagnetic, green crystalline complex **B**.

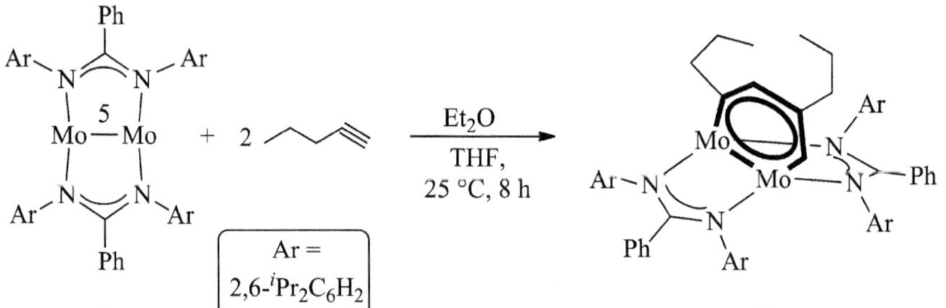

(i) Give a mechanism for this reaction.

(ii) Predict the bond order of the Mo–Mo linkage in **B**.

REFERENCE: H.-Z. Chen, S.-C. Liu, C.-H. Yen, J.-S. K. Yu, Y.-J. Shieh, T.-S. Kuo, and Y.-C. Tsai, Reactions of metal–metal quintuple bonds with alkynes: [2+2+2] and [2+2] cycloadditions. *Angew. Chem. Int. Ed.* **51**, 10342–6 (2012).

12.13 Why does the di-rhenium(III) complex $[Re_2Cl_8]^{2-}$ exhibit an eclipsed conformation (D_{4h} symmetry), whereas the di-osmium(III) complex $[Os_2Cl_8]^{2-}$ has a staggered structure (D_{4d} symmetry)?

12.14 Synthesis of the first benzene-ring-bridged dimolybdenum complex $[(MoL)_2(C_6H_6)]$ (**A**, where L denotes ArNCH=CHNAr) is outlined below:

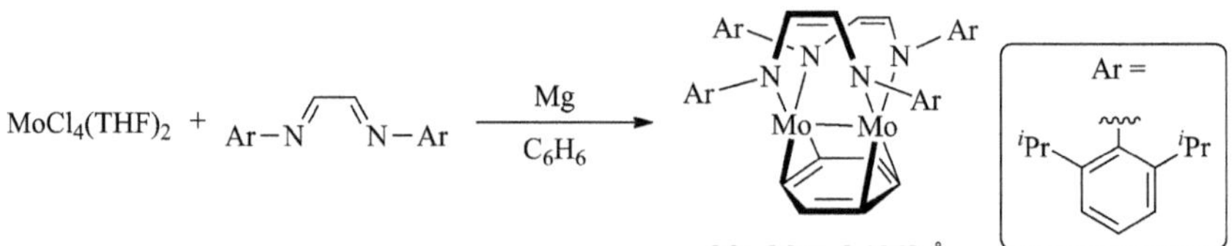

Mo–Mo = 2.1968 Å

A

An X-ray crystallographic analysis revealed that **A** has a very short Mo–Mo bond distance of 2.1968(4) Å.

(i)   What is the role of magnesium in this reaction?

(ii) Describe a bonding scheme for the Mo_2 core. Deduce the metal-metal bond order in **A**.

REFERENCE: Y. Kajita, T. Ogawa, J. Matsumoto, and H. Masuda, Synthesis and characterization of a benzene-dimolybdenum complex with a new bridging mode. *Inorg. Chem.* **48**, 9069–71 (2009).

12.15 The coordination chemistry of a tricobalt cluster carboxylate ligand, $[\{(CO)_3Co\}_3(\mu^3\text{-}COO)]^-$, was examined. As shown in the following figure, complex **A** is a hybrid cluster compound which contains both Mo–Mo and Co–Co bonds. The molecule consists of a Mo_2 core, which is bound by four equatorial $[\{(CO)_3Co\}_3(\mu^3\text{-}COO)]^-$ ligands and two axial tricobalt cluster carboxylic acid molecules $[\{(CO)_3Co\}_3(\mu^3\text{-}COOH)]$.

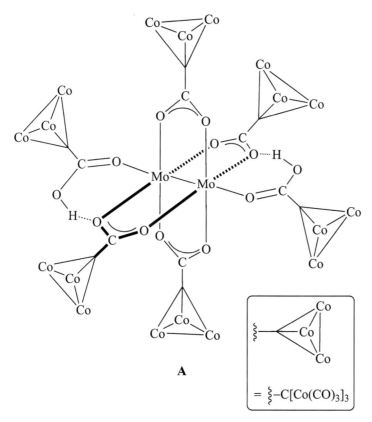

A

$$= \text{\textonequarter}-C[Co(CO)_3]_3$$

(i)   Predict the formal Mo–Mo bond order in **A**.

(ii) The IR spectrum of **A** showed two symmetric A_1 stretching bands for the carbonyl ligands at 2105 and 2111 cm^{-1} (in a 2:1 ratio). For comparison, the A_1 stretching frequency of the carbonyl ligands of free tricobalt cluster carboxylic acid $[\{(CO)_3Co\}_3(\mu^3\text{-COOH})]$ was found to be 2111 cm^{-1}. Comment on the lower carbonyl stretching frequency for the equatorial cluster ligands in **A**.

REFERENCE: W. Cen, P. Lindenfeld, and T. P. Fehlner, On the interface of metal-metal multiple bond compounds and organometallic clusters: synthesis and structure of $Mo_2\{\mu-[(CO)_9Co_3(\mu_3-CCO_2)]\}_4[(CO)_9Co_3(\mu_3-CCO_2H)]_2$ and related compounds. *J. Am. Chem. Soc.* **114**, 5451–2 (1992).

12.16 The structure of a dimolybdenum complex is shown below. This complex has a staggered, ethane-like Mo_2N_6 moiety. The Mo–Mo bond distance is short, namely 2.214(2) Å (average), which is suggestive of a Mo–Mo triple bond. Each Mo–NC$_2$ unit is planar with a short Mo–N bond distance of 1.98 Å.

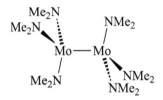

(i)   Describe a bonding scheme for the Mo$_2$ core in $[Mo_2(NMe_2)_6]$.

(ii) Do you expect $[Mo_2(NMe_2)_6]$ to be paramagnetic or diamagnetic? Rationalize your answer.

(iii) The structurally related complex $[Mo_2(CH_2SiMe_3)_6]$ also has a similar ethane-like geometry. The Mo–Mo bond distance in this complex was found to be 2.167 Å. Comment on the longer Mo–Mo bond distance in the $[Mo_2(NMe_2)_6]$ complex. (Assume the steric requirements of the NMe$_2$ and CH$_2$SiMe$_3$ ligands do not impose significant effects on the Mo–Mo bond distances.)

REFERENCE: M. H. Chisholm, F. A. Cotton, B. A. Frenz, W. W. Reichert, L. W. Shive, and B. R. Stults, The molybdenum-molybdenum triple bond. 1. Hexakis(dimethylamido)dimolybdenum and some homologues: Preparation, structure, and properties. *J. Am. Chem. Soc.* **98**, 4469–76 (1976).

12.17 Illustrated below are the structures of $Re_3Cl_{12}^{3-}$ (left, D_{3h} symmetry) and $Re_2Cl_8^{2-}$ (right, D_{4h}).

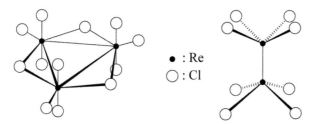

● : Re
○ : Cl

(i) Determine the bond order of the Re–Re linkage in each ion.

(ii) Surprisingly, the Cl atoms in $Re_2Cl_8^{2-}$ adopt an eclipsed conformation, contrary to that found in C_2H_6. Apply the VSEPR theory to rationalize the conformation found in $Re_2Cl_8^{2-}$.

12.18 A metal centered Lewis base has an electron rich metal center which can act as a formal two-electron donor. Two platinum-beryllium adducts, $[(Cy_3P)_2Pt–BeClX]$ ($Cy =$ cyclohexyl; $X = Cl$ or Me), containing two-center two-electron Pt–Be bonds were prepared as outlined in the following reaction scheme.

$$
\begin{array}{c}
PCy_3 \\
| \\
Pt \\
| \\
PCy_3
\end{array}
\;+\; BeCl_2 \longrightarrow
\begin{array}{c}
PCy_3 \\
| \\
Pt-Be \cdots Cl \\
| \quad\quad Cl \\
PCy_3
\end{array}
\;\xrightarrow{\;A\;}\;
\begin{array}{c}
PCy_3 \\
| \\
Pt-Be \cdots Cl \\
| \quad\quad Me \\
PCy_3
\end{array}
$$

$$Cy = \text{⸽}\text{—}\bigcirc$$

(i) Suggest a reagent (**A**) for the preparation of $[(Cy_3P)_2Pt–BeCl_2]$ from $[(Cy_3P)_2Pt–BeClMe]$.

(ii) The crystal structures of these two platinum-beryllium adducts were determined by X-ray diffraction analysis. The Pt–Be bond lengths and Cl–Be–X bond angles of the two complexes are tabulated below.

	$[(Cy_3P)_2Pt–BeCl_2]$	$[(Cy_3P)_2Pt–BeClMe]$
Pt–Be (Å)	2.168(4)	2.195(3)
Cl–Be–X (°)	118.6(1)	123.7(1)

Account for a longer Pt–Be bond length and a larger Cl–Be–X bond angle for the $[(Cy_3P)_2Pt–BBeClMe]$ complex

REFERENCE: H. Braunschweig, K. Gruss, and K. Radacki, Complexes with dative bonds between d- and s-block metals: Synthesis and structure of [(Cy$_3$P)$_2$Pt–Be(Cl)X] (X = Cl, Me). *Angew. Chem. Int. Ed.* **48**, 4239–41 (2009).

12.19 The reaction of mixed-metal cluster H$_2$FeRu$_3$(CO)$_{13}$ with diphenylacetylene (PhC≡CPh) led to insertion of the alkyne across a metal-metal bond, yielding an alkyne insertion product FeRu$_3$(CO)$_{12}$(PhC≡CPh). It consists of two isomeric forms, namely "axial" and "equatorial" isomers, which differ only in the positioning of the Fe atom. The "FeRu$_3$C$_2$" cores of these two isomeric forms of FeRu$_3$(CO)$_{12}$(PhC≡CPh) are shown below:

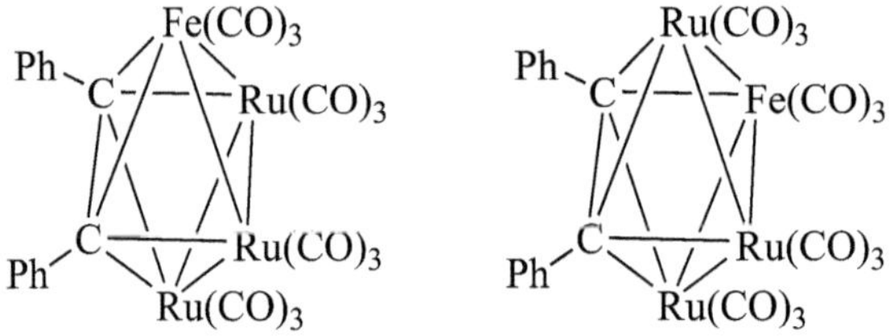

"Axial" isomer   "Equatorial" isomer

(i) The "axial" isomer was found to be thermodynamically more stable than the "equatorial" isomer. Explain.

(ii) Does the FeRu$_3$(CO)$_{12}$(PhC≡CPh) cluster have a *closo*, *nido*, or *arachno* structure?

(iii) The H$_2$FeRu$_3$(CO)$_{13}$ cluster reacted with MeC≡CPh in an analogous manner, giving three isomeric products of formula FeRu$_3$(CO)$_{12}$(MeC≡CPh). Draw the structures of these three isomeric products.

REFERENCE: G. L. Geoffroy, Synthesis, molecular dynamics, and reactivity of mixed-metal clusters. *Acc. Chem. Res.* **13**, 469–76 (1980).

12.20 Diplumbenes, with a general formula of R$_2$Pb=PbR$_2$ (R = variety of bulky organic substituents), are considered as lead analogues of alkenes. The Pb=Pb double bonds in diplumbenes are relatively weak. It is noted that diplumbenes generally have a high tendency to dissociate in solution according to the following equilibrium:

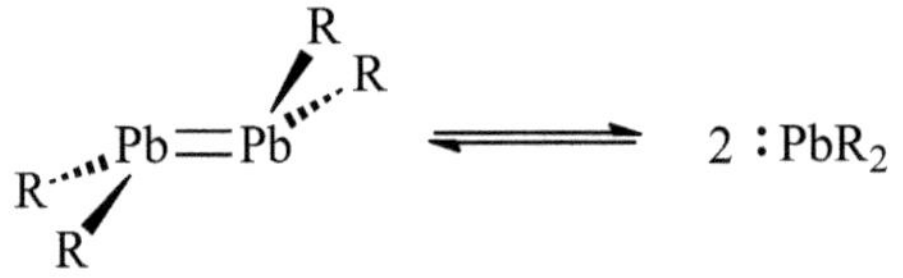

In fact, only few examples of stable diplumbenes were isolated and structurally characterized. Account for the weakness of the Pb=Pb double bonds in diplumbenes.

REFERENCE: M. Stürmann, W. Saak, H. Marsmann, and M. Weidenbruch, Tetrakis(2,4, 6-triisopropylphenyl)diplumbene: a molecule with a lead–lead double bond. *Angew. Chem. Int. Ed.* **38**, 187–9 (1999).

SOLUTIONS

A12.1 (i) **A** has two Cr(II) centers. The electronic configuration of the Cr_2 core is $\sigma^2\pi^4\delta^2$ (bond order $= 4$; a quadruple bond).

(ii) The two axially coordinated CH_3CN ligands in **B** donate electron density to the antibonding orbitals of the Cr_2 core. This weakens the Cr–Cr bond.

A12.2 For **A**, the total number of valence electrons is 6×2 (two Mo atoms) $+ 1 \times 6$ (six $O^i Pr$ ligands) $= 18$. After we use 12 electrons for the six Mo–O bonds, six electrons are left for Mo–Mo bonding. Thus the Mo–Mo bond order is 3.

For **B**, the total number of valence electrons is 6×2 (two Mo atoms) $+ 1 \times 6$ (six terminal $O^i Pr$ ligands) $+ 3 \times 2$ (two bridging $O^i Pr$ ligands) $= 24$. The ten Mo–O bonds use up $2 \times 10 = 20$ electrons. Therefore, four electrons are left for Mo–Mo bonding, giving rise to a Mo=Mo double bond.

A12.3 (i) Total number of valence electrons in the Cr_2 core in **A**: 6×2 (two Cr atoms) $+ 3 \times 2$ (two aryl ligands) $+ 3 \times 2$ (two carboxylate ligands) $= 24$. Sixteen electrons are used to form metal-ligand bonds, leaving eight electrons for Cr–Cr bonding (bond order $= 4$).

(ii) The difference in the Cr–Cr bond distances in the two dichromium complexes may be ascribed to the difference in the steric bulkiness of the supporting ligands.

A12.4 (i) Total number of valence electrons surrounding the V_2^{4+} core: 5×2 (two V atoms) $+ 3 \times 4$ (four ligands) $= 22$. The metal-ligand bonds take up $2 \times 8 = 16$ electrons. Therefore, 6 electrons are left for the V–V bond. The electronic configuration of the V_2^{4+} core is $\sigma^2\pi^4$ (bond order $= 3$).

(ii) The "V_2" core in the $[V_2L_4]^-$ complex has an electronic configuration of $\sigma^2\pi^4\delta^1$ (bond order $= 3.5$). Therefore, the V–V bond distance in the $[V_2L_4]^-$ complex is expected to be shorter than that of the $[V_2L_4]$ complex.

(iii) In general, more basic ligands are better electron donors, which may enhance the electron density in the V_2^{4+} core. Hence the V–V bond is expected to be shorter.

A12.5 (i) Total number of valence electrons in the Cr_2 core in the complex: 6×2 (two Cr atoms) $+ 3 \times 4$ (four ligands) $= 24$. Number of electrons used to form metal-ligand bonds is $2 \times 8 = 16$. Therefore, eight electrons are left for Cr–Cr bonding, leading to a bond order of 4.

(ii) The "Cr_2" core in the $[Cr(L^{Me})_2]_2$ complex has an electronic configuration of $\sigma^2\pi^4\delta^2$. The complex is, thus, diamagnetic. This can be verified by magnetic susceptibility measurements.

(iii) The elongation of the Cr–Cr bond distances is attributed to the effect of intramolecular donation by the *ortho* substituent X to the σ^* orbital of the Cr_2 core. For X $=$ Me, there is no significant axial donation to the Cr_2 core. The Cr–Cr bond distance increases for X $=$ OMe $<$ Cl $<$ Br. As the size and "softness" of the substituents X increases, more electron density can be donated to the axial σ^* orbital of the Cr_2 core.

A12.6 (i) Total number of valence electrons in the Cr_2 core in **B**: 6×2 (2 Cr atoms) $+ 3 \times 2$ (2 pyridyl amido ligands) $+ 1$ (1 Me ligand) $+ 1$ (1 $AlMe_2$ ligand) $= 20$. Number of electrons used to form metal-ligand bonds is $2 \times 6 = 12$. Therefore, eight electrons are left for the Cr–Cr bond. Bond order $= 4$.

(ii) The Cr–Cr quintuple bond in **A** undergoes cycloaddition reaction with the C–C triple bond of diphenyl acetylene, forming dichromium(II) **C**:

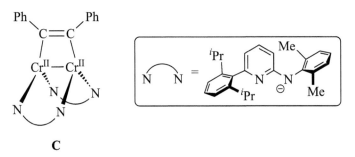

C

The Cr–Cr bond order is reduced from 5 (**A**) to 4 (**C**).

A12.7 (i) & (ii)

Molecular orbital	Symmetry species	Nature	Energy level diagram*
(d)	$1b_{1g}$	nonbonding	
(e)	$2a_{1g}$	nonbonding	
(f)	$1a_{2u}$	π bonding	
(c)	$1b_{2g}$	bonding	
(b) and (g)	$1e_u$	bonding	
(a)	$1a_{1g}$	bonding	

* The ordering of the molecular orbitals is based on qualitative arguments.

(iii) The most important resonance structures of Al_4^{2-} should be:

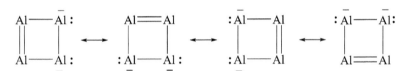

Hence the Al–Al bond should have a bond order of $1\frac{1}{4}$. In addition, there are two π electrons around the ring, satisfying the $(4n + 2)$ rule. Thus the Al_4^{2-} anion has been called the first all-metal aromatic species. Finally, the bonding picture put forth by the resonance theory: four σ bonds, one π bond, and two lone pairs, is in agreement with the description provided by the molecular orbital theory: five occupied delocalied bonding orbitals (one of which is π-type) and two filled delocalized nonbonding orbitals.

(iv) The (partially filled or vacant) $3d_{z^2}$, 4s, and $4p_z$ atomic orbitals of Cu^+ can interact with the filled $1a_{2u}$ (or π) orbital of Al_4^{2-} to form $CuAl_4^-$.

A12.8 (i) The electronic configuration of Mo is $4d^5 5s^1$. Thus there are $6 \times 6 + 8 \times 1 - 4 = 40$ valence electrons in $Mo_6Cl_8^{4+}$. On each triangular face, one Cl atom forms a 4c–2e Mo–Cl bond with three Mo atoms. These eight 4c–2e bonds use up 16 of the valence electrons, leaving 24 electrons for the Mo_6 skeleton, with 12 Mo–Mo linkages. Hence each Mo–Mo linkage is a 2c–2e bond, which is a single bond with bond order equal to 1. On the other hand, the electronic configuration of Ta is $5d^3 4s^2$. So there are also 40 valence electrons in $Ta_6 Cl_{12}^{2+}$. On each edge of the Ta_6 octahedron, one Cl atom forms a 3c–2e Ta–Cl bond with two Ta atoms. These 12 3c–2e bonds use up 24 of the valence electrons, leaving 16 (eight pairs of) electrons for the Ta_6 skeleton, which has 12 Ta–Ta linkages. So each Ta–Ta linkage has a bond order of $8/12 = 2/3$.

(ii) Based on the bonding picture described above, it is clear that there are eight electron pairs around each metal atom. In accordance with the VSEPR model, these electron pairs are arranged in a square antiprismatic way, with four pairs above the four edges and the other four pairs above the four faces.

A12.9 (i) $[Mo_2(SO_4)_4]^{3-}$ has 23 valence electrons $[6 \times 2$ (two Mo atoms) $+ 2 \times 4$ (four SO_4^{2-} ligands) $+ 3$ (−3 charges)]. Seven of them participate in the Mo–Mo bond, giving a bond order of $3^1/2$.

(ii) The Mo–Mo bond has an electronic configuration $\sigma^2 \pi^4 \delta^1$. There is an unpaired electron in the δ orbital.

A12.10 (i) The total number of valence electrons in **B**: 5×2 (two Ta atoms) $+ 1 \times 4$ (four Cl atoms) $+ 2 \times 4$ (four PMe_3 ligands) $+ 1 \times 2$ (two H atoms) $= 24$. After we subtract 20 electrons for metal-ligand bonding, four electrons are left for Ta–Ta bonding. Thus the Ta–Ta bond order is 2.

(ii)

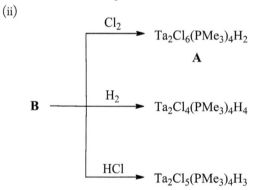

A12.11

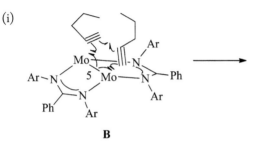

B

A12.12 (i)

B

B1   **B2**

(ii)   The structure of **B** can be represented by two resonance structures, namely **B1**
and **B2**, which have Mo–Mo bond orders of 4 and 3 respectively. Therefore,
B has a Mo–Mo bond order of $3^1/_2$.

A12.13   The $[Re_2Cl_8]^{2-}$ complex has a quadruple Re–Re bond with a $\sigma^2\pi^4\delta^2$ electronic
configuration. The eclipsed conformation of this complex is attributed to the pres-
ence of the δ bond. On the other hand, the Os–Os bond in $[Os_2Cl_8]^{2-}$ contains
10 electrons, giving rise to a $\sigma^2\pi^4\delta^2\delta^{*2}$ electronic configuration (i.e., a bond order
of 3). The absence of a net δ component in the Os–Os linkage of $[Os_2Cl_8]^{2-}$ results
in a less strained staggered conformation.

A12.14 (i) Magnesium acts as a reducing agent.

(ii) Ligand L is dianionic. Thus **A** formally contains a $[Mo_2]^{4+}$ core. The electronic configuration of a Mo atom is $4d^5 5s^1$. Therefore, the $[Mo_2]^{4+}$ core contains eight valence electrons with an electronic configuration of $\sigma^2 \pi^4 \delta^2$. The Mo–Mo bond order in **A** is 4.

A12.15 (i) **A** contains four anionic cluster carboxylate ligands and two free cluster acid ligands. It formally contains a $[Mo_2]^{4+}$ core. Because the electronic configuration of a Mo atom is $4d^5 5s^1$, the $[Mo_2]^{4+}$ core contains eight valence electrons with an electronic configuration of $\sigma^2 \pi^4 \delta^2$. **A** has a formal Mo–Mo bond order of 4.

(ii) Based on the intensity ratio, the stretching bands at 2105 and 2111 cm^{-1} are assignable to the equatorial and axial tricobalt cluster ligands, respectively. The carbonyl stretching frequencies are sensitive to the electron density on the tricobalt units. A lower carbonyl frequency for the tricobalt units of the equatorial $[\{(CO)_3Co\}_3(\mu^3\text{-}COO)]^-$ ligands ($\nu_{CO} = 2105$ cm^{-1}) indicates an increase in the electron density on the tricobalt clusters. On the other hand, the carbonyl stretching bands for the tricobalt cluster units of free $[\{(CO)_3Co\}_3(\mu^3\text{-}COOH)]$ acid and the axially coordinated acid ligands are identical ($\nu_{CO} = 2111$ cm^{-1}), suggesting that the Mo_2 core exerts very little effect on the electron density of the axially coordinated cluster acid ligands.

A12.16 (i) The electronic configuration of a Mo atom is $4d^5 5s^1$. $[Mo_2(NMe_2)_6]$ formally contains a $[Mo_2]^{6+}$ core. Therefore, the $[Mo_2]^{6+}$ core contains six valence electrons with an electronic configuration of $\sigma^2 \pi^4$ (a formal Mo–Mo bond order of 3).

(ii) $[Mo_2(NMe_2)_6]$ is diamagnetic as it does not contain any unpaired electrons.

(iii) The amido ligand $-NMe_2$ in $[Mo_2(NMe_2)_6]$ shows a nearly trigonal planar geometry. Thus the amido nitrogen atom is considered to be sp^2 hybridized. Besides, the Mo–N bond distances are short. These two structural features suggest a considerable N to Mo π-bonding interaction in $[Mo_2(NMe_2)_6]$. The π-electron density is donated to the δ and δ^* orbitals of the metal-metal bond, resulting in the lengthening of the Mo–Mo triple bond in $[Mo_2(NMe_2)_6]$ relative to that in $[Mo_2(CH_2SiMe_3)_6]$.

A12.17 (i) The electronic configuration of Re is $5d^5 6s^2$. Thus there are $3 \times 7 + 12 \times 1 + 3 = 36$ valence electrons in $Re_3Cl_{12}^{3-}$. On each Re–Re linking, each bridging Cl atom forms a 3c–2e Re–Cl bonds with two Re atoms. The three 3c–2e and nine terminal 2c–2e Re–Cl bonds use up 24 of the valence electrons, leaving 12 electrons for Re–Re bonding. Thus each Re–Re linkage is a double bond, with bond order equal to 2. Analogously, there are 24 valence electrons in $Re_2Cl_8^{2-}$. After using up 16 of these electrons for the eight Re–Cl bonds, there are eight electrons for the Re–Re linkage. Hence we expect the Re–Re linkage to be a quadruple bond, with bond order equal to 4.

(ii) Based on the bonding picture of $Re_2Cl_8^{2-}$ described above, each Re has eight electron pairs in its valence shell and they are arranged in a square antiprismatic manner. Such an arrangement for both metal atoms will result in an eclipsed conformation for the eight Cl atoms. This is in agreement with the crystallographic results.

A12.18 (i) Methyllithium (LiMe).

(ii) Both a shorter Pt–Be bond length and a smaller Cl–Be–X angle in $[(Cy_3P)_2Pt-BeCl_2]$ are indicative of a stronger Pt–Be bond. This is attributed to a more Lewis acidic $BeCl_2$ moiety compared with $BeCl(Me)$.

A12.19 (i) The atomic size of Fe is smaller than that of Ru. According to VSEPR theory, the Fe atom prefers to occupy an "axial" position.

(ii) Using Wade's skeletal electron counting rules, the total number of electrons available for cluster bonding in $FeRu_3(CO)_{12}(PhC \equiv CPh)$:

$$1 \times Fe(CO)_3 = 1 \times 2 = 2$$
$$3 \times Ru(CO)_3 = 3 \times 2 = 6$$
$$2 \times CPh \quad\quad = 2 \times 3 = 6$$

Total number of electrons $= 2 + 6 + 6 = 14$, corresponding to seven electron pairs. Hence, the cluster has $2n + 2$ electrons for cluster bonding, i.e. a *closo* structure.

(iii)

A12.20 A diplumbene can be considered as a plumbylene ($:PbR_2$) dimer. Unlike the C=C double bonds in alkenes, the Pb=Pb double bonds in diplumbenes make use of valence 6s electrons in bonding through a 6s-to-6p donor-acceptor interaction:

The intrinsic weakness of this donor-acceptor interaction is attributed to a relativistic contraction of the valence 6s orbital of lead, which increases the energy gap between the 6s and 6p orbitals. This reduces the stability of the 6s $\rightarrow$ 6p donor-acceptor interactions. In fact, the X=X (where X denotes Si, Ge, Sn, and Pb) bond strength decreases on going from silicon to lead.

Bioinorganic Chemistry | **13**

PROBLEMS

13.1 Cryptands are "three-dimensional" analogues of crown ethers, both of which are macrocyclic ligands for metal cations. The binding of K^+ by [2.2.2]cryptand in methanol is $\sim 10^4$ times stronger than 18-crown-6, though these two macrocyclic ligands have similar ring size. Account for this dramatic enhancement of K^+ binding ability of [2.2.2]cryptand over its crown ether analogue.

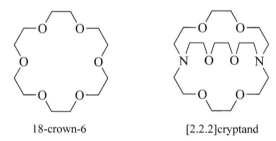

18-crown-6   [2.2.2]cryptand

REFERENCE: J.-M. Lehn, Supramolecular chemistry: concepts and perspectives. VCH, Weinheim (1995).

13.2 The binding free energies of [2.2.2]cryptand for alkali metal ions are given below:

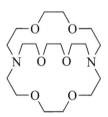

[2.2.2]cryptand

Problems in Structural Inorganic Chemistry. Second edition. Wai-Kee Li, Hung Kay Lee, Dennis Kee Pui Ng, Yu-San Cheung, Kendrew Kin Wah Mak, and Thomas Chung Wai Mak. © Oxford University Press 2019. Published in 2019 by Oxford University Press. DOI: 10.1093/oso/9780198823902.001.0001

Alkali metal ions	Li^+	Na^+	K^+	Rb^+
$\Delta G^\ominus$ (kJ mol^{-1})	-41.8	-74.1	-64.0	-53.1

Explain the variation of these Gibbs binding free energies.

REFERENCE: D. J. Cram, Preorganization—from solvents to spherands. *Angew. Chem. Int. Ed. Engl.* **25**, 1039–57 (1986).

13.3 The binding constants (K) of $[2.2.2]$cryptand $(\mathbf{A})$ and a series of its nitrogen analogues, $\mathbf{B}$ to $\mathbf{D}$, for K^+ and Ag^+ in water are shown below.

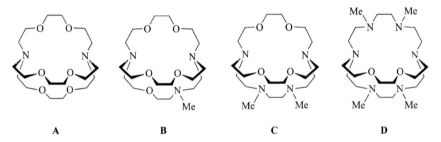

| | $\mathbf{A}$ | $\mathbf{B}$ | $\mathbf{C}$ | $\mathbf{D}$ |

Ligand	Log K	
	K^+	Ag^+
A	5.4	9.6
B	4.2	10.8
C	2.7	11.5
D	1.1	13.0

The macrocyclic ligands $\mathbf{A}$ to $\mathbf{D}$ exhibit a smooth change in the binding constants as the oxygen atoms are successively replaced by nitrogen atoms. Explain this variation of the binding constants of ligands $\mathbf{A}$ to $\mathbf{D}$ for (i) K^+ and (ii) Ag^+.

REFERENCE: J. W. Steed and J. L. Atwood, Supramolecular chemistry. 2nd ed., Wiley, Chichester (2009).

13.4 Valinomycin, an antibiotic isolated from *Streptomyces*, is a dodecadepsipeptide macrocyclic ionophore with a high selectivity for K^+ and Na^+ ions. This ionophore binds the alkali metal ions utilizing six carbonyl ligands. The binding constant of the K^+-valinomycin complex is $\sim$5 orders of magnitude higher than that of the Na^+-valinomycin counterpart.

Valinomycin

(i) How can valinomycin facilitate transport of K^+ ions through the hydrophobic lipid bilayer of cellular membranes?

(ii) Suggest one possible explanation which accounts for the high K^+ and Na^+ selectivity of valinomycin.

REFERENCES: M. Dobler, *Ionophores and their structures*, John Wiley and Sons, New York (1981); M. C. Rose and R. W. Henkens, Stability of sodium and potassium complexes of valinomycin. *Biochim. Biophys. Acta* **372**, 426–35 (1974); S. Varma, D. Sabo, and S. B. Rempe, K^+/Na^+ selectivity in K channels and valinomycin: over-coordination *versus* cavity-size constraints. *J. Mol. Biol.* **376**, 13–22 (2008).

13.5 Enterobactin is a siderophore produced by enteric bacteria such as *Escherichia coli* for effective chelation and transport of Fe(III). The molecular structure of enterobactin consists of a trilactone backbone with three catechol pendants. Enterobactin has the highest known affinity for ferric ion $(K_f = 10^{52} \text{ L mol}^{-1})$, forming a very stable Fe(III)–enterobactin complex through the six catecholate oxygen atoms.

Enterobactin

(i) In order to investigate the solution-state structure of the Fe(III)-enterobactin complex by NMR, a gallium(III) derivative was prepared for comparison. Why are gallium(III) derivatives often used in this type of study?

(ii) Both Δ- and Λ-complexes are possible upon coordination of the six catecholate oxygen atoms of enterobactin with Fe(III). In fact, the Δ-complex forms preferentially. Suggest one possible reason for this observation. (Hint: the trilactone backbone of natural enterobactin is derived from three ℓ-serine units.)

REFERENCES: M. Llinis, D. M. Wilson, and J. B. Neilands, Effect of metal binding on the conformation of enterobactin. A proton and carbon-13 nuclear magnetic resonance study. *Biochemistry* **12**, 3836–43 (1973); B. F. Matzanke, G. Mtiller-Matzanke, and K. N. Raymond, in: Iron carriers and iron proteins (T. M. Loehr, Ed.), VCH, New York, 1989, p. 1; T. D. P. Stack, T. B. Karpishin, and K. N. Raymond, Structural and spectroscopic characterization of chiral ferric tris-catecholamides: unraveling the design of enterobactin. *J. Am. Chem. Soc.* **114**, 1512–4 (1992).

13.6 A group of students examined the binding of enterobactin with iron. The complex $[CrL_3]^{3-}$, where L is the dianion of $1,2\text{-}(HO)_2C_6H_4$, was used as a model compound in their research work.

(i) Why was a chromium complex, instead of an iron derivative, used in this research project?

(ii) Besides circular dichroism, name one spectroscopic method which is commonly employed to examine the binding interactions of siderophores with metal ions. Explain your answers.

13.7 A tris(catechol) compound, MECAM, was prepared as a synthetic analogue for enterobactin (ent). MECAM is a very strong iron complexant. It takes up Fe^{3+} to form a very stable ferric complex, $[Fe(MECAM)]^{3-}$. The binding constant of $[Fe(MECAM)]^{3-}$ is on the order of $\sim 10^{46}$, which is almost 10^6 times lower than that of $[Fe(ent)]^{3-}$

$(K_f = 10^{52} \text{ L mol}^{-1})$. Suggest an explanation for the difference in the binding affinity of enterobactin and MECAM for Fe^{3+}.

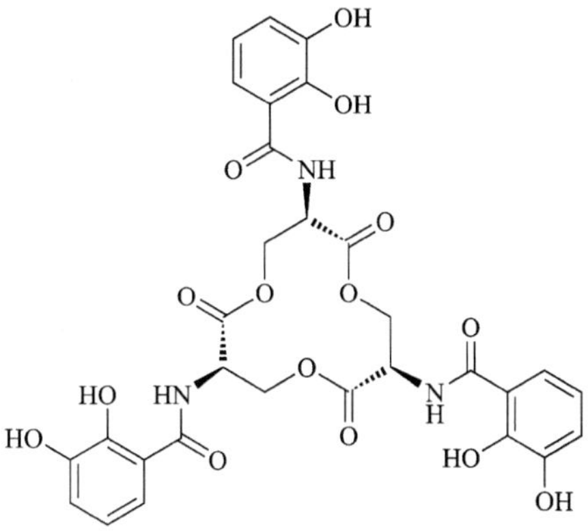

Enterobactin 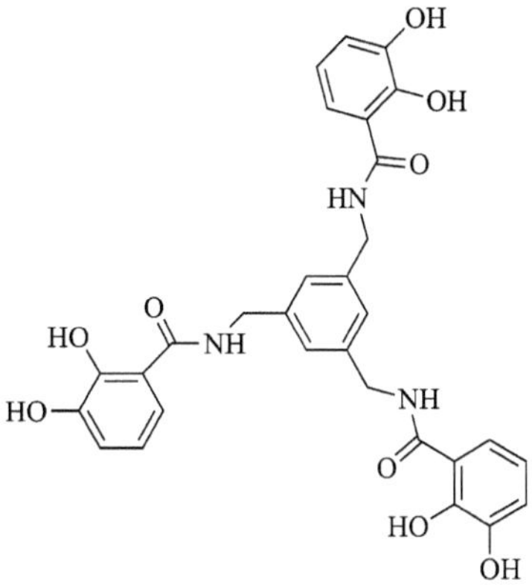   MECAM

REFERENCE: M. S. Mitchell, D.-L. Walker, J. Whelan, and B. Bosnich, Biological analogues. Synthetic iron(III)-specific chelators based on the natural siderophores. *Inorg. Chem.* **26**, 396-400 (1987).

13.8 The macrocyclic molecules **A** and **B** belong to a class of compounds known as *siderands* (synthetic siderophores). These compounds provide large molecular cavities for metal cations. Compound **A** displays an exceptionally high binding constant ($\sim 10^{59}$) for Fe^{3+}. What do you expect the binding constant of Fe^{3+}-**B** might be compared with that of Fe^{3+}-**A**?

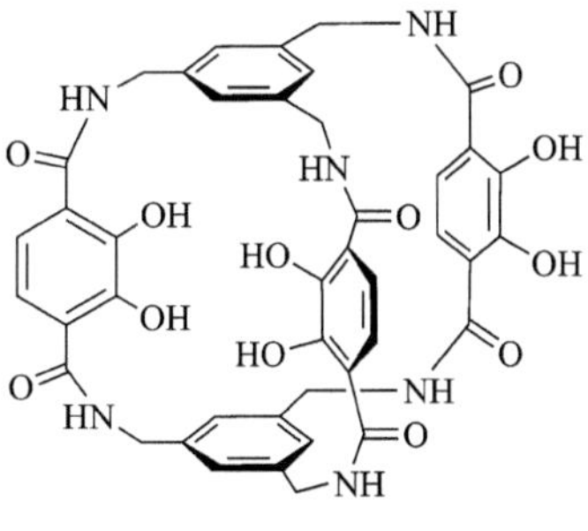

A 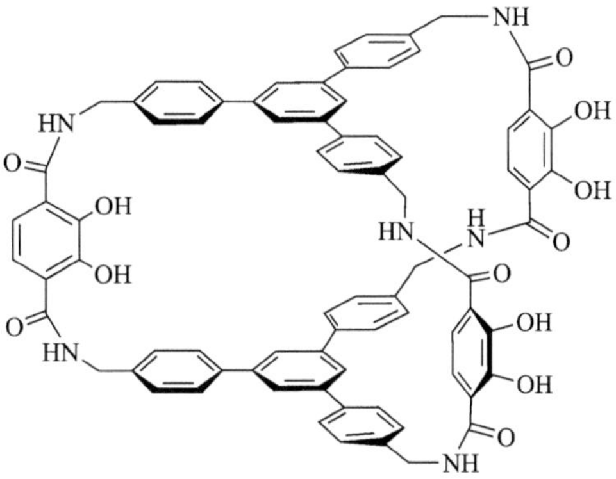   B

REFERENCE: P. Stutte, W. Kiggen, and F. Vogtle, Large molecular cavities bearing siderophore type functions. *Tetrahedron* **43**, 2065–74 (1987).

13.9 Describe one general mechanism of heavy metal poisoning.

13.10 Tetraphenylporphyrin H_2TPP and its metalated derivatives have been extensively studied. The electronic absorption spectra of H_2TPP, $Ni(TPP)$ and $Zn(TPP)$ are shown below:

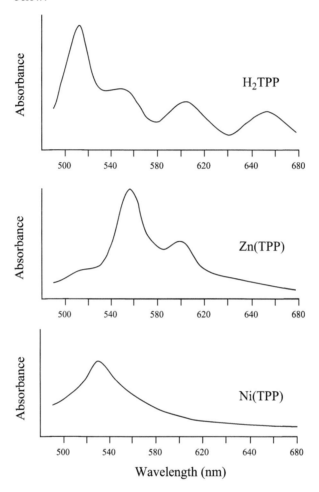

(Note: The absorbance of these spectra is not shown on the same scale.)

(i) Why does the absorption maximum of $Ni(TPP)$ show a blue shift as compared to that of $Zn(TPP)$?

(ii) Why do the electronic spectra of the two metal porphyrins $M(TPP)$ ($M = Ni, Zn$) show fewer absorption features as compared to that of the free porphyrin H_2TPP?

REFERENCE: D. F. Marsh and L. M. Mink, Microscale synthesis and electronic absorption spectroscopy of tetraphenylporphyrin $H_2(TPP)$ and metalloporphyrins $Zn^{II}(TPP)$ and $Ni^{II}(TPP)$. *J. Chem. Edu.* **73**, 1188-90 (1996).

13.11 Complexes **A** and **B** (see figure below) are five-coordinate porphyrin complexes in which one of the axial coordinate sites is occupied by an imidazole base. They can bind dioxygen in a reversible manner. Thus, they are considered as synthetic analogues for the five-coordinate heme iron sites in hemoglobin and myoglobin. The binding of **A** and **B** with carbon monoxide has also been studied. Both complexes have greater affinity for CO than O_2, though the CO-binding affinity of **B** is much lower than that of **A**.

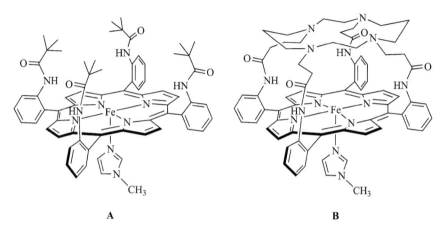

A   B

(i) What is an advantage of using these two complexes as synthetic models for oxygenation reactivity studies as compared with the use of the simple $Fe^{II}(TPP)$ (TPP = tetraphenylporphyrin) complex?

(ii) Why does CO bind tightly to iron(II) porphyrins?

(iii) Why do complexes **A** and **B** show different CO binding affinity?

REFERENCE: J. P. Collman and L. Fu, Synthetic models for hemoglobin and myoglobin. *Acc. Chem. Res.* **32**, 455-63 (1999).

13.12 Catalase is a heme protein which catalyzes the disproportionation of H_2O_2 to H_2O and O_2. The resting state of catalase consists of a ferric heme active center, which is oxidized by H_2O_2 to form a high-valent iron oxo species during catalytic turnovers. This high-valent iron oxo species can reduce another molecule of H_2O_2 to form H_2O and O_2 with regeneration of the ferric heme center.

$$2\,H_2O_2 \xrightarrow{\text{Catalase}} 2\,H_2O + O_2$$

The mechanism is proposed as below:

$$Fe(III) + H_2O_2 \rightarrow \text{``Por}^{\cdot+}-Fe(IV)=O\text{''}$$
$$\text{``Por}^{\cdot+}-Fe(IV)=O\text{''} + H_2O_2 \rightarrow Fe(III) + H_2O + O_2$$

(i) Suggest one single physical method for identifying the oxidation state of the heme center of catalase during enzymatic turnovers. Explain your answers.

(ii) Interestingly, carbon monoxide does not affect the function of catalase but an aqueous solution of sodium cyanide inhibits the function of this enzyme. Account for this observation.

REFERENCES: J. T. Groves, R. C. Haushalter, M. Nakamura, T. E. Nemo, and B. J. Evans, High-valent iron-porphyrin complexes related to peroxidase and cytochrome P-450. *J. Am. Chem. Soc.* **103**, 284–6 (1981); M. Alfonso-Prieto, X. Biarnés, P. Vidossich, and C. Rovira, The molecular mechanism of the catalase reaction. *J. Am. Chem. Soc.* **131**, 11751–61 (2009).

13.13 Carbonic anhydrases are metalloenzymes, which catalyze the interconversion of carbon dioxide to bicarbonate in biological systems.

$$CO_2(aq) + H_2O(l) \xrightleftharpoons[\text{anhydrase}]{\text{Carbonic}} H^+(aq) + HCO_3^-(aq)$$

The active site of human carbonic anhydrase contains a $Zn(II)$ ion

(i) Is it possible for the Zn-containing carbonic anhydrases to act as dioxygen carriers in biological systems? Why?

(ii) In order to obtain structural and mechanistic information of carbonic anhydrases, the $Zn(II)$ ion in the native enzymes was often replaced by $Co(II)$ (with retention of activity) in various studies. Explain the principles of using $Co(II)$-substituted enzymes, instead of the native enzymes, in those studies.

13.14 **A** is an electron-transfer protein which contains two redox cofactors, namely a Type I copper center and a flavin mononucleotide. In addition, **A** also contains a quinone binding site, where reduction of quinone to quinol occurs.

(i) A group of students intended to study **A** by measuring its UV-Vis absorption spectra. However, their teacher suggested that UV-Vis spectroscopy might not be an appropriate physical method for characterizing this protein. Why?

(ii) Suggest, with explanation, one appropriate physical method which can provide valuable information for understanding the electron-transfer mechanism of **A**.

13.15 Cytochrome *c* is a heme iron protein that plays a vital role in electron-transfer in a wide range of organisms. The coordination sphere of the iron center in cytochrome *c* is occupied by a heme group, an imidazole group of a histidine residue, and a thioether group of a methionine residue. The iron center cycles between the Fe(II) and Fe(III) oxidation states during electron-transfer reactions.

(i) Suggest one physical method that best suits for determination of the oxidation state of the iron center in cytochrome c. Explain your answer.

(ii) Besides electron-transfer, can cytochrome c also take up the function as a dioxygen carrier? Explain your answer.

(iii) Compare and explain any difference you expect on the rate of electron-transfer if the iron center in cytochrome c is replaced by cobalt.

13.16 Comment on the following observations:

(i) Although both cytochrome c and hemoglobin are heme proteins, the iron center in cytochrome c has a six-coordinate geometry whereas that in deoxyhemoglobin is five-coordinated.

(ii) The reduction potential of the enzyme plastocyanin from *P. vulgaris* is found to be +360 mV (vs. NHE) at pH 7. Comment on the relevance of this reduction potential for the understanding of the structure and redox chemistry of this enzyme.

13.17 Plastocyanin (CuPc) is a blue copper protein which plays an important role in electron transport in photosynthesis. This protein undergoes reversible oxidation and reduction reactions, carrying electrons from cytochrome f to $P700^+$ of Photosystem I (PSI):

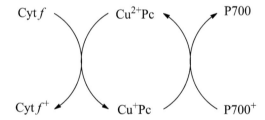

(i) The enzyme plastocyanin is blue in color in its oxidized state, with a characteristic absorption band at $\lambda_{max} \sim 600$ nm ($\varepsilon \sim 5000$ L mol^{-1} cm^{-1}). The intensity of this absorption is two orders of magnitude greater than those reported for common tetragonal copper(II) complexes ($\varepsilon \sim 100$ L mol^{-1} cm^{-1}). Account for the different absorption intensities of plastocyanin and common copper(II) complexes.

(ii) Besides electronic absorption spectroscopy, suggest one single spectroscopic method which can be used to monitor the redox state of plastocyanin. Explain your answers.

(iii) Treatment of chloroplasts with a high dosage of potassium cyanide (KCN) resulted in inhibition of electron-transfer from cytochrome *f* to P700$^+$. Account for this phenomenon.

REFERENCES: J. M. Berg and S. J. Lippard, *Principles of bioinorganic chemistry*. University Science Books, Sausalito, California (1994); A. A. Gewirth and E. I. Solomon, Electronic structure of plastocyanin: excited state spectral features. *J. Am. Chem. Soc.* **110**, 3811–9 (1988); S. Izawa, R. Kraayenhof, E. K. Ruuge, and D. Devault, The site of KCN inhibition in the photosynthetic electron transport pathway. *Biochim. Biophys. Acta* **314**, 328–39 (1973).

13.18 Plastocyanin was the first blue copper protein to be structurally characterized by X-ray diffraction analysis. In the oxidized form, the active site of plastocyanin consists of a highly distorted "tetrahedral" Cu(II) ion, which is coordinated by two histidine nitrogen atoms, one cysteine thiolate group and one methionine thioether group (see the figure below). Very little structural changes around the copper site were observed upon reduction of the protein.

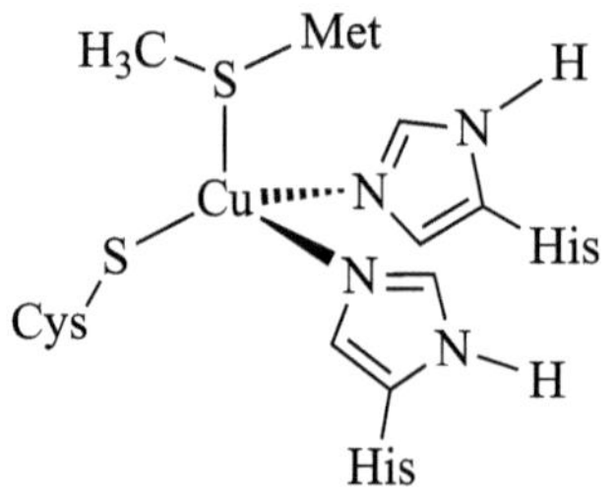

Explain how the structures of the copper site in both the oxidized and reduced forms of plastocyanin are related to its function as an electron-transfer protein.

REFERENCES: P. M. Colman, H. C. Freeman, J. M. Guss, M. Murata, V. A. Norris, J. A. M. Ramshaw, and M. P. Venkatappa, X-ray crystal structure analysis of plastocyanin at 2.7 Å resolution. *Nature* **272** 319–24 (1978); B. L. Vallee and R. J. P. Williams, Metalloenzymes: the entatic nature of their active sites. *Proc. Natl. Acad. Sci. U.S.A.* **59**, 498–505 (1968); R. A. Marcus and N. Sutin, Electron transfers in chemistry and biology. *Biochim. Biophys. Acta* **811**, 265–322 (1985).

13.19 Succinate:ubiquinone oxidoreductase (SQR) is a multi-subunit enzyme in the tricarboxylic acid cycle. This enzyme catalyzes the two-electron reduction of succinate to fumarate, while the electron-transfer to quinone to yield quinol. A schematic diagram of the enzyme is shown below.

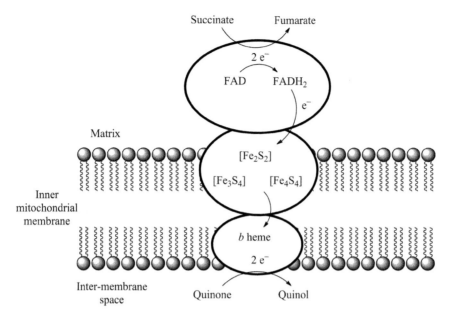

The SQR enzyme isolated from *Paracoccus denitrificans* contains one covalently bound flavin cofactor (FAD), three iron-sulfur clusters of the type $[Fe_2S_2]$, $[Fe_3S_4]$, and $[Fe_4S_4]$, and one mono-heme cytochrome *b*. SQR from *P. denitrificans* has been studied by EPR experiments at different levels of reduction, namely the air-oxidized, succinate-reduced, and dithionite-reduced forms.

(i) Suggest one physical method for determination of the FAD content of this enzyme in the air-oxidized form (the resting state). Briefly describe the design of the experiment. (Hint: the FAD moiety is a chromophore.)

(ii) Upon reduction of the enzyme with succinate, the FAD cofactor turned out to be EPR active. However, the FAD moiety became EPR silent when the enzyme was treated with dithionite. Account for these observations.

(iii) The air-oxidized SQR enzyme showed one nearly isotropic EPR signal centered at $g = 2.006$ (at 4 K). This signal was attributed to the $[Fe_3S_4]$ center. Deduce the oxidation state of the iron centers and the overall spin state of the $[Fe_3S_4]$ cluster in the air-oxidized enzyme.

(iv) The $[Fe_2S_2]$ cluster of the air-oxidized SQR enzyme was EPR silent, while it became EPR active in both the succinate- and dithionite-reduced enzyme. Predict the oxidation state of the two iron centers in this iron-sulfur cluster in both the oxidized and reduced forms of the enzyme.

REFERENCE: A. R. Waldeck, M. H. B. Stowell, H. K. Lee, S.-C. Hung, M. Matsson, L. Hederstedt, B. A. C. Ackrell, and S. I. Chan, Electron paramagnetic resonance studies of succinate:ubiquinone oxidoreductase from *Paracoccus denitrificans. J. Biol. Chem.* **272**, 19373–82 (1997).

SOLUTIONS

A13.1 The K^+ cation is encapsulated entirely within a pre-organized, three-dimensional cavity of the [2.2.2]cryptand host molecule. This strongly favors spherical recognition of metal cations by cryptands to take place.

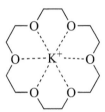

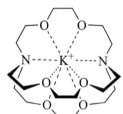

K^+-18-crown-6 K^+-[2.2.2]cryptand

A13.2 Cryptands are host molecules which are conformationally organized (pre-organized) to favor the selective binding of one cation over another. The macrobicyclic cavity of [2.2.2]cryptand is rigid such that it cannot constrict or expand sufficiently to fit cations of too small or too large ionic radii. See the figure in the answer to the previous question.

A13.3 (i) The binding constants of the macrocyclic ligands for K^+ decrease in the order **A > B > C > D**. Because the binding of K^+ by these macrocyclic ligands is largely ionic in nature, K^+ is more stabilized by oxygen donor ligands.

(ii) On the other hand, the Ag^+–N bonding interaction has a relatively high covalent character and is, thus, more stabilized. Therefore, the binding constants of the macrocyclic ligands for Ag^+ increase in the order **A < B < C < D** as the oxygen atoms on the ligands are replaced by nitrogen.

A13.4 (i) An X-ray diffraction analysis revealed that the K^+-valinomycin complex adopts a configuration in which the six carbonyl oxygen atoms point inward, whereas its amino acid and hydroxy acid side-chains point outward. This forms a cavity for ion complexation and a hydrophobic exterior, facilitating the movement of the K^+-valinomycin complex across biological membranes.

(ii) It was first believed that the ring size of valinomycin plays an important role in achieving K^+/Na^+ selectivity. Results of later studies revealed that the selectivity is also determined by hydration energies of the metal ions because both metal ions must lose their solvation shell before being taken up by valinomycin. Given that the ionic radius of Na^+ and K^+ is 1.16 Å and 1.52 Å, respectively. Therefore, the q^2/r ratio of the smaller Na^+ ion is larger than that of K^+ ion. Note that hydration energies are related to the q^2/r ratio of the metal ions. Thus, the hydrated form of Na^+ is energetically more stable as compared to hydrated K^+ ions. On the basis of this energetic consideration, the binding of K^+ to valinomycin is favored.

A13.5 (i) NMR studies can provide valuable information regarding the conformation of metal-enterobactin complexes in solution. Substitution of the paramagnetic iron(III) ion by a diamagnetic gallium(III) ion can avoid line broadening in the NMR spectra. In addition, gallium(III) is non-reducible as compared to iron(III).

(ii) Because the trilactone backbone of enterobactin is chiral, two diasteroisomers, namerly Δ-*cis* and Λ-*cis*, are possible. These diasteroisomers have different thermodynamic and physical properties.

A13.6 (i) The crystal field stabilization energy (CFSE) of an octahedral chromium(III) complex (d^3) is 12 Dq. Therefore, octahedral chromium(III) complexes are kinetically more stable than an analogous high-spin iron(III) complex (d^5), which has a CFSE value $= 0$.

(ii) UV/Vis spectroscopy can provide important information on the binding of ligands to metal ions.

A13.7 Enterobactin has a trilactone backbone, whilst the mesityl spacer group in MECAM is smaller in size. Compared to $[Fe(ent)]^{3-}$, the $[Fe(MECAM)]^{3-}$ complex, presumably, has a more strained molecular geometry.

A13.8 Compound **A** has rigid and pre-organized structure with a molecular cavity of an appropriate size for encapsulation of Fe^{3+} ion. The larger spacer units (1,3,5-triphenylmesityl) in compound **B** result in an over-sized cavity for Fe^{3+} ion. Therefore, we expect that the binding constant of Fe^{3+}-**B** is lower than that of Fe^{3+}-**A**.

A13.9 Heavy metal ions are Lewis acids. They have strong binding interactions with most of the biological molecules, such as proteins, polypeptides and nucleic acids, which

A13.10 (i) An intense absorption observed in the 500-550 nm region in the electronic absorption spectra of porphyrin compounds is attributed to a $\pi \rightarrow \pi^*$ transition of the porphyrin ring. In transition-metal porphyrins, a metal-to-porphyrin back bonding due to interactions of the empty π^* orbital of the porphyrin ligand and the filled d_{xz} and d_{yz} (d_π) orbitals of the metal may be present.

For metal ions with a closed shell electronic configuration (e.g. Zn^{2+}, d^{10}), and the energy of their d_π orbitals must be relatively low to have much influence on the $\pi \rightarrow \pi^*$ energy gap. On the other hand, $[Ni^{II}(TPP)]$ is a hyposoporphyrin in which the metal ion has a d^8 electronic configuration. A more significant $d_\pi \rightarrow \pi^*$ back bonding interaction is present in the case of $[Ni^{II}(TPP)]$ which results in a larger $\pi \rightarrow \pi^*$ energy gap.

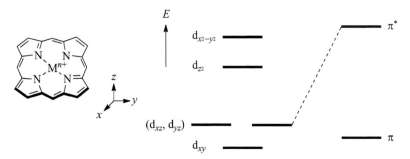

(ii) H_2TPP (D_{2h}) has a relatively lower molecular symmetry than the metalated derivatives $[M(TPP)]$ ($M = $ Ni, Zn; D_{4h}). In general, a more symmetrical molecule gives a simpler spectrum with less spectral features.

A13.11 (i) Complex **A** is a "picket fence" iron porphyrin complex and complex **B** is a "capped" iron porphyrin. $[Fe^{II}(TPP)]$ does not bind O_2 reversibly due to formation of a μ-peroxo diiron(III) complex. The sterically congested environment of **A** and **B** can prevent the formation of undesirable μ-peroxo dimer.

(ii) CO is a good σ-donor and π-acceptor ligand which binds in a linear manner to heme iron centers. A metal-to-ligand back bonding enhances the iron carbonyl ($Fe \leftarrow C \equiv O$) bond.

(iii) A small binding cavity in the pocket of "capped" porphyrin **B** does not favor a linear binding mode of CO to the heme iron center.

A13.12 (i) Mössbauer spectroscopy is a useful physical method for studying ^{57}Fe nuclei. The oxidation state and coordination environment of an iron center in a system can be deduced by analysis of the isomer shift and splitting pattern of its Mössbauer signal.

EPR spectroscopy can also be used as the ferric heme [Fe(III)] and the "Por$^{\cdot+}$–Fe(IV)=O" species generally show quite distinctive EPR signals. Besides, UV-visible and extended X-ray absorption fine structure (EXAFS) spectroscopy can provide valuable information on the coordination environment around the heme iron center.

(ii) Cyanide anion (CN$^-$) has a high affinity for Fe(III) and, hence, it inhibits the enzymatic activity of catalase. In general, carbon monoxide has a very strong affinity for Fe(II) heme centers but not Fe(III) heme compounds.

A13.13 (i) The zinc(II) center (with a d^{10} configuration) in carbonic anhydrases does not undergo further oxidation. Therefore, the zinc-containing carbonic anhydrases do not bind oxygen molecules.

(ii) Cobalt(II) has a d^7 electronic configuration and, thus, it is paramagnetic. This facilitates their complexes to be studied by UV/Vis spectroscopy, EPR spectroscopy, and magnetic moment measurements.

A13.14 (i) Both the Type I copper center and the flavin mononucleotide cofactor exhibit absorptions in the UV/Vis region with plausible (or maybe likely) overlapping absorption peaks. Therefore, UV/Vis spectroscopy is not the best physical method for studying protein **A**.

(ii) Electron paramagnetic resonance (EPR) spectroscopy is an excellent physical method to study **A**. Redox titrations of **A** can give distinct EPR signals for the copper(II) center, flavin mononucleotide radical, and semi-quinone at different measurement temperatures.

A13.15 (i) Mössbauer spectroscopy. EPR spectroscopy is also an acceptable answer as Fe(III) species are EPR active whereas Fe(II) species generally do not elicit EPR signals in X-band EPR spectra.

(ii) No, as there is no available vacant coordination site for binding oxygen molecules on the iron center in cytochrome c.

(iii) Compared with the oxidation of Fe(II) $\rightarrow$ Fe(III), the Co(II) $\rightarrow$ Co(III) process is kinetically more difficult to occur.

A13.16 (i) Cytochrome c is an electron-transfer protein. The iron center in cytochrome c switches between the +II and +III oxidation states during electron-transfer processes. Meanwhile, hemoglobin is a carrier of oxygen molecule, for which a vacant site in the iron's coordination sphere is necessary for binding oxygen molecule.

(ii) Plastocyanin is a blue copper protein which is also an important electron carrier in the chloroplasts of higher plants and algae. The reduction potential of the copper center is regulated by the protein environment. The reduction potential plays a very important role in controlling the transfer of electrons to and from this protein in an electron-transfer chain.

A13.17 (i) The characteristic feature of a blue copper absorption spectrum is attributed to an intense ligand-to-metal charge-transfer transition. It is associated with the ligand to metal (L→M) variety. Specifically, the CT is assigned as a $S_{p\pi} \rightarrow Cu_{x^2-y^2}$ transition.

(ii) The copper site in plastocyanin shuttles between the +I and +II oxidation states during the electron-transfer processes: Cu(I) (d^{10}) and Cu(II) (d^9) are diamagnetic and paramagnetic, respectively. Therefore, the redox state of the protein can be readily monitored by EPR spectroscopy.

(iii) Copper forms stable complexes with CN^-. This leads to removal of copper from plastocyanin.

A13.18 In general, Cu(I) (d^{10}) prefers a tetrahedral geometry with soft ligands (such as S-donor ligands), whereas Cu(II) (d^9) prefers a square planar structure with hard ligands (such as O- or N-donor ligands). The distorted "trigonal pyramidal" geometry of the copper site of plastocyanin has an intermediate geometry between a tetrahedral and a square planar structure. This represents an entatic state situation, which can help to minimize the reorganization energy and facilitate electron-transfer. In addition, Marcus Theory requires little change in geometry during electron-transfer processes.

A13.19 (i) The FAD content can be determined using a fluorometric method. For experimental details, please see D. F. Wilson and D. E. King, The determination of the acid-nonextractable flavin in mitochondrial preparations from heart muscle. *J. Biol. Chem.* **239**, 2683–90 (1964).

(ii)

$$\text{FAD} \xrightarrow{e^-} \text{FAD}^{\cdot-} \xrightarrow{e^-} \text{FAD}^{2-}$$
$$S = 0 \qquad\quad S = 1/2 \qquad\quad S = 0$$

(iii) In the air-oxidized enzyme, the $[Fe_3S_4]$ cluster contains three Fe(III) centers. Two of these Fe(III) centers in the cluster undergo anti-ferromagnetic coupling. Therefore, the total spin state of the cluster is $S_T = 5/2$.

(iv) In the air-oxidized form: $[Fe_2S_2]$ contains two Fe(III) centers, $S_T = 0$ (EPR silent).

In the reduced forms: $[Fe_2S_2]$ contains one Fe(III) center $(S = 5/2)$ and one Fe(II) center $(S = 2)$, $S_T = 1/2$ (EPR active).

APPENDIX 1

Periodic Table

1																		18
1 **H** 1.0079	2											13	14	15	16	17		2 **He** 4.0026
3 **Li** 6.941	4 **Be** 9.0122											5 **B** 10.811	6 **C** 12.011	7 **N** 14.007	8 **O** 15.999	9 **F** 18.998		10 **Ne** 20.180
11 **Na** 22.990	12 **Mg** 24.305	3	4	5	6	7	8	9	10	11	12	13 **Al** 26.982	14 **Si** 28.086	15 **P** 30.974	16 **S** 32.065	17 **Cl** 35.453		18 **Ar** 39.948
19 **K** 39.098	20 **Ca** 40.078	21 **Sc** 44.956	22 **Ti** 47.867	23 **V** 50.942	24 **Cr** 51.996	25 **Mn** 54.938	26 **Fe** 55.845	27 **Co** 58.933	28 **Ni** 58.693	29 **Cu** 63.546	30 **Zn** 65.38	31 **Ga** 69.723	32 **Ge** 72.64	33 **As** 74.922	34 **Se** 78.96	35 **Br** 79.904		36 **Kr** 83.798
37 **Rb** 85.468	38 **Sr** 87.62	39 **Y** 88.906	40 **Zr** 91.224	41 **Nb** 92.906	42 **Mo** 95.96	43 **Tc** –	44 **Ru** 101.07	45 **Rh** 102.91	46 **Pd** 106.42	47 **Ag** 107.87	48 **Cd** 112.41	49 **In** 114.82	50 **Sn** 118.71	51 **Sb** 121.76	52 **Te** 127.60	53 **I** 126.90		54 **Xe** 131.29
55 **Cs** 132.91	56 **Ba** 137.33	57–71	72 **Hf** 178.49	73 **Ta** 180.95	74 **W** 183.84	75 **Re** 186.21	76 **Os** 190.23	77 **Ir** 192.22	78 **Pt** 195.08	79 **Au** 196.97	80 **Hg** 200.59	81 **Tl** 204.38	82 **Pb** 207.2	83 **Bi** 208.98	84 **Po** –	85 **At** –		86 **Rn** –
87 **Fr** –	88 **Ra** –	89–103	104 **Rf** –	105 **Db** –	106 **Sg** –	107 **Bh** –	108 **Hs** –	109 **Mt** –	110 **Ds** –	111 **Rg** –								

57 **La** 138.91	58 **Ce** 140.12	59 **Pr** 140.91	60 **Nd** 144.24	61 **Pm** –	62 **Sm** 150.36	63 **Eu** 151.96	64 **Gd** 157.25	65 **Tb** 158.93	66 **Dy** 162.50	67 **Ho** 164.93	68 **Er** 167.26	69 **Tm** 168.93	70 **Yb** 173.05	71 **Lu** 174.97
89 **Ac** –	90 **Th** 232.04	91 **Pa** 231.04	92 **U** 238.03	93 **Np** –	94 **Pu** –	95 **Am** –	96 **Cm** –	97 **Bk** –	98 **Cf** –	99 **Es** –	100 **Fm** –	101 **Md** –	102 **No** –	103 **Lr** –

Adapted from:
http://old.iupac.org/reports/periodic_table/

APPENDIX 2

Energy Conversion Factors, Physical Constants, and Atomic Units

ENERGY CONVERSION FACTORS

	cm^{-1}	$kJ\,mol^{-1}$	eV	Hartree
cm^{-1}	1	1.19627×10^{-2}	1.23985×10^{-4}	4.55635×10^{-6}
$kJ\,mol^{-1}$	83.5934	1	1.03643×10^{-2}	3.80880×10^{-4}
eV	8065.51	96.4851	1	3.67493×10^{-2}
Hartree	219474	2625.50	27.2114	1

PHYSICAL CONSTANTS

Constant	Symbol	SI Value
Speed of light	c	$2.99792458 \times 10^8\,m\,s^{-1}$
Electron charge	e	$1.602177 \times 10^{-19}\,C$
Permittivity of vacuum	ε_0	$8.8541878 \times 10^{-12}\,J^{-1}\,C^2\,m^{-1}$
Avogadro's number	N_0	$6.02214 \times 10^{23}\,mol^{-1}$
Electron mass*	m_e	$9.10939 \times 10^{-31}\,kg$
Planck constant	h	$6.62608 \times 10^{-34}\,J\,s$
Bohr radius	a_0	$5.291772 \times 10^{-11}\,m$

* In some problems in this book, electron mass is denoted as m.

Appendix 2

ATOMIC UNITS

Length	1 a.u. = a_0 = 5.29177 × 10^{-11} m (Bohr radius)
Mass	1 a.u. = m_e = 9.109382 × 10^{-31} kg (rest mass of electron)
Charge	1 a.u. = $|e|$ = 1.6021764 × 10^{-19} C (charge of electron)
Energy	1 a.u. = $\dfrac{e^2}{4\pi\varepsilon_0 a_0}$ = 27.2114 eV (potential energy between two electrons when they are 1a_0 apart) $\equiv$ 1 hartree
Angular momentum	1 a.u. = $\dfrac{h}{2\pi}$ $\equiv$ $\hbar$ = 1.05457 × 10^{-34} J s
Scale factor	$4\pi\varepsilon_0 = 1$

REFERENCE: B. K. Teo and W.-K. Li, The scales of time, length, mass, energy, and other fundamental physical quantities in the atomic world and the use of atomic units in quantum mechanical calculations. *J. Chem. Educ.* **88**, 921–8 (2011).

APPENDIX 3

Useful Integrals

$$\int_0^\infty x^n e^{-ax} dx = \frac{n!}{a^{n+1}};$$

$$\int_0^\infty x^{2n} e^{-ax^2} dx = \frac{(2n-1)(2n-3)\ldots 5 \times 3 \times 1}{2^{n+1} a^n} \sqrt{\frac{\pi}{a}};$$

$$\int_0^\infty x e^{-ax^2} dx = \frac{1}{2a};$$

$$\int_0^\infty x^3 e^{-ax^2} dx = \frac{1}{2a^2};$$

$$\int \sin x \, dx = -\cos x + C;$$

$$\int \cos^2 x \, dx = \frac{1}{2}(x + \sin x \cos x) + C;$$

$$\int \sin^3 x \, dx = \frac{1}{3}[-\cos x(\sin^2 x + 2)] + C;$$

$$\int \cos^n x \sin x \, dx = -\frac{1}{n+1}(\cos^{n+1} x) + C;$$

$$\int \sin^n x \cos x \, dx = \sin^{n+1} x + C;$$

$$\int \rho^2 e^{-\rho} d\rho = -(\rho^2 + 2\rho + 2)e^{-\rho} + C;$$

$$\int \rho^3 e^{-\rho} d\rho = -(\rho^3 + 3\rho^2 + 6\rho + 6)e^{-\rho} + C;$$

$$\int \rho^4 e^{-\rho} d\rho = -(\rho^4 + 4\rho^3 + 12\rho^2 + 24\rho + 24)e^{-\rho} + C.$$

APPENDIX 4

Slater's Rules for Screening Constants

In an alkali metal atom such as sodium, the 3s electron penetrates the neon core, i.e., it moves into the field of attraction of the nucleus, being only partially screened by the K and L shells. In an excited sodium atom the electron in a 3p orbital penetrates the electron cloud to a lesser extent, and the electron raised to a 3d orbital is practically non-penetrating. Thus the 3s, 3p, and 3d orbitals in a many-electron atom have different energies whereas these orbitals in a hydrogen atom have the same energy.

An alternative way of discussing penetration effects is to speak of the effective nuclear charge Z_{eff} for a particular electron. The following rules for estimating the screening (or shielding) constant σ (where $Z_{eff} = Z - \sigma$) were proposed by Slater. [J. C. Slater, Atomic shielding constants. *Phys. Rev.* **36**, 57–64 (1930)].

(a) Electrons are divided into the following groups separated by "/":

$$1s/2s, 2p/3s, 3p/3d/4s, 4p/4d/4f/5s, 5p/ \text{ etc.}$$

Note that the s and p subshells are grouped together, the d and f subshells are separate, and the subshells are arranged in the order of increasing n, e.g., 3d comes before 4s. Note that this is the approximate order of the average distances of the subshells from the nucleus.

(b) The shielding constant σ is obtained from the sum of the following contributions:
   (1) nothing from any shell outside the one considered;
   (2) an amount 0.35 from each other electron in the group considered (except for a 1s electron which contributes 0.30 to shielding of the other 1s electron); and
   (3) for d and f electrons the shielding is 1.00 for each electron in the underlying groups; for s and p electrons an amount 0.85 from each electron in the next inner shell, and 1.00 from each electron still further in.

As a first example, we take nitrogen, $Z = 7$. Here we have two /1s/ electrons, and five /2s, 2p/ electrons. The effective nuclear charges are:

$$Z_{\text{eff}}(1s) = 7 - 1(0.30) = 6.70.$$

$$Z_{\text{eff}}(2s) = Z_{\text{eff}}(2p) = 7 - 4(0.35) - 2(0.85) = 3.90.$$

As a second example, we take Co, $Z = 27$. There are two /1s/, eight /2s, 2p/, eight /3s, 3p/, seven /3d/, and two /4s/ electrons. The effective nuclear charges are

$$Z_{\text{eff}}(1s) = 27 - 1(0.30) = 26.70.$$

$$Z_{\text{eff}}(2s) = Z_{\text{eff}}(2p) = 27 - 7(0.35) - 2(0.85) = 22.85.$$

$$Z_{\text{eff}}(3s) = Z_{\text{eff}}(3p) = 27 - 7(0.35) - 8(0.85) - 2(1.00) = 15.75.$$

$$Z_{\text{eff}}(3d) = 27 - 6(0.35) - 18(1.00) = 6.90.$$

$$Z_{\text{eff}}(4s) = 27 - 1(0.35) - 15(0.85) - 10(1.00) = 3.90.$$

APPENDIX 5

Wade's Rules

A5.1 CARBORANES

For a carborane system with the general formula $[(CH)_a(BH)_pH_q]^{c-}$, we can readily deduce its structural classification–*closo*, *nido*, or *arachno*, using Wade's $2n + 2$ rules. These rules are easily "derived" in the following manner.

With the aforementioned general molecular formula, we can see that the number of vertices of the polyhedron or polyhedral fragment is $a + p = n$, and there are q hydrogens involved in bridging B–H–B or extra B–H bonding. Now let us count the atomic orbitals and electrons taking part in the bonding of the skeleton: each BH group contributes three orbitals and two electrons, and each CH group contributes three orbitals and three electrons. So the number of framework electrons is

$$3a \text{ [from } (CH)_a] + 2p \text{ [from } (BH)_p] + q + c$$
$$= (2a + 2p) + a + q + c$$
$$= 2n + a + q + c.$$

So the number of electron pairs is $\frac{1}{2}(2n + a + q + c)$. On the orbital side, there are $3a + 3p + q = 3n + q$ orbitals for skeletal bonding. Hence at most there are $\frac{1}{2}(3n + q)$ bonding molecular orbitals. We may assume that a stable structure is obtained when all the bonding orbitals are filled.

Next we consider the orbitals involved in the skeletal bonding. Let us assume that there are n cluster atoms (B or C in the present case) at the vertices of a regular deltahedron. The four valence orbitals on each vertex atom may be divided into one external sp hybrid orbital, two equivalent tangential p orbitals and one internal hybrid orbital pointing at the center of the polyhedron. This orbital arrangement is shown below.

The external sp hybrid orbital is used to form a σ bond to an external hydrogen atom. The tangential orbitals are then used to bind the cluster atoms together. Pairwise interaction of these $2n$ tangential orbitals will form n bonding and n antibonding molecular orbitals. These orbitals are delocalized over the surface of the polyhedron. Finally, the n internal sp hybrid orbitals overlap at the center of the polyhedron to generate one strongly bonding molecular orbital[†] and $n - 1$ non-bonding orbitals. Thus the cluster has $n + 1$ bonding orbitals which can accommodate $2n + 2$ electrons. These are the well known $2n + 2$ rules of polyhedral clusters. These rules were first formulated by K. Wade and hence are also known as Wade's rules.

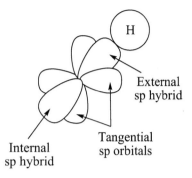

The orbitals used by a polyhedral vertex atom (B or C) in cluster bonding.

[†] Take $n = 4$ as an example. The four internal sp hybrids are labeled on the right. The four linear combinations are $a + b + c + d$, $a + b - c - d$, $a - b + c - d$, and $a - b - c + d$. Only the first one is (strongly) bonding.

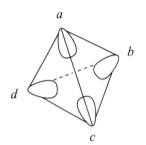

Hence we can easily arrive at the following conclusions:

(a) When $C = 2n + 2$, the carborane has a *closo* structure with n vertices. Examples include borane anions $(BH)_n^{2-}$, carboranes $(CH)_2(BH)_{n-2}$, etc.

(b) When $C = 2n + 4$, the carborane has a *nido* structure with n vertices, which is derived from a *closo* structure with $n + 1$ vertices. Examples include boranes $(BH)_n H_4$ (or $B_n H_{n+4}$) and carboranes $(BH)_5(CH)H_3$ and $(BH)_2(CH)_4$, etc.

(c) When $C = 2n + 6$, the carborane has an *arachno* structure with n vertices, which is derived from a *closo* structure with $n + 2$ vertices. Examples include boranes $(BH)_4 H_6$ (or $B_4 H_{10}$) and carboranes $(CH)_2(BH)_8^{4-}$ and $(BH)_7(CH)_2 H_4$, etc.

A5.2 MAIN-GROUP CLUSTERS

The $2n + 2$ rules appear to be also applicable to clusters with main-group elements. In these clusters, atoms with no substituents are assumed to have a lone pair each, directed outward from the surface of the clusters. Some examples are discussed below.

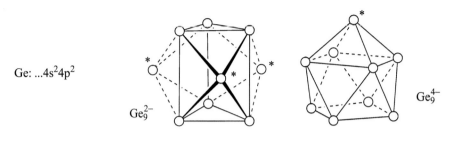

Ge: ...$4s^24p^2$

Ge_9^{2-}

Ge_9^{4-}

* capping atoms

Germanium has four valence electrons. Thus there are 38 valence electrons for Ge_9^{2-}. After using 18 of them as nine lone pairs, there remain 20 electrons for the Ge_9 skeleton. With $n = 9$, we have $C = 20 = 2n + 2$. Hence Ge_9^{2-} is a *closo* species with a tricapped trigonal prismatic structure, as shown above. Similarly, for Ge_9^{4-}, we have $n = 9$ and $C = 22 = 2n + 4$. So this anion has a *nido* structure derived from a bicapped square antiprism. This structure is also shown above.

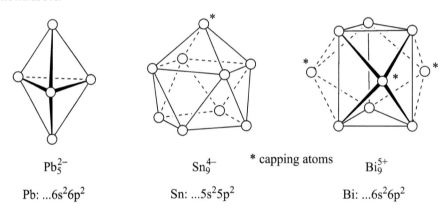

Pb_5^{2-}

Pb: ...$6s^26p^2$

Sn_9^{4-}

Sn: ...$5s^25p^2$

* capping atoms

Bi_9^{5+}

Bi: ...$6s^26p^2$

For Pb_5^{2-}, we have $n = 5$ and $C = 12 = 2n + 2$. Hence this is a *closo* compound, with a trigonal bipyramidal structure, as shown above. For Sn_9^{4-}, $n = 9$ and $C = 22 = 2n + 4$, so this is a *nido* compound; its monocapped square antiprismatic structure, shown above, is derived from the *closo* structure of a bicapped square antiprism. For Bi_9^{5+}, again we have $n = 9$ and $C = 22 = 2n + 4$. So Bi_9^{5+} is isoelectronic with Sn_9^{4-}; they should have the same structure. However, it is found that Bi_9^{5+} has the *closo* structure of a tricapped trigonal prism. So the $2n + 2$ rules do not hold here. Still, it should be pointed out that only a small distortion in Bi_9^{5+} would convert it to the geometry observed for Sn_9^{4-}.

A5.3 METALLOCARBORANES

When one or more of the vertices of a carborane are replaced by metal (M) atoms or metallo (ML_n) groups, a metallocarborane is formed. Transition metal atoms in these cluster compounds are considered to be pseudo-octahedrally coordinated with valence shells having 18 electrons, corresponding to noble gas configuration. When this assumption is combined with Wade's $2n+2$ rules, we can readily arrive at a general formula for the number of valence electrons C for the skeleton of a metallocarborane:

$$C = V + L + m - 12t - q,$$

where V = total number of electrons of all the transition metals in the cluster;

L = total number of ligand electrons donated to the transition metals;

m = total number of electrons contributed by the non-transition elements in the cluster;

t = number of transition metals in the cluster; and

q = charge of the entire cluster molecule.

Upon determining C, we can then apply the $2n+2$ rules to arrive at the structure of the cluster, where n is the number of cluster atoms:

(a) When $C = 2n + 2$, the compound has the *closo* structure of a regular polyhedron with n vertices. An example is $Os_5(CO)_{16}$ with $n = 5$ and $C = 12$, and the Os_5 skeleton is a trigonal bipyramid.

(b) When $C = 2n + 4$, the compound has a *nido* structure, corresponding to an $(n + 1)$-vertexed polyhedron lacking a vertex. An example is $Fe_5(CO)_{15}C$ with $n = 5$ and $C = 14$, and the Fe_5 skeleton is a square pyramid with a C atom slightly below the base. Note that the Fe_5 square pyramid is a *nido* structure derived from an octahedron.

(c) When $C = 2n + 6$, the structure is an $(n + 2)$-vertexed polyhedron missing two vertices, i.e., an *arachno* structure. An example is $Re_4(CO)_{16}^{2-}$ with $n = 4$ and $C = 14 = 2n + 6$, and the Re_4 has a disphenoidal (or butterfly) structure which is an octahedron minus two vertices.

(d) When $C < 2n+2$, the cluster is expected to be a capped polyhedron: $C = 2n$ corresponds to mono capping, $C = 2n - 2$ indicates bicapping, etc. Examples include $Os_7(CO)_{21}$ (with $n = 7$ and $C = 14 = 2n$, and the Os_7 skeleton is a monocapped octahedron) and $Os_8(CO)_{22}^{2-}$ (with $n = 8$ and $C = 14 = 2n - 2$, and the Os_8 skeleton is a bicapped octahedron).

REFERENCE: W. L. Jolly, *Modern Inorganic Chemistry*, McGraw-Hill, New York, 1984, pp. 269–73; pp. 416–8.

APPENDIX 6

Character Tables of Point Groups

Note: In the point groups containing C_5, the following relationships may be useful:

$$\eta^+ = -2\cos 144° = (1+\sqrt{5})/2 = 1.61803\ldots; \; \eta^- = -2\cos 72°$$
$$= (1-\sqrt{5})/2 = -0.61803\ldots; \; (\eta^\pm)^2 = 1 + \eta^\pm; \; \eta^+\eta^- = -1 = -(\eta^+ + \eta^-).$$

1. The non-axial groups

C_1	E
A	1

C_s	E	σ_h		
A'	1	1	x, y, R_z	x^2, y^2, z^2, xy
A''	1	-1	z, R_x, R_y	xz, yz

C_i	E	i		
A_g	1	1	R_x, R_y, R_z	$x^2, y^2, z^2, xy, xz, yz$
A_u	1	-1	x, y, z	

2. The C_n groups

C_2	E	C_2		
A	1	1	z, R_z	x^2, y^2, z^2, xy
B	1	-1	x, y, R_x, R_y	xz, yz

C_3	E	C_3	C_3^2		$\epsilon = \exp(2\pi i/3)$
A	1	1	1	z, R_z	$x^2 + y^2, z^2$
E	$\left\{\begin{array}{c} 1 \\ 1 \end{array}\right.$	$\begin{array}{c}\epsilon \\ \epsilon^*\end{array}$	$\left.\begin{array}{c}\epsilon^* \\ \epsilon\end{array}\right\}$	$(x, y), (R_x, R_y)$	$(x^2 - y^2, xy), (xz, yz)$

C_4	E	C_4	C_2	C_4^3		
A	1	1	1	1	z, R_z	x^2+y^2, z^2
B	1	-1	1	-1		x^2-y^2, xy
E	$\begin{cases} 1 \\ 1 \end{cases}$	$\begin{matrix} i \\ -i \end{matrix}$	$\begin{matrix} -1 \\ -1 \end{matrix}$	$\begin{matrix} -i \\ i \end{matrix}$	$(x,y), (R_x, R_y)$	(xz, yz)

C_5	E	C_5	C_5^2	C_5^3	C_5^4		$\epsilon = \exp(2\pi i/5)$
A	1	1	1	1	1	z, R_z	x^2+y^2, z^2
E_1	$\begin{cases} 1 \\ 1 \end{cases}$	$\begin{matrix} \epsilon \\ \epsilon^* \end{matrix}$	$\begin{matrix} \epsilon^2 \\ \epsilon^{2*} \end{matrix}$	$\begin{matrix} \epsilon^{2*} \\ \epsilon^2 \end{matrix}$	$\begin{matrix} \epsilon^* \\ \epsilon \end{matrix}$	$(x,y), (R_x, R_y)$	(xz, yz)
E_2	$\begin{cases} 1 \\ 1 \end{cases}$	$\begin{matrix} \epsilon^2 \\ \epsilon^{2*} \end{matrix}$	$\begin{matrix} \epsilon^* \\ \epsilon \end{matrix}$	$\begin{matrix} \epsilon \\ \epsilon^* \end{matrix}$	$\begin{matrix} \epsilon^{2*} \\ \epsilon^2 \end{matrix}$		(x^2-y^2, xy)

C_6	E	C_6	C_3	C_2	C_3^2	C_6^5		$\epsilon = \exp(2\pi i/6)$
A	1	1	1	1	1	1	z, R_z	x^2+y^2, z^2
B	1	-1	1	-1	1	-1		
E_1	$\begin{cases} 1 \\ 1 \end{cases}$	$\begin{matrix} \epsilon \\ \epsilon^* \end{matrix}$	$\begin{matrix} -\epsilon^* \\ -\epsilon \end{matrix}$	$\begin{matrix} -1 \\ -1 \end{matrix}$	$\begin{matrix} -\epsilon \\ -\epsilon^* \end{matrix}$	$\begin{matrix} \epsilon^* \\ \epsilon \end{matrix}$	$(x,y), (R_x, R_y)$	(xz, yz)
E_2	$\begin{cases} 1 \\ 1 \end{cases}$	$\begin{matrix} -\epsilon^* \\ -\epsilon \end{matrix}$	$\begin{matrix} -\epsilon \\ -\epsilon^* \end{matrix}$	$\begin{matrix} 1 \\ 1 \end{matrix}$	$\begin{matrix} -\epsilon^* \\ -\epsilon \end{matrix}$	$\begin{matrix} -\epsilon \\ -\epsilon^* \end{matrix}$		(x^2-y^2, xy)

C_7	E	C_7	C_7^2	C_7^3	C_7^4	C_7^5	C_7^6		$\epsilon = \exp(2\pi i/7)$
A	1	1	1	1	1	1	1	z, R_z	x^2+y^2, z^2
E_1	$\begin{cases} 1 \\ 1 \end{cases}$	$\begin{matrix} \epsilon \\ \epsilon^* \end{matrix}$	$\begin{matrix} \epsilon^2 \\ \epsilon^{2*} \end{matrix}$	$\begin{matrix} \epsilon^3 \\ \epsilon^{3*} \end{matrix}$	$\begin{matrix} \epsilon^{3*} \\ \epsilon^3 \end{matrix}$	$\begin{matrix} \epsilon^{2*} \\ \epsilon^2 \end{matrix}$	$\begin{matrix} \epsilon^* \\ \epsilon \end{matrix}$	$(x,y), (R_x, R_y)$	(xz, yz)
E_2	$\begin{cases} 1 \\ 1 \end{cases}$	$\begin{matrix} \epsilon^2 \\ \epsilon^{2*} \end{matrix}$	$\begin{matrix} \epsilon^{3*} \\ \epsilon^3 \end{matrix}$	$\begin{matrix} \epsilon^* \\ \epsilon \end{matrix}$	$\begin{matrix} \epsilon \\ \epsilon^* \end{matrix}$	$\begin{matrix} \epsilon^3 \\ \epsilon^{3*} \end{matrix}$	$\begin{matrix} \epsilon^{2*} \\ \epsilon^2 \end{matrix}$		(x^2-y^2, xy)
E_3	$\begin{cases} 1 \\ 1 \end{cases}$	$\begin{matrix} \epsilon^3 \\ \epsilon^{3*} \end{matrix}$	$\begin{matrix} \epsilon^* \\ \epsilon \end{matrix}$	$\begin{matrix} \epsilon^2 \\ \epsilon^{2*} \end{matrix}$	$\begin{matrix} \epsilon^{2*} \\ \epsilon^2 \end{matrix}$	$\begin{matrix} \epsilon \\ \epsilon^* \end{matrix}$	$\begin{matrix} \epsilon^{3*} \\ \epsilon^3 \end{matrix}$		

C_8	E	C_8	C_4	C_8^3	C_2	C_8^5	C_4^3	C_8^7		$\in = \exp(2\pi i/8)$
A	1	1	1	1	1	1	1	1	z, R_z	x^2+y^2, z^2
B	1	-1	1	-1	1	-1	1	-1		
E_1 $\begin{cases} \\ \\ \end{cases}$	1	$\in$	i	$-\in^*$	-1	$-\in$	$-i$	$-\in^*$	$(x,y), (R_x, R_y)$	(xz, yz)
	1	$\in^*$	$-i$	$-\in$	-1^*	$-\in^*$	i	$\in$		
E_2 $\begin{cases} \\ \\ \end{cases}$	1	i	-1	$-i$	1	i	-1	$-i$		(x^2-y^2, xy)
	1	$-i$	-1	i	1	$-i$	-1	i		
E_3 $\begin{cases} \\ \\ \end{cases}$	1	$-\in$	i	$\in^*$	-1	$\in$	$-i$	$-\in^*$		
	1	$-\in^*$	$-i$	$\in$	-1	$\in^*$	i	$-\in$		

3. The D_n groups

D_2	E	$C_2(z)$	$C_2(y)$	$C_2(x)$		
A	1	1	1	1		x^2, y^2, z^2
B_1	1	1	-1	-1	z, R_z	xy
B_2	1	-1	1	-1	y, R_y	xz
B_3	1	-1	-1	1	x, R_x	yz

D_3	E	$2C_3$	$3C_2$		(x-axis coincident with C_2)	
A_1	1	1	1		x^2+y^2, z^2	
A_2	1	1	-1	z, R_z		
E	2	-1	0	$(x,y), (R_x, R_y)$	$(x^2-y^2, xy), (xz, yz)$	

D_4	E	$2C_4$	$C_2(=C_4^2)$	$2C_2'$	$2C_2''$	(x-axis coincident with C_2')	
A_1	1	1	1	1	1		x^2+y^2, z^2
A_2	1	1	1	-1	-1	z, R_z	
B_1	1	-1	1	1	-1		x^2-y^2
B_2	1	-1	1	-1	1		xy
E	2	0	-2	0	0	$(x,y), (R_x, R_y)$	(xz, yz)

D_5	E	$2C_5$	$2C_5^2$	$5C_2$	$\eta^\pm = (1\pm\sqrt{5})/2$ (x-axis coincident with C_2)	
A_1	1	1	1	1		x^2+y^2, z^2
A_2	1	1	1	-1	z, R_z	
E_1	2	$-\eta^-$	$-\eta^+$	0	$(x,y), (R_x, R_y)$	(xz, yz)
E_2	2	$-\eta^+$	$-\eta^-$	0		(x^2-y^2, xy)

D_6	E	$2C_6$	$2C_3$	C_2	$3C_2'$	$3C_2''$	(x-axis coincident with C_2')	
A_1	1	1	1	1	1	1		$x^2 + y^2,\ z^2$
A_2	1	1	1	1	−1	−1	z, R_z	
B_1	1	−1	1	−1	1	−1		
B_2	1	−1	1	−1	−1	1		
E_1	2	1	−1	−2	0	0	$(x, y),\ (R_x, R_y)$	(xz, yz)
E_2	2	−1	−1	2	0	0		$(x^2 - y^2, xy)$

4. The C_{nv} groups

C_{2v}	E	C_2	$\sigma_v(xz)$	$\sigma_v'(yz)$		
A_1	1	1	1	1	z	$x^2,\ y^2,\ z^2$
A_2	1	1	−1	−1	R_z	xy
B_1	1	−1	1	−1	x, R_y	xz
B_2	1	−1	−1	1	y, R_x	yz

C_{3v}	E	$2C_3$	$3\sigma_v$		
A_1	1	1	1	z	$x^2 + y^2,\ z^2$
A_2	1	1	−1	R_z	
E	2	−1	0	$(x, y),\ (R_x, R_y)$	$(x^2 - y^2, xy),\ (xz, yz)$

C_{4v}	E	$2C_4$	C_2	$2\sigma_v$	$2\sigma_d$		
A_1	1	1	1	1	1	z	$x^2 + y^2,\ z^2$
A_2	1	1	1	−1	−1	R_z	
B_1	1	−1	1	1	−1		$x^2 - y^2$
B_2	1	−1	1	−1	1		xy
E	2	0	−2	0	0	$(x, y),\ (R_x, R_y)$	(xz, yz)

C_{5v}	E	$2C_5$	$2C_5^2$	$5\sigma_v$	$\eta^{\pm} = (1 \pm \sqrt{5})/2$	
A_1	1	1	1	1	z	$x^2 + y^2,\ z^2$
A_2	1	1	1	−1	R_z	
E_1	2	$-\eta^-$	$-\eta^+$	0	$(x, y),\ (R_x, R_y)$	(xz, yz)
E_2	2	$-\eta^+$	$-\eta^-$	0		$(x^2 - y^2, xy)$

C_{6v}	E	$2C_6$	$2C_3$	C_2	$3\sigma_v$	$3\sigma_d$		
A_1	1	1	1	1	1	1	z	$x^2+y^2,\ z^2$
A_2	1	1	1	1	−1	−1	R_z	
B_1	1	−1	1	−1	1	−1		
B_2	1	−1	1	−1	−1	1		
E_1	2	1	−1	−2	0	0	$(x,\ y),\ (R_x,\ R_y)$	$(xz,\ yz)$
E_2	2	−1	−1	2	0	0		$(x^2-y^2,\ xy)$

5. The C_{nh} groups

C_{2h}	E	C_2	i	σ_h		
A_g	1	1	1	1	R_z	$x^2,\ y^2,\ z^2,\ xy$
B_g	1	−1	1	−1	$R_x,\ R_y$	$xz,\ yz$
A_u	1	1	−1	−1	z	
B_u	1	−1	−1	1	$x,\ y$	

C_{3h}	E	C_3	C_3^2	σ_h	S_3	S_3^5		$\epsilon = \exp(2\pi i/3)$
A'	1	1	1	1	1	1	R_z	$x^2+y^2,\ z^2$
E'	$\begin{cases}1\\1\end{cases}$	$\begin{matrix}\epsilon\\\epsilon^*\end{matrix}$	$\begin{matrix}\epsilon^*\\\epsilon\end{matrix}$	$\begin{matrix}1\\1\end{matrix}$	$\begin{matrix}\epsilon\\\epsilon^*\end{matrix}$	$\begin{matrix}\epsilon^*\\\epsilon\end{matrix}$	(x,y)	$(x^2-y^2,\ xy)$
A''	1	1	1	−1	−1	−1	z	
E''	$\begin{cases}1\\1\end{cases}$	$\begin{matrix}\epsilon\\\epsilon^*\end{matrix}$	$\begin{matrix}\epsilon^*\\\epsilon\end{matrix}$	$\begin{matrix}-1\\-1\end{matrix}$	$\begin{matrix}-\epsilon\\-\epsilon^*\end{matrix}$	$\begin{matrix}-\epsilon^*\\-\epsilon\end{matrix}$	$(R_x,\ R_y)$	$(xz,\ yz)$

C_{4h}	E	C_4	C_2	C_4^3	i	S_4^3	σ_h	S_4		
A_g	1	1	1	1	1	1	1	1	R_z	$x^2+y^2,\ z^2$
B_g	1	−1	1	−1	1	−1	1	−1		$x^2-y^2,\ xy$
E_g	$\begin{cases}1\\1\end{cases}$	$\begin{matrix}i\\-i\end{matrix}$	$\begin{matrix}-1\\-1\end{matrix}$	$\begin{matrix}-i\\i\end{matrix}$	$\begin{matrix}1\\1\end{matrix}$	$\begin{matrix}i\\-i\end{matrix}$	$\begin{matrix}-1\\-1\end{matrix}$	$\begin{matrix}-i\\i\end{matrix}$	$(R_x,\ R_y)$	$(xz,\ yz)$
A_u	1	1	1	1	−1	−1	−1	−1	z	
B_u	1	−1	1	−1	−1	1	−1	1		
E_u	$\begin{cases}1\\1\end{cases}$	$\begin{matrix}i\\-i\end{matrix}$	$\begin{matrix}-1\\-1\end{matrix}$	$\begin{matrix}-i\\i\end{matrix}$	$\begin{matrix}-1\\-1\end{matrix}$	$\begin{matrix}-i\\i\end{matrix}$	$\begin{matrix}1\\1\end{matrix}$	$\begin{matrix}i\\-i\end{matrix}$	(x,y)	

C_{5h}	E	C_5	C_5^2	C_5^3	C_5^4	σ_h	S_5	S_5^7	S_5^3	S_5^9		$\epsilon = \exp(2\pi i/5)$
A'	1	1	1	1	1	1	1	1	1	1	R_z	$x^2+y^2,\ z^2$
E_1'	$\begin{cases}1\\1\end{cases}$	$\begin{matrix}\epsilon\\\epsilon^*\end{matrix}$	$\begin{matrix}\epsilon^2\\\epsilon^{2*}\end{matrix}$	$\begin{matrix}\epsilon^{2*}\\\epsilon^2\end{matrix}$	$\begin{matrix}\epsilon^*\\\epsilon\end{matrix}$	$\begin{matrix}1\\1\end{matrix}$	$\begin{matrix}\epsilon\\\epsilon^*\end{matrix}$	$\begin{matrix}\epsilon^2\\\epsilon^{2*}\end{matrix}$	$\begin{matrix}\epsilon^{2*}\\\epsilon^2\end{matrix}$	$\begin{matrix}\epsilon^*\\\epsilon\end{matrix}$	(x,y)	
E_2'	$\begin{cases}1\\1\end{cases}$	$\begin{matrix}\epsilon^2\\\epsilon^{2*}\end{matrix}$	$\begin{matrix}\epsilon^*\\\epsilon\end{matrix}$	$\begin{matrix}\epsilon\\\epsilon^*\end{matrix}$	$\begin{matrix}\epsilon^{2*}\\\epsilon^2\end{matrix}$	$\begin{matrix}1\\1\end{matrix}$	$\begin{matrix}\epsilon^2\\\epsilon^{2*}\end{matrix}$	$\begin{matrix}\epsilon^*\\\epsilon\end{matrix}$	$\begin{matrix}\epsilon\\\epsilon^*\end{matrix}$	$\begin{matrix}\epsilon^{2*}\\\epsilon^2\end{matrix}$		$(x^2-y^2,\ xy)$
A''	1	1	1	1	1	-1	-1	-1	-1	-1	z	
E_1''	$\begin{cases}1\\1\end{cases}$	$\begin{matrix}\epsilon\\\epsilon^*\end{matrix}$	$\begin{matrix}\epsilon^2\\\epsilon^{2*}\end{matrix}$	$\begin{matrix}\epsilon^{2*}\\\epsilon^2\end{matrix}$	$\begin{matrix}\epsilon^*\\\epsilon\end{matrix}$	$\begin{matrix}-1\\-1\end{matrix}$	$\begin{matrix}-\epsilon\\-\epsilon^*\end{matrix}$	$\begin{matrix}-\epsilon^2\\-\epsilon^{2*}\end{matrix}$	$\begin{matrix}-\epsilon^{2*}\\-\epsilon^2\end{matrix}$	$\begin{matrix}-\epsilon^*\\-\epsilon\end{matrix}$	(R_x, R_y)	(xz, yz)
E_2''	$\begin{cases}1\\1\end{cases}$	$\begin{matrix}\epsilon^2\\\epsilon^{2*}\end{matrix}$	$\begin{matrix}\epsilon^*\\\epsilon\end{matrix}$	$\begin{matrix}\epsilon\\\epsilon^*\end{matrix}$	$\begin{matrix}\epsilon^{2*}\\\epsilon^2\end{matrix}$	$\begin{matrix}-1\\-1\end{matrix}$	$\begin{matrix}-\epsilon^2\\-\epsilon^{2*}\end{matrix}$	$\begin{matrix}-\epsilon^*\\-\epsilon\end{matrix}$	$\begin{matrix}-\epsilon\\-\epsilon^*\end{matrix}$	$\begin{matrix}-\epsilon^{2*}\\-\epsilon^2\end{matrix}$		

C_{6h}	E	C_6	C_3	C_2	C_3^2	C_6^5	i	S_3^5	S_6^5	σ_h	S_6	S_3		$\epsilon = \exp(2\pi i/6)$
A_g	1	1	1	1	1	1	1	1	1	1	1	1	R_z	$x^2+y^2,\ z^2$
B_g	1	-1	1	-1	1	-1	1	-1	1	-1	1	-1		
E_{1g}	$\begin{cases}1\\1\end{cases}$	$\begin{matrix}\epsilon\\\epsilon^*\end{matrix}$	$\begin{matrix}-\epsilon^*\\-\epsilon\end{matrix}$	$\begin{matrix}-1\\-1\end{matrix}$	$\begin{matrix}-\epsilon\\-\epsilon^*\end{matrix}$	$\begin{matrix}\epsilon^*\\\epsilon\end{matrix}$	$\begin{matrix}1\\1\end{matrix}$	$\begin{matrix}\epsilon\\\epsilon^*\end{matrix}$	$\begin{matrix}-\epsilon^*\\-\epsilon\end{matrix}$	$\begin{matrix}-1\\-1\end{matrix}$	$\begin{matrix}-\epsilon\\-\epsilon^*\end{matrix}$	$\begin{matrix}\epsilon^*\\\epsilon\end{matrix}$	(R_x, R_y)	(xz, yz)
E_{2g}	$\begin{cases}1\\1\end{cases}$	$\begin{matrix}-\epsilon^*\\-\epsilon\end{matrix}$	$\begin{matrix}-\epsilon\\-\epsilon^*\end{matrix}$	$\begin{matrix}1\\1\end{matrix}$	$\begin{matrix}-\epsilon^*\\-\epsilon\end{matrix}$	$\begin{matrix}-\epsilon\\-\epsilon^*\end{matrix}$	$\begin{matrix}1\\1\end{matrix}$	$\begin{matrix}-\epsilon^*\\-\epsilon\end{matrix}$	$\begin{matrix}\epsilon\\-\epsilon^*\end{matrix}$	$\begin{matrix}1\\1\end{matrix}$	$\begin{matrix}-\epsilon^*\\-\epsilon\end{matrix}$	$\begin{matrix}-\epsilon\\-\epsilon^*\end{matrix}$		$(x^2-y^2,\ xy)$
A_u	1	1	1	1	1	1	-1	-1	-1	-1	-1	-1	z	
B_u	1	-1	1	-1	1	-1	-1	1	-1	1	-1	1		
E_{1u}	$\begin{cases}1\\1\end{cases}$	$\begin{matrix}\epsilon\\\epsilon^*\end{matrix}$	$\begin{matrix}-\epsilon^*\\-\epsilon\end{matrix}$	$\begin{matrix}-1\\-1\end{matrix}$	$\begin{matrix}-\epsilon\\-\epsilon^*\end{matrix}$	$\begin{matrix}\epsilon^*\\\epsilon\end{matrix}$	$\begin{matrix}-1\\-1\end{matrix}$	$\begin{matrix}-\epsilon\\-\epsilon^*\end{matrix}$	$\begin{matrix}\epsilon^*\\\epsilon\end{matrix}$	$\begin{matrix}1\\1\end{matrix}$	$\begin{matrix}\epsilon\\\epsilon^*\end{matrix}$	$\begin{matrix}-\epsilon^*\\-\epsilon\end{matrix}$	(x,y)	
E_{2u}	$\begin{cases}1\\1\end{cases}$	$\begin{matrix}-\epsilon^*\\-\epsilon\end{matrix}$	$\begin{matrix}-\epsilon\\-\epsilon^*\end{matrix}$	$\begin{matrix}1\\1\end{matrix}$	$\begin{matrix}-\epsilon^*\\-\epsilon\end{matrix}$	$\begin{matrix}-\epsilon\\-\epsilon^*\end{matrix}$	$\begin{matrix}-1\\-1\end{matrix}$	$\begin{matrix}\epsilon^*\\\epsilon\end{matrix}$	$\begin{matrix}\epsilon\\\epsilon^*\end{matrix}$	$\begin{matrix}-1\\-1\end{matrix}$	$\begin{matrix}\epsilon^*\\\epsilon\end{matrix}$	$\begin{matrix}\epsilon\\\epsilon^*\end{matrix}$		

6. The D_{nh} groups

D_{2h}	E	$C_2(z)$	$C_2(y)$	$C_2(x)$	i	$\sigma(xy)$	$\sigma(xz)$	$\sigma(yz)$		
A_g	1	1	1	1	1	1	1	1		x^2, y^2, z^2
B_{1g}	1	1	-1	-1	1	1	-1	-1	R_z	xy
B_{2g}	1	-1	1	-1	1	-1	1	-1	R_y	xz
B_{3g}	1	-1	-1	1	1	-1	-1	1	R_x	yz
A_u	1	1	1	1	-1	-1	-1	-1		
B_{1u}	1	1	-1	-1	-1	-1	1	1	z	
B_{2u}	1	-1	1	-1	-1	1	-1	1	y	
B_{3u}	1	-1	-1	1	-1	1	1	-1	x	

D_{3h}	E	$2C_3$	$3C_2$	σ_h	$2S_3$	$3\sigma_v$		(x-axis coincident with C_2)
A_1'	1	1	1	1	1	1		$x^2+y^2,\ z^2$
A_2'	1	1	-1	1	1	-1	R_z	
E'	2	-1	0	2	-1	0	(x,y)	$(x^2-y^2,\ xy)$
A_1''	1	1	1	-1	-1	-1		
A_2''	1	1	-1	-1	-1	1	z	
E''	2	-1	0	-2	1	0	$(R_x,\ R_y)$	$(xz,\ yz)$

D_{4h}	E	$2C_4$	C_2	$2C_2'$	$2C_2''$	i	$2S_4$	σ_h	$2\sigma_v$	$2\sigma_d$		(x-axis coincident with C_2')
A_{1g}	1	1	1	1	1	1	1	1	1	1		$x^2+y^2,\ z^2$
A_{2g}	1	1	1	-1	-1	1	1	1	-1	-1	R_z	
B_{1g}	1	-1	1	1	-1	1	-1	1	1	-1		x^2-y^2
B_{2g}	1	-1	1	-1	1	1	-1	1	-1	1		xy
E_g	2	0	-2	0	0	2	0	-2	0	0	$(R_x,\ R_y)$	$(xz,\ yz)$
A_{1u}	1	1	1	1	1	-1	-1	-1	-1	-1		
A_{2u}	1	1	1	-1	-1	-1	-1	-1	1	1	z	
B_{1u}	1	-1	1	1	-1	-1	1	-1	-1	1		
B_{2u}	1	-1	1	-1	1	-1	1	-1	1	-1		
E_u	2	0	-2	0	0	-2	0	2	0	0	(x,y)	

$$\eta^{\pm} = (1 \pm \sqrt{5})/2$$

D_{5h}	E	$2C_5$	$2C_5^2$	$5C_2$	σ_h	$2S_5$	$2S_5^3$	$5\sigma_v$		(x-axis coincident with C_2)
A_1'	1	1	1	1	1	1	1	1		$x^2+y^2,\ z^2$
A_2'	1	1	1	-1	1	1	1	-1	R_z	
E_1'	2	$-\eta^-$	$-\eta^+$	0	2	$-\eta^-$	$-\eta^+$	0	(x,y)	
E_2'	2	$-\eta^+$	$-\eta^-$	0	2	$-\eta^+$	$-\eta^-$	0		$(x^2-y^2,\ xy)$
A_1''	1	1	1	1	-1	-1	-1	-1		
A_2''	1	1	1	-1	-1	-1	-1	1	z	
E_1''	2	$-\eta^-$	$-\eta^+$	0	-2	η^-	η^+	0	$(R_x,\ R_y)$	$(xz,\ yz)$
E_2''	2	$-\eta^+$	$-\eta^-$	0	-2	η^+	η^-	0		

D_{6h}	E	$2C_6$	$2C_3$	C_2	$3C_2'$	$3C_2''$	i	$2S_3$	$2S_6$	σ_h	$3\sigma_d$	$3\sigma_v$		(x-axis coincident with C_2')
A_{1g}	1	1	1	1	1	1	1	1	1	1	1	1		x^2+y^2, z^2
A_{2g}	1	1	1	1	-1	-1	1	1	1	1	-1	-1	R_z	
B_{1g}	1	-1	1	-1	1	-1	1	-1	1	-1	1	-1		
B_{2g}	1	-1	1	-1	-1	1	1	-1	1	-1	-1	1		
E_{1g}	2	1	-1	-2	0	0	2	1	-1	-2	0	0	(R_x, R_y)	(xz, yz)
E_{2g}	2	-1	-1	2	0	0	2	-1	-1	2	0	0		(x^2-y^2, xy)
A_{1u}	1	1	1	1	1	1	-1	-1	-1	-1	-1	-1		
A_{2u}	1	1	1	1	-1	-1	-1	-1	-1	-1	1	1	z	
B_{1u}	1	-1	1	-1	1	-1	-1	1	-1	1	-1	1		
B_{2u}	1	-1	1	-1	-1	1	-1	1	-1	1	1	-1		
E_{1u}	2	1	-1	-2	0	0	-2	-1	1	2	0	0	(x, y)	
E_{2u}	2	-1	-1	2	0	0	-2	1	1	-2	0	0		

D_{8h}	E	$2C_8$	$2C_8^3$	$2C_4$	C_2	$4C_2'$	$4C_2''$	i	$2S_8$	$2S_8^3$	$2S_4$	σ_h	$4\sigma_v$	$4\sigma_d$	(x-axis coincident with C_2')
A_{1g}	1	1	1	1	1	1	1	1	1	1	1	1	1	1	x^2+y^2, z^2
A_{2g}	1	1	1	1	1	-1	-1	1	1	1	1	1	-1	-1	R_z
B_{1g}	1	-1	-1	1	1	1	-1	1	-1	-1	1	1	1	-1	
B_{2g}	1	-1	-1	1	1	-1	1	1	-1	-1	1	1	-1	1	
E_{1g}	2	$\sqrt{2}$	$-\sqrt{2}$	0	-2	0	0	2	$\sqrt{2}$	$-\sqrt{2}$	0	-2	0	0	(R_x, R_y) (xz, yz)
E_{2g}	2	0	0	-2	2	0	0	2	0	0	-2	2	0	0	(x^2-y^2, xy)
E_{3g}	2	$-\sqrt{2}$	$\sqrt{2}$	0	-2	0	0	2	$-\sqrt{2}$	$\sqrt{2}$	0	-2	0	0	
A_{1u}	1	1	1	1	1	1	1	-1	-1	-1	-1	-1	-1	-1	
A_{2u}	1	1	1	1	1	-1	-1	-1	-1	-1	-1	-1	1	1	z
B_{1u}	1	-1	-1	1	1	1	-1	-1	1	1	-1	-1	-1	1	
B_{2u}	1	-1	-1	1	1	-1	1	-1	1	1	-1	-1	1	-1	
E_{1u}	2	$\sqrt{2}$	$-\sqrt{2}$	0	-2	0	0	-2	$-\sqrt{2}$	$\sqrt{2}$	0	2	0	0	(x, y)
E_{2u}	2	0	0	-2	2	0	0	-2	0	0	2	-2	0	0	
E_{3u}	2	$-\sqrt{2}$	$\sqrt{2}$	0	-2	0	0	-2	$\sqrt{2}$	$-\sqrt{2}$	0	2	0	0	

7. The D_{nd} groups

D_{2d}	E	$2S_4$	C_2	$2C_2'$	$2\sigma_d$	x-axis coincident with C_2'	
A_1	1	1	1	1	1		x^2+y^2, z^2
A_2	1	1	1	-1	-1	R_z	
B_1	1	-1	1	1	-1		x^2-y^2
B_2	1	-1	1	-1	1	z	xy
E	2	0	-2	0	0	$(x, y), (R_x, R_y)$	(xz, yz)

D_{3d}	E	$2C_3$	$3C_2$	i	$2S_6$	$3\sigma_d$		x-axis coincident with C_2	
A_{1g}	1	1	1	1	1	1		x^2+y^2, z^2	
A_{2g}	1	1	-1	1	1	-1	R_z		
E_g	2	-1	0	2	-1	0	(R_x, R_y)	$(x^2-y^2, xy), (xz, yz)$	
A_{1u}	1	1	1	-1	-1	-1			
A_{2u}	1	1	-1	-1	-1	1	z		
E_u	2	-1	0	-2	1	0	(x, y)		

D_{4d}	E	$2S_8$	$2C_4$	$2S_8^3$	C_2	$4C_2'$	$4\sigma_d$		x-axis coincident with C_2'	
A_1	1	1	1	1	1	1	1		x^2+y^2, z^2	
A_2	1	1	1	1	1	-1	-1	R_z		
B_1	1	-1	1	-1	1	1	-1			
B_2	1	-1	1	-1	1	-1	1	z		
E_1	2	$\sqrt{2}$	0	$-\sqrt{2}$	-2	0	0	(x, y)		
E_2	2	0	-2	0	2	0	0		(x^2-y^2, xy)	
E_3	2	$-\sqrt{2}$	0	$\sqrt{2}$	-2	0	0	(R_x, R_y)	(xz, yz)	

									$\eta^{\pm} = (1 \pm \sqrt{5})/2$	
D_{5d}	E	$2C_5$	$2C_5^2$	$5C_2$	i	$2S_{10}^3$	$2S_{10}$	$5\sigma_d$	(x-axis coincident with C_2)	
A_{1g}	1	1	1	1	1	1	1	1		x^2+y^2, z^2
A_{2g}	1	1	1	-1	1	1	1	-1	R_z	
E_{1g}	2	$-\eta^-$	$-\eta^+$	0	2	$-\eta^-$	$-\eta^+$	0	(R_x, R_y)	(xz, yz)
E_{2g}	2	$-\eta^+$	$-\eta^-$	0	2	$-\eta^+$	$-\eta^-$	0		(x^2-y^2, xy)
A_{1u}	1	1	1	1	-1	-1	-1	-1		
A_{2u}	1	1	1	-1	-1	-1	-1	1	z	
E_{1u}	2	$-\eta^-$	$-\eta^+$	0	-2	η^-	η^+	0	(x, y)	
E_{2u}	2	$-\eta^+$	$-\eta^-$	0	-2	η^+	η^-	0		

D_{6d}	E	$2S_{12}$	$2C_6$	$2S_4$	$2C_3$	$2S_{12}^5$	C_2	$6C_2'$	$6\sigma_d$	(x-axis coincident with C_2')	
A_1	1	1	1	1	1	1	1	1	1		x^2+y^2, z^2
A_2	1	1	1	1	1	1	1	-1	-1	R_z	
B_1	1	-1	1	-1	1	-1	1	1	-1		
B_2	1	-1	1	-1	1	-1	1	-1	1	z	
E_1	2	$\sqrt{3}$	1	0	-1	$-\sqrt{3}$	-2	0	0	(x, y)	
E_2	2	1	-1	-2	-1	1	2	0	0		(x^2-y^2, xy)
E_3	2	0	-2	0	2	0	-2	0	0		
E_4	2	-1	-1	2	-1	-1	2	0	0		
E_5	2	$-\sqrt{3}$	1	0	-1	$\sqrt{3}$	-2	0	0	(R_x, R_y)	(xz, yz)

8. The S_n groups

S_4	E	S_4	C_2	S_4^3		
A	1	1	1	1	R_z	x^2+y^2, z^2
B	1	-1	1	-1	z	x^2-y^2, xy
E	$\left\{\begin{array}{c}1\\1\end{array}\right.$	$\begin{array}{c}i\\-i\end{array}$	$\begin{array}{c}-1\\-1\end{array}$	$\left.\begin{array}{c}-i\\i\end{array}\right\}$	$(x,y), (R_x, R_y)$	(xz, yz)

S_6	E	C_3	C_3^2	i	S_6^5	S_6		$\epsilon = \exp(2\pi i/3)$
A_g	1	1	1	1	1	1	R_z	x^2+y^2, z^2
E_g	$\left\{\begin{array}{c}1\\1\end{array}\right.$	$\begin{array}{c}\epsilon\\\epsilon^*\end{array}$	$\begin{array}{c}\epsilon^*\\\epsilon\end{array}$	$\begin{array}{c}1\\1\end{array}$	$\begin{array}{c}\epsilon\\\epsilon^*\end{array}$	$\left.\begin{array}{c}\epsilon^*\\\epsilon\end{array}\right\}$	(R_x, R_y)	$(x^2-y^2, xy), (xz, yz)$
A_u	1	1	1	-1	-1	-1	z	
E_u	$\left\{\begin{array}{c}1\\1\end{array}\right.$	$\begin{array}{c}\epsilon\\\epsilon^*\end{array}$	$\begin{array}{c}\epsilon^*\\\epsilon\end{array}$	$\begin{array}{c}-1\\-1\end{array}$	$\begin{array}{c}-\epsilon\\-\epsilon^*\end{array}$	$\left.\begin{array}{c}-\epsilon^*\\-\epsilon\end{array}\right\}$	(x,y)	

S_8	E	S_8	C_4	S_8^3	C_2	S_8^5	C_4^3	S_8^7		$\epsilon = \exp(2\pi i/8)$
A	1	1	1	1	1	1	1	1	R_z	x^2+y^2, z^2
B	1	-1	1	-1	1	-1	1	-1	z	
E_1	$\left\{\begin{array}{c}1\\1\end{array}\right.$	$\begin{array}{c}\epsilon\\\epsilon^*\end{array}$	$\begin{array}{c}i\\-i\end{array}$	$\begin{array}{c}-\epsilon^*\\-\epsilon\end{array}$	$\begin{array}{c}-1\\-1\end{array}$	$\begin{array}{c}-\epsilon\\-\epsilon^*\end{array}$	$\begin{array}{c}-i\\i\end{array}$	$\left.\begin{array}{c}\epsilon^*\\\epsilon\end{array}\right\}$	$(x,y), (R_x, R_y)$	
E_2	$\left\{\begin{array}{c}1\\1\end{array}\right.$	$\begin{array}{c}i\\-i\end{array}$	$\begin{array}{c}-1\\-1\end{array}$	$\begin{array}{c}-i\\i\end{array}$	$\begin{array}{c}1\\1\end{array}$	$\begin{array}{c}i\\-i\end{array}$	$\begin{array}{c}-1\\-1\end{array}$	$\left.\begin{array}{c}-i\\i\end{array}\right\}$		(x^2-y^2, xy)
E_3	$\left\{\begin{array}{c}1\\1\end{array}\right.$	$\begin{array}{c}-\epsilon^*\\-\epsilon\end{array}$	$\begin{array}{c}-i\\i\end{array}$	$\begin{array}{c}\epsilon\\\epsilon^*\end{array}$	$\begin{array}{c}-1\\-1\end{array}$	$\begin{array}{c}\epsilon^*\\\epsilon\end{array}$	$\begin{array}{c}i\\-i\end{array}$	$\left.\begin{array}{c}-\epsilon\\-\epsilon^*\end{array}\right\}$		(xz, yz)

9. The cubic groups

T	E	$4C_3$	$4C_3^2$	$3C_2$		$\epsilon = \exp(2\pi i/3)$
A	1	1	1	1		$x^2+y^2+z^2$
E	$\left\{\begin{array}{c}1\\1\end{array}\right.$	$\begin{array}{c}\epsilon\\\epsilon^*\end{array}$	$\begin{array}{c}\epsilon^*\\\epsilon\end{array}$	$\left.\begin{array}{c}1\\1\end{array}\right\}$		$(2z^2-x^2-y^2, x^2-y^2)$
T	3	0	0	-1	$(R_x, R_y, R_z), (x,y,z)$	(xy, xz, yz)

T_h	E	$4C_3$	$4C_3^2$	$3C_2$	i	$4S_6$	$4S_6^5$	$3\sigma_h$		$\epsilon = \exp(2\pi i/3)$
A_g	1	1	1	1	1	1	1	1		$x^2 + y^2 + z^2$
E_g	$\begin{cases}1\\1\end{cases}$	$\begin{matrix}\epsilon\\\epsilon^*\end{matrix}$	$\begin{matrix}\epsilon^*\\\epsilon\end{matrix}$	$\begin{matrix}1\\1\end{matrix}$	$\begin{matrix}1\\1\end{matrix}$	$\begin{matrix}\epsilon\\\epsilon^*\end{matrix}$	$\begin{matrix}\epsilon^*\\\epsilon\end{matrix}$	$\begin{matrix}1\\1\end{cases}$		$(2z^2 - x^2 - y^2,\ x^2 - y^2)$
T_g	3	0	0	−1	3	0	0	−1	(R_x, R_y, R_z)	(xy, xz, yz)
A_u	1	1	1	1	−1	−1	−1	−1		
E_u	$\begin{cases}1\\1\end{cases}$	$\begin{matrix}\epsilon\\\epsilon^*\end{matrix}$	$\begin{matrix}\epsilon^*\\\epsilon\end{matrix}$	$\begin{matrix}1\\1\end{matrix}$	$\begin{matrix}-1\\-1\end{matrix}$	$\begin{matrix}-\epsilon\\-\epsilon^*\end{matrix}$	$\begin{matrix}-\epsilon^*\\-\epsilon\end{matrix}$	$\begin{matrix}-1\\-1\end{cases}$		
T_u	3	0	0	−1	−3	0	0	1	(x, y, z)	

T_d	E	$8C_3$	$3C_2$	$6S_4$	$6\sigma_d$		
A_1	1	1	1	1	1		$x^2 + y^2 + z^2$
A_2	1	1	1	−1	−1		
E	2	−1	2	0	0		$(2z^2 - x^2 - y^2, x^2 - y^2)$
T_1	3	0	−1	1	−1	(R_x, R_y, R_z)	
T_2	3	0	−1	−1	1	(x, y, z)	(xy, xz, yz)

O	E	$8C_3$	$3C_2(= C_4^2)$	$6C_4$	$6C_2$		
A_1	1	1	1	1	1		$x^2 + y^2 + z^2$
A_2	1	1	1	−1	−1		
E	2	−1	2	0	0		$(2z^2 - x^2 - y^2, x^2 - y^2)$
T_1	3	0	−1	1	−1	$(R_x, R_y, R_z), (x, y, z)$	
T_2	3	0	−1	−1	1		(xy, xz, yz)

O_h	E	$8C_3$	$6C_2$	$6C_4$	$3C_2(= C_4^2)$	i	$6S_4$	$8S_6$	$3\sigma_h$	$6\sigma_d$		
A_{1g}	1	1	1	1	1	1	1	1	1	1		$x^2 + y^2 + z^2$
A_{2g}	1	1	−1	1	1	1	−1	1	1	−1		
E_g	2	−1	0	0	2	2	0	−1	2	0		$(2z^2 - x^2 - y^2, x^2 - y^2)$
T_{1g}	3	0	−1	1	−1	3	1	0	−1	−1	(R_x, R_y, R_z)	
T_{2g}	3	0	1	−1	−1	3	−1	0	−1	1		(xy, xz, yz)
A_{1u}	1	1	1	1	1	−1	−1	−1	−1	−1		
A_{2u}	1	1	−1	−1	1	−1	1	−1	−1	1		
E_u	2	−1	0	0	2	−2	0	1	−2	0		
T_{1u}	3	0	−1	1	−1	−3	−1	0	1	1	(x, y, z)	
T_{2u}	3	0	1	−1	−1	−3	1	0	1	−1		

10. The groups $C_{\infty v}$ and $D_{\infty h}$ linear molecules

$C_{\infty v}$	E	$2C_\infty^\phi$	...	$\infty\sigma_v$		
$A_1 \equiv \Sigma^+$	1	1	...	1	z	$x^2+y^2,\ z^2$
$A_2 \equiv \Sigma^-$	1	1	...	-1	R_z	
$E_1 \equiv \Pi$	2	$2\cos\phi$	...	0	$(x,y),\ (R_x,R_y)$	$(xz,\ yz)$
$E_2 \equiv \Delta$	2	$2\cos 2\phi$	...	0		$(x^2-y^2,\ xy)$
$E_3 \equiv \Phi$	2	$2\cos 3\phi$	...	0		
$\vdots$	$\vdots$	$\vdots$	$\vdots$	$\vdots$		

$D_{\infty h}$	E	$2C_\infty^\phi$	...	$\infty\sigma_v$	i	$2S_\infty^\phi$	...	∞C_v		
$A_{1g} \equiv \Sigma_g^+$	1	1	...	1	1	1	...	1		$x^2+y^2,\ z^2$
$A_{2g} \equiv \Sigma_g^-$	1	1	...	-1	1	1	...	-1	R_z	
$E_{1g} \equiv \Pi_g$	2	$2\cos\phi$	...	0	2	$-2\cos\phi$	...	0	(R_x,R_y)	$(xz,\ yz)$
$E_{2g} \equiv \Delta_g$	2	$2\cos 2\phi$	...	0	2	$2\cos 2\phi$	...	0		$(x^2-y^2,\ xy)$
...	...	...	...	...	...	...	...	...		
$A_{1u} \equiv \Sigma_u^+$	1	1	...	1	-1	-1	...	-1	z	
$A_{2u} \equiv \Sigma_u^-$	1	1	...	-1	-1	-1	...	1		
$E_{1u} \equiv \Pi_u$	2	$2\cos\phi$	...	0	-2	$2\cos\phi$	...	0	(x,y)	
$E_{2u} \equiv \Delta_u$	2	$2\cos 2\phi$	...	0	-2	$-2\cos 2\phi$	...	0		
$\vdots$	$\vdots$	$\vdots$	$\vdots$	$\vdots$	$\vdots$	$\vdots$	$\vdots$	$\vdots$		

11. The icosahedral groups

I	E	$12C_5$	$12C_5^2$	$20C_3$	$15C_2$	$\eta^\pm = (1\pm\sqrt{5})/2$	
A	1	1	1	1	1		$x^2+y^2+z^2$
T_1	3	η^+	η^-	0	-1	$(x,y,z),\ (R_x,R_y,R_z)$	
T_2	3	η^-	η^+	0	-1		
G	4	-1	-1	1	0		
H	5	0	0	-1	1		$(2z^2-x^2-y^2,\ x^2-y^2,\ xy,\ xz,\ yz)$

I_h	E	$12C_5$	$12C_5^2$	$20C_3$	$15C_2$	i	$12S_{10}$	$12S_{10}^3$	$20S_6$	15σ		$\eta^\pm = (1\pm\sqrt{5})/2$
A_g	1	1	1	1	1	1	1	1	1	1		$x^2+y^2+z^2$
T_{1g}	3	η^+	η^-	0	-1	3	η^-	η^+	0	-1	(R_x, R_y, R_z)	
T_{2g}	3	η^-	η^+	0	-1	3	η^+	η^-	0	-1		
G_g	4	-1	-1	1	0	4	-1	-1	1	0		
H_g	5	0	0	-1	1	5	0	0	-1	1		$(2z^2-x^2-y^2, x^2-y^2, xy, xz, yz)$
A_u	1	1	1	1	1	-1	-1	-1	-1	-1		
T_{1u}	3	η^+	η^-	0	-1	-3	$-\eta^-$	$-\eta^+$	0	1	(x, y, z)	
T_{2u}	3	η^-	η^+	0	-1	-3	$-\eta^+$	$-\eta^-$	0	1		
G_u	4	-1	-1	1	0	-4	1	1	-1	0		
H_u	5	0	0	-1	1	-5	0	0	1	-1		

APPENDIX 7

Electrocyclic Reactions and Cycloadditions

A7.1 PREDICTING THE COURSE OF A REACTION BY CONSIDERING THE SYMMETRY OF THE RELEVANT ORBITALS: ELECTROCYCLIC REACTIONS

The molecular orbitals of butadiene can be used to predict, or at least rationalize, the course of reactions this compound undergoes. For instance, experimentally it is known that

In 1965, American chemists R. B. Woodward and R. Hoffmann ("conservation of orbital symmetry" – Woodward–Hoffmann Rules) and Japanese chemist K. Fukui ("frontier orbital theory") proposed theories to explain these results as well as those for other related reactions. Woodward won the Nobel Prize in Chemistry in 1965 for his synthetic work. In 1981, after the death of Woodward, Hoffmann and Fukui shared the same prize for the theories discussed here.

These theories assert that the course of an electrocyclic reaction that a compound undergoes is controlled by its "highest occupied molecular orbital (HOMO)". Referring to reactions A7.1 and A7.2, we pictorially illustrate the four π MOs of butadiene in Fig. A7.1.

For a thermal reaction of butadiene [reaction (A7.1)] ("ground state chemistry"), the HOMO is ψ_2 (reaction pathways shown in Fig. A7.2). For a photochemical reaction of butadiene [reaction (A.7.2)] ("excited state chemistry"), the HOMO is ψ_3 (reaction pathways shown in Fig. A7.3).

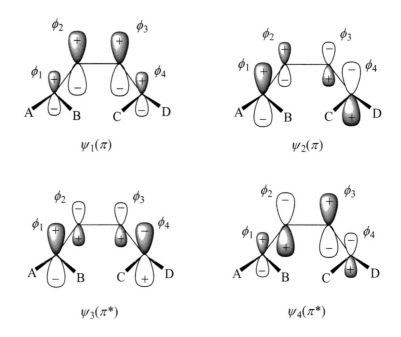

Fig. A7.1 The four π molecular orbitals of butadiene.

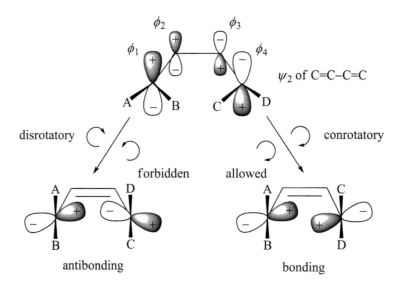

Fig. A7.2 The thermal electrocyclic reaction of butadiene.

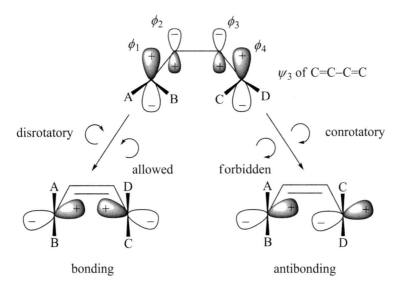

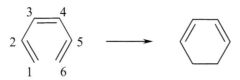

bonding antibonding

Fig. A7.3 The photochemical electrocyclic reaction of butadiene.

For wavefunction ψ_2, the terminal atomic orbitals ϕ_1 and ϕ_4 have the relative orientations shown in Fig. A7.1. It is now clear from Fig. A7.2 that a conrotatory process leads to a bonding interaction between ϕ_1 and ϕ_4, while a disrotatory process leads to an antibonding interaction between ϕ_1 and ϕ_4. In other words, the conrotatory pathway is allowed, while the disrotatory one is forbidden.

Conversely, for wavefunction ψ_3, the terminal atomic orbitals ϕ_1 and ϕ_4 have the relative orientations also shown in Fig. A7.1. Now, as illustrated in Fig. A7.3, a conrotatory pathway yields an antibonding interaction between the terminal atomic orbitals, while a disrotatory step leads to a stabilizing bonding interaction. Hence now the disrotatory process wins out.

Now let us apply this theory to the electrocyclic reaction of hexatriene:

The wavefunctions of the six π molecular orbitals of hexatriene are summarized in the following table:

Energies	Wavefunctions					
$E_6 = \alpha - 1.802\beta$	$\psi_6 = 0.232\phi_1$	$-0.418\phi_2$	$+0.521\phi_3$	$-0.521\phi_4$	$+0.418\phi_5$	$-0.232\phi_6$
$E_5 = \alpha - 1.247\beta$	$\psi_5 = 0.418\phi_1$	$-0.521\phi_2$	$+0.232\phi_3$	$+0.232\phi_4$	$-0.521\phi_5$	$+0.418\phi_6$
$E_4 = \alpha - 0.445\beta$	$\psi_4 = 0.521\phi_1$	$-0.232\phi_2$	$-0.418\phi_3$	$+0.418\phi_4$	$+0.232\phi_5$	$-0.521\phi_6$
$E_3 = \alpha + 0.445\beta$	$\psi_3 = 0.521\phi_1$	$+0.232\phi_2$	$-0.418\phi_3$	$-0.418\phi_4$	$+0.232\phi_5$	$+0.521\phi_6$
$E_2 = \alpha + 1.247\beta$	$\psi_2 = 0.418\phi_1$	$+0.521\phi_2$	$+0.232\phi_3$	$-0.232\phi_4$	$-0.521\phi_5$	$-0.418\phi_6$
$E_1 = \alpha + 1.802\beta$	$\psi_1 = 0.232\phi_1$	$+0.418\phi_2$	$+0.521\phi_3$	$+0.521\phi_4$	$+0.418\phi_5$	$+0.232\phi_6$

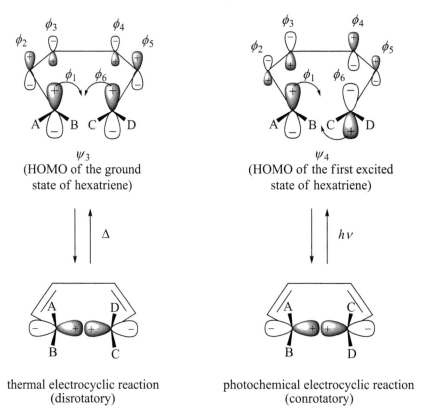

ψ_3
(HOMO of the ground
state of hexatriene)

ψ_4
(HOMO of the first excited
state of hexatriene)

Δ

$h\nu$

thermal electrocyclic reaction
(disrotatory)

photochemical electrocyclic reaction
(conrotatory)

Fig. A7.4 The thermal and photochemical electrocyclic reactions of hexatriene.

For the thermal and photochemical electrocyclic reactions of hexatriene, the controlling HOMOs are ψ_3 and ψ_4, respectively. As Fig. A7.4 shows, the allowed pathway for the thermal reaction is disrotatory. On the other hand, the allowed pathway for the photochemical reaction is conrotatory. These results are just the opposite to those found for the electrocyclic reactions of butadiene (Figs A7.2 and A7.3).

To summarize, for a linear polyene with 4, 8, 12, ... π electrons involved in its thermal electrocyclic reaction, the reaction will follow a conrotatory pathway; conversely, its photochemical electrocyclic reaction will adopt a disrotatory pathway. On the other hand, for a linear polyene electrocyclic reaction with 6, 10, 14, ... π electrons involved, the pathways of its thermal and photochemical electrocyclic reactions will be disrotatory and conrotatory, respectively.

A7.2 PREDICTING THE COURSE OF A REACTION BY CONSIDERING THE SYMMETRY OF THE RELEVANT ORBITALS: CYCLOADDITIONS

The best known cycloaddition may well be the Diels–Alder reaction between butadiene and ethylene (under thermal conditions):

According to the frontier orbital theory, the orbitals that control these reactions are the aforementioned HOMO of one reactant and the LUMO (lowest unoccupied molecular orbital) of the other reactant. So, for this reaction, we have two possible scenarios: interaction between HOMO ψ_2 of butadiene and LUMO π^* of ethylene or that between HOMO π of ethylene and LUMO ψ_3 of butadiene. As the following illustration indicates, both possibilities lead to bonding overlap between the interacting orbitals of the reactants; so the reaction is allowed, as we all know. In addition, as found by theory, between these two possible scenarios, we favor the interaction between the HOMO of the electron-rich reactant (butadiene in this case) with the LUMO of the electron-poor reactant (ethylene).

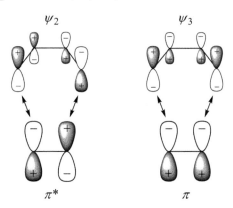

On the other hand, the dimerization of ethylene is symmetry-forbidden, as illustrated below:

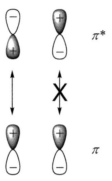

In order for this $[2 + 2]$ cycloaddition (each of the reactants has two π electrons participating in the interaction) to proceed, we need to carry out the experiment under photochemical conditions, where both the HOMO of the excited ethylene [with configuration $(\pi)^1(\pi^*)^1$] and the LUMO of the ground-state ethylene [configuration $(\pi)^2(\pi^*)^0$] are π^*. As the following illustration reveals, now we have a symmetry-allowed situation:

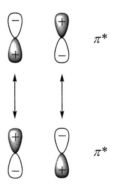

To summarize, the $[2 + 2]$ cycloadditions will proceed photochemically, but not thermally. On the other hand, the $[2 + 4]$ cycloadditions such as Diels–Alder reactions are allowed thermally and forbidden photochemically.

REFERENCE: I. Fleming, *Molecular Orbitals and Organic Chemical Reactions: Student Edition*, Wiley, Chichester, 2009, pp. 215–6; pp. 304–7.

APPENDIX 8

Selected Topics in Bioinorganic Chemistry

Bioinorganic chemistry is a frontier area that spans the boundary between the classical areas of chemistry and biology. It deals with the chemistry of metals in various biological systems. As metals play key roles in maintaining life, the major part of bioinorganic chemistry is focused on the study of natural occurring inorganic elements in biology.

A8.1 BIOLOGICAL LIGANDS FOR METAL IONS

Metal-ligand interactions can be considered as Lewis acid-base interactions. In coordination compounds, metal ions are surrounded by electron-donating ligands such as water, amines, halide anions and the carbonyl ligand (CO). In biological systems, metal ions are most commonly bound by various biological ligands (biomolecules), which can be broadly classified into three groups:

(a) peptides (or proteins) with appropriate amino acid side chain groups usable for metal coordination,

(b) biosynthesized macrocyclic ligands (such as porphyrinoids and ionophores), and

(c) nucleic acids.

A8.1.1 Coordination by Peptides and Proteins

Proteins are natural polymers formed by various combinations of 20 natural occurring amino acids. A general structure of an amino acid molecule in its un-ionized form is depicted in Fig. A8.1(a). The α carbon atom (C_α) in each amino acid molecule is bonded to one amino group $(-NH_2)$, one carboxylic acid group $(-COOH)$, one hydrogen atom and one organic side chain (R). Thus, all amino acids (except glycine) are chiral. The C_α atom of naturally occurring amino acids has an S configuration (called L-amino acids). In aqueous solutions, amino acids readily ionize, forming the corresponding zwitterions [Fig. A8.1(b)]. It is the side chain R which determines the identity of the amino acids (Fig. A8.2). An amino acid is abbreviated by a standard three-letter code.

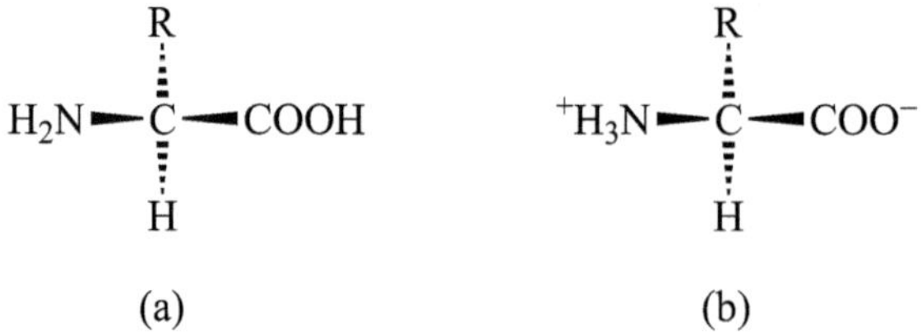

$$H_2N \text{---} \overset{R}{\underset{H}{C}} \text{---} COOH \qquad\qquad {}^+H_3N \text{---} \overset{R}{\underset{H}{C}} \text{---} COO^-$$

(a)   (b)

Fig. A8.1 The general structure of an amino acid in its (a) un-ionized form and (b) zwitterionic form.

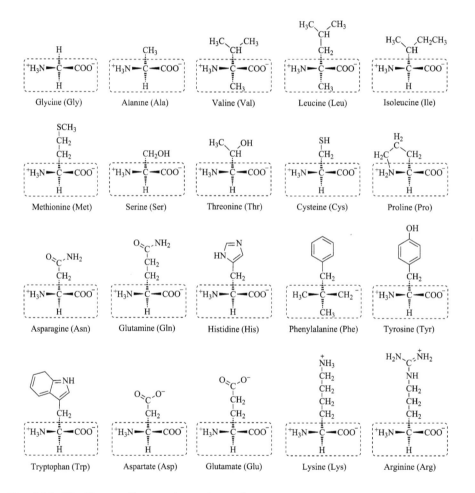

Fig. A8.2 The 20 naturally occurring amino acids.

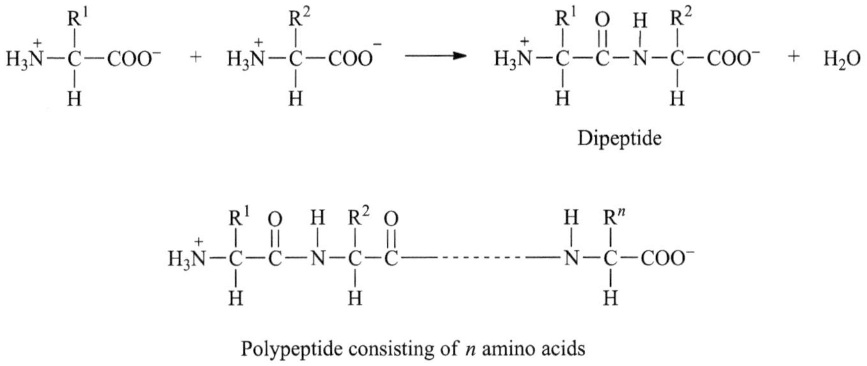

$$H_3\overset{+}{N}-\underset{\underset{H}{|}}{\overset{\overset{R^1}{|}}{C}}-COO^- \;+\; H_3\overset{+}{N}-\underset{\underset{H}{|}}{\overset{\overset{R^2}{|}}{C}}-COO^- \;\longrightarrow\; H_3\overset{+}{N}-\underset{\underset{H}{|}}{\overset{\overset{R^1}{|}}{C}}-\overset{\overset{O}{\|}}{C}-\underset{\underset{H}{|}}{\overset{\overset{H}{|}}{N}}-\underset{\underset{H}{|}}{\overset{\overset{R^2}{|}}{C}}-COO^- \;+\; H_2O$$

Dipeptide

$$H_3\overset{+}{N}-\underset{\underset{H}{|}}{\overset{\overset{R^1}{|}}{C}}-\overset{\overset{O}{\|}}{C}-\underset{\underset{H}{|}}{\overset{\overset{H}{|}}{N}}-\underset{\underset{H}{|}}{\overset{\overset{R^2}{|}}{C}}-\overset{\overset{O}{\|}}{C}------\underset{\underset{H}{|}}{\overset{\overset{H}{|}}{N}}-\underset{\underset{H}{|}}{\overset{\overset{R^n}{|}}{C}}-COO^-$$

Polypeptide consisting of n amino acids

Fig. A8.3 Formation of peptides through condensation of amino and carboxylate groups of amino acids.

Two amino acid molecules can undergo condensation reaction to form a dipeptide. Further condensation reactions lead to formation of polypeptides (Fig. A8.3). Amino-acids in proteins are numbered sequentially, starting from the first residue at the amino terminus.

Many amino acids possess donor atoms on the side chain groups (the R substituents), making them suitable for coordination to metals. For example, the thiolate group of cysteine, the imidazole group of histidine, the carboxylate groups of glutamate and aspartate, and the phenolate group of tyrosine are common protein donor groups which readily bind metal ions (Fig. A8.4). It is worth noting that different donor groups on the amino acid side chains show different affinity towards a particular metal ion. Beside the donor groups on the protein side chains, peptide carbonyl groups and deprotonated N-terminus amino groups of proteins can also coordinate to metal ions, though examples of the latter complexation are rare.

A8.1.2 Biosynthesized Macrocyclic Ligands

A8.1.2.1 Macrocyclic tetrapyrrole systems

Porphyrins are macrocyclic tetrapyrrole molecules which, in their deprotonated forms, can bind a wide range of metal ions to form metalloporphyrin complexes (Fig. A8.5). A metal ion is bonded to the four pyrrole nitrogens of a porphyrin ligand in a planar or nearly planar configuration. Metalloporphyrins are important functional units which are widely found in various biological systems such as chlorophyll, hemoglobin, and a wide range of heme-containing enzymes. Corrins and corroles are closely related macrocyclic tetrapyrrolic molecules which have a skeletal structure very similar to that of porphyrins. Vitamin B_{12} is a cobalt corrinoid complex. These porphyrinoid systems perform important functions in various biological processes such as light harvesting, electron transfer, dioxygen activation, and transport (Fig. A8.6).

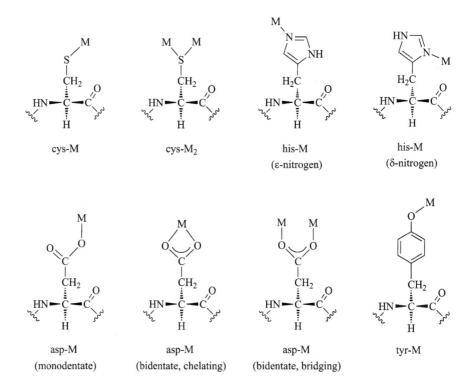

Fig. A8.4 Examples of metal binding by amino acid side chain groups.

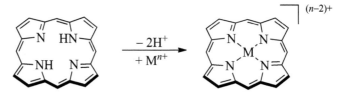

Fig. A8.5 Porphyrins are dianionic tetrapyrrole ligands.

Porphyrinoid ligands and their metal complexes contain an extensive conjugated π system. They fulfill the Hückel rule for aromaticity with $4n + 2 = 18$ π electrons in the macrocyclic ring. The electronic absorption (UV-vis) spectra of most porphyrin complexes show a characteristic intense absorption band at $\lambda_{max} \approx 400 - 450$ nm which is attributed to a $\pi \rightarrow \pi^*$ transition. This absorption is called the Soret band, named after its discoverer Jacques-Louis Soret. A weaker absorption feature at $\sim$550 nm (the Q band) is also observed in some porphyrin

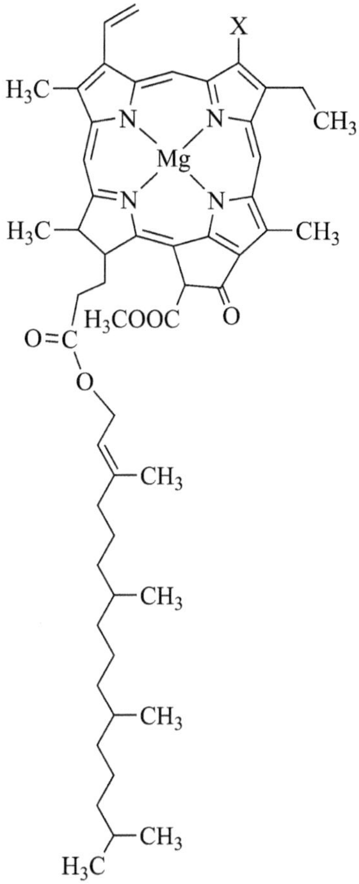

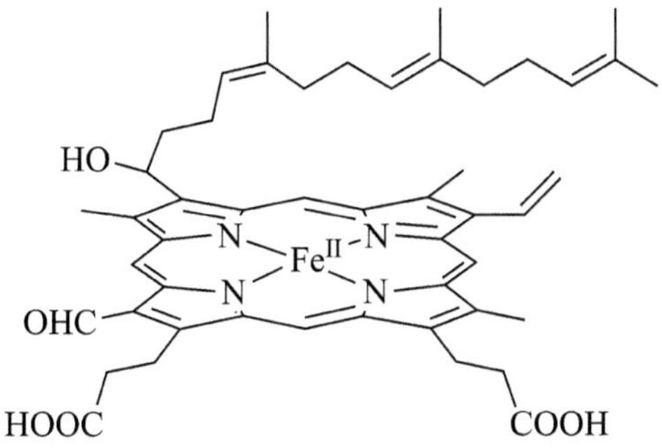

(b) Heme *a* (ferrous form)

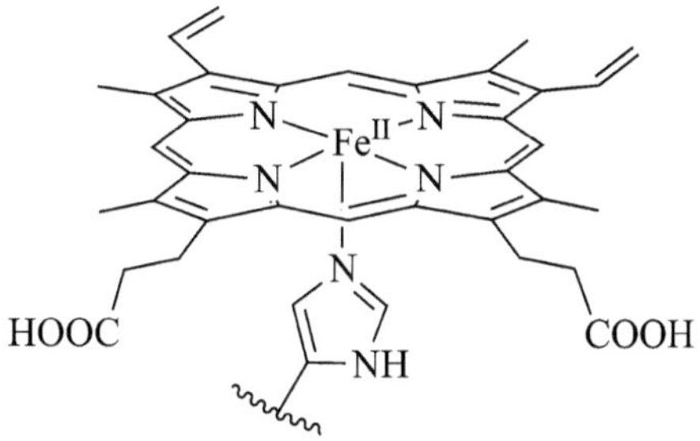

(c) The prosthetic heme group in deoxyhemoglobin. The fifth coordination site of the Fe(II) ion is occupied by an imidazole ring of a histidine residue.

(a) Chlorophyll *a* (X = CH$_3$) and chlorophyll *b* (X = CHO)

Fig. A8.6 Examples of porphyrinoid complexes: (a) chlorophyll *a* and *b*, (b) heme *a*, (c) the prosthetic heme group in deoxyhemogloblin, and (d) vitamin B$_{12}$.

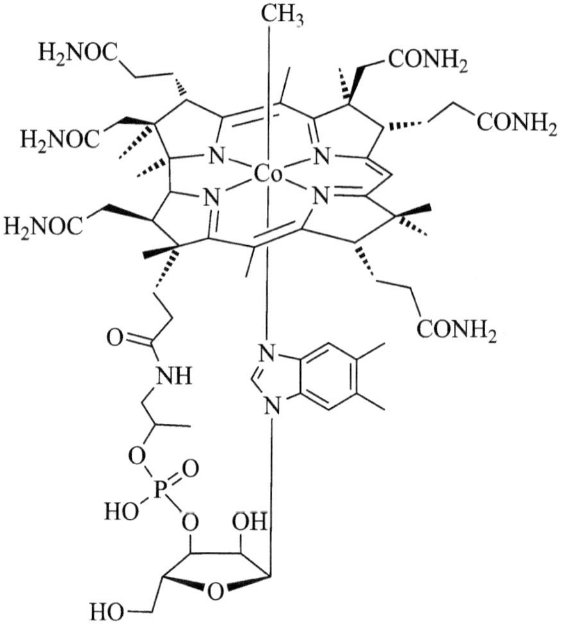

(d) Methylcobalamine (a form of vitamin B_{12})

Fig. A8.6 Continued

compounds. Another noteworthy property of porphyrinoid compounds is a narrow π frontier orbital gap, facilitating their involvement in the uptake and release of electrons (i.e. one-electron oxidation-reduction processes). As a consequence of these light absorption and redox properties, porphyrin compounds play important roles in photosynthesis and respiration.

A8.1.2.2 Ionophores and siderophores

Ionophores are biologically important macrocyclic host molecules that can reversibly bind metal ions. They act as ion carriers, facilitating the transport of metal ions (e.g. Na^+ and K^+) across cell membranes. Many antibiotics (e.g. valinomycin) are ionophores synthesized by microorganisms, whereas crown ethers, cryptands and calixarenes are synthetic ionophores. Depending on the size of the macrocyclic rings, ionophores show high selectivity towards binding of specific metal cations.

Siderophores are iron-binding small molecules secreted by microorganisms. The iron-binding properties of siderophores will be discussed in detail in a later section of this appendix.

A8.1.3 Nucleic Acids as Ligands

Nucleic acids are natural polymers of nucleotides. A nucleotide is composed of one pentose sugar unit, one nucleotide base, and one phosphate group. A nucleotide base is bonded to the pentose sugar unit through a glycosidic bond. Adjacent nucleotide monomers are joined together by phosphodiester linkages, resulting in formation of a sugar-phosphate polymer backbone (Fig. A8.7).

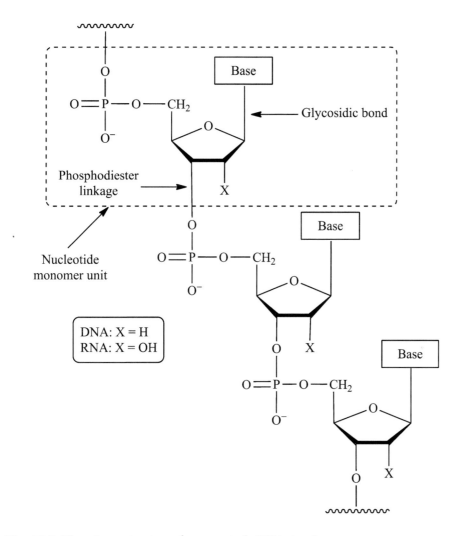

Fig. A8.7 The primary structure of a segment of a DNA strand.

Appendix 8

There are two types of nucleic acids, namely deoxyribonucleic acid (DNA) and ribonucleic acid (RNA). Both DNA and RNA are important biomolecules which carry genetic information, yet they are structurally different:

(a) DNA and RNA are polymers of deoxyribonucleotides and ribonucleotides, respectively. The pentose sugar in deoxyribonucleotides is deoxyribose, which does not contain a –OH group attached to the 2′-carbon of the pentose ring. Ribonucleotides contain ribose which has one more –OH group than deoxyribose.
(b) DNA has a double-stranded structure while RNA is a single-stranded molecule.
(c) The nitrogenous bases found in DNA are cytosine (C), thymine (T), guanine (G), and adenine (A), while RNA contains cytosine (C), uracil (U), guanine (G), and adenine (A) (Fig. A8.8). Two DNA strands form a double helical structure through Watson-Crick base-pairing interactions (Fig. A8.9).

There are three potential donor sites on nucleotides for metal binding:

(a) The nitrogen bases (primarily the endocyclic nitrogen atoms) can form strong complexes with transition-metal ions such as Cu(II), Cr(III), and Pt(II). By contrast, the exocyclic amino groups on cytosine, guanine and adenine are relatively poor metal-binding groups because their lone pairs largely delocalize into the ring.

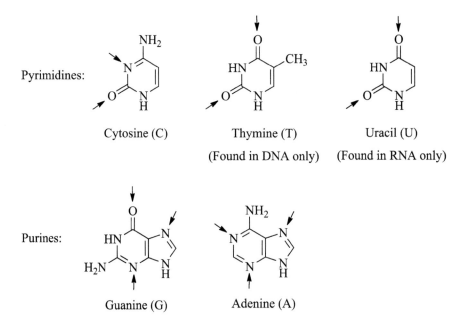

Fig. A8.8 The five nucleotide bases. The arrows show the common metal-binding sites.

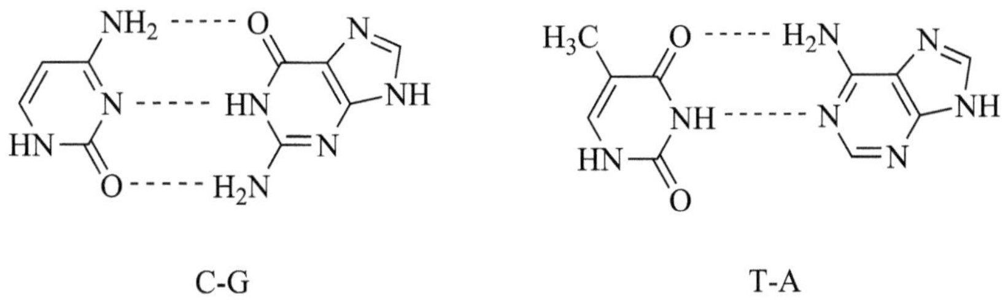

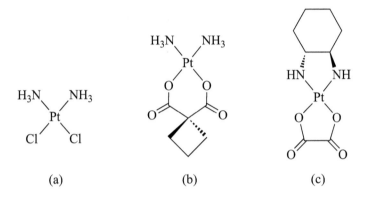

Fig. A8.9 In the Watson-Crick DNA base pairing model, the base pairs (C-G and T-A) are bound together by hydrogen bonds.

Fig. A8.10 Platinum-containing compounds. (a) Cisplatin, (b) carboplatin, and (c) oxaliplatin.

(b) The phosphodiester groups can bind metal ions, particularly hard alkali and alkaline-earth metal ions such as Na^+, K^+, and Mg^{2+}.

(c) The hydroxyl groups on the pentose sugar units can also bind metal ions, though complexation to these hydroxyl groups have been rarely observed.

cis-$[Pt(NH_3)_2Cl_2]$ (cisplatin) is a well-known chemotherapeutic drug that has been known since the 1960s. It is effective against various types of tumors such as bladder, ovarian, and testicular tumors, though it induces many undesirable side effects such as kidney damage and reduction in immunity to infections. The mode of action of cisplatin is based on its interaction with purine bases on DNA, causing DNA damage and replication inhibition, and subsequently inducing cell death. A generally accepted mechanism involves aquation of cisplatin inside cells to form an aquo complex of the type cis-$[Pt(NH_3)_2Cl(H_2O)]^+$. The aqua ligand is labile and readily displaced by a nucleotide base. Other platinum-based drugs such as carboplatin and oxaliplatin have also been developed for treatment of cancers with fewer side effects (Fig. A8.10).

A8.2 IRON STORAGE AND TRANSPORT

Living organisms have developed various ways to store and transport metal ions. Let us take iron as an example. Iron is one of the most important metals in the human body. It has a rich redox chemistry and plays vital roles in various biological systems in our body, and the most important iron-containing biological molecules are hemoglobin (for dioxygen transport), and cytochromes and Fe-S proteins (for electron-transfer processes).

The uptake, transport and storage of iron in mammals involve complex regulatory mechanisms. These processes may consist of the following steps:

(a) active or passive absorption of iron from foods,
(b) selective transport of iron through membranes into cells,
(c) incorporation of iron into a protein (a process which happens inside the cells), and
(d) elimination of iron through excretion or temporary storage.

An excess of free iron, particularly Fe(II), is fatal as reactive oxygen species (superoxide and hydroxyl radicals) can be generated in the presence of dioxygen and hydrogen peroxide, respectively, through the Fenton and Fenton-like oxidation reactions:

$$Fe(II) + O_2 \rightarrow Fe(III) + O_2^-$$
$$Fe(II) + H_2O_2 \rightarrow Fe(III) + OH^- + OH$$

In addition, a continuous supply of iron also can promote the growth and multiplication of pathogenic microorganisms. Therefore, efficient ways for regulating iron contents in all organisms are vital.

A8.2.1 Ferritin and Transferrin

In mammals, iron is taken up from dietary sources in the form of Fe(II) by absorption through the stomach and intestines. It is taken into the blood in the form of a Fe(III)-containing metalloprotein (transferrin). In mammals, iron is mainly stored in ferritin, a water-soluble iron storage protein found in liver, bone marrow and spleen.

In order to understand how iron is being stored inside ferritin, the structure of the protein has been studied in detail. Apoferritin from horse spleen was isolated and characterized. An X-ray crystallographic analysis showed that the protein consists of 24 equivalent subunits (each subunit consists of a polypeptide chain of 163 amino acid residues). These subunits are arranged to form a hollow shell with a diameter of about 80 Å (Fig. A8.11). Iron is stored in the core of ferritin as Fe(III) ions in the form of an oxohydroxophosphate of composition $(FeO \cdot OH)_8 (FeO \cdot H_2PO_4)$. It is believed that iron is taken up in the form of Fe(II) through channels in the spherical structure,

$$\text{Fe(O}_2\text{CCH}_3)_2 + \text{LiOCH}_3 \xrightarrow[\text{CH}_3\text{OH}]{\text{O}_2} [\text{Fe}^{\text{III}}_4\text{Fe}^{\text{II}}_8 (\text{O})_2(\text{OCH}_3)_{18}(\text{O}_2\text{CCH}_3)_6(\text{CH}_3\text{OH})_{4.67}]$$

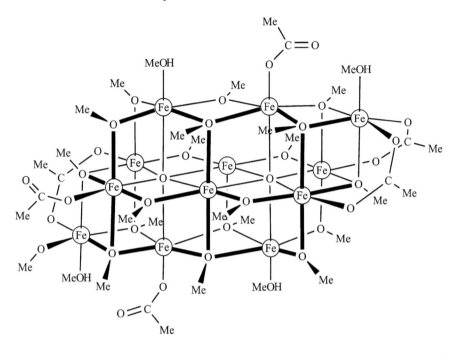

Fig. A8.11 Preparation and structure of the polyiron oxo cluster complex $[\text{Fe}^{\text{III}}_4$ $\text{Fe}^{\text{II}}_8(\text{O})_2(\text{OCH}_3)_{18}(\text{O}_2\text{CCH}_3)_6(\text{CH}_3\text{OH})_{4.67}]$, which resembles the core of ferritin. The iron atoms are highlighted with circles. [Adapted from K. L. Taft, G. C. Papaefthymiou, and S. J. Lippard, A mixed-valent polyiron oxo complex that models the biomineralization of the ferritin core, *Science* **259**, 1302–5 (1993).]

followed by oxidation and storage in the form of Fe(III) ions inside ferritin. When the body needs iron, it is released as Fe(II) ions in a controlled fashion.

Studies of model compounds can provide valuable insights into the related bioinorganic systems. Several polyiron oxo cluster complexes have been reported as mimics to the three-dimensional layered structure of the inorganic core of ferritin. Two examples are illustrated in Figs. A8.12 and A8.13, respectively.

Transferrins are iron-transport glycoproteins found primarily in mammals. Several types of transferrins, namely serum transferrin, lactoferrin and ovotransferrin, have been identified. The human serum transferrin is composed of a single polypeptide chain of molecular weight ∼80 kDa. This polypeptide chain folds to form two protein pockets where hard N- and O-donor atoms with a high affinity for Fe(III) ions are found. Besides these, the presence of $[\text{HCO}_3^-]$ and

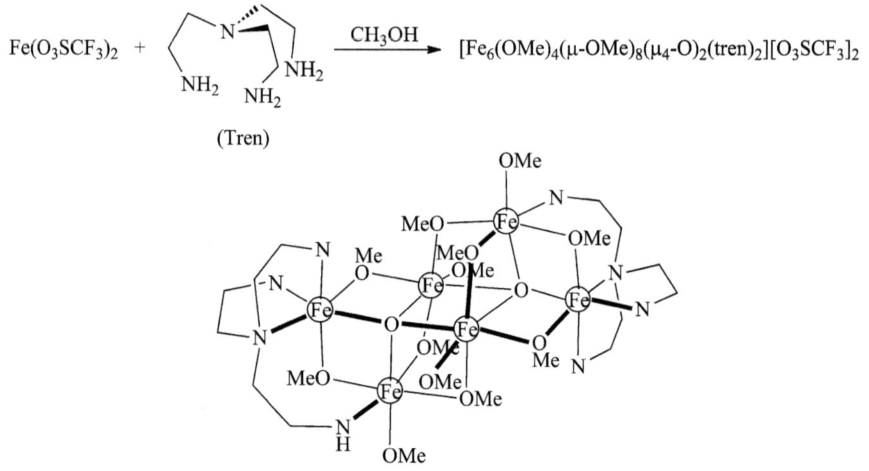

Fe(O₃SCF₃)₂ + [Tren structure] $\xrightarrow{\text{CH}_3\text{OH}}$ [Fe₆(OMe)₄(μ-OMe)₈(μ₄-O)₂(tren)₂][O₃SCF₃]₂

(Tren)

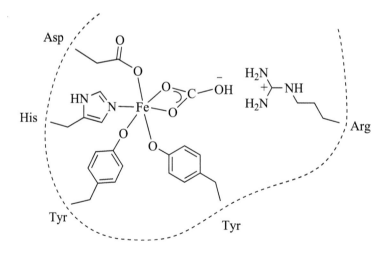

Fig. A8.12 Synthesis and structure of hexanuclear $[\text{Fe}_6(\mu_4\text{-O})_2(\mu_2\text{-OMe})_8(\text{OMe})_4(\text{tren})_2]^{2+}$ cluster, where tren is tridentate 2,2′,2″-triaminotriethylamine. A notable feature of this polyiron oxo cluster is the presence of two unusual $\mu_4\text{-O}^{2-}$ ions. The iron atoms are highlighted with circles. [Adapted from V. A. Nair and K. S. Hagen, Iron oxo aggregation: Fe₃ to Fe₆. Synthesis, structure, and magnetic properties of the hexanuclear dication $[\text{Fe}_6(\mu_4\text{-O})_2(\mu_2\text{-OMe})_8(\text{OMe})_4(\text{tren})_2]^{2+}$, a soluble, crystalline model of iron oxo hydroxo nanoparticles, the core of ferritin and rust formation, *Inorg. Chem.* **31**, 4048–50 (1992).]

[Sketch of transferrin Fe binding site with Asp, His, Tyr, Tyr, Arg residues and central Fe]

Fig. A8.13 A sketch of the Fe^{3+} binding site in human transferrin.

$[CO_3^{2-}]$ ligands also facilitates the binding of Fe(III) (Fig. A8.13). Indeed, Fe(III)-transferrin complexes have very high binding constants ($\beta \approx 10^{23}$-10^{28} M^{-1} at pH $= 7.4$). As a consequence, transferrins have an extremely high efficiency in iron transport. The Fe(III) binding affinity of transferrins was found to be pH dependent; the binding affinity decreases when the pH of the medium becomes lower. It is generally believed that transferrins loaded with iron are brought into a cell across the cell membrane via transferrin receptors. Reduction of pH values (by variation of $[H^+]$) inside the cell causes transferrins to release their iron content.

A8.2.2 Siderophores

Uptake of iron by microorganisms (bacteria and fungi) is limited by the insolubility of Fe(OH)$_3$ in an aqueous medium [K_{sp} of Fe(OH)$_3$ $= 2.64 \times 10^{-39}$ M^4]. Microorganisms can overcome this problem by making use of polydentate O-donor ligands called siderophores to take up iron. Siderophores are low molecular weight metal-chelating compounds that form stable complexes with Fe(III). Several examples of siderophores are shown in Fig. A8.14.

On the other hand, the release of iron requires the reduction of Fe(III) to Fe(II) since the complex formation constants of siderophores and Fe(II) are orders of magnitude lower than those of Fe(III).

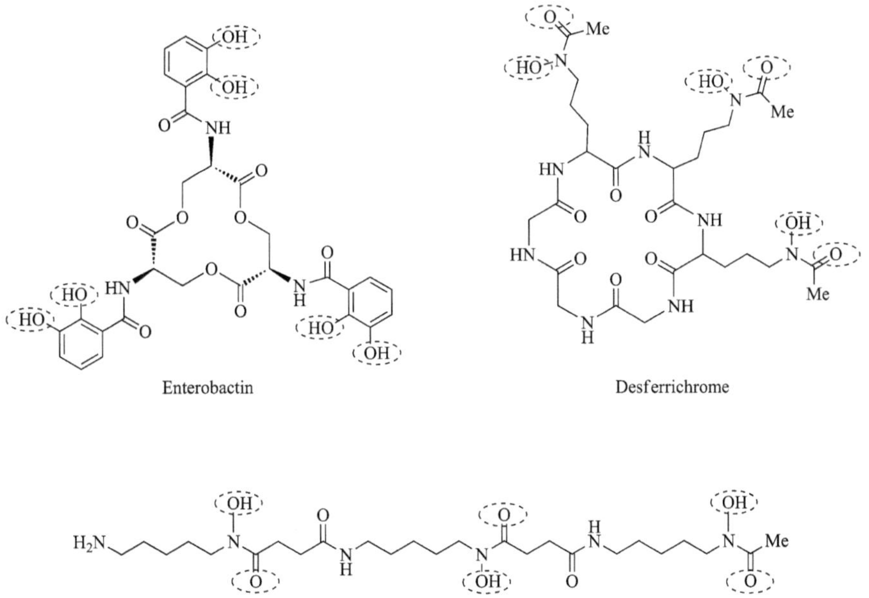

Enterobactin

Desferrichrome

Desferrioxamine

Fig. A8.14 Examples of siderophores. The iron-binding groups are circled in dashed lines.

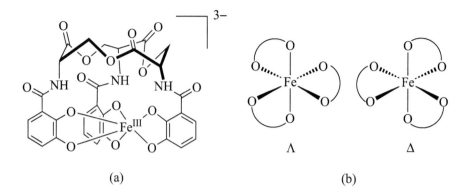

Fig. A8.15 (a) Enterobactin coordinates to a Fe(III) ion to form a pseudo-octahedral complex. (b) The Λ and Δ configurations of an octahedral complex. [Adapted from: K. N. Raymond, E. A. Dertz, and S. S. Kim, Enterobactin: An archetype for microbial iron transport, *Proc. Natl. Acad. Sci. U.S.A.* **100**, 3584–8 (2003).]

Most of the siderophores form chiral complexes with Fe(III) ions. These Fe(III)-siderophore complexes are absorbed by microorganisms through specific receptors on the cell membrane. Because these receptors are chiral, they can recognize the absolute configuration of the Fe(III)-siderophore complexes.

Let us consider the siderophore enterobactin. Enterobactin is derived from three L-serine residues, each of which carries a 2,3-dihydroxybenzoyl substituent. The deprotonated form of enterobactin is a hexadentate triscatecholate ligand which readily binds Fe(III) to form an pseudo-octahedral complex (Fig. A8.15). Based on UV/Vis and circular dichroism (CD) data, the Δ-complex is formed diastereoselectively.

Circular Dichroism (CD)

Opposite enantiomers show difference in absorption of left- and right-circularly polarized light. In other words, the absorption of right-circularly polarized light (A_R) is different from that of left-circularly polarized light (A_L) for a chiral molecule. Thus, the difference in absorption intensity $\Delta A = A_L - A_R \neq 0$. This effect is known as circular dichroism. This different absorption property provides important information on the structures of chiral molecules. Circular dichroism is widely used to characterize chiral organic and biological compounds.

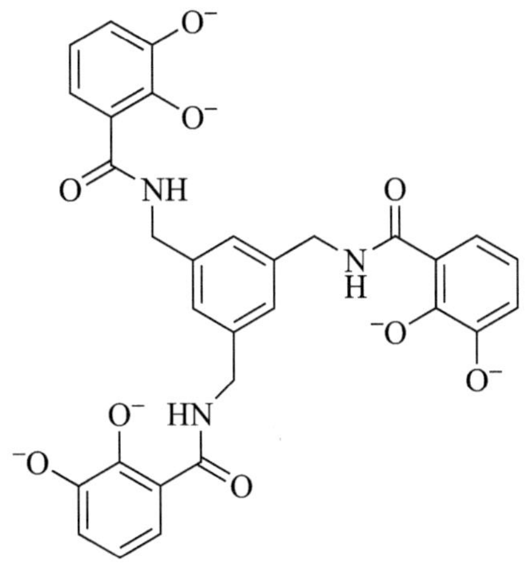

Fig. A8.16 A synthetic analog of enterobactin.

In order to understand the coordination chemistry of enterobactin, a structural mimic consisting of three triscatecholate pendants has been synthesized (Fig. A8.16), and its complexation with Fe(III) ions has been studied in detail. Interestingly, this model compound also forms a Fe(III) complex with a binding constant close to that of the native enterobactin. The crystal structure of this model Fe(III) complex was also determined.

It is noted that high-spin Fe(III)-siderophore complexes are kinetically labile. Therefore, other metal ions are often used as substitutes for Fe(III) ion in model studies. For example, Cr(III) ions form kinetically inert complexes with siderophores. This makes the studies of their complexation chemistry in solution easier to carry out.

A8.3 BIOLOGICAL DIOXYGEN CARRIERS

Aerobic organisms require dioxygen to carry out respiration. Nature has developed a few elegant systems to transport dioxygen from regions of high abundance (air or water) to regions of relatively low abundance and high demand (e.g. tissue cells). Dioxygen-carrier proteins are key components of these systems. Dioxygen-carriers can bind and release O_2 reversibly at a rapid rate.

Although cobalt complexes are well known to perform dioxygen-binding in inorganic chemistry, the process appears to be carried out exclusively by iron and copper complexes in biological systems.

Three biological dioxygen transport and storage systems have been identified:

(a) the hemoglobin family - found in all vertebrates and many invertebrates;
(b) the hemerythrin family - found in some marine invertebrates; and
(c) the hemocyanin family - found in arthropods and mollusks.

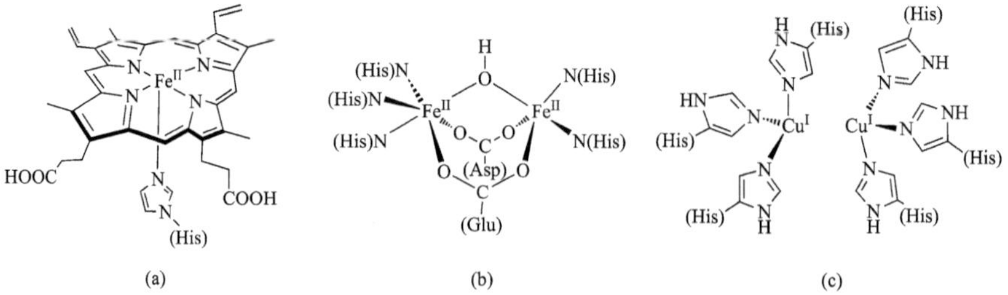

Fig. A8.17 The O_2-binding site in (a) hemoglobin, (b) hemerythrin, and (c) hemocyanin.

Table A8.1 Some physical properties of hemoglobin, hemerythrin, and hemocyanin.

Property	Hemoglobin	Hemerythrin	Hemocyanin
Metal	Fe	Fe	Cu
Oxid. state of the metals in deoxy protein	+2	+2	+1
Metal:O_2	Fe:O_2	2Fe:O_2	2Cu:O_2
Color (oxy form)	Red	Violet-pink	Blue
Color (deoxy form)	Red-purple	Colorless	Colorless
Coordination to metal center	Heme ring	Protein side chains	Protein side chains
Molecular weight	65 kDa	108 kDa	400 to 20,000 kDa
Number of subunits	4	8	Many

These three dioxygen transport systems employ metalloproteins that contain either iron or copper in their O_2-binding sites (Fig. A8.17). Some physical properties of these dioxygen carrier proteins are summarized in Table A8.1.

A8.3.1 Hemoglobin (Hb) and Myoglobin (Mb)

Both hemoglobin and myoglobin are O_2-binding heme-iron proteins. Hemoglobin is a metal complex found in the red blood cells of vertebrates.[1] Hemoglobin takes up dioxygen from air in the lungs and delivers the bound dioxygen to various parts of the body through the blood stream.

Structures:

The crystal structure of the deoxy form of human hemoglobin was determined by Max Perutz and co-workers in 1959.[2] Human hemoglobin is an $\alpha_2\beta_2$ tetramer, consisting of four polypeptide chains (two α and two β subunits). Each subunit contains an iron-protoporphyrin IX group. In deoxyhemoglobin, each iron-protoporphyrin IX cofactor has a high-spin five-coordinate Fe(II) center, which is bound by four pyrrolic nitrogen atoms of the porphyrin and a histidine residue of

---

[1] It is noted that hemoglobin is also found in the tissues of some invertebrates.

[2] M. F. Perutz, M. G. Rossmann, A. F. Cullis, H. Muirhead, G. Will and A. C. T. North, Structure of haemoglobin: a three-dimensional Fourier synthesis at 5.5-Å resolution, obtained by X-ray analysis. *Nature* **185**, 416–22 (1960).

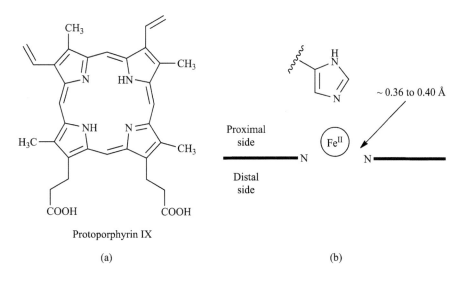

Fig. A8.18 (a) Molecular structure of protoporphyrin IX. (b) A schematic diagram showing the coordination environment of a heme cofactor in hemoglobin.

the polypeptide chain. The coordination geometry around the Fe(II) center can be considered as square pyramidal. The Fe(II) center is located at ≈ 0.36 - 0.40 Å above the plane of the porphyrin ring (Fig. A8.18). Upon oxygenation, the iron center [now becomes Fe(III)] moves toward the FeN_4 plane.

Myoglobin is a monomeric heme protein for dioxygen storage found mainly in muscle tissues of vertebrates. It consists of an iron-protoporphyrin IX cofactor, which is bonded to a single polypeptide chain. The coordination environment of the iron center in the heme group is very similar to that of hemoglobin.[3] In deoxymyoglobin, the Fe(II) center lies 0.42 Å out of the plane of the porphyrin ring.

The dioxygen-binding reaction:

The O_2-binding properties of hemoglobin and myoglobin have been examined by spectroscopic, thermodynamic and kinetic studies. The electronic spectra of Hb and Mb are dominated by intense absorptions at 400-500 nm (the Soret band) and 500-600 nm (α and β bands), which are assignable to $\pi \rightarrow \pi^*$ transitions of the heme group. It is noted that these optical absorption bands are sensitive to the state of oxygenation.

The binding of dioxygen to deoxy-hemoglobin and -myoglobin has been studied by vibrational spectroscopy. Resonance Raman spectra of oxy-hemoglobin and -myoglobin show an O–O stretching frequency at ~ 1105 cm^{-1}, which is close to that of superoxide anion O_2^-

---

[3] J. C. Kendrew, R. E. Dickerson, B. E. Strandberg, R. G. Hart, D. R. Davies, D. C. Phillips and V. C. Shore, Structure of myoglobin: a three-dimensional Fourier synthesis at 2 Å resolution. *Nature* **185**, 422–7 (1960).

Table A8.2 Vibrational and structural properties of dioxygen species.

Species	ν_{O-O} (cm^{-1})	r_{O-O} (Å)
O_2^+	1,905	1.12
O_2	1,580	1.21
O_2^-	1,097	1.33
O_2^{2-}	802	1.49

(Table A8.2). The vibrational spectroscopic data suggest that coordination of dioxygen to deoxy-hemoglobin and -myoglobin is accompanied by electron-transfer to form an Fe(III)-superoxide species (Fe^{III}-O_2^-).

In deoxyhemoglobin, the Fe(II) center has a high-spin electronic configuration ($S = 2$). Fe(II) has a covalent radius too large to fit in the plane of the four nitrogen atoms of the heme ring. Addition of O_2 leads to formation of a Fe^{III}-O_2^- adduct with a diminished covalent radius of the iron center, which then moves into the ring of the heme group.[4] This gives rise to a substantial change in electronic configuration and structural reorganization of the iron coordination sphere (Fig. A8.19). The movement of iron towards the center of the heme ring results in a change in the quaternary structure of the tetrameric hemoglobin protein and triggers a cooperative binding of dioxygen by the four heme groups.

The O_2-binding behavior of the four heme groups in hemoglobin is not independent of each other. After a dioxygen molecule binds to the first heme group of deoxyhemoglobin, the O_2-binding affinity of the remaining heme groups becomes enhanced. When the second and third heme groups become oxygenated, the O_2-binding affinity of the protein is further strengthened. In other words, the dioxygen affinity of hemoglobin rises with increasing oxygen saturation. Thus, hemoglobin in the lung tissues, where oxygen tension is high, becomes fully oxygenated. In tissue cells where oxygen tension is low, the bound dioxygen is readily released and the affinity of hemoglobin for dioxygen is reduced.

Fig. A8.20 shows oxygen saturation curves for hemoglobin and myoglobin. Hemoglobin exhibits a characteristic sigmoidal saturation curve due to the cooperative dioxygen binding behavior. On the other hand, the monomeric myoglobin contains only one heme prosthetic group and, hence, it shows no cooperative O_2-binding property. Myoglobin displays a hyperbolic curve, indicating a rapid dioxygen saturation behavior.

Synthetic model complexes for hemoglobin and myoglobin:
A number of synthetic models for hemoglobin and myoglobin have been reported. An earlier study involved the reactions of five-coordinate cobalt(II) Schiff base complexes with dioxygen, which led to formation of the corresponding cobalt(III) compounds with an O_2 molecule being

---

[4] According to Pauling's Rules, the ionic radii of Fe^{2+} and Fe^{3+} with C.N. $= 6$ are 0.78 Å and 0.65 Å, respectively.

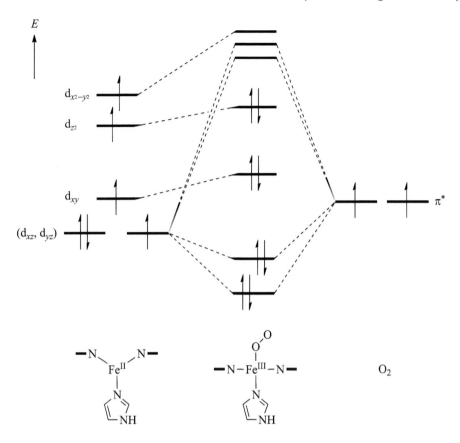

Fig. A8.19 Electronic structures of the heme cofactor in deoxy- and oxy-hemoglobin.

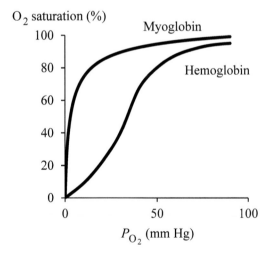

Fig. A8.20 Oxygen saturation curves for hemoglobin and myoglobin. Hemoglobin: four hemes found in RBC O_2/CO_2 transport. Myoglobin: one heme found in muscle O_2 storage.

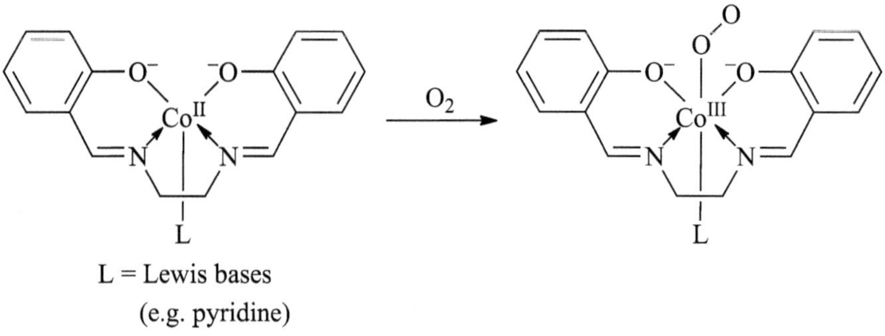

L = Lewis bases
(e.g. pyridine)

Fig. A8.21 Oxygenation reactions of five-coordinate cobalt(II) Schiff base complexes. [References: D. Chen and A. E. Martel, Dioxygen affinities of synthetic cobalt Schiff base complexes. *Inorg. Chem.* **26**, 1026–30 (1987). D. Chen, A. E. Martel, and Y. Sun, New synthetic cobalt Schiff base complexes as oxygen carriers. *Inorg. Chem.* **28**, 2647–52 (1989).]

bound "end-on" to the metal center (Fig. A8.21). The O–O bond distances in these Co(III)-O_2 adducts were found to be ~1.26 Å [*c.f.* O–O distances in O_2 (1.21 Å) and O_2^- (1.34 Å)]. The Lewis base L was also found to be crucial in the stabilization of these mononuclear Co(III)-O_2 adducts.

Iron porphyrin complexes are close structural analogs for the heme cofactors of hemoglobin and myoglobin. In order to mimic the coordination environment of the heme groups in the native proteins, a number of synthetic models containing a five-coordinate, high spin Fe(II) porphyrin system with a proximal base were prepared. One obstacle in these studies is the formation of undesirable μ-peroxo di-Fe(III) complexes upon exposure of the Fe(II) porphyrin precursors to dioxygen which renders reversible binding of dioxygen not feasible.

In order to suppress the formation of peroxo-bridged Fe(III) dimer, iron porphyrin complexes containing sterically bulky substituents were studied. A "picket-fence" porphyrin system has been developed by Collman and co-workers (Fig. A8.22). This bulky porphyrin system has proved to be successful in preventing the formation of the peroxo-bridged dimers. In addition, this "picket-fence" porphyrin model shows reversible oxygen binding.

A8.3.2 Hemerythrins (Hr)
Hemerythrin is a non-heme iron dioxygen carrier found in some marine invertebrates (such as annelids, and some species of molluscs and arthropods) for dioxygen transport and storage.

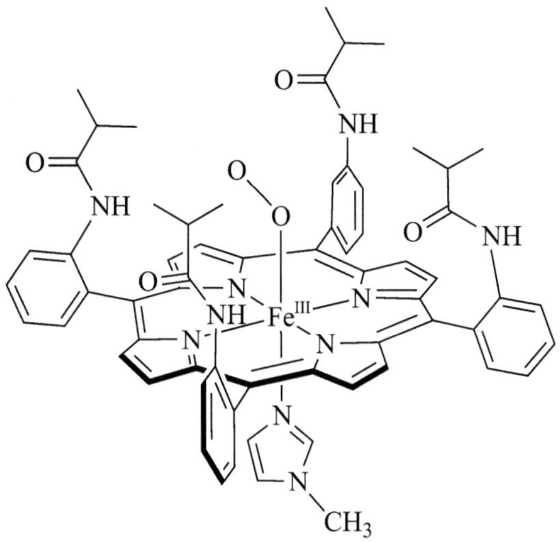

Fig. A8.22 A dioxygen-bound "picket-fence" iron(III) porphyrin complex. [Reference: J. P. Collman and L. Fu, Synthetic models for hemoglobin and myoglobin, *Acc. Chem. Res.* **32**, 455–63 (1999).]

Structures:

The structure of hemerythrin was determined by X-ray crystallography. Deoxyhemerythrin contains a diiron(II) core which is the O_2-binding site. The two Fe(II) centers have different coordination geometries. They are joined by one bridging hydroxyl group (μ-OH^-) and two bridging carboxylate groups. The carboxylate groups are contributed by glutamate and aspartate amino-acid residues of the protein core. In addition, one of the Fe(II) centers is ligated by three histidine nitrogens whilst the other one is bound by two histidine residues (Fig. A8.23). In other words, one of the Fe(II) centers is coordinatively saturated (six-coordinate), while the other one is five-coordinate. A vacant site on the latter iron center is available for dioxygen binding.

The dioxygen–binding reaction:

Deoxyhemerythrin consists of a (μ-hydroxo)*bis*(μ-carboxylato)diiron(II) core in which a vacant terminal coordination position on one iron atom is available for dioxygen binding. Upon oxygenation, the diiron(II) core transfers two electrons and a proton (from the bridging OH group) to dioxygen, forming a terminal hydroperoxy (–OOH) ligand. It is believed that this oxidative addition of dioxygen to the diiron(II) core may occur by two sequential one-electron steps, which involve a mixed-valence $\{Fe^{II}Fe^{III}\}$ intermediate.

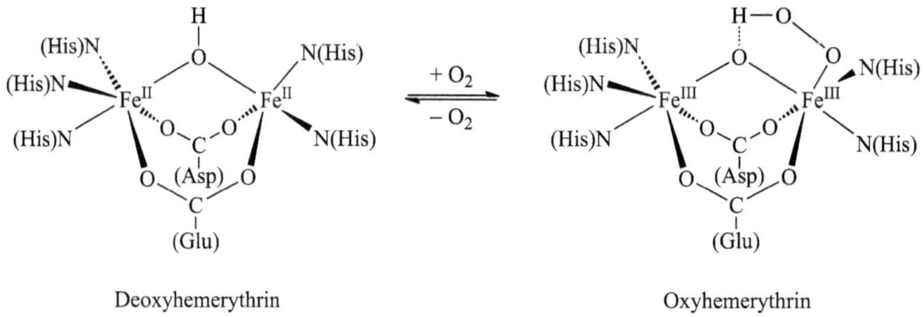

Fig. A8.23 A schematic diagram showing the redox and structural changes that occur in the diiron core of hemerythrin upon reversible dioxygen binding.

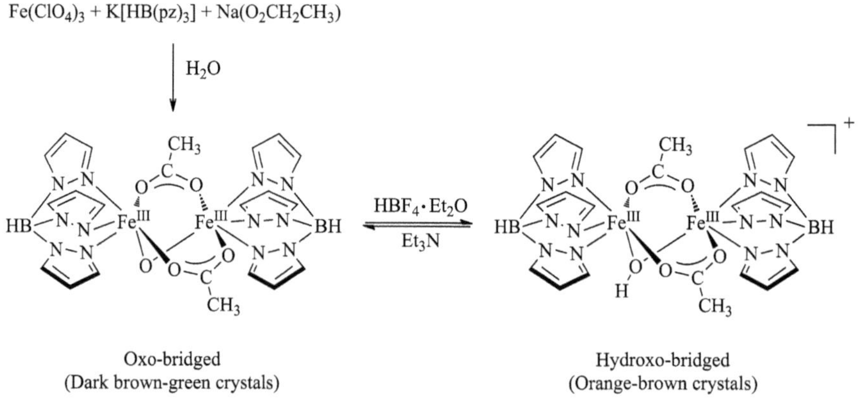

Fig. A8.24 Preparation of the μ-oxo bridged complex $[\{HB(pz)_3\}Fe(\mu\text{-}O)(\mu\text{-}OAc)_2 Fe\{HB(pz)_3\}]$ as a synthetic model for methemerythrin. [References: W. H. Armstrong and S. J. Lippard, $(\mu\text{-}Oxo)bis(\mu\text{-}acetato)bis(tri\text{-}1\text{-}pyrazolylborato)diiron(III)$, $[(HBpz_3)FeO(CH_3CO_2)_2Fe(HBpz_3)]$: Model for the binuclear iron center of hemerythrin, *J. Am. Chem. Soc.* **105**, 4837-8 (1983); W. H. Armstrong and S. J. Lippard, Reversible protonation of the oxo bridge in a hemerythrin model compound. Synthesis, structure, and properties of $(\mu\text{-}hydroxo)bis(\mu\text{-}acetato)bis[hydrotris(1\text{-}pyrazolyl)borato]diiron(III)$, $[(HB(Pz)_3)Fe(OH)(O_2CCH_3)_2Fe(HB(pz)_3)]^+$, *J. Am. Chem. Soc.* **106**, 4632-3 (1984).]

Synthetic model complexes for hemerythrin:

A number of dinuclear iron complexes supported by various types of nitrogen-donor ligands were reported as synthetic analogs for the diiron core of hemerythrin. An early example is a μ-oxo bridged diiron(III) complex supported by two tris(pyrazolyl)borate $([HB(C_3H_3N_2)_3]^-)$ ligands (Fig. A8.24). The two capping $[HB(C_3H_3N_2)_3]^-$ ligands closely mimic the histidine residues coordinating the diiron core in the native enzyme. Interestingly,

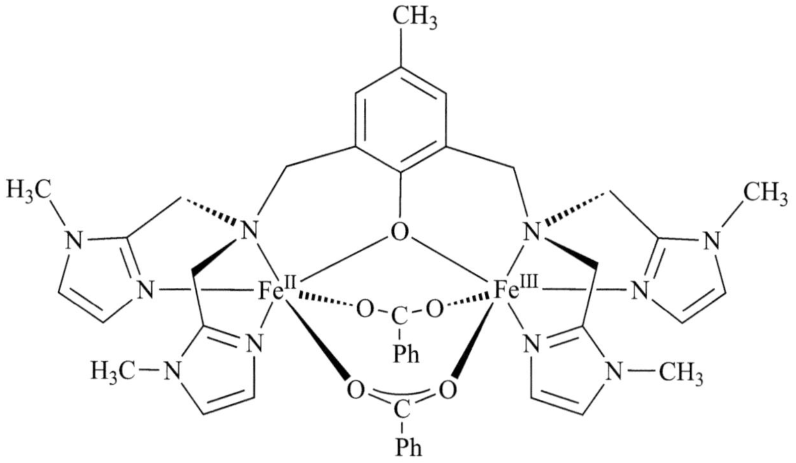

Fig. A8.25 A mixed-valence $Fe^{II}Fe^{III}$ complex supported by the dinucleating bimp⁻ ligand and two bridging benzoate ligands as a structural model of the $\{Fe^{II}Fe^{III}\}$ intermediate species in oxygenation of hemerythrin. [Reference: M. S. Mashuta, R. J. Webb, J. K. McCusker, E. A. Schmitt, K. J. Oberhausen, J. F. Richardson, R. M. Buchanan, and D. N. Hendrickson, Electron transfer in $Fe^{II}Fe^{III}$ model complexes of iron-oxo proteins, *J. Am. Chem. Soc.* **114**, 3815–27 (1992).]

this μ-oxo bridged diiron(III) complex undergoes a reversible protonation reaction to give the corresponding μ-hydroxo bridged complex.

Besides, a series of μ-phenoxo-bis(carboxylate)-bridged $Fe^{II}Fe^{III}$ complexes with the general formula $[Fe^{II}Fe^{III}(bimp)(\mu\text{-}O_2CR)_2]^{2+}$ (where bimp⁻ is a binucleating ligand as shown in Fig. A8.25) were reported as structural models of the mixed-valence $\{Fe^{II}Fe^{III}\}$ intermediate species in oxygenation of hemerythrin.

A8.3.3 Hemocyanins (Hc)

Hemocyanins are large, multi-subunit oxygen carrier proteins found in some species of molluscs (e.g. octopus) and arthropods (e.g. lobsters and horseshoe crabs). The deoxy form of hemocyanins contains a dicopper(I) oxygen binding site, and hence it is colorless. Upon oxygenation, the dicopper core is oxidized to a dicopper(II) form, giving the proteins a blue color (cyanin from cyanos, Greek for blue).

Structures:

The binuclear site of the deoxy form of hemocyanins contains two three-coordinate Cu(I) ions, each of which is bound by three histidine residues. The Cu$\cdots$Cu distance ($\sim$3.5 - 3.7 Å) is long and non-bonded. An empty cavity between the two Cu(I) centers is available for O_2-binding [Fig. A8.26(a)].

431

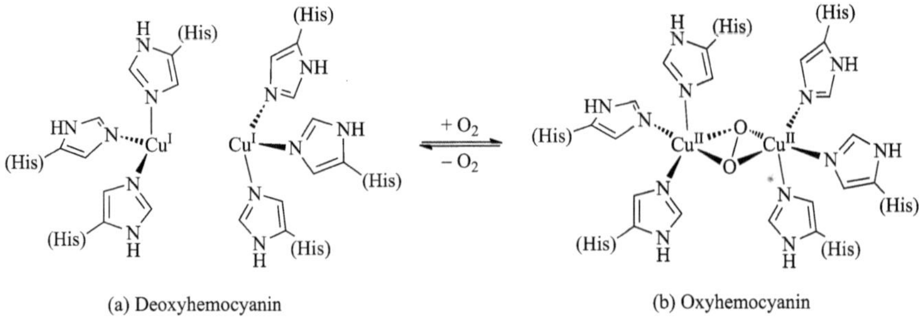

(a) Deoxyhemocyanin   (b) Oxyhemocyanin

Fig. A8.26 A schematic diagram showing the structure of the dicopper core in (a) deoxyhemo-cyanin, and (b) oxyhemocyanin. The dioxygen unit binds in a $\mu{:}\eta^2,\eta^2$-peroxy mode.

The dioxygen-binding reaction:

Upon oxygenation, the dicopper(I) core is oxidized to the dicopper(II) form, while an O_2 molecule is reduced to a peroxide. Oxyhemocyanins show a characteristic, intense optical absorption at $\sim$345 nm, which is assignable to a ligand-to-metal charge-transfer. Resonance Raman spectroscopy shows an O–O stretching frequency of $\sim$800 cm^{-1}, which is close to that of a peroxide anion. Data obtained from resonance Raman spectroscopy and isotopic $^{16}O{-}^{18}O$ labeling studies are suggestive of symmetric binding of a peroxide ligand to the dicopper(II) core to give a (μ-peroxo)dicopper(II) form [Fig. A8.26(b)]. At room temperature, the two Cu(II) ions (d^9) are so strongly antiferromagnetically coupled that the Cu(II)Cu(II) center is essentially diamagnetic. An X-ray crystallographic structure of an oxyhemocyanin was finally determined. Interestingly, the dioxygen unit binds in a side-on $\mu{:}\eta^2,\eta^2$-peroxy form to both copper(II) centers.

Synthetic model complexes for hemocyanin:

A variety of dicopper complexes supported by various types of nitrogen donor ligands were reported as synthetic analogs for the active site of hemocyanin. One noteworthy example is the dinuclear $[\{Cu(Tpz^{iPr})\}_2(O_2)]$ complex which contains a side-on $\mu\text{-}\eta^2{:}\eta^2$ peroxide ion coordinating to two Cu(II) centers with a Cu$\cdots$Cu distance of $\sim$3.56 Å (Fig. A8.27). Moreover, spectroscopic data of this kind of side-on peroxo complexes closely resemble those of oxyhemo-cyanin.

A8.4 ELECTRON TRANSFER IN BIOLOGICAL SYSTEMS

Oxidation-reduction reactions play key roles in many biological processes, such as respiration, photosynthesis and nitrogen fixation. Nature utilizes some elegant redox components to transfer electrons from one site to another in these processes. For example, transition metals like iron and copper that have variable oxidation states can undergo facile redox reactions, and hence

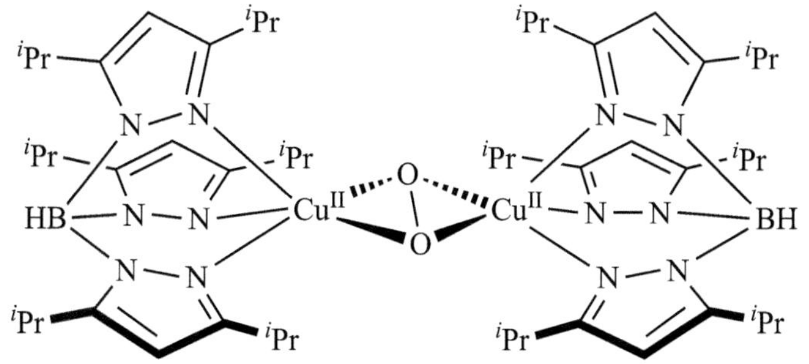

Fig. A8.27 Structure of dinuclear $[\{Cu(Tpz^{iPr})\}_2(O_2)]$ $(Tpz^{iPr} = [HB(3, 5-Pr^i_2C_3HN_2)_3]^-)$ complex in which the peroxide ion binds a side-on μ-η^2:η^2 mode. [Reference: N. Kitajima, K. Fujisawa, C. Fujimoto, Y. Moro-oka, S. Hashimoto, T. Kitagawa, K. Toriumi, K. Tatsumi, and A. Nakamura, *J. Am. Chem. Soc.* **114**, 1277–91 (1992).]

their complexes can act as electron donor and acceptor centers in biological systems. The outer-sphere and inner-sphere mechanisms are well-established in the understanding of electron-transfer reactions between simple inorganic complexes. In biological systems, electron transfer may occur over relatively long distances (e.g. from one protein to another). The Marcus-Hush theory applies to long-distance electron transfer processes in biology.

Let us take cellular respiration and photosynthesis as two examples to understand the roles of redox components in biological electron transfer. Aerobic cellular respiration involves a series of enzymatic reactions in which carbohydrates are broken down to CO_2, while O_2 is being reduced to water. The overall reaction is an exothermic process and the free energy released is used to produce ATP. Cellular respiration is composed of three main stages, namely glycolysis, the Kreb's cycle and oxidative phosphorylation (Fig. A8.28). Glucose (a 6-carbon sugar compound) is first converted to two molecules of pyruvate (a 3-carbon compound) through the process of glycolysis, which takes place in the cytoplasm of cells. The breakdown of pyruvate to CO_2 (the Kreb's cycle) occurs in the mitochondrion matrix. Water-soluble reductants, such as nicoti-namide adenine dinucleotide (NADH) and succinate, are produced. These reductants donate electrons to NADH dehydrogenase (also known as Complex I) and succinate:quinone reductase (Complex II) of the mitochondrial electron-transfer chain, respectively. Ubiquinones pick up the electrons from Complexes I and II to form Complex III. The electrons are eventually transferred to a "terminal electron acceptor", namely cyctochrome c oxidase (Complex IV), where O_2 is being reduced to H_2O (Fig. A8.29).

Complexes I to IV are components of the mitochondrial electron-transfer chain. They are multi-subunit electron-carrier proteins, which contain various redox cofactors for the electron-transfer processes. Except Complex II, Complexes I, III, and IV are proton pumps. The free

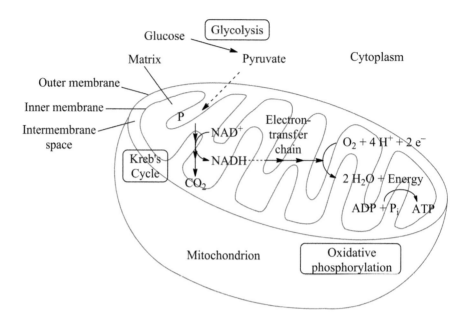

Fig. A8.28 Aerobic cellular respiration is composed of three main processes. Glycolysis takes place in the cytoplasm of a cell whilst the Kreb's cycle and oxidative phosphorylation occur inside mitochrondria.

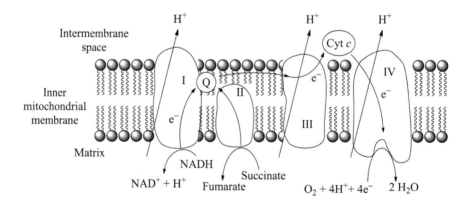

Fig. A8.29 A schematic diagram showing the flow of electrons through the electron-transfer chain in the mitochondrion. (Cyt c: cytochrome c; Q: ubiquinone.)

energy released is used to drive these proton pumps, pumping H^+ out of the matrix to the intermediate space of mitochrondria. This process generates a proton gradient and drives the phosphorylation of ADP to ATP.

In the process of photosynthesis, absorption of a photon by a pigment (e.g. carotenoids) promotes the chlorophyll molecule into an electronically excited state. The energy of the excited state is, then, converted into electrochemical potential energy leading to electron transfers. This is known as the light reactions of photosynthesis. A number of electron carriers are involved in these electron-transfer processes (Fig. A8.30). Two photosystems, namely Photosystem I and Photosystem II, are involved in the light reaction. Water is oxidized to dioxygen by the oxygen evolving complex (a Mn_4 cluster) associated with Photosystem II. In other words, water is the source of electrons. In eukaryotes, the light reactions take place in the thylakoid membranes of the chloroplasts. The electrons are used for fixation of CO_2 to produce carbohydrates in the dark reactions.

Biological nitrogen fixation is a process in the nitrogen cycle which is vital to all organisms in nature. Nitrogen is a major element found in a variety of biomolecules, and hence it is an essential component of all living organisms. Although dinitrogen constitutes $\sim$79% of the atmosphere, it is an inert gas and not readily available for use by most organisms because of its strong N$\equiv$N triple bond energy of $\approx$943 kJ mol^{-1}. As a consequence, most organisms obtain nitrogen in usable ("fixed") forms such as ammonia, nitrate, proteins or other smaller organo-nitrogen molecules. Nitrogen fixation is a process by which atmospheric nitrogen is converted to a bioaccessible form of nitrogen. Biological nitrogen fixation is carried out by

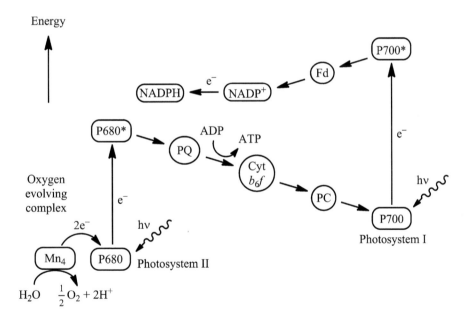

Fig. A8.30 A schematic diagram showing the flow of electrons along the two electron-transfer chains in the photosynthetic systems. The two electron-transfer chains are called the Z-scheme.

a special group of nitrogen-fixing microorganisms (such as *Rhizobium* and *Azobacter*) which utilize the enzyme nitrogenase to catalyze the fixation of atmospheric nitrogen to ammonia. Compared with the fixation of dinitrogen accomplished by non-biological processes such as lightning and the industrial Haber-Bosch process (the latter requires the use of an iron-based catalyst under reaction conditions of high temperature and pressure), nitrogenase can catalyze the transformation of atmospheric nitrogen to ammonia with high efficiency under ambient conditions. Indeed, nitrogenase is the most elaborate biological catalyst and its chemistry has attracted considerable research interest.

Nitrogenase is a two-component metalloprotein composed of a dinitrogenase and a dinitrogenase reductase. The reduction of atmospheric nitrogen by the nitrogenase enzyme involves a transient interaction of these two component proteins. The dinitrogenase reductase protein delivers electrons to the dinitrogenase component; the latter contains an active site where reduction of dinitrogen to ammonia occurs. As summarized in the following equation, the whole process of biological nitrogen fixation involves eight electrons together with an energy input through hydrolysis of ATP:

$$N_2 + 16\,ATP + 16\,H_2O + 8\,H^+ + 8\,e^- \rightarrow 2\,NH_3 + 16\,ADP + 16\,P_i + H_2$$

Several types of nitrogenase containing different combinations of metals in their active site cofactors have been found: the molybdenum-dependent nitrogenase contains a FeMo-cofactor, the vanadium-dependent nitrogenase uses a FeV-cofactor, and the iron-only nitrogenase possesses a FeFe-cofactor in their respective active sites. Among these nitrogenase groups, the molybdenum-dependent enzyme is the most abundant one which has been widely investigated.

The molybdenum-dependent enzyme contains a molybdenum-iron (MoFe) protein (the dinitrogenase component) and an iron (Fe) protein (the dinitrogenase reductase component). The Fe protein contains a single [4Fe-4S] cluster, whereas the MoFe protein contains four metal clusters of two unique types, namely two [MoFe$_7$S$_8$] clusters (the FeMo-cofactors) and two [8Fe-7S] clusters (the P clusters). The structures of these metal clusters are shown in Fig. A8.31. Electrons are delivered from the Fe protein via the P cluster to the FeMo-cofactor where substrate binding and reduction occurs. It is interesting to note that some of the Fe sites in the FeMo-cofactor are three-coordinated. Conceivably, one or more of these low-coordinate Fe sites may be the substrate binding sites. Another interesting issue regarding the structure of the MoFe protein is the identification of a light interstitial atom (X) within the interior of the FeMo-cofactor. Recent reports on high-resolution X-ray structures of nitrogenase have provided evidence for an assignment of carbon to this interstitial atom. Although a substantial amount of structural and spectroscopic data on the molybdenum-dependent nitrogenases have been reported, the exact binding site for dinitrogen on the FeMo-cofactor remains unclear.

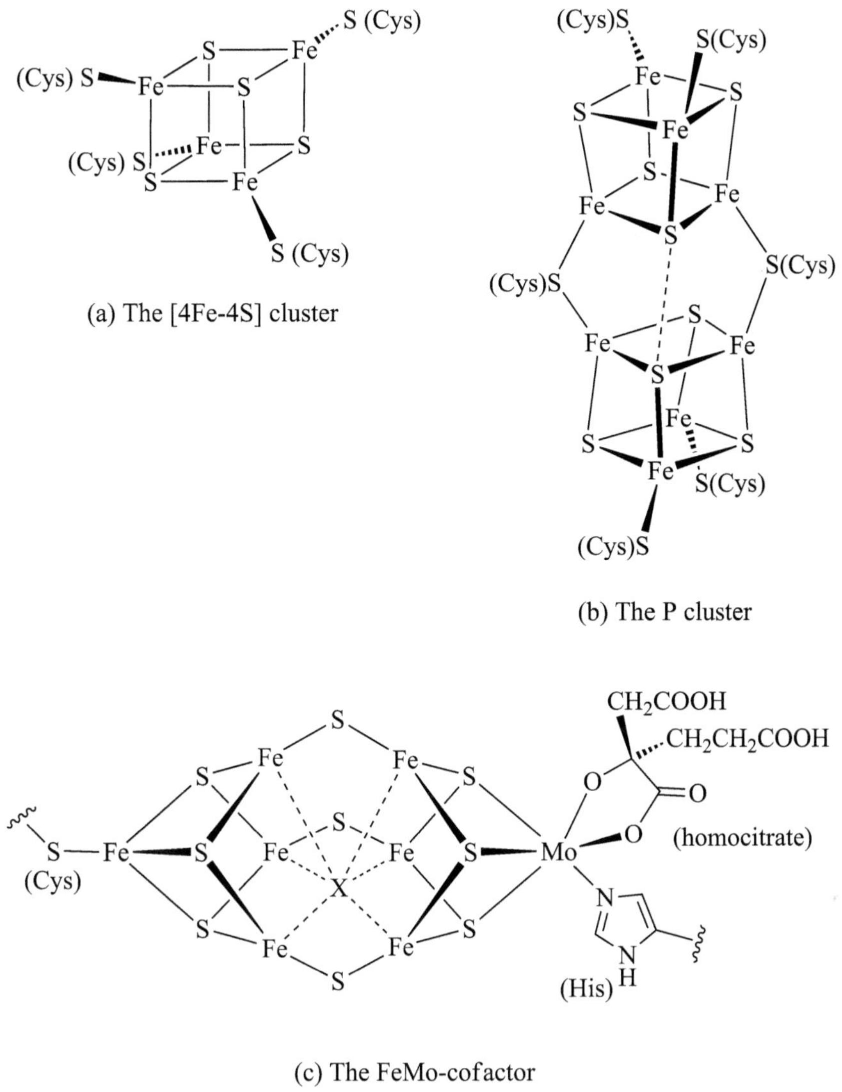

(a) The [4Fe-4S] cluster

(b) The P cluster

(c) The FeMo-cofactor

Fig. A8.31 Metal clusters in the molybdenum-dependent nitrogenases. (a) The [4Fe-4S] cluster in the Fe protein. (b) The P cluster and (c) the FeMo-cofactor in the MoFe protein. Results of recent studies have suggested that an interstitial C atom is present within the core of the FeMo-cofactor. [Reference: L.-M. Zhang, C. N. Morrison, J. T. Kaisera, and D. C. Rees, Nitrogenase MoFe protein from Clostridium pasteurianum at 1.08 Å resolution: comparison with the *Azotobacter vinelandii* MoFe protein. *Acta Cryst.* **D71**, 274–82 (2015).]

An overview of the chemistry of several types of biological electron-transfer components is presented in the following section.

Three types of biological redox components are known. They are redox active protein side chains, small organic molecules (such as NADH and ubiquinone), and redox cofactors (such as copper ions, iron-sulfur clusters and cytochromes). They participate in electron-transfer reactions by acting as electron donors and acceptors.

A8.4.1 Redox Active Protein Side Chains

Some amino acids contain redox active side chains which enable them to undergo reversible oxidation-reduction reactions. For instance, the sulfhydryl group of cysteine is readily oxidized to yield the corresponding disulfide compound (cystine). Another example is the oxidation of tyrosine to give the corresponding tyrosyl radical (Fig. A8.32).

A8.4.2 Redox Active Organic Molecules

Several organic molecules are known to play key roles in electron transfer in biological systems. Nicotinamide adenine dinucleotide (NADH) and nicotinamide adenine dinucleotide phosphate (NADPH) are two well known examples. Nature utilizes the NAD^+/NADH and $NADP^+$/NADPH redox couples for electron transfer among biological systems (Fig. A8.33).

Ubiquinone and plastoquinone are organic compounds which can function as either one- or two-electron carriers in a wide variety of biological systems. As shown in Fig. A8.34, 1,4-benzoquinone is readily reduced to form the corresponding semiquinone radical. This semiquinone radical species can undergo either a reversible oxidation reaction to yield the original benzoquinone compound, or a subsequent reduction to form the corresponding

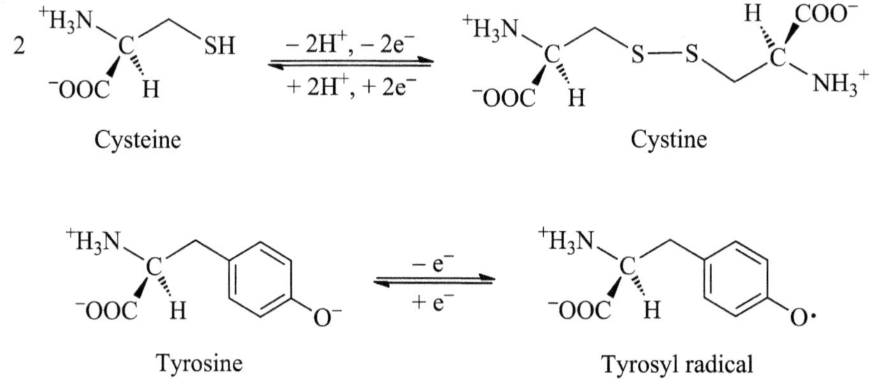

Fig. A8.32 Cysteine and tyrosine are readily oxidized to give cystine and tyrosyl radical, respectively.

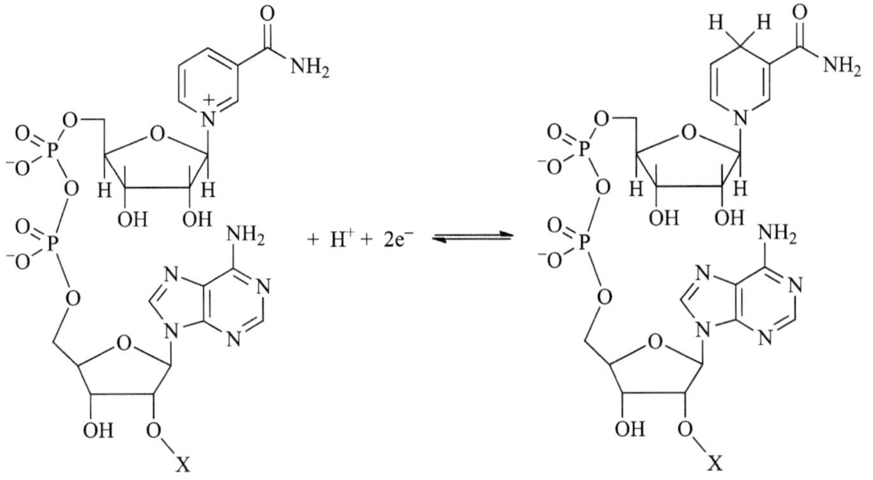

Fig. A8.33 NAD^+ and $NADP^+$ pick up two electrons and a proton to become NADH and NADPH, respectively.

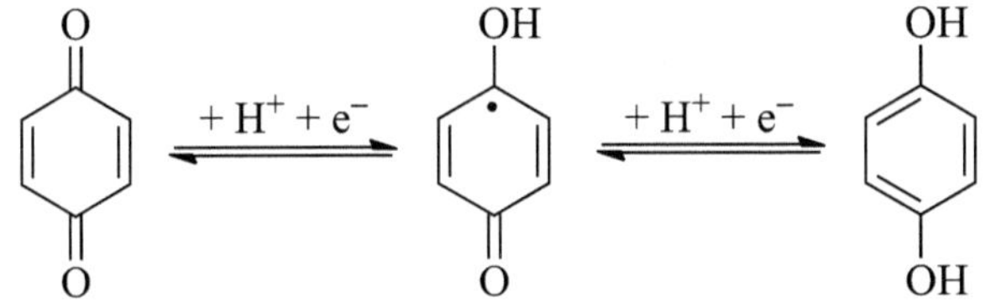

Fig. A8.34 Quinone can pick up one electron to yield a semiquinone radical. Subsequent reduction of the latter produces quinol.

ubiquinol as the final reduction product. Semiquinone is a free radical intermediate species which can be detected by electron paramagnetic resonance (EPR) spectroscopy.

Ubiquinone and plastoquinone contain a 1,4-benzoquinone moiety associated with a long isoprenoid tail. The latter enables the molecule to diffuse rapidly through biological membranes. Ubiquinone is also known as coenzyme Q_{10}. The abbreviation "Q" represents the quinone chemical group, and the number "10" refers to the number of isoprenyl units in the hydrocarbon tail (Fig. A8.35).

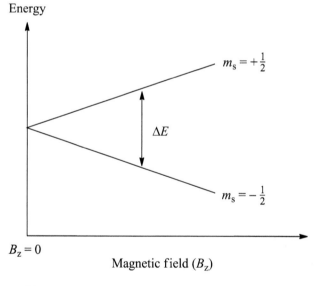

$(n = 6 - 10)$

Fig. A8.35 Ubiquinone (coenzyme Q_{10}) contains a long isoprenoid tail which enables the molecule to diffuse freely within biological membranes.

Electron Paramagnetic Resonance (EPR) Spectroscopy
(Electron Spin Resonance (ESR) Spectroscopy)

Some atoms and molecules may contain one or more unpaired electrons. For example, semiquinone is an organic radical with one unpaired electron, and a Cu(II) ion (d^9) and its complexes are also paramagnetic. These paramagnetic species are easily studied by electron paramagnetic resonance (EPR) spectroscopy. Interaction of the magnetic moment of an unpaired electron with an applied magnetic field gives rise to two energy levels:

Energy

$m_s = +\frac{1}{2}$

ΔE

$m_s = -\frac{1}{2}$

$B_z = 0$

Magnetic field (B_z)

The splitting of the energy levels in an applied magnetic field B_z is given by the equation:

$$\Delta E = g\beta B_z$$

where β is the Bohr magneton $(9.273 \times 10^{-24}\,\mathrm{J\,T^{-1}})$ and g is the Landé splitting factor. A resonance absorption occurs when an appropriate electromagnetic radiation of frequency ν is applied to the system:

$$h\nu = g\beta B_z$$

where h is the Planck constant $(6.626 \times 10^{-34}\,\mathrm{J\,s})$. The g value provides valuable information on the electronic structure and geometry of the radical species. For example, organic radicals have g values of 1.99-2.01, whereas the $Cu(acac)_2$ complex has a g value of $\sim$2.13.

Other important information which can be obtained from EPR spectral data regards hyperfine splitting patterns. As many atoms have their own nuclear spin (I), this will generate an additional local magnetic field interacting with the nearby unpaired electron spin. For instance, the nuclear spin of hydrogen and copper is $\frac{1}{2}$ and $\frac{3}{2}$, respectively. As a consequence, a characteristic hyperfine splitting pattern due to this nuclear spin interaction will be observed on the EPR signal.

A8.4.3 Redox Cofactors

Four classes of redox cofactors, namely flavins, copper ions (in enterobactins), iron-sulfur clusters and cytochromes, have been extensively studied.

A8.4.2.1 Flavins

Flavins are organic redox coenzymes that may serve as either one- or two-electron carriers (Fig. A8.36).

Let us consider the redox properties of flavin mononucleotide (FMN). The reduction of FMN to $FMNH_2$ involves two one-electron reduction processes (Fig. A8.37). It is noted that different oxidation states of this flavin chromophore are characterized by different absorption maxima in the UV-visible spectrum. Moreover, the flavin radical intermediate can be unambiguously identified using EPR spectroscopy.

A8.4.2.2 Copper ions

Copper plays an important role in biological redox processes through an involvement of a reversible Cu(II)/Cu(I) redox couple. In general, copper centers in biological systems can be categorized into three main types, namely Type I, Type II, and Type III copper centers (Fig. A8.38). This classification is based on the structures and spectroscopic properties of the copper centers in the proteins.

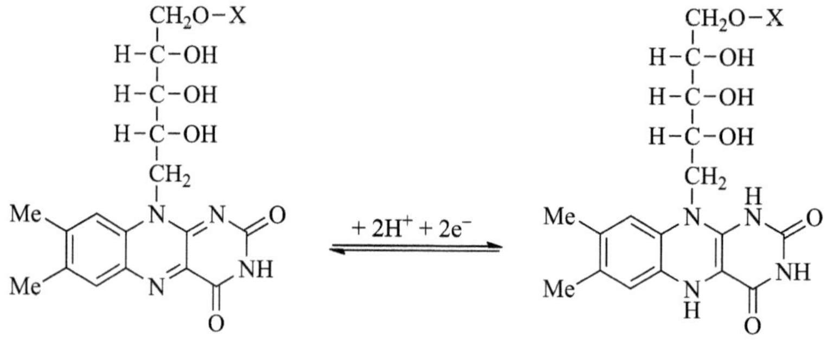

Riboflavin (vit. B$_2$): X = H
Flavin mononucleotide (FMN): X = PO$_3{}^{2-}$
Flavin adenine dinucleotide (FAD):
X = adenosin disphosphate

5,10-dihydroriboflavin: X = H
FMNH$_2$: X = PO$_3{}^{2-}$
FADH$_2$: X = adenosin disphosphate

Fig. A8.36 Structures and redox reactivity of riboflavin (vitamine B$_2$), flavin mononucleotide (FMN) and flavin adenine dinucleotide (FAD).

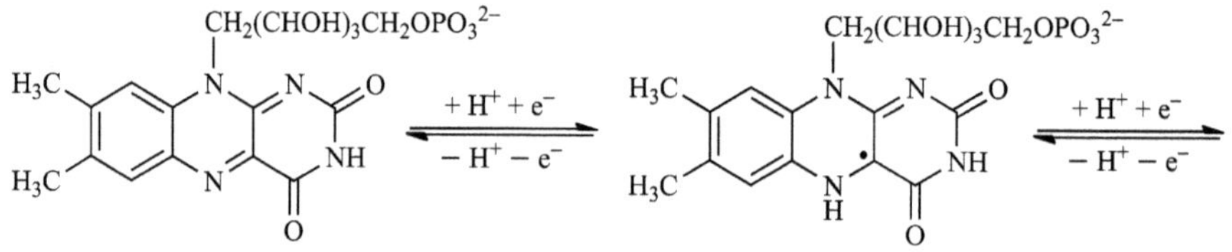

FMN
$\lambda_{max} \sim 337, 445$ nm
$S = 0$

$\lambda_{max} \sim 565$ nm
$S = \frac{1}{2}$

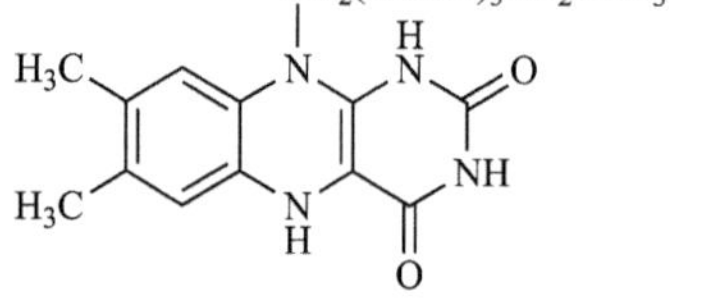

FMNH$_2$
Colorless
$S = 0$

Fig. A8.37 The flavin mononucleotide (FMN) cofactor is a chromophore which shows different absorption maxima at different redox levels. The FMN radical intermediate is EPR active.

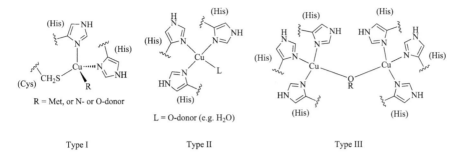

Fig. A8.38 The three classes of copper centers found in copper proteins.

Type I copper centers are also known as "blue" copper centers because of their unusually intense blue color. A Type I copper center has a distorted tetrahedral structure. The ligand environment consists of N- and S-donor atoms of the amino acid residues of the corresponding protein. The intense blue color of Type I copper proteins is assignable to a symmetry-allowed cysteine → Cu(II) charge-transfer absorption (a LMCT band) at $\lambda_{max} \sim 600$ nm ($\varepsilon > 3000$ $M^{-1}cm^{-1}$), which is much more intense than that of a simple Cu(II) complex in an aqueous solution (e.g. $[Cu(H_2O)_6]^{2+}, \varepsilon \sim 5\text{-}10\, M^{-1}cm^{-1}$). The Cu(II) center ($d^9, I = \frac{3}{2}$) also shows a characteristic EPR signal with a narrow hyperfine splitting pattern.

Plastocyanin is a "blue" copper protein found in the chloroplasts of higher plants and algae. The structure of poplar plastocyanin was determined using X-ray crystallography. The oxidized form of this protein contains a Cu(II) active site which is bound by one cysteine, two histidines, and one methionine ligand to form a distorted tetrahedral structure.

Type II copper centers are the most common type of copper centers found in biology. A Type II copper center is usually bound by N- or O-donor ligands, which form a square planar or tetragonal geometry around the metal center. They have electronic and EPR spectral characteristics typical of simple Cu(II) complexes.

A Type III copper center contains a pair of copper ions bound by histidine residues of the respective protein. The dinuclear center can function as a two-electron transfer center. In its oxidized state, a Type III copper center consists of a pair of antiferromagnetically coupled Cu(II) ions and, hence, it is EPR silent. The dinuclear copper center in hemocyanin is a typical example of a Type III copper center.

A8.4.2.3 Iron-sulfur Clusters

Iron-sulfur ($[Fe\text{–}S]$) clusters are versatile cofactors found in a variety of electron-transfer proteins in all living organisms. They are polynuclear species containing iron and sulfur atoms. The most common iron-sulfur clusters are $[2Fe\text{-}2S]$, $[3Fe\text{-}4S]$, and $[4Fe\text{-}4S]$.

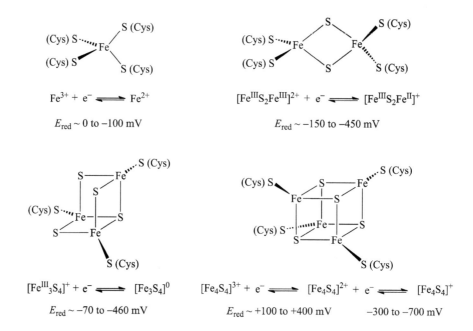

Fig. A8.39 Iron-sulfur clusters are one-electron transfer centers found in all living organisms. Their reduction potentials are tuned by the protein environment that the clusters are associated with.

Iron-sulfur clusters contain high-spin Fe(II) or Fe(III) ions coordinated tetrahedrally by four S-donors such as S^{2-} (sulfido) ions or thiolate groups of cysteine residues of the protein core. Four classes of iron-sulfur clusters are known. They are named according to the number of iron and sulfur atoms that the clusters contain (Fig. A8.39).

Iron-sulfur proteins usually show a dark reddish-brown color. However, attempts to characterize the iron-sulfur clusters in these proteins using UV-visible spectroscopy are difficult. Their electronic absorption spectra generally show broad and featureless absorption bands, probably due to those overlapping charge-transfer transitions of the clusters. On the other hand, iron-sulfur clusters show distinctive EPR spectra. Therefore, EPR spectroscopy provides valuable information on both structures and redox levels of iron-sulfur proteins.

A8.4.2.4 Cytochromes

Cytochromes are heme iron proteins containing one or more heme cofactors. They are the most thoroughly characterized electron carriers in biological systems. Cytochromes are facile one-electron carriers by virtue of the reversible Fe(III)/Fe(II) redox couple. They exhibit an intense absorption in the 400–500 nm region due to the presence of the heme cofactors (the Soret band). Cytochromes are classified according to the structures of the heme groups and the axial ligands

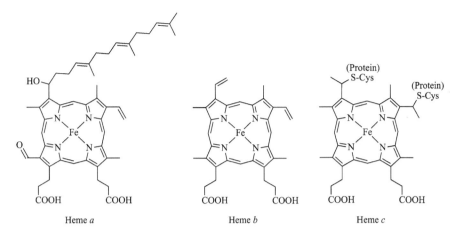

Fig. A8.40 The structure of hemes *a*, *b*, and *c*.

coordinate to the heme iron center. Hemes *a*, *b*, and *c* differ in the substituents attached to the porphyrin unit (Fig. A8.40).

Cytochrome *c* is a water-soluble heme protein which has widespread occurrence in biological systems. In eukaryotic cells, cytochrome *c* is responsible for shuttling electrons from Complex III to Complex IV in the mitochondrial electron-transfer chain. The protein contains a six-coordinate, low-spin iron center which is bonded to a heme *c* group and two axial ligands (one histidine and one methionine residue). The reduction potential of the protein is tunable by the microenvironment of the protein core. Compared to the five-coordinate heme iron centers in hemoglobin, the heme cofactor in cytochrome *c* does not contain any vacant coordination site for O_2-binding. That explains why cytochrome *c* can only act as an electron carrier by switching between Fe(II) and Fe(III) oxidation states, but not as a dioxygen carrier.

Further Reading

1. C. E. Housecroft and A. G. Sharpe, *Inorganic Chemistry*, 2nd ed., Pearson Prentice Hall, Harlow (2005).
2. S. J. Lippard and J. M. Berg, *Principles of Bioinorganic Chemistry*, University Science Books, Mill Valley (1994).
3. I. Bertini, H. B. Gray, E. I. Stiefel and J. S. Valentine, *Biological Inorganic Chemistry – Structure and Reactivity*, University Science Books, Sausalito (2007).
4. P. C. Wilkins and R. G. Wilkins, *Inorganic Chemistry in Biology*, Oxford University Press, New York (1997).
5. J. A. Cowan, *Inorganic Biochemistry - An Introduction*, 2nd ed., VCH, New York (1997).
6. J. J. R. Fraústo da Silva and R. J. P. Williams, *The Biological Chemistry of the Elements - The Inorganic Chemistry of Life*, 2nd ed., Oxford University Press, Oxford (2001).

7. W. Kaim, B. Schwederski and A. Klein, *Bioinorganic Chemistry: Inorganic Elements in the Chemistry of Life*, 2nd ed., Wiley, Chichester (2013).

8. J. J. Stephanos and A. W. Addison *Chemistry of Metalloproteins: Problems and Solutions in Bioinorganic Chemistry* Wiley New Jersey (2014).

9. K. D. Karlin, S. J. Lippard, J. S. Valentine and C. J. Burrows, "*Solving 21st Century Problems in Biological Inorganic Chemistry Using Synthetic Models*", Accounts of Chemical Research, **48**, 2659-60 (2015).

10. M. Dobler, *Ionophores and Their Structures*, Wiley, New York (1981).

11. J.-M. Lehn, *Supramolecular Chemistry: Concepts and Perspectives*, VCH, Weinheim (1995).

12. J. W. Steedand and J. L. Atwood, *Supramolecular Chemistry*, 2nd ed., Wiley, Chichester (2009).

13. B. F. Matzanke, G. Mtiller-Matzanke and K. N. Raymond, in: *Iron Carriers and Iron Proteins*, T. M. Loehr, Ed., VCH, New York (1989).

14. D. C. Rees and J. B. Howard, "Nitrogenase: Standing at the Crossroads", *Current Opinion in Chemical Biology*, **4**, 559–66 (2000).

15. J. W. Peters and R. K Szilagyi, "*Exploring New Frontiers of Nitrogenase Structure and Mechanism*", Current Opinion in Chemical Biology, **10**, 101–8 (2006).

16. B. K. Burgess and D. J. Lowe, "*Mechanism of Molybdenum Nitrogenase*", Chemical Reviews, **96**, 2983-3011 (1996).

17. B. M. Hoffman, D. Lukoyanov, Z.-Y. Yang, D. R. Dean and L. C. Seefeldt, "*Mechanism of Nitrogen Fixation by Nitrogenase: The Next Stage*", Chemical Reviews, **114**, 4041–62 (2014).

BIBLIOGRAPHY

The reference list in this Bibliography consists of 157 titles. It is safe to say that all the topics that are dealt with in this problem book are covered in one or more of these reference books. In this list there are recently published books which allow readers of the present volume to study the subjects from a modern perspective. On the other hand, there are also books that are decades old; it is hoped that readers will find them as classics, rather than merely outdated.

The titles in the reference list are broadly divided into three categories in increasing levels of difficulty. **Textbooks** offer a logically organized presentation of the subject matter from which readers can study at their own pace to gain a broad understanding. **Problem books** are suggested for readers who wish to have more drilling on solving problems. **Monographs and Advanced Texts** are graduate-level or specialized books for readers who wish to delve deeper into specific topics. Clearly, some books may be viewed as either a textbook or a monograph, so the categorization is not clearly defined.

TEXTBOOKS

Inorganic Chemistry

1. G. L. Miessler, P. J. Fisher and D. A. Tarr, *Inorganic Chemistry*, 5th edn., Pearson, Upper Saddle River, NJ, 2013.
2. J. E. House and K. A. House, *Descriptive Inorganic Chemistry*, 2nd edn., Elsevier/Academic Press, London, 2010.
3. J. E. House, *Inorganic Chemistry*, 2nd edn., Elsevier/Academic Press, London, 2012.
4. W. W. Porterfield, *Inorganic Chemistry: A Unified Approach*, 2nd edn., Academic Press, San Diego, CA, 1993.
5. B. E. Douglas, D. H. McDaniel and J. J. Alexander, *Concepts and Models of Inorganic Chemistry*, 3rd edn., Wiley, New York, 1994.
6. J. E. Huheey, E. A. Keiter and R. L. Keiter, *Inorganic Chemistry: Principles of Structure and Reactivity*, 4th edn., Harper Collins, New York, 1993.
7. C. E. Housecroft and A. G. Sharpe, *Inorganic Chemistry*, 4th edn., Pearson Education, Essex, 2012.

8. D. M. P. Mingos, *Essential Trends in Inorganic Chemistry*, Oxford University Press, Oxford, 1998.

9. W.-K. Li, G.-D. Zhou and T. C. W. Mak, *Advanced Structural Inorganic Chemistry*, Oxford University Press, Oxford, 2008.

10. F. A. Cotton, C. A. Murillo, M. Bochmann and R. N. Grimes, *Advanced Inorganic Chemistry*, 6th edn., Wiley, New York, 1999.

11. N. N. Greenwood and A Earnshaw, *Chemistry of the Elements*, 2nd edn., Butterworth-Heinemann, Oxford, 1997.

Physical Chemistry

12. P. W. Atkins and J. de Paula, *Atkins' Physical Chemistry*, 10th edn., Oxford University Press, Oxford, 2014.

13. I. N. Levine, *Physical Chemistry*, 6th edn., McGraw-Hill, Boston, 2008.

14. R. J. Silbey, R. A. Alberty and M. G. Bawendi, *Physical Chemistry*, 4th edn., Wiley, New York, 2004.

15. M. Karplus and R. N. Porter, *Atoms and Molecules: An Introduction for Students of Physical Chemistry*, Benjamin, New York, 1970.

16. R. S. Berry, S.A. Rice and J. Ross, *Physical Chemistry*, 2nd edn., Oxford University Press, New York, 2000.

Chemical Bonding

17. E. Cartmell and G. W. A. Fowles, *Valency and Molecular Structure*, 4th edn., Butterworth, London, 1977.

18. R. L. DeKock and H. B. Gray, *Chemical Structure and Bonding*, 2nd edn., University Science Books, Sausalito, 1989.

19. J. N. Murrell, S. F. A. Kettle and J. M. Tedder, *The Chemical Bond*, 2nd edn., Wiley, Chichester, 1987.

20. R. McWeeny, *Coulson's Valence*, 3rd edn., Oxford University Press, Oxford, 1979.

21. N. W. Alcock, *Bonding and Structure: Structural Principles in Inorganic and Organic Chemistry*, Ellis Horwood, New York, 1990.

22. A. Haaland, *Molecules and Models: The Molecular Structures of Main Group Element Compounds*, Oxford University Press, Oxford, 2008.

23. R. J. Gillespie and P. L. A. Popelier, *Chemical Bonding and Molecular Geometry from Lewis to Electron Densities*, Oxford University Press, New York, 2001.

24. P. F. Bernath, *Spectra of Atoms and Molecules*, 2nd edn., Oxford University Press, New York, 2005.

25. I. Fleming, *Molecular Orbitals and Organic Chemical Reactions: Student Edition*, 2009, *Reference Edition*, 2010, Wiley, London.

26. T. Shida, *The Chemical Bond: A Fundamental Quantum-Mechanical Picture*, Springer, New York, 2004.

27. V. M. S. Gil, *Orbitals in Chemistry: A Modern Guide for Students*, Cambridge University Press, Cambridge, 2000.

28. Y. Jean and F. Volatron (translated by J. Burdett), *An Introduction to Molecular Orbitals*, Oxford University Press, New York, 1993.

29. I. Fleming, *Pericyclic Reactions*, Oxford University Press, Oxford, 1999.

30. A. Rauk, *Orbital Interaction Theory of Organic Chemistry*, 2nd edn., Wiley, New York, 2001.

Symmetry and Group Theory

31. A. Vincent, *Molecular Symmetry and Group Theory: A Programmed Introduction to Chemical Applications*, 2nd edn., Wiley, Chichester, 2001.

32. Y. Öhrn, *Elements of Molecular Symmetry*, Wiley, New York, 2000.

33. R. L. Carter, *Molecular Symmetry and Group Theory*, Wiley, New York, 1998.

34. G. Davison, *Group Theory for Chemists*, Macmillan, London, 1991.

35. R. L. Flurry, Jr., *Symmetry Groups: Theory and Chemical Applications*, Prentice-Hall, Englewood Cliffs, NJ, 1980.

36. A. M. Lesk, *Introduction to Symmetry and Group Theory for Chemists*, Kluwer, Dordrecht, 2004.

37. K. C. Molloy, *Group Theory for Chemists: Fundamental Theory and Applications*, 2nd edn., Woodhead Publishing, Cambridge, 2011.

38. M. Ladd, *Symmetry of Crystals and Molecules*, Oxford University Press, New York, 2014.

39. P. R. Bunker and P. Jensen, *Fundamentals of Molecular Symmetry*, Institute of Physics Publishing, Bristol, 2005.

40. S. F. A. Kettle, *Symmetry and Structure: Readable Group Theory for Chemists*, 3rd edn., Wiley, Chichester, 2007.

41. K. V. Reddy, *Symmetry and Spectroscopy of Molecules*, 2nd edn., New Age Science, Tunbridge Wells, 2009.

42. D. C. Harris and M. D. Bertolucci, *Symmetry and Spectroscopy: An Introduction to Vibrational and Electronic Spectroscopy*, Oxford University Press, New York, 1978.

43. B. D. Douglas and C. A. Hollingsworth, *Symmetry in Bonding and Spectra: An Introduction*, Academic Press, Orlando, 1985.

44. F. A. Cotton, *Chemical Applications of Group Theory*, 3rd edn., Wiley, New York, 1990.

Molecular Quantum Mechanics

45. S. M. Blinder, *Introduction to Quantum Mechanics*, Elsevier/Academic Press, Amsterdam, 2004.

46. M. A. Ratner and G. C. Schatz, *Introduction to Quantum Mechanics in Chemistry*, Prentice Hall, Upper Saddle River, NJ, 2001.

47. V. Magnasco, *Elementary Methods of Molecular Quantum Mechanics*, Elsevier, Amsterdam, 2007.

48. V. Magnasco, *Methods of Molecular Quantum Mechanics: An Introduction to Electronic Molecular Structure*, Wiley, Hoboken, NJ, 2009.

49. J. E. House, *Fundamentals of Quantum Chemistry*, 2nd edn., Elsevier/Academic Press, San Diego, CA, 2004.

50. T. Veszprémi and M. Fehér, *Quantum Chemistry: Fundamentals to Applications*, Kluwer/Plenum, New York, 1999.

51. D. D. Fitts, *Principles of Quantum Mechanics as Applied to Chemistry and Chemical Physics*, Cambridge University Press, Cambridge, 1999.

52. D. B. Cook, *Quantum Chemistry: A Unified Approach*, 2nd edn., Imperial College Press, London, 2012.

53. P. W. Atkins and R. S. Friedman, *Molecular Quantum Mechanics*, 5th edn., Oxford University Press, New York, 2011.

54. J. P. Lowe and K. Peterson, *Quantum chemistry*, 3rd edn., Elsevier Academic Press, London, 2006.

55. I. N. Levine, *Quantum Chemistry*, 7th edn., Prentice-Hall, Upper Saddle River, NJ, 2013.

56. J. Simons and J. Nicholls, *Quantum Mechanics in Chemistry*, Oxford University Press, New York, 1997.

57. F. L. Pilar, *Elementary Quantum Chemistry*, 2nd edn., McGraw-Hill, New York, 1990.

58. R. Grinter, *The Quantum in Chemistry: An Experimentalist's View*, Wiley, Chichester, 2005.

59. A. Hinchliffe, *Molecular Modelling for Beginners*, 2nd edn., Wiley, Chichester, 2008.

60. E. G. Lewars, *Computational Chemistry: Introduction to the Theory and Applications of Molecular and Quantum Mechanics*, 2nd edn., Springer, Dordrecht, 2011.

61. F. Jensen, *Introduction to Computational Chemistry*, 2nd edn., Wiley, Chichester, 2007.

X-Ray Crystallography

62. T. Hahn (ed.), *Brief Teaching Edition of International Tables for Crystallography, Volume A: Space-group symmetry (International Tables for Crystallography)*, 5th edn. (corrected reprint), Springer, Chichester, 2005.

63. W. Borchardt-Ott, *Crystallography*, 2nd edn., Springer-Verlag, Berlin, 1995.

64. J.-J. Rousseau, *Basic Crystallography*, Wiley, Chichester, 1998.

65. A. M. Glazer, *Crystallography: A Very Short Introduction*, Oxford University Press, New York, 2016.

66. C. Hammond, *The Basics of Crystallography and Diffraction*, 3rd edn., Oxford University Press, New York, 2009.

67. J. A. K. Tareen and T. R. N. Kutty, *A Basic Course in Crystallography*, Universities Press, Hyderabad, 2001.

68. M. M. Woolfson, *An Introduction to X-Ray Crystallography*, 2nd edn., Cambridge University Press, 1997.

69. L. E. Smart and E. A. Moore, *Solid State Chemistry: An Introduction*, 3rd edn., CRC Press, Boca Raton, 2005.

70. A. R. West, *Solid State Chemistry and its Applications*, 2nd edn., Wiley, Chichester, 2013.

71. J. P. Glusker and K. N. Trueblood, *Crystal Structure Analysis: A Primer,* 3rd edn., Oxford University Press, New York, 2010.

72. W. Clegg, *Crystal Structure Determination*, Oxford University Press, New York, 1998.

73. W. Massa, *Crystal Structure Determination* 2nd edn., Springer-Verlag, Berlin, 2004, corrected 5th printing, 2010

74. M. Ladd and R. Palmer, *Structure Determination by X-Ray Crystallography: Analysis by X-rays and Neutrons*, 5th edn., Springer, New York, 2013.

75. W. Clegg (ed.), A. J. Blake, W. Clegg, J. M. Cole, J. S. O. Evans, P. Main, S. Parsons and D. J. Watkin, *Crystal Structure Analysis: Principles and Practice*, 2nd edn., Oxford University Press, New York, 2009.

76. C. Giacovazzo (ed.), C. Giacovazzo, H. L. Monaco, D. Viterbo, M. Milanesio, G. Ferraris, G. Gilli, P. Gilli, G. Zanotti and M. Catti, *Fundamentals of Crystallography*, 3rd edn., Oxford University Press, 2011.

Problem Books

77. M. Cini, F. Fucito and M. Sbragaglia, *Solved Problems in Quantum and Statistical Mechanics*, Springer, Milan, 2012.

78. E. d'Emilio and L. E. Picasso, *Problems in Quantum Mechanics: With Solutions*, Springer, Milan, 2011.

79. K. Tamvakis, *Problems and Solutions in Quantum Mechanics*, Cambridge University Press, Cambridge, 2005.

80. Y.-K. Lim (ed.), *Problems and Solutions on Quantum Mechanics*, World Scientific Publishing, Singapore, 1998.

81. C. S. Johnson, Jr. and L. G. Pedersen, *Problems in Quantum Chemistry and Physics*, Addison-Wesley, Reading, MA, 1974; repr. Dover, New York, 1987.

82. G. L. Squires, *Problems in Quantum Mechanics: with Solutions*, Cambridge University Press, Cambridge, 1995.

83. T. A. Albright and J. K. Burdett, *Problems in Molecular Orbital Theory*, Oxford University Press, New York, 1992.

Monographs and Advanced Texts

84. J. Emsley, *The Elements*, 3rd edn., Clarendon Press, Oxford, 1998.

85. D. H. Rouvray and R. B. King (eds.), *The Periodic Table: Into the 21st Century*, Research Studies Press Ltd., Baldock, Hertfordshire, 2004.

86. E. U. Condon and H. Odabasi, *Atomic Structure*, Cambridge University Press, Cambridge, 1980.

87. J. A. McCleverty and T. J. Meyer (eds.), *Comprehensive Coordination Chemistry II: From Biology to Nanotechnology*, Elsevier Pergamon, Amsterdam, 2004.

88. B. N. Figgis and M. A. Hitchman, *Ligand Field Theory and Its Applications*, Wiley-VCH, New York, 2000.

89. J. Mulak and Z. Gajek, *The Effective Crystal Field Potential*, Elsevier, New York, 2000.

90. C. K. Jørgenson, *Modern Aspects of Ligand Field Theory*, North-Holland, Amsterdam, 1971.

91. T. M. Dunn, D. S. McClure and R. G. Pearson, *Some Aspects of Crystal Field Theory*, Harper and Row, New York, 1965.

92. C. J. Ballhausen, *Introduction to Ligand Field Theory*, McGraw-Hill, New York, 1962.

93. J. S. Griffith, *The Theory of Transition-Metal Ions*, Cambridge University Press, Cambridge, 1961.

94. I. B. Bersuker, *Electronic Structure and Properties of Transition Metal Compounds: Introduction to the Theory*, 2nd edn., Wiley, New Jersey, 2010.

95. J. R. Ferraro and K. Nakamoto, *Introductory Raman Spectroscopy*, 2nd edn., Academic Press, San Diego, CA, 2003.

96. K. Nakamoto, *Infrared and Raman Spectra of Inorganic and Coordination Compounds (Part A: Theory and Applications in Inorganic Chemistry; Part B: Applications in Coordination, Organometallic, and Bioinorganic Chemistry)*, 6th edn., Wiley, Hoboken, NJ, 2009.

97. R. A. Nyquist, *Interpreting Infrared, Raman, and Nuclear Magnetic Resonance Spectra, Vols. 1 and 2*, Academic Press, San Diego, CA, 2001.

98. D. W. H. Rankin, N. Mitzel and C. Morrison, *Structural Methods in Molecular Inorganic Chemistry*, Wiley, Chichester, 2013.

99. R. S. Drago, *Physical Methods for Chemists*, 2nd edn., Saunders, Fort Worth, TX, 1992.

100. J. P. Fackler, Jr., and L. R. Falvello (eds.), *Techniques in Inorganic Chemistry*, CRC Press, Taylor & Francis, Boca Raton, FL, 2011.

101. R. J. Gillespie and I. Hargittai, *The VSEPR Model of Molecular Geometry*, Allyn and Bacon, Boston, 1991.

102. J. Demaison, J. E. Boggs and A. G. Csaszar (eds.), *Equilibrium Molecular Structures: From Spectroscopy to Quantum Chemistry*, CRC Press, Boca Raton, FL, 2011.

103. V. Magnasco, *Models for Bonding in Chemistry*, Wiley, Chichester, 2010.

104. W. J. Hehre, L. Radom, P. v. R. Schleyer and J. A. Pople, *Ab initio Molecular Orbital Theory*, Wiley, New York, 1986.

105. F. Weinhold and C. R. Landis, *Valency and Bonding: A Natural Bond Orbital Donor-Acceptor Perspective*, Cambridge University Press, Cambridge, 2005.

106. A. A. Levin and P. N. D'Yachkov (translated by V. A. Sipachev), *Heteroligand Molecular Systems: Bonding, Shapes and Isomer Stabilities*, Taylor & Francis, London, 2002.

107. C. J. Ballhausen and H. B. Gray, *Molecular Orbital Theory*, Benjamin, New York, 1964.

108. J. G. Verkade, *A Pictorial Approach to Molecular Bonding and Vibrations*, 2nd edn., Springer, New York, 1997.

109. J. K. Burdett, *Molecular Shapes: Theoretical Models of Inorganic Stereochemistry*, Wiley, New York, 1980.

110. B. M. Gimarc, *Molecular Structure and Bonding: The Quantitative Molecular Orbital Approach*, Academic Press, New York, 1979.

111. L. Pauling, *The Nature of the Chemical Bond and the Structure of Molecules and Crystals: An Introduction to Modern Structural Chemistry*, 3rd edn., Cornell University Press, Ithaca, NY, 1960.

112. T. A. Albright, J. K. Burdett and M.-H. Whangbo, *Orbital Interactions in Chemistry*, Wiley, New York, 1985.

113. I. Hargittai and M. Hargittai, *Symmetry through the Eyes of a Chemist*, 3rd edn., Springer, Dordrecht, 2009.

114. P. W. M. Jacobs, *Group Theory with Applications in Chemical Physics*, Cambridge University Press, Cambridge. 2005.

115. S. K. Kim, *Group Theoretical Methods and Applications to Molecules and Crystals*, Cambridge University Press, Cambridge, 1999.

116. B. S. Tsukerblat, *Group Theory in Chemistry and Spectroscopy: A Simple Guide to Advanced Usage,* Academic Press, London, 1994.

117. M. Reiher and A. Wolf, *Relativistic Quantum Chemistry: The Fundamental Theory of Molecular Science,* Wiley-VCH, Weinheim, 2009.

118. D. S. Sholl and J. A. Steckel, *Density Functional Theory: A Practical Introduction,* Wiley, Hoboken, NJ, 2009.

119. H.-D. Höltje, W. Sippl, D. Rognan and G. Folkers, *Molecular Modeling: Basic Principles and Applications,* 3rd edn, Wiley-VCH, Weinheim, 2008.

120. I. P. Grant, *Relativistic Quantum Theory of Atoms and Molecules: Theory and Computation,* Springer, New York, 2007.

121. I. Mayer, *Simple Theorems, Proofs, and Derivations in Quantum Chemistry,* Kluwer/Plenum, New York, 2003.

122. J. Cioslowski (ed.), *Quantum-Mechanical Prediction of Thermochemical Data,* Kluwer, Dordrecht, 2001.

123. A. R. Leach, *Molecular Modeling: Principles and Applications,* 2nd edn., Prentice Hall, Harlow, 2001.

124. T. M. Klapötke and A. Schulz, *Quantum Chemical Methods in Main-Group Chemistry,* Wiley, Chichester, 1998.

125. K. Balasubramanian (ed.), *Relativistic Effects in Chemistry, Part A and Part B,* Wiley, New York, 1997.

126. A. Szabo and N. S. Ostlund, *Modern Quantum Chemistry,* Dover, Mineola, 1996.

127. J. B. Foresman and A. Frisch, *Exploring Chemistry with Electronic Structure Methods,* 2nd edn., Gaussian Inc., Pittsburgh, 1996.

128. R. G. Parr and W. Yang, *Density-Functional Theory of Atoms and Molecules,* Oxford University Press, New York, 1989.

129. G. Burns and A. M. Glazer, *Space Groups for Solid State Scientists,* 2nd edn., Academic Press, New York, 1990.

130. T. Hahn (ed.), *International Tables for Crystallography, Volume A: Space-group symmetry,* 5th edn. (corrected reprint), Kluwer Academic, Dordrecht, 2005.

131. E. Prince (ed.), *International Tables for Crystallography, Volume C: Mathematical, physical and chemical tables,* 3rd edn., Kluwer Academic, Dordrecht, 2004.

132. M. O'Keeffe and B. G. Hyde, *Crystal Structures. I. Patterns and Symmetry,* Mineralogical Society of America, Washington, DC, 1996.

133. R. C. Buchanan and T. Park, *Materials Crystal Chemistry,* Marcel Dekker, New York, 1997.

134. B. Douglas and S.-M. Ho, *Structure and Chemistry of Crystalline Solids,* Springer, New York, 2006.

135. J. F. Nye, *Physical Properties of Crystals,* Oxford University Press, Oxford, 1957.

136. J. Bernstein, *Polymorphism in Molecular Crystals,* Oxford University Press, New York, 2002.

137. G. A. Jeffrey, *An Introduction to Hydrogen Bonding,* Oxford University Press, New York, 1997.

138. J. D. Dunitz, *X-Ray Analysis and the Structure of Organic Molecules,* 2nd Corrected Reprint, VCH Publishers, New York, 1995.

139. J. P. Glusker, M. Lewis and M. Rossi, *Crystal Structure Analysis for Chemists and Biologists,* VCH Publishers, New York, 1994.

140. I. D. Brown, *The Chemical Bond in Inorganic Chemistry: The Valence Bond Model*, Oxford University Press, New York, 2002.

141. B. K. Vainshtein, V. M. Fridkin and V. L. Indenbom, *Structures of Crystals*, 2nd edn., Springer Verlag, Berlin, 1995.

142. R. J. D. Tilley, *Crystals and Crystal Structures*, Wiley, Chichester, 2006.

143. B. G. Hyde and S. Andersson, *Inorganic Crystal Structures*, Wiley, New York, 1989.

144. F. S. Galasso, *Structure and Properties of Inorganic Solids*, Pergamon, Oxford, 1970.

145. A. K. Cheetham and P. Day (eds.), *Solid State Chemistry: Techniques*, Oxford University Press, New York, 1987.

146. A. K. Cheetham and P. Day (eds.), *Solid State Chemistry: Compounds*, Oxford University Press, New York, 1991.

147. G. S. Rohrer, *Structure and Bonding in Crystalline Materials*, Cambridge University Press, Cambridge, 2001.

148. U. Müller, *Inorganic Structural Chemistry*, 2nd edn., Wiley, Chichester, 2006.

149. U. Müller, *Symmetry Relationships between Crystal Structures*, Oxford University Press, Oxford, 2013.

150. A. F. Wells, *Structural Inorganic Chemistry*, 5th edn., Oxford University Press, Oxford, 1984.

151. M.-C. Hong and L. Chen (eds.), *Design and Construction of Coordination Polymers*, Wiley, Hoboken, New Jersey, 2009.

152. R. Xu, W. Pang, J. Yu, Q. Huo and J. Chen (eds.), *Chemistry of Zeolites and Related Porous Materials: Synthesis and Structure*, Wiley(Asia), Singapore, 2007.

153. R. Xu, W. Peng and Q. Huo (eds.), *Modern Inorganic Synthetic Chemistry*, Elsevier, Amsterdam, 2011.

154. E. W. T. Tiekink and J. J. Vittel (eds.), *Frontiers in Crystal Engineering*, Wiley, Chichester, 2006.

155. E. W. T. Tiekink, J. J. Vittel and M. J. Zaworotko (eds.), *Organic Crystal Engineering* (Frontiers in Crystal Engineering II), Wiley, Chichester, 2010.

156. E. W. T. Tiekink and J. Zukerman-Schpector (eds.), *The Importance of Pi-Interactions in Crystal Engineering* (Frontiers in Crystal Engineering III), Wiley, Chichester, 2012.

157. T. C. W. Mak and G.-D. Zhou, *Crystallography in Modern Chemistry: A Resource Book of Crystal Structures*, Wiley-Interscience, New York, 1992; Wiley Professional Paperback Edition, 1997.

INDEX

This index shows the numbers of the Problems where a given keyword or topic appears, instead of listing the page numbers. The Question numbers are written in the form of *Chapter.Question*: for example, 3.11 denotes Problem 11 in Chapter 3. It is likely the reader may also wish to refer to the corresponding Answers, even though the numbers of the Answers are not redundantly listed.